Fundamentals of Queueing-Game Models

Jinting Wang

Fundamentals of Queueing-Game Models

Science Press
Beijing

Springer

Jinting Wang
School of Management Science
and Engineering
Central University of Finance
and Economics
Beijing, China

ISBN 978-981-95-0260-8   ISBN 978-981-95-0261-5 (eBook)
https://doi.org/10.1007/978-981-95-0261-5

Jointly published with Science Press
The print edition is not for sale in China (Mainland). Customers from China (Mainland) please order the print book from: Science Press.

This Springer imprint is published by the registered company Springer Nature Singapore Pte Ltd.
The registered company address is: 152 Beach Road, #21-01/04 Gateway East, Singapore 189721, Singapore

To My Beloved Parents
Jiquan Wang and Daxi Li

To My Loving Wife and Son
Tongyan Jiang, and Yingdong Wang

Preface

This book arose out of several graduate courses that the author has taught during the past years, the first one being on queueing theory followed by the second one on the game-theoretic analysis of queueing models. During the past decade since the publication of the first edition (in Chinese) of this book, queueing-game (or called queueing economics, rational queueing) theory has developed substantially.

Queueing theory has a very general scope, encompassing problems that are basic and also important to various service systems, telecommunication systems, transportation systems, healthcare systems, etc. The word queue comes from the Latin word cauda, a sixteenth-century heraldic term, which means the tail of an animal. It vividly describes a waiting line or a sequence of people awaiting their turn to proceed in service systems, such as telecommunications, manufacturing, or on fairness and scheduling in computer science. Subsequently, the classical queueing theory mainly studies the objective law of the queueing system itself, statistically models the distribution law of customer arrival and service time in a variety of queueing models, and obtains the queue length, waiting time, busy period and other indexes of queueing systems by using the theories of probability, statistics, and stochastic process.

Nowadays, it becomes evident that the service sector is dominating the economy globally, and as this has happened, queueing theorists have become more interested in the human side of queues. Actually, being the indispensable constituent of queueing systems, customers, service providers, system managers and their behavior indeed play a non-negligible role in a queueing system. In hubs of transportation, hospitals, shops and restaurants, lines comprise people, not inanimate objects. People are at the front of lines, too, distributing passengers to scheduled trips, routing patients to hospital beds, ringing up purchases, or leading hungry patrons to tables. Their autonomous behavior will directly affect the performance of the whole system, making the queueing strategy flexible.

These considerations are more in line with the actual situation, but it also makes the queueing analysis more complex. On the one hand, in the classical queueing theory, the strategic decision-making behavior of individual customers is often ignored, and the economic utility of customers (or service providers, system managers) and its impact on the system performance are rarely considered in the performance analysis, optimal design and control of various service systems. On the other hand, with the progress in behavioral operations management, many studies have shown that people's behavior shows some unique properties, such as risk aversion (preference), loss aversion, reference dependence, and bounded rationality. As a result, people do not necessarily follow the "maximum utility" principle from the perspective of overall optimization of the system, but make "self satisfaction" according to different information of the system and their own preference or characteristics. Therefore, for the service systems with personalized customer decision-making behavior, many new research problems emerge and they can hardly be handled by the classical methodologies of operations research, statistics and economics directly.

Queueing-game theory is a rapidly developing research field stepping from behavioral operations management, statistics and economics and quickly grows to be an important research topic. Specifically, for the game-theoretic aspect of queueing systems, on the one hand, customers hope to get services in time. Customers can get certain utility benefits after receiving services, but under a certain cost structure, generally speaking, the longer the customer's queueing time, the greater the corresponding loss. On the other hand, service facilities also have their own operational cost or losses, such as preventive maintenance, supply chain management, equipments replacement, and logistics, on which service providers can price the services provided. Sometimes customers are allowed to join the system after charging admission fees, by which means service providers used to control customer flow. In this process, customers and service providers pursue the maximization of their own interests. Customers choose queueing strategies independently, while their behavior is affected by the system information and other customer behavior strategies. In short, both of them have to consider the impact of each other's behavior on themselves while making decisions, so a game between both sides and even more players is formed. Furthermore, the cost structure of the system can change according to different practical problems, different priorities of different customers or service providers, different service capabilities of service providers, and different changes of queueing rules, which make the problem diverse and complicated. But, despite all this, the game-theoretic analysis of queueing systems has a wide range of important applications in the fields of banking services, enterprise order production, communication networks and so on.

I attempt to briefly introduce the foundation of queueing game or queueing economics in the book. When preparing to write this book, I was trying to include the basic contents, fundamental models and methodologies as far as possible. The content includes brief queueing theory and basic knowledge of game theory which are the foundations to equilibrium analysis of various continuous-time queueing systems, including observable queueing systems, unobservable queueing systems, priority queueing game, repairable queueing game, vacation queueing game and retrial queueing game, as well as the applications of queueing game in various fields. Much of this book is the author's research results in recent years and contains some new unpublished results. Due to the limitation of space and the universality of readers, the author abandons many more in-depth theoretical problems and some important but special systematic discussions.

This book is positioned as an introductory text to the fundamental models and methodologies that are nowadays widely used and accepted in the queueing-game literature. It is intended to start from a beginner level and accelerate to research level, targeting upper undergraduate operations research students and graduate students in the areas of operations research, management science, system science, transportation, wireless communication, computer science. The book will also be suitable for OR/MS researchers who wish to build a background and advance to research level in queueing game. In brief, this book is expected to be a useful introductory text for readers interested in the interdisciplinary research on queueing theory and game theory.

In preparing the book, I relied on help from many people. Indeed, this book could not have been written without the selfless help and great support of my family members. First and foremost I am in debt to my family for their continuous support. My wife, Tongyan, and my son, Yingdong, were very supportive and patiently endured my preoccupation while this book was being written.

I sincerely thank my Ph.D. supervisor, Prof. Jinhua Cao, for his long-term encouragement and support. I have also gained enormously from the long-term cooperation and guidance with Wei Wayne Li of Southern University of Texas, Zhe George Zhang of Western Washington University and Simon Fraser University, Yiqiang Zhao of Carlton University, Hanqin Zhang of National University of Singapore, Jesús R. Artalejo of Complutense University of Madrid, Attahiru S. Alfa of University of Manitoba, Raj Srinivasan of University of Saskatchewan, Bin Liu of Anhui Jianzhu University, Dequan Yue of Yanshan University, Quanlin Li of Beijing University of Technology, and other friends and colleagues.

I am grateful to Profs. Refael Hassin, Antonis Economou, Naishuo Tian, Pengfei Guo, Yaxiang Yuan, Shouyang Wang, Yuhong Dai, Zhidong Liu for their great support and encouragement in the past decades of years. I have been particularly fortunate in having excellent doctoral students and postdoctoral students work with me on different aspects of queueing game models. Shan Gao, Xuelu Zhang, Fang Wang, Yu Zhang, Ruopeng Wang, Sheng Zhu, Zhongbin Wang, Kaili Li, Lingjiao Zhang, Yilin Wang, Yujie Xie, Qin Chen, Yuying Zhang have freely shared their ideas as well as their enthusiasm for queueing game studies and they have also made detailed comments on the manuscript; I would thank them for their input and some proofreading work.

I am grateful to Jingke Li of Science Press (Beijing) for her editorial help and advice in preparing and organizing the manuscript. She put in hard work for the publication of this book.

The continuous support from the research grants of National Natural Science Foundation of China (Nos. 72371259, 71871008, 71571014) and the support of New Century Excellent Talents Program of the Ministry of Education (No. NCET-11-0568) are gratefully acknowledged. The monograph is also financially supported by a research start-up funds from Central University of Finance and Economics, China.

This book is dedicated to the memory of my father Jiquan Wang and mother Daxi Li. I wish we could spend more time in my hometown, drink tea and talk about everyday life, and share more wonderful times.

Beijing, China
May 2024

Jinting Wang

Contents

Chapter 1
Preliminaries

The present chapter, which is of reference nature, presents in concise terms and without any proof the relatively general knowledge of the basic concepts of game theory and queueing theory. Game theory has been viewed as a framework on the construction of rigorous models that deal with situation of conflict and cooperation between rational decision makers. It also shed light on other disciplines, such as social and economic networks, evolutionary biology and psychology. Queueing theory has been regarded as an operations management technique that is commonly used to determine and streamline staffing needs, scheduling, and inventory in order to improve overall customer service. This chapter is mainly based on the monographs of Shortle et al. (2018), Hassin and Haviv (2003), Nisan et al. (2007), and there are many excellent textbooks on introduction to game theory and queueing theory such as Fudenberg and Tirole (1991), Myerson (1997), Tadelis (2013), Cooper (1981), Kleinrock (1975, 1976), Neuts (1989), Takagi (1991), Bocharov et al. (2004), He (2014), Zhang (2023), to mention only a few.

1.1 Basics from Game Theory

1.1.1 Definition of Game

In general, a game is an interactive process in which some individuals, teams or other organizations, facing certain environmental conditions and under certain rules, choose and implement from their allowed behaviors or strategies simultaneously or successively, one or more times, and obtain their corresponding strategic results. By the above description, the following four aspects need to be set up to specify or define a game.

J. Wang, *Fundamentals of Queueing-Game Models*,
https://doi.org/10.1007/978-981-95-0261-5_1

1. **Players**. Players are individuals or organizations that make independent decisions and bear the results independently in the defined game. No matter how large an organization is, even a country or even the United Nations composed of many countries, as long as they make unified decisions, take unified actions and bear the results in a game, the organization can be a participant in the game. And, once the rules of the game are well-defined and determined, all participants are equal, and everyone must act in strict accordance with the rules. For the sake of convenience, each independent participant in the game is called a "player".
2. **A strategy profile or the set of all actions of players**. The action set specifies the methods, ways of doing a thing or the level and value of economic activities that each player can choose when making decisions. In different games, the number of strategies or behaviors that can be selected by the player is quite different. In the same game, the content and quantity of optional strategies or behaviors of different players are often different, sometimes there are only a finite set, or even only one, and sometimes there may be many, even unlimited, alternative strategies or behaviors.
3. **Orders**. In various decision-making activities in reality, when there are multiple independent decision-makers to make decisions, sometimes these game players need to make choices at the same time to ensure fairness and rationality of the game. While many times, the decisions of each player have a sequential order, and sometimes a player has to make more than one decision and choice. Therefore, it is necessary to specify the order of a game playing. Different orders are generally mean different games, even if other aspects of the game are the same.
4. **Payoffs**. Each set of possible decision choices corresponding to each player should have a result representing the gains or losses of each player under the profile of strategy. Game-theoretic analysis is usually carried out through the comparison of quantitative relations, and therefore most of the games have quantitative results or can be characterized as quantitative results, such as income, profit, loss, personal utility, social utility, economic welfare, etc. The quantitative values of these possible results in the game are called the "benefits" of each player under the corresponding circumstances. The benefits, which can be positive or negative, should be specified in a formulated game. They are the standard and basis for analyzing various game models. It is worth noting that, although the benefits of each game player in various situations should exist objectively, this does not mean that each game player knows the benefits of each other.

The above four aspects must be set first when defining a game. That is, a game is defined once the above four aspects are determined. In this sense game theory refers to the systematic study of various game problems that can be defined using the above methods, seeking reasonable strategy selection and the results of the game under the condition that each player has sufficient or bounded rationality and ability, and analyzing the economic and efficiency significance of these results.

A game in which all players simultaneously choose or can be considered as simultaneously choosing strategies is called a *static game*. The game with sequential decision-making is called a *dynamic game*, which is also known as *multi-stage game*. In a dynamic game, the player who fully knows the process of the game upon his turn to choose is called the player with perfect information. If all players in the dynamic game have perfect information, it is called the *dynamic game with perfect information*. In a dynamic game, when the player whose turn is to act does not fully understand all the previous game processes, it is called the player with imperfect information, and the dynamic game with this player is called the *dynamic game with imperfect information*.

1.1.2 Non-cooperative Game

Generally, we call games that allow binding agreements as cooperative games. In contrast, we call a game that does not allow binding agreements as a noncooperative game. Due to the significant differences in the basic behavioral logic and research methods of the players in cooperative and noncooperative games, they are two very different types of games. In fact, "cooperative game theory" and "noncooperative game theory" are the most fundamental classifications of game theory. The generation and development of "cooperative game theory" and "noncooperative game theory", and their status, role and influence in economics vary greatly in many aspects. Noncooperative game theory is currently dominant and widely studied and applied to various problems arising from economics, finance, engineering, operations research, statistics, and computer science.

We often use G to represent a non-cooperative game. If G has n players, the set of all optional strategies of each player is called "strategy space", denoted by S, and s_i is the action chosen by the player i, where $i = 1, 2, \ldots, n$. As mentioned above, we call the game in which all players fully understand each other in various situations as complete information game, and the game in which at least some players do not fully understand the benefits of other players as incomplete information game or Bayesian game. Accordingly, we have the following concepts.

Definition (Pure Strategy): In a complete information game, if only one specific strategy can be chosen under each given information, this strategy is called a pure strategy.

Definition (Mixed Strategy): If there are different strategies can be chosen with a certain probability under each given information, the strategy is called a mixed strategy.

Mixed strategy can be understood as the probability distribution of pure strategies in a probability space, and pure strategy can be regarded as a special case of the mixed strategy.

- **Payoff Function**: The benefits that each player can obtain when participating in a game according to the game's rule and the strategy adopted by the player.

Denote $s = (s_1, \ldots, s_n)$ by a combination of strategies or an action profile, $s_i \in S_i$, the payoff function of each player is $F_i(s)$ Let s_{-i} denote the strategy combination of everyone except the player i. If s_i is a mixed strategy, choose s_i^1 with probability α and s_i^2 with probability $1 - \alpha$, then the benefit of the i-th player is

$$F_i(s_i, s_{-i}) = \alpha F_i(s_i^1, s_{-i}) + (1 - \alpha) F_i(s_i^2, s_{-i}).$$

For player i, a strategy s_i^1 is called as a weakly dominant strategy over strategy s_i^2 if for any s_{-i}, we have $F_i(s_i^1, s_{-i}) \geq F_i(s_i^2, s_{-i})$ and the inequality is strictly hold for at least one s_{-i}. A strategy s_i is said to be weakly dominant if it weakly dominates all other strategies in S_i. Further, a strategy s_i^* is said to be a best response for player i against the profile s_{-i} if we have

$$s_i^* \in \arg \max_{s_i \in S_i} F_i(s_i, s_{-i}).$$

1.1.3 Nash Equilibrium

Definition In the n-player game G, the strategies $s^e = (s_1^e, \ldots, s_n^e)$ are a Nash equilibrium if, for each player i, s_i^e is (at least tied for) player i's best response to the strategies specified for the $n - 1$ other players, $s_{-i}^e = (s_1^e, \ldots, s_{i-1}^e, s_{i+1}^e, \ldots, s_n^e)$:

$$F_i(s_i^e, s_{-i}^e) \geq F_i(s_i, s_{-i}^e).$$

If we simply consider that all players are of the same type, then we get a symmetric equilibrium. In addition, a commonly recognized drawback of the equilibrium concept is the possibility that the solution is not unique.

A subgame of a dynamic game consisting of subsequent game stages starting from a stage other than the first stage, with an initial information set and all the information required for the game, that can form a part of the original game.

If in a dynamic game with perfect information, a strategy combination composed of the strategies of each player satisfies the Nash equilibrium in the entire dynamic game and all its subgames, then this strategy combination is called a subgame of the dynamic game with perfect Nash equilibrium. In short, this set of strategies must be optimal in every subgame.

A subgame perfect Nash equilibrium is stable and stronger than the Nash equilibrium (excluding untrusted threats or commitments in the equilibrium strategy). If it is not the subgame perfect Nash equilibrium, it may not be in the interests of the players in some subgames.

1.1.4 Evolutionarily Stable Strategies

Definition: Let y be an equilibrium strategy. There may be another best response strategy $z(z \neq y)$. Therefore, when people originally began to choose strategy y, some people may choose z. If z is the best response against themselves and better than y, then everyone will choose z. In this case, y is unstable in the sense that when starting with y, it may be that the players adopt the best response z, and then a new equilibrium, at z, will be reached. If no such z exists then y is said to be an evolutionarily stable strategy and often abbreviated to ESS (see Smith 1982).

Let $F(x, y)$ be the payoff for a customer who selects a strategy x when all other choose strategy y. An equilibrium strategy y is called an evolutionarily stable strategy when it satisfies the following conditions.

$$\begin{aligned} &F(y, y) \geq F(x, y), \quad x \in S, \\ &F(z, y) \geq F(x, y), \quad x \in S, z \neq y, \\ &F(y, z) > F(z, z). \end{aligned}$$

1.1.5 Follow the Crowd and Avoid the Crowd

Suppose that for any strategy y there exists a unique best response $x(y)$ such that

$$x(y) = \arg\max_x F(x, y).$$

If $x(y)$ is monotone decreasing in y, we call the situation as *avoid the crowd* (ATC).

If $x(y)$ is monotone increasing in y, we call the situation as *follow the crowd* (FTC).

It can be shown that in the ATC situation at most one equilibrium exists whereas multiple equilibria are possible in the FTC case (see Fig. 1.1 in Hassin and Haviv 2003)

1.2 Basics from Queueing Theory

Queueing theory, also known as stochastic service system theory, is a mathematical discipline to study the phenomenon of congestion. It is an important branch of operations research, and also an important branch of applied probability as well as applied mathematics. It focuses on the optimal design and optimal control of systems by studying the probability characteristics of various service systems with queueing/congestion. Apparently, the queueing problems studied in the literature have a strong practical background. Historically, it originated from the research of Danish telecommunication engineer A. K. Erlang on telecommunication system in the early twentieth century. After that, queueing theory has become a rather mature one with the efforts of many scholars from mathematics and operations research societies. With the rapid development of computer and information technology, the scientific research of queueing theory is fast-developing and its application field is also expanding. At present, various theoretic results in queueing theory have been widely used in many fields, such as telecommunication, transportation, production and inventory management, computer system design, computer communication network, military operations, flexible manufacturing system and system reliability, etc., and queueing theory and its applications have yielded fruitful application achievements. In summary, queueing theory has played a direct and important role in the aforementioned fields and it has become one of the essential mathematical tools for system design and optimization.

1.2.1 Basics of Queueing Systems

1. **Input process**: The input process is to describe the arrival pattern of customers, the source of customers and how customers arrive in the queueing system.
   (1) **Source of customers in the system**: The source of customers may be finite or infinite.
   (2) **Type of arrival**: Customers arrive to the system individually or in batches.
   (3) **Arrival process**: To explain what probability distribution does the arrival interval of successive customers obey, what are the parameters of the distribution, and whether the arrival interval is independent.
2. **Queue discipline**: Queueing disciplines refer to whether the service is allowed to queue or not, and whether the customer is willing to queue. In the case of queueing, what is the service order, which is divided into:
   (1) **Loss system**: When the customer arrives, if all the service facilities are occupied and the service system does not allow the customer to wait, the customer will leave automatically.
   (2) **Waiting system**: When customers arrive, if all service facilities are occupied, they queue up for service. In the waiting system, the service orders are usually categorised as the following types.
      - **First Come, First Served (FCFS)**: Customers receive services in the order of their arrivals.
      - **Last Come, First Served (LCFS)**: Customers who arrive later receive service first.
      - **Random Selection for Service (RSS)**: The service facility randomly selects one of the waiting customers for service.

- **Priority Queues**: Among the customers waiting in line, some types of customers have priorities. Special service should be rendered in the service order to let these special customers get their service first.

(3) **Mixed systems**: It is the mixture of loss system and waiting system, which is classified into mixed systems with limited capacity, mixed systems with limited waiting time and mixed systems with limited sojourn time.

3. **Service facility**: The main aspects of describing the service facilities are:

(1) **Number of servers.** In the case of multiple-server queueing systems, these servers are connected in series or in parallel or other type of structure.
(2) **Service pattern.** To demonstrate what probability distribution does the service time required by customers obey, whether the service time required by each customer is independent of each other, whether it is a batch service or a single service, etc.

The service time distributions of customers are differently fitted depending on practical problems and data, and they usually be assumed as: Deterministic distribution, negative exponential distribution, Hyper-exponential distribution, k-order Erlang distribution, Geometric distribution, General distribution, among others.

1.2.2 Notations for Classic Queueing Systems

A queueing system consists of several elements. For the sake of simplicity, 3–5 English letters are often used in the queueing literature to represent a queueing system, which are separated by slashes. It is due to the most part to Kendall (1953) as shorthand for describing queueing processes.

More specifically, a queueing process is described by a series of symbols and slashes A/B/X/Y/Z, where A denotes the interarrival time distribution, B denotes the service time distribution, X denotes the number of parallel servers, Y denotes the system capacity, and Z denotes the queue discipline. For example,

- $M/M/c/\infty$ represents a queueing system with exponential interarrival times, exponential service times, c parallel servers, infinite system capacity. Therefore, $M/M/c/\infty$ queueing system is a waiting system.
- $M/G/1/\infty$ indicates a queueing system with exponential interarrival times, independent and generally distributed service times, one server, infinite system capacity. It is a waiting system.
- $GI/M/1/\infty$ indicates a queueing system with independent and generally distributed interarrival times, independent and exponentially distributed service times, one server, infinite system capacity. It is also a waiting system.

- $E_k/G/1/K$ is a mixed queueing system in which the interval of successive arrivals is independent and follows the k-order Erlangian distribution, the service time is independent and follows a general probability distribution, one server, and a capacity of k.
- $D/M/c/K$ presents a mixed queueing system with deterministic interarrival times, exponential service times, c parallel servers, finite k system capacity.
- $M^r/M/1/\infty$ is a waiting queueing system in which customers arrive in batches, each batch contains a fixed r customers, and the arrival interval between batches is independent and follows a negative exponential distribution, the service time of customers is independent and follows a negative exponential distribution. There is only one server and infinite system capacity.
- $M^X/M^r/1/\infty$ indicates a waiting queueing system in which the arrival stream forms a batched Poisson process, the actual number of customers in any arriving module is a random variable X, which takes on the value n with probability c_n, where n is a positive integer $(0 < n < \infty)$. Service is provided in batches, each batch contains a fixed r customers. There is one server in the system and the system capacity is infinite.

For more details on the notations readers can refer to the first chapter of Shortle et al. (2018).

1.2.3 Main Measures of Queueing Systems

1. **System size and queue length**

The system size refers to the number of customers in the system (including customers receiving service), while the waiting queue length refers to the number of customers waiting in the system. They are crucial random variables and also quantitative indicators which are of great concern to both customers and service providers. In practice, it is very important to obtain the distribution functions and relevant moments of system size and queue length (at least the expected value). Obviously, the system size is equal to the waiting queue length plus the number of customers being served.

2. **Waiting time and sojourn time**

The customer's waiting time refers to the period from the moment when the customer enters the system to the moment when he starts receiving service, and the sojourn time is the sum of the customer's waiting time and service time in the system. Under the assumption that arrival process and service process are independent of each other, waiting time and service time are independent of each other. Waiting time and sojourn time are both the important quantitative indicators most concerned by customers. In application, we are concerned about their distribution and expected average value in steady state.

3. **Busy period and idle period**

When the customer arrives at the idle system, the service starts immediately until the system becomes idle again. This period is the continuous busy time of the system, which is called the busy period of the system. It reflects the work intensity of the servers in the system.

The concept corresponding to busy period in a queueing system is the idle period of the system, which refers to the length of time the system continuously remains idle. In a queueing system, busy and idle periods alternate under stationary equilibrium.

The busy period cycle refers to the interval between the start of two adjacent busy periods. Obviously, it is equal to the sum of the current busy period length and idle period length.

4. **Output process**

The output process, also known as the departure process, refers to the process in which the customers who have received the service leave the system one after another. The main characteristics describing an output process are the time interval of successive departure and the number of customers who have left in a known period of time. These measures also reflect the work efficiency of the system from another side.

1.2.4 $M/M/1$ Queueing Systems

Assume that customers arrive in the system according to a Poisson process with parameter λ, that is, the interval time sequences of successive arrivals are independent and follow a negative exponential distribution of the same parameter λ. The service time sequences required by customers are independent, exponentially distributed random variables with a common parameter μ. There is, only one server in the system with infinite waiting capacity, and the arrival process and service process are mutually independent. The existence of steady-state solution of the system depends on the condition that $\rho \equiv \lambda/\mu < 1$, or equivalently, $\lambda < \mu$. Let N represent the random variable "number of customers in the system in the steady state" and L represent its expected value. Similarly, denoted by N_q the random variable "number in queue in steady state" and by L_q its expected value, then in the steady-state we have the following results.

1. The probability of n customers in the system is

$$p_n = (1-\rho)\rho^n.$$

2. The expected number of customers in the system is

$$L = \mathbb{E}[N] = \frac{\rho}{1-\rho} = \frac{\lambda}{\mu-\lambda}.$$

3. The mean queue length is

$$L_q = \mathbb{E}[N_q] = \frac{\rho^2}{1-\rho} = \frac{\lambda^2}{\mu - \lambda}.$$

4. The average length of sojourn time of customers W is given by

$$W = \frac{1}{\mu - \lambda}.$$

5. The average busy period length T_{BP} is given by

$$T_{BP} = \frac{1}{\mu - \lambda}.$$

6. The average number of customers served in a busy period is

$$\frac{1}{1-\rho}.$$

7. The average time from the customer's arrival at the system to the first time the server is idle is

$$\frac{1}{\mu(1-\rho)^2}.$$

1.2.5 M/M/1/K Model: System with Limited Waiting Space

We now take up the M/M/1/K model that there is a limit K placed on the number of customers allowed to join the system at any time. Customers arrive in accordance with a Poisson process with rate λ; a customer will join the system whenever he finds less than K in the system and a customer who arrives when there are K in the system leaves the system and is lost to the system. Service times follow an i.i.d. exponential distribution with rate μ. A queue with limited waiting space is called as a queue with finite buffer. The derivation of main performance measures is similar to those of M/M/1 model, except that now the waiting space is finite and thus there is no requirement that the traffic intensity ρ be less than 1. So we have the following results.

1. The probability of n customers in the system is

$$p_0 = \begin{cases} \frac{1-\rho}{1-\rho^{K+1}}, & \rho \neq 1; \\ \frac{1}{K+1}, & \rho = 1. \end{cases}$$

$$p_n = \begin{cases} \frac{(1-\rho)\rho^n}{1-\rho^{K+1}}, & \rho \neq 1; \\ \frac{1}{K+1}, & \rho = 1. \end{cases}$$

2. The expected number of customers in the system is

$$L_K = \begin{cases} \frac{K}{2}, & \rho = 1; \\ \frac{\rho}{1-\rho} - \frac{(K+1)\rho^{K+1}}{1-\rho^{K+1}}, & \rho \neq 1. \end{cases}$$

3. The effective arrival rate $\hat{\lambda}$ to the system M/M/1/K is the expected number of customers joining the system in unit time and is given by

$$\hat{\lambda} = \lambda(1 - p_K) = \begin{cases} \frac{\lambda K}{K+1}, & \rho = 1; \\ \frac{\lambda(1-\rho^K)}{1-\rho^{K+1}}, & \rho \neq 1. \end{cases}$$

4. The average number of customers diverted from the system in unit time equals

$$\lambda - \lambda(1 - p_K) = \lambda p_K = \begin{cases} \frac{\lambda}{K+1}, & \rho = 1; \\ \frac{\lambda(1-\rho^K)\rho^K}{1-\rho^{K+1}}, & \rho \neq 1. \end{cases}$$

5. The utilization factor of the server is not equal to ρ as in usual M/M/1 model, but equals

$$\hat{\rho} = \frac{\hat{\lambda}}{\mu} = \rho(1 - p_K) = \begin{cases} \frac{\rho K}{K+1}, & \rho = 1; \\ \frac{(1-\rho^K)\rho}{1-\rho^{K+1}}, & \rho \neq 1. \end{cases}$$

6. Upon completion of service, the average number of customers leaving the system in unit time equals

$$\mu\hat{\rho} = \begin{cases} \frac{\lambda K}{K+1}, & \rho = 1; \\ \frac{\lambda(1-\rho^K)}{1-\rho^{K+1}}, & \rho \neq 1. \end{cases}$$

7. The expected waiting time in queue equals $L_K/\hat{\lambda}$.
8. It should be noted that the actual input for the M/M/1/K model is not truly Poisson, although the original arrival process is Poisson. The property PASTA does not hold in this case. In fact, define $\hat{\lambda}_n$ as the probability that an effective arrival (joining the system upon arrival) finds n in the system in an infinitesimal interval $(t, t+h]$, we can derive it by using Bayes theorem. That is,

$$\begin{aligned} \hat{\lambda}_n &= \Pr\{n \text{ in system} | \text{an arrival is about to occur}\} \\ &= \frac{p_n}{1 - p_K}, \quad n = 0, 1, \ldots, K-1. \end{aligned}$$

1.2.6 Little's Law

Comparing the stationary mean number L of customers in the system and the stationary mean customer sojourn time W, the system, we observe the relation $L = \lambda W$ holds, which is called the Little's law or Little's formula. It is one of the most fundamental and useful relationships in queueing theory named after J. D. C. Little, who gave its first formal proof (1961). It relates the long-term mean number of customers to the mean amount of time customers spend in the system provided the number of customers entering the system is equivalent to the number of customers departing from it. Little's law is a quite general result holding even for G/G/1 queues, and it also holds with other service disciplines than FIFO. Little's law establishes a relationship between the average number of customers in the system, the mean arrival rate and the mean customer response time (time between entering and leaving the system after getting service) in the steady state.

Intuitively, the reasoning is natural and obvious. Indeed, let the system be in the steady state over a sufficiently long time interval T. Then, on the average, λT customers arrive at the system during this time. In turn, each customer, on the average, stays in the system for the time W; therefore, the mean time V spent in the system by all λT customers is λWT, and at each instant the system, on the average, has $L = V/T = \lambda W$ customers.

The above reasoning can be made more rigorous if one makes use of the ergodic theory of stationary random processes. It is also clear that the area of application of the Little formula is much wider than the M/M/l/∞ system under consideration. This proves useful for analysing systems for which only one characteristic—either L or W—can be calculated. If we are looking at the number of customers waiting, a similar equality is valid for L_q and W_q: $L_q = \lambda W_q$. Note that the principle of customer conservation is needed, which means the frequency of entries into a service channel equals the frequency of departures from the channel. For example, it is not correct if one replaces either L_q or W_q by L or W respectively.

The above results hold not only for the consideration of customers in the system, but also for the servers. In this scenario, L can be considered as the average number of busy servers, and W be the average service time. As such, in a c-server system $c\rho$ equals λ/μ (with $\rho = \lambda/(c\mu)$).

A more general formula $H = \lambda G$ presented in Heyman and Stidham (1980) showed a similar relationship between more general customer and time averages. In this formula, H represents the total average cost per time incurred by the system, G represents the average "cost" or "work" associated with a customer.

Due to context and space limitation we omit the proofs for Little's law here. These proofs can be found, for example, in Stidham (1974) and Wolff (2011), see also Whitt (1991). Theoretically, the Little's law type relationship can be proved with slightly different assumptions on the underlying stochastic processes. For instance, the original proof by Little requires the underlying processes to be strictly stationary, as does the theorem in Brumelle (1971), while some other versions require the existence of regeneration points when the system empties out and "starts over" (e.g., Jewell,

1967). Some variants of the theorem in finite time are given by Little (2011) in a retrospective article. See Shortle et al. (2018) and the references therein for more details on Little's law.

1.2.7 PASTA: Poisson Arrivals See Time Averages

In practice we usually face the following question: when we look at the number of customers in a queueing system at "random points" in time, do then have all "random points" the same properties or do there exist points where the results differ fundamentally? Image that you are running a simulation of some system, and you want to calculate the fraction of time that the system has n jobs. To this end, you may record each arrival into the system and check whether it sees n jobs or not. By doing so, the fraction of arrivals that witnessed n jobs is tracked when arriving into the system. The question in this case is whether this method gives you the right answer?

Keeping the M/M/1 queue in mind, let's assume an ergodic continuous-time system. Let $\pi_n = p_n$ be the limiting probability that there are n jobs in the system (or equivalently the long-run fraction of time that there are n jobs in the system). Let a_n be the limiting probability that an arrival sees n jobs in the system (or equivalently the long-run fraction of arrivals that see n jobs). Let d_n be the limiting probability that a departure leaves behind n jobs in the system when it departs (or equivalently the long-run fraction of departures that leave behind n jobs). It can be shown that a_n is not necessarily equal to p_n (see Bertsekas and Gallager 1992 or Harchol-Balter 2013).

For completeness, we recap the example therein. Consider a single queue whose customers have an interarrival time distributed uniformly between 1 and 2 (i.e., $\sim U(1, 2)$). Assume that all service times are constant with a value of 1. Then, $a_0 = 1$ and $d_0 = 1$ because a customer will complete service before the next customer arrives. However, $p_0 \neq 1$ because the system is not always empty.

Evidently, an arriving customer sees n customers already in the system whenever the number of customers in the system goes from n to $n + 1$. Also, a departure leaves n customers in the system whenever the number of customers in the system goes from $n + 1$ to n. That is, there is no bulk arrival nor bulk service in the system. Since we consider the system in stationary state ($\rho < 1$), the overall arrival rate equals the overall departure rate. Given that customers arrive one at a time and are served one at a time, the proportion of arrivals that find n customers in the system is equal to the proportion of departures that leave n customers in the system. It is readily seen that $a_n = d_n$.

The property of PASTA—"Poisson Arrivals See Time Averages"—states that if the arrival process to the system is a Poisson process, then $a_n = p_n$ and consequently, $d_n = p_n$. That is, in a queue with Poisson arrivals, the limiting proportion of arrivals that find the system in some state n is equal to the limiting proportion of time the system spends in that state n. By PASTA is meant the equality of these two limiting fractions.

Intuitively, a brief proof is provided as follows. Viewing p_n and a_n as limiting probabilities, we have $p_n = \lim_{t\to\infty} \Pr\{N(t) = n\}$ and $a_n = \lim_{t\to\infty} \Pr\{N(t) = n|\text{an arrival occurred just after time } t\}$. We show that $a_n = p_n$.

Let $A(t, t+\delta)$ be the event that an arrival occurred just after time t. Then, we have

$$\begin{aligned}
a_n &= \lim_{t\to\infty}\lim_{\delta\to 0} \Pr\{N(t) = n|A(t, t+\delta)\} = \lim_{t\to\infty}\lim_{\delta\to 0} \frac{\Pr\{N(t) = n, A(t, t+\delta)\}}{\Pr\{A(t, t+\delta)\}} \\
&= \lim_{t\to\infty}\lim_{\delta\to 0} \frac{\Pr\{A(t, t++\delta)|N(t) = n\}}{\Pr\{N(t) = n\}\Pr\{A(t, t+\delta)\}} \\
&= \lim_{t\to\infty}\lim_{\delta\to 0} \frac{\Pr\{A(t, t+\delta)\}\Pr\{N(t) = n\}}{\Pr\{A(t, t+\delta)\}} \\
&= \lim_{t\to\infty} \Pr\{N(t) = n\} = p_n.
\end{aligned}$$

In the above derivation, the key step where the Poisson process assumption is used is in going from the third equals sign to the fourth line (note that $A(t, t+\delta)$ is independent of $N(t)$). It is the independent-increments property of the Poisson process that really accounts for PASTA.

Note that the above results hold also in the transient state—that is, $a_n(t) = p_n(t)$, for $n \geq 0, t \geq 0$. In other words, the PASTA property may be generalized to the case of a nonstationary Poisson arrival process with rate $\lambda(t)$. It also holds whether the system is a delay system or a loss system, for example, M/G/1/K. However, it does not hold in case of quasi-random input. That is, when the arrivals occur from a finite source, such as the machine repair problem (MRP), though the interarrival times are exponential random variables, $a_n(t) \neq p_n(t)$. In addition, PASTA can be generalized to the property of ASTA (Arrivals See Time Average), RSTA (Retrials See Time Average). Furthermore, there is another related question of interest: When does ASTA imply that the associated counting process is Poisson? It is called anti-PASTA and interested readers are referred to Green and Melamed (1990), Melamed and Whitt (1990), Melamed and Yao (1995), and we omit the details here.

1.3 Basic Concepts for Queueing Game Models

1.3.1 Costs and Objective Functions

We usually assume that the customer is risk neutral (that is, the expected payoff is equivalent to the realized payoff). The customer will get a reward after receiving the service, and needs to pay a fee in waiting. The customer's total cost refers to those direct or indirect cost incurred, and the customer's income is the difference between the service reward and the total cost. In a non-cooperative game, the individual's goal is to maximize his benefit.

From the perspective of social administration, the goal is to maximize the social benefits. The social benefits here refer to the total benefits of all members of the society (including customers and service providers). It should be noted that the benefit conversion among social members (between customers and service providers) has no impact on social benefits.

1.3.2 Information of System

When customers arrive at the system, they will determine their own joining strategy according to his reward-cost structure along with the system information available to them upon arrival. Taking the M/M/1 queueing systems as an example, according to the degree of information about the server's status and the queue length disclosed to customers by the system, these can usually be categorized into four cases as follows.

1. **Fully Observable Case (FO)**: In this case, the status of the server and the queue length are both observable when the customer arrives at the system.
2. **Almost Observable Case (AO)**: In this case, when the customer arrives at the system, the server's status is unobservable but the queue length is observable.
3. **Almost Unobservable Case (AU)**: In this case, when the customer arrives at the system, the server's status is observable, but the queue length is unobservable.
4. **Fully Unobservable Case (FU)**: In this case, when the customer arrives at the system, the server's status and queue length are both not observable.

1.3.3 Threshold Strategies

Suppose that the customer arriving at the system has two behavior choices according to the non-negative integer variable describing the system state, a_1 and a_2. For example, the status of the system can be described by the queue length, and the customer's behavior can be to enter the queue or balk after arriving at the system.

1. **Pure threshold strategy**
   Pure threshold strategy with threshold n indicates that one of the actions, say, A_1, for every state in $\{0, 1, \cdots, n-1\}$ and the other action, A_2, otherwise
2. **Mixed threshold strategy**
   The strategies $x = n + p, n \in \mathbb{N}, p \in [0, 1)$ specified in the mixed threshold strategy are:

$$\begin{cases} A_1, & \text{if } 0 \le i \le n-1, \\ pA_1 + (1-p)A_2, & \text{if } i = n, \\ A_2, & \text{if } i > n. \end{cases}$$

When x is an integer ($p = 0$), the mixed threshold strategy degenerates into a pure threshold strategy. Generally, in the observable (queue length observable) queueing system, customers mostly adopt the threshold strategy.

1.3.4 Price of Anarchy

The concept of Price of Anarchy (PoA) in economics and game theory measures how the efficiency of a system degrades due to selfish behavior of the players. It is a quite general notion that can be used in diverse systems as notions of efficiency. As an illustration, consider the system of transportation of a city and many commuters trying to go from some initial location to a destination. Here, efficiency means the average time for a commuter to reach the destination. In the "centralized" solution, a central authority can tell each commuter which path to take in order to minimize the average travel time. However, in the "decentralized" version, each commuter chooses its own path. The Price of Anarchy measures the ratio between average travel time in the two cases.

A variety of service systems can be modeled as a game and the efficiency of these systems is some function of the outcomes (e.g. congestion in a transportation system, maximum delay in a network, social welfare in an auction, etc.). Different concepts of equilibrium can be used to model the selfish behavior of the players, among which the most common is the Nash equilibrium. Different flavors of Nash equilibrium lead to variations of the notion of PoA as Pure Price of Anarchy (for deterministic equilibria), Mixed Price of Anarchy (for randomized equilibria), and Bayes-Nash Price of Anarchy (for games with incomplete information). Solution concepts other than Nash equilibrium lead to variations such as the Price of Sinking (see, for example, Goemans et al. 2005).

Consider a game defined by a set of players N, strategy sets S_i for each player and utilities $u_i : S \to \mathbb{R}$ (where $S = S_1 \times \cdots \times S_n$ also called set of outcomes). We can define a measure of efficiency of each outcome which we call welfare function $\text{Welf} : S \to \mathbb{R}$. Natural candidates include the sum of players utilities (utilitarian objective) $\text{Welf}(s) = \sum_{i \in N} u_i(s)$, minimum utility (fairness or egalitarian objective) $\text{Welf}(s) = \min_{i \in N} u_i(s)$, ..., or any function that is meaningful for the particular game being analyzed and is desirable to be maximized.

We can define a subset $\text{Equil} \subseteq S$ to be the set of strategies in equilibrium (for example, the set of Nash equilibria). The Price of Anarchy is then defined as the ratio between the optimal "centralized" solution and "worst equilibrium":

$$\text{PoA} = \frac{\max_{s \in S} \text{Welf}(s)}{\min_{s \in Equil} \text{Welf}(s)}.$$

If, instead of a "welfare" which we want to "maximize", the function measure efficiency is a "cost function" Cost:$S \to \mathbb{R}$ which we want to "minimize" (e.g. delay in a network) we use (following the convention in approximation algorithms):

$$\text{PoA} = \frac{\max_{s\in\text{Equil}} \text{Cost}(s)}{\min_{s\in S} \text{Cost (s)}}.$$

A related notion is that of the Price of Stability (PoS) which measures the ratio between the optimal "centralized" solution and the "best equilibrium":

$$\text{PoS} = \frac{\max_{s\in S} \text{Welf}(s)}{\max_{s\in\text{Equil}} \text{Welf}(s)},$$

or in the case of cost functions:

$$\text{PoS} = \frac{\min_{s\in\text{Equil}} \text{Cost}(s)}{\min_{s\in S} \text{Cost}(s)}.$$

It is evident that $1 \leq \text{PoS} \leq \text{PoA}$ by the definition. Accordingly, it is expected that the loss in efficiency due to game-theoretical constraints is somewhere between "PoS" and "PoA".

References

Bertsekas, Dimitri, and Robert G. Gallager. 1992. *Data Networks*. Prentice Hall.

Bocharov, P., C. D'Apice, A. V. Pechinkin, and S. Salerno. 2004. *Modern Probability and Statistics: Queueing Theory*. Utrecht, Boston: VSP Publisher.

Brumelle, S. L. 1971. On the relation between customer and time averages in queues. *Journal of Applied Probability* 8 (3): 508–520.

Cooper, Robert B. 1981. *Introduction to Queueing Theory*. New York: North Holland.

Fudenberg, Drew, and Jean Tirole. 1991. *Game Theory*. Cambridge, Massachusetts: The MIT Press.

Goemans, M., V. Mirrokni, and A. Vetta. 2005. Sink equilibria and convergence. In *FOCS 05*.

Green, L., and B. Melamed. 1990. An anti-PASTA result for Markovian systems. *Operations Research* 38 (1): 173–175.

Harchol-Balter, Mor. 2013. *Performance Modeling and Design of Computer Systems: Queueing Theory in Action*. New York, USA: Cambridge University Press.

Hassin, Refael, and Moshe Haviv. 2003. *To Queue or Not to Queue: Equilibrium Behavior in Queueing Systems*. Boston: Kluwer Academic Publishers.

He, Qi-Ming. 2014. *Fundamentals of Matrix-Analytic Methods*. New York, NY: Springer.

Heyman, D.P., and S. Stidham Jr. 1980. The relation between customer and time averages in queues. *Operations Research* 28 (4): 983–994.

Jewell, W. S. 1967. A simple proof of $L = \lambda W$. *Operations Research* 15 (6): 1109–1116.

Kendall, D. G. 1953. Stochastic processes occurring in the theory of queues and their analysis by the method of imbedded Markov chains. *The Annals of Mathematical Statistics* 24: 338–354.

Kleinrock, L. 1975. *Queueing Systems—Vol. 1: Theory*. New York, London, Sydney, Toronto: John Wiley & Sons Inc.

Kleinrock, L. 1976. *Queueing Systems—Vol. 2: Computer Applications*. New York, London, Sydney, Toronto: John Wiley & Sons Inc.

Little, J. D. C. 1961. A proof for the queueing formula: $L = \lambda W$. *Operations Research* 9: 383–387.

Little, J. D. C. 2011. Little's law as viewed on its 50th anniversary. *Operations Research* 59 (3): 536–549.

Melamed, B., and David D. Yao. 1995. The ASTA property. In *Advances in Queueing Theory, Methods, and Open Problems*, 195–224. CRC Press.
Melamed, B., and Ward Whitt. 1990. On arrivals that see time averages. *Operations Research* 38 (1): 156–172.
Myerson, Roger B., and Game Theory. 1997. *Analysis of Conflict*. Cambridge, Massachusetts: Harvard University Press.
Neuts, M. F. 1989. *Structured Stochastic Matrices of M/G/1 Type and Their Applications*. New York and Basel: Marcel Dekker.
Nisan, N., T. Roughgarden, E. Tardos, and V. V. Vazirani. 2007. *Algorithmic Game Theory, 2007*. New York: Cambridge University Press.
Shortle, J.F., J.M. Thompson, D. Gross, and C.M. Harris. 2018. *Fundamentals of Queueing Theory*. 5th ed. John Wiley & Sons.
Smith, J.M. 1982. *Evolution and the Theory of Games*. Cambridge University Press.
Stidham, S., Jr. 1974. A last word on $L = \lambda W$. *Operations Research* 22 (2): 417–421.
Tadelis, Steven. 2013. *Game Theory: An Introduction*. Princeton and Oxford: Princeton University Press.
Takagi, H. 1991. *Queueing Analysis: A Foundation of Performance Analysis, Volume 1: Vacation and Priority Systems, Part 1*, North-Holland, Amsterdam: Elsevier Science Publishers B.V.
Whitt, Ward. 1991. A review of $L = \lambda W$ and extensions. *Queueing Systems* 9 (3): 235–268.
Wolff, R. 2011. Little's law and related results. In *Wiley Encyclopedia of Operations Research and Management Science*, vol. 4, ed. J. Cochran. New York: Wiley. 2929-2841.
Zhang, Zhe George. 2023. *Fundamentals of Stochastic Models*. Taylor & Francis Group, Boca Raton: CRC Press.

Chapter 2
Observable Queueing Systems

In this chapter we focus on the queueing-game models with observable information provision. Here observable queues refer to the queueing systems in which customers can observe the state of the server and queue length upon arrival, i.e., the "FO" case mentioned in Sect. 1.3.2. Customers make their decision that maximizes their benefit based on these available system information in hand. The study of strategic behavior in observable queues dates back to the semina work of Naor (1969) . This pioneering work studies customer decisions in an observable M/M/1 queue in which joining-customers utility is linearly decreasing with the joining position. Naor derives the optimal threshold strategies for the individuals, social planner, and monopolist and proves that the monopoly optimal threshold is (weakly) smaller than the socially optimal threshold, which is (weakly) smaller than the individually optimal one. He also designs a procedure that computes the socially optimal joining strategy. Since then, a lot of research work has been expanded on the basis of this model.

Hassin (1985) extends the model of Naor (1969) in two aspects: (1) The service discipline is changed from FCFS to last-come, first-served with preemptive-resume (LCFS-PR), that is, a newly arrived customer accepts the service and is immediately served even if another customer is being serving at the moment. Preempted customers join a queue where later arrivals receive priority over earlier arrivals. When the preempted customer begins his service, it is resumed from the point of interruption. (2) Every customer is allowed to renege the system at any instant without any cost but never return. At any instant the queue is fully observable such that a customer can base his decision on the queue length and on his position in the queue. Hassin (1985) observes that this LCFS-PR discipline leads to a socially optimal behavior by the customers even without charging them But obviously, this discipline lacks of fairness among customers from a general perspective.

Subsequently, Nalebuff (1989) presents illustrative descriptions of the LCFS-PR model and its consequences. Larsen (1998) generalizes Naor's model from another perspective. He assumes that customers differ by their service values, that is, it is supposed that the service value R of an arriving customer is a random variable

J. Wang, *Fundamentals of Queueing-Game Models*,
https://doi.org/10.1007/978-981-95-0261-5_2

with a distribution function F. An arriving customer knows his value, but this is his private information and hence it cannot be used by the server to discriminate among customers. It is consistent with Naor's conclusion that the toll charged to customers in order to maximize profits is higher than or equal to the toll charged to customers when the social welfare is optimal.

Debo and Veeraraghavan (2013) further assume that the arriving customers do not know the information of service rate and service reward on the basis of Naor (1969) . They prove that there is no joining threshold strategy for customers in some cases. Besides, Schroeter (1982) modifies Naor's model by assuming that the time value is uniformly distributed over a finite interval, which simplifies the derivation of the profit-maximizing price. Hassin and Henig (1986) give the general conditions of the optimal threshold strategy for controlling the throughput of the system. De Vany (1976) considers an observable queue in which the demand for service is a function of the admission fee. Accordingly, the conclusion under the assumption of this model is that the actual equilibrium arrival rate of customers is much lower than the socially-optimal arrival rate because the service providers charge too much for customers. Miller and Buckman (1987) consider an M/M/s/s model in which there is no queue and an arriving customer who finds all servers busy must balk. The model assumes heterogeneous service values, so that for any given value of service fee, the demand consists of only those customers whose service value exceeds the fee. Miller and Buckman (1987) obtain the optimal fee to customers under the socially-optimal condition when the customer service value is a random variable. Guo and Zhang (2013) study the behavior of strategic customers in a queueing system with multiple servers. In this system, the number of servers with working mode on changes with the number of customers in the system.

In addition, there are some more observable queueing economic models. Chen and Frank (2004) extend Naor's model by assuming that both the customers and the server maximize their expected discounted utility using a common discount rate. They compute the profit and social welfare maximizing pricing schemes for this modified model. They also study the state dependent pricing issue of queueing economics models. They point out that when the service provider can set the price according to the system state (on the premise that the customers are the same, that is, the service value obtained after the completion of the service is the same), the result of maximizing the interests of the service provider and the result of maximizing social benefits can be agreed. The situation of different customer service values is also investigated. Borgs et al. (2014) discuss the optimal admission threshold in observable queues with state dependent pricing. Under dynamic pricing, they study the social welfare maximization issue in Naor's model and the profit maximization problem in Chen and Frank (2004).

It is evident that in the observable M/M/1 queueing system studied by Naor, the service times and the interarrival times of customers are memoryless (due to Markovian property). However, when these random variables no longer obey exponential distribution, the analysis of this kind of queueing system becomes difficult and complex. Altman and Hassin (2002) investigate the M/G/1 observable queues where the service times obey Bernoulli distribution and customers can decide to join or to balk the queueing system according to the queue length information they observed upon arrival. It is shown that the equilibrium strategy prescribes randomization when there is a customer in service and no queue, joining if the queue size is greater than one but less than some threshold, and balking otherwise. Whitt (1999) further discusses the customer equilibrium strategy of the observable M/G/s queueing systems. Kerner (2011) proposes a recursive algorithm to obtain the Nash equilibrium joining strategy of M/G/1 queueing systems. Zhu, Wang and Liu (2020) study the equilibrium joining strategy of M/G/1 queueing systems with server breakdowns and repairs. Mandelbaum and Yechiali (1983) suppose that in the M/G/1 queueing system, except for a special customer, all other arriving customers must join the system queue. The special customer can choose to enter the system, balk or wait until the next service is completed. They obtain the individually-optimal strategy of the special customer. Stidham (1978) considers the observable GI/M/1 queueing system where the inter-arrival times of customers follow a general distribution. In this model, the reward of service is a random variable, and the waiting time of customers is a convex function of the number of customers in the system. Consistent with Naor's conclusion, the socially-optimal strategy is smaller than the customer's individually-optimal strategy. Yechiali (1971, 1972) also generalize the results of Naor (1969) to more general GI/M/1 and GI/M/s queueing systems.

2.1 M/M/1 Queueing Systems

2.1.1 Model Description

Consider a first-come, first-served (FCFS) single-server queueing system in which customers arrive in the system according to a Poisson process with parameter λ and service times for customers are assumed to be independent, exponentially distributed with parameter μ. Customers are strategic in the sense they can decide whether to join the queue or to balk the system based on the information of the queue length observed upon arrival. A customer's benefit from completed service is R and his cost for staying in the system (either while waiting in queue or while being served) is C per unit of time. Customers are risk neutral and they maximize the expected value of their net benefit. Once a decision is made to join or to balk, it is irrevocable, and reneging during the waiting process is not allowed. A customer who balks leaves the system and never returns. Assume that $R > \frac{C}{\mu}$ holds, which ensures that when the system is empty, customers will choose to enter the queueing system to receive services.

2.1.2 Individually Optimal Strategy

When a customer arrives and sees i customers (including the served customers) in the system, the expected payoff after he chooses to join the team is

$$U_i = R - \frac{C}{\mu}(i + 1).$$

When his payoff is nonnegative, i.e., $i + 1 \leq \frac{R\mu}{C}$, the customer chooses to join the queueing system, otherwise, he chooses to balk. Accordingly, a pure threshold strategy can be obtained as

$$n_e = \left\lfloor \frac{R\mu}{C} \right\rfloor,$$

such that when the customer arrives and sees i customers in the system, if $i \leq n_e - 1$, he decides to join the system, where $\lfloor x \rfloor$ denotes the largest integer which is less than or equal to x. Otherwise, if $i \geq n_e$, he balks. Therefore, in the case of individual optimization, the maximum number of customers in the system does not exceed n_e, so it can also be regarded as an M/M/1/n_e queueing system.

2.1.3 Socially-Optimal Strategy

Suppose a maximum queue length of n, that is, an arriving customer will balk the system if he observe n customers queued in front of him. Denote the probability of this event (observing n customers in the system) by q_n, we can derive the following balance equations:

$$\begin{aligned} \lambda q_0 &= \mu q_1, \\ (\lambda + \mu) q_i &= \lambda q_{i-1} + \mu q_{i+1}, \quad 1 \leq i \leq n - 1, \\ \lambda q_{n-1} &= \mu q_n. \end{aligned}$$

By solving the equations and in virtue of the normalization condition, we can obtain

$$q_n = \frac{\rho^n (1 - \rho)}{1 - \rho^{n+1}}.$$

Therefore, by the results presented in Sect. 1.2.5, the effective arrival rate of customers is given by

$$\lambda(1-q_n) = \frac{\lambda(1-\rho^n)}{1-\rho^{n+1}},$$

and the expected queue length of customers in the system is given by

$$L_n = \sum_{i=0}^{n} i q_i = \frac{\rho}{1-\rho} - \frac{(n+1)\rho^{n+1}}{1-\rho^{n+1}}.$$

Also, the average social welfare per unit time is given by

$$S(n) = \lambda R(1-q_n) - CL_n = \frac{\lambda R(1-\rho^n)}{1-\rho^{n+1}} - C\Big(\frac{\rho}{1-\rho} - \frac{(n+1)\rho^{n+1}}{1-\rho^{n+1}}\Big).$$

Since $S(n)$ is a unimodel function, by virtue of the property that

$$\begin{cases} S(n-1) \le S(n), \\ S(n+1) < S(n), \end{cases}$$

we can obtain the solution n_{soc} that maximizes the expected social welfare per unit time. According to the relationship $n_{soc} \le n_e$, the individually-optimal decision deviates from the socially-optimal decision. This difference is caused by the negative externality effect of the individual optimal behavior. To make customers accept the overall optimal threshold of n_{soc}, an admission fee of p can be imposed to make customers' income satisfies (see Naor 1969)

$$\begin{cases} U_{n_{\rm soc}} = R - p - \frac{C}{\mu} n_{soc} \ge 0, \\ U_{n_{\rm soc}+1} = R - p - \frac{C}{\mu}(n_{\rm soc}+1) < 0, \end{cases}$$

i.e., the fee p satisfies

$$n_{\rm soc} = \left\lfloor \frac{(R-p)\mu}{C} \right\rfloor.$$

Therefore, the optimal admission fee p^* meets:

$$R - \frac{C}{\mu}(n_{\rm soc}+1) < p^* \le R - \frac{C}{\mu} n_{\rm soc}.$$

In fact, it is also possible to increase the waiting cost per unit time by C instead of reducing the reward R, so as to urge customers to adopt the overall optimal threshold of n_{soc}. That is, increase the waiting cost per unit time by t, so that the individually-optimal threshold is reduced to n_{soc}, and then t satisfies

$$n_{\rm soc} = \left\lfloor \frac{R\mu}{C+t} \right\rfloor.$$

2.1.4 Admission Revenue Maximization Strategy

Suppose that the server charges customers an admission fee of p, and every customer decides whether to join the queueing system according to such a fee along with available queue length information. That is, when a customer arrives at the system and sees i customers, if and only if $R \geq p + \frac{C(i+1)}{\mu}$, he will choose to join the queueing system. Therefore, when the maximum number of customers in the system is n, the maximum value of the admission fee of p charged is $R - \frac{Cn}{\mu}$. Subsequently, for the profit-maximizing service provider, the total profit per unit time is

$$Z(n) = \lambda(1 - q_n)p = \frac{\lambda(1 - \rho^n)}{1 - \rho^{n+1}}\left(R - \frac{Cn}{\mu}\right).$$

Therefore, the threshold to maximize total revenue needs to be satisfied

$$\begin{cases} Z(n-1) \leq Z(n), \\ Z(n+1) < Z(n). \end{cases}$$

The threshold for maximizing service provider's profit can be derived (see Naor 1969) and we denote it by n_m. By comparison, the threshold relationship can be obtained as $n_m \leq n_{soc} \leq n_e$, as well as $p_m > p^*$ This shows that if the social benefit is taken as the objective function, it is beneficial to levy a certain admission fee. However, if the service provider is to maximize its own benefit, it will impose an excessive admission fee of p_m on customers entering the system, but such a result can not achieve social optimization.

2.2 M/G/1 Queues

2.2.1 Model Description

Consider a first-come, first-served M/G/1 queueing system in which customers arrive according to a Poisson flow with a parameter of λ. The service times are independent and identically distributed (i.i.d.) with a general distribution function $G(\cdot)$ whose Laplace-Stieltjes transform (LST) is $G^*(s) = \int_{x=0}^{\infty} e^{-sx} \mathrm{d}G(x)$ Denote by $\bar{x}$ and $\bar{x}^2$ the first and second moments of the waiting time distribution, respectively. Upon arrival, customers can decide whether to enter the queueing system according to the queue length information. Once they make the decision to join the queue or balk, the decision is irrevocable. After the service is completed, each customer will receive a reward of R. The customer's stay fee per unit time in the system is C. Suppose $R > C\bar{x}$, so as to ensure that customers who find that the system is empty will choose to join the team.

2.2.2 Equilibrium Strategy

We first introduce some notations:

1. Let p_n denote the probability that a newly arrived customer will choose to join the system when he sees that there are $n$customers in the system, and define $P_n = (p_1, \ldots, p_n)P = P_\infty$
2. Define L and L_a represents the number of customers in the system at any time and arrival epoch, respectively.
3. Denote by S_R the remaining service time of the customer currently receiving service, use $\bar{r}_n(p)$ represents the expected remaining service time when the joining probability set is p and the number of customers in the system is n, i.e. $\bar{r}_n(p) = \mathbb{E}_p(S_R|L_a = n)$
4. $F_n^*(s)$ represents the Laplace-Stieltjes transformation of the remaining service time of the customer receiving service when there are n customers in the system.

By virtue of Corollaries 2.2.1 and 2.2.2 in Kerner (2008), we can obtain the following recursive formula of Laplace-Stieltjes transformation for the remaining service time:

$$F_n^*(s) = \frac{\lambda p_n}{s - \lambda p_n}\left(G^*(\lambda p_n)\frac{1 - F_{n-1}^*(s)}{1 - F_{n-1}^*(\lambda p_n)} - G^*(s)\right), \ n \geq 2, \tag{2.2.1}$$

with the initial condition

$$F_1^*(s) = \frac{\lambda p_1}{\lambda p_1 - s}\frac{G^*(s) - G^*(\lambda p_1)}{1 - G^*(\lambda p_1)}. \tag{2.2.2}$$

Similarly, we have

$$\bar{r}_n(P) = \frac{G^*(\lambda p_n)}{1 - F_{n-1}^*(\lambda p_n)}\bar{r}_{n-1}(p) - \frac{1}{\lambda p_n} + \bar{x}, \ n \geq 2, \tag{2.2.3}$$

with the initial condition

$$\bar{r}_1(P) = \frac{\bar{x}}{1 - G^*(\lambda p_1)} - \frac{1}{\lambda p_1}. \tag{2.2.4}$$

When a newly arrived customer sees n customers in the system, given a joining probability set of $\mathbf{P}$, his expected waiting time (including the time served) is:

$$p_n^e \in \arg\max_{0 \leq p \leq 1} p(R - C(\bar{r}_n(P^e) + n\bar{x})), \quad n \geq 1. \tag{2.2.5}$$

Notice that $F_n^*(s)$ and $\bar{r}_n(P_n)$ are both functions of $p_1, \ldots, p_n$, we can obtain (2.2.5) by the recursive approach with the help of the initial value $n = 1$. And therefore,

$$p_n^e = \begin{cases} 1, & R > C(\bar{r}_n(P_{n-1}^e, 1) + n\bar{x}), \\ p, & R = C(\bar{r}_n(P_n^e) + n\bar{x}), \qquad \forall p \in [0, 1]; \\ 0, & R < C(\bar{r}_n(P_{n-1}^e, 0) + n\bar{x}). \end{cases}$$

If $R = C(\bar{r}_n(P_n^e) + n\bar{x})$, the customer's payoff is equal to zero, and in this case she can choose to join or to balk the system, or with a certain probability she joins the system. It is readily seen that if n is large enough, e.g., $n \geq \frac{R}{C\bar{x}}$, we have $p_n^e = 0$

To obtain the equilibrium strategy P^e, we need to find the initial value p_1^e at first.

Theorem 2.2.1 Kerner (2011)

Assume that $L_a = 1$, then we have at least one of the following cases:

***Case 1** If*

$$\bar{x} + \max_{0 \leq p \leq 1} \left\{ \frac{\bar{x}}{1 - G^*(\lambda p)} - \frac{1}{\lambda p} \right\} \leq \frac{R}{C},$$

then joining is a dominant strategy

***Case 2** If*

$$\bar{x} + \max_{0 \leq p \leq 1} \left\{ \frac{\bar{x}}{1 - G^*(\lambda p)} - \frac{1}{\lambda p} \right\} \geq \frac{R}{C},$$

then not joining is dominant strategy

***Case 3** If a dominant strategy does not exist, then at least one of the following holds*

3a If $\bar{x} + \frac{\bar{x}}{1-G^*(\lambda p)} - \frac{1}{\lambda} \leq \frac{R}{C}$, then the choice of joining with probability one is an equilibrium.

3b If $\lim_{p \to 0} \left(\bar{x} + \frac{\bar{x}}{1-G^*(\lambda p)} - \frac{1}{\lambda p} \right) = \bar{x} + \frac{\bar{x}^2}{2\bar{x}} \geq \frac{R}{C}$, then the choice of balking with probability one is an equilibrium

3c There exists a (not necessarily unique) mixed Nash equilibrium p_1^e, for $0 < p_1^e < 1$, where $\bar{x} + \frac{\bar{x}}{1-G^*(\lambda p_1^e)} - \frac{1}{\lambda p_1^e} = \frac{R}{C}$. Moreover, any p_1^e that solves this equation is an equilibrium.

Proof If Case 1 (Case 2, respectively) holds, then the individual reward from joining is positive (negative) regardless of the behavior of other customers. Hence, a dominant strategy for him is to join (to not join).

If the inequality in Case 3a (3b, respectively) holds, and all join (do not join), then the individual s net gain from joining is positive (negative). Therefore, one's best response is to join (not to join) and hence $p_1^e = 1(p_1^e = 0)$

If there exists a p_1^e as defined in Case 3c, and if all those who find one customer in the system upon their arrival join with probability p_1^e, then the individual's net gain from joining is zero. Hence, this individual has no better response against p_1^e than the same probability. Therefore, p_1^e is a Nash equilibrium. Finally, as $\bar{r}_1(p)$ is a continuous function in p, if neither Case 1 nor Case 2 hold, then Case 3c must hold for some (not necessarily unique) p_1^e, $0 < p_1^e < 1$. □

Once $P^e_{n-1}, n \geq 1$, is available (no matter if it is unique or not), the value(s) of P^e_n can be derived and determined. Similar to the proof of Theorem 2.2.1, we can get the following theorem.

Theorem 2.2.2 Kerner (2011)

For any equilibrium probabilities P^e_{n-1}, the (not necessarily unique) equilibrium joining probability P^e_n is determined by the not necessarily mutually exclusive (but comprehensive) cases:

Case 1 *If $\bar{r}_n(P^e_{n-1}, 1) + n\bar{x} \leq \frac{R}{C}$, then $p^e_n = 1$ is an equilibrium.*

Case 2 *If $\bar{r}_n(P^e_{n-1}, 0) + n\bar{x} \geq \frac{R}{C}$, then $p^e_n = 0$ is an equilibrium*

Case 3 *For $\forall p$, $0 < p < 1$, such that $\bar{r}_n(P^e_{n-1}, p) + n\bar{x} = \frac{R}{C}$, then $p^e_n = p$ is an equilibrium*

When the probability distribution of system service time is different, the customer's equilibrium strategy may or may not be unique.

If the service distribution belongs to the class of decreasing mean residual life (DMRL), then for any $n \geq 1$ and for any $P_{n-1} \in (0, 1]^{n-1}$, $\bar{r}_n(P_{n-1}, p)$ is increasing with p. In this case the customer's Nash equilibrium strategy is unique.

If the service distribution belongs to the class of increasing mean residual life (IMRL), then for any $n \geq 1$ and for any $P_{n-1} \in (0, 1]^{n-1}$, $\bar{r}_n(P_{n-1}, p)$ is decreasing with p. In this case a pure Nash equilibrium exists. Furthermore, the Nash equilibrium is not necessarily unique, and all equilibria are of the structure given in Proposition 3.1 of Kerner (2011).

2.3 GI/M/c Queueing Systems

2.3.1 Model Description

We consider a GI/M/s queueing system where customers arrive at instants $\tau_0, \tau_1, \tau_2, \ldots, \tau_m, \ldots$, and the interarrival times $\tau_{m+1} - \tau_m$, $m = 0, 1, 2, \ldots$, are independent, identically distributed positive random variables with common distribution function $H(\cdot)$ and finite mean $\frac{1}{\lambda}$. There are $c \geq 1$ identical servers in the system and the distribution of service times is exponential with parameter μ. Denote by $\eta(t)$ the queue size (customers waiting or being served) at instant t, $\eta_m = \eta(\tau_m - 0)$ the queue size just before the mth arrival. The system is said to be in state i at the mth step if $\eta_m = i$. Suppose that upon arrival each customer can take one of two actions: either join the queue or balk. Let $\{\Delta_m, m = 0, 1, 2, \ldots\}$ denote the sequence of decisions (i.e., actions) made by the arriving customers. Let $\Delta_m = 1$ stand for the decision to join and $\Delta_m = 0$ for the decision to balk. Therefore, regarding the system as making its transitions at instants of arrival, we obtain a Markovian Decision Process $\{\eta_m, \Delta_m\}$, $m = 0, 1, 2, \ldots$, whose stationary transition probabilities $q_{ij}(k) = \Pr\{\eta_{m+1} = j \mid \eta_m = i, \Delta_m = k\} i, j = 0, 1, 2, \ldots, k = 0, 1$ are calculated as follows:

(1) If $j > i + 1$, $q_{ij}(1) = 0$ for all $i = 0, 1, 2, \ldots$
(2) If $j \le i + 1 \le c$, then we have

$$q_{ij}(1) = \binom{i+1}{j} \int_0^\infty e^{-j\mu x}(1 - e^{-\mu x})^{i+1-j} \mathrm{d}H(x).$$

(3) If $i + 1 \ge j \ge c$ and $i \ge c$, then

$$q_{ij}(1) = \int_0^\infty e^{-c\mu x} \frac{(c\mu x)^{i+1-j}}{(i+1-j)!} \mathrm{d}H(x).$$

(4) If $i + 1 > c > j$,

$$\begin{aligned} q_{ij}(1) &= \int_0^\infty \left[\int_0^x \frac{(c\mu)^{i+1-c}}{(i-c)!} t^{i-c} e^{-c\mu t} \binom{c}{j} e^{-\mu(x-t)j} (1 - e^{-\mu(x-t)})^{c-j} \mathrm{d}t \right] \mathrm{d}H(x) \\ &= \binom{c}{j} \int_0^\infty e^{-j\mu x} \left[\int_0^x \frac{(c\mu t)^{i-c}}{(i-c)!} (e^{-\mu t} - e^{-\mu x})^{c-j} c\mu \mathrm{d}t \right] \mathrm{d}H(x). \end{aligned}$$

It is readily seen that:

$$q_{ij}(0) = q_{ij}(1), \quad i = 1, 2, \ldots, j = 0, 1, 2, \ldots, \tag{2.3.1}$$

and $q_{00}(0) = 1$

Let D_{ik} denote the stationary conditional probability of taking action k when state i is observed. That is, $D_{ik} = \Pr\{\Delta_m = k \,|\eta_m = i\}$, $m = 0, 1, 2, \ldots$ Since $k = 0$ or 1 let $D_{i1} = D_i$ and let $D_{i0} = 1 - D_i$. The sequence D_i will be called a balking sequence. For any given balking sequence, D_i, the sequence of random variables $\{\eta_m\}$ forms a homogeneous Markov chain (imbedded at instants of arrival) with transition probabilities

$$p_{ij} = q_{ij}(1)D_i + q_{ij}(0)(1 - D_i), \quad i, j = 0, 1, 2, \ldots. \tag{2.3.2}$$

Let ${p_{ij}}^{(t)}$ denote the t-step transition probability from state i to state j. Then it is well known that the limiting probabilities $\pi_j = \lim_{t\to\infty} {p_{ij}}^{(t)}$, $i, j = 0, 1, 2, \ldots$ always exist, they are nonnegative, and for an ergodic chain they form a distribution which uniquely satisfies the set of linear equations

$$\pi_j = \sum_i \pi_i p_{ij}, \quad \sum_j \pi_j = 1. \tag{2.3.3}$$

Suppose now that the service facility has a limited waiting room of size $n \geq 0$, which means that there could be at most $n + c$ customers in the system. An arrival balks with probability one if he finds that all the c servers and all the n waiting positions are occupied. Customers who find the system in state $i < c + n$ join the the queue with a positive probability. The system is equivalent to a balking sequence $\{D_i\}$ for which $0 < D_i \leq 1(i < c + n)$ for $i < c + n$ and $D_i = 0(i \geq c + n)$ for all $i \geq c + n$. A GI/M/c queueing process with such a balking sequence $\{D_i\}$ will be denoted as a GI/M/c/n queue. It is well-known that such a GI/M/c/n process may be imbedded at instants of arrival to yield a homogeneous Markov chain whose transition probabilities $\{p_{ij}(n)\}$ are given by

$$p_{ij}(n) = q_{ij}(1)D_i + q_{ij}(0)(1 - D_i), \quad i < c + n, j \leq c + n, \tag{2.3.4}$$

$$p_{c+n,j}(n) = q_{c+n,j}(0), \quad j \leq c + n. \tag{2.3.5}$$

Since this is a finite Markov chain and thus obtained is aperiodic and irreducible, the limiting probabilities of the $GI/M/c/n$ process, $\pi_j = \lim_{t\to\infty} p_{ij}{}^{(t)}$, are all positive and satisfy the equation of (2.3.3)

2.3.2 *Individually Optimal Strategy*

We consider a linear cost-reward structure associated with the queueing process. A reward R is obtained by each customer upon successful completion of his service. A service charge (toll) θ is paid by the customer to the service agency for the service rendered. The waiting costs for each arrival who joins the queue are incurred at the rate of $C > 0$ per unit time. The decision to balk is accompanied with a fixed penalty l. Let $g = R - \theta$ be the net reward of a customer who has been served, and let $b = g + l$ be the benefit of a customer who joins the queue. In order to eliminate trivialities we assume that $b \geq \frac{C}{\mu}$

Regarding a policy S, it can be interpreted as a set of functions $\left\{D_k^S(H_{m-1}, \eta_m)\right\}$ $(m \geq 0)$, where $H_{m-1} = \{\eta_0, \Delta_0, \ldots, \eta_{m-1}, \Delta_{m-1}\}$ is to be interpreted as the collection of the first $M - 1$ customer behaviors, $D_k^S(H_{m-1}, \eta_m)$ is the probability of implementing decision k $(k = 0, 1)$ at time m given the past information or history H_{m-1} and the present state η_m, $D_k^S(\cdot) \geq 0$ $\sum_{k=0}^{1} D_k^S(\cdot) = 1$ Note that in the preceding sections we have assumed that

$$D_k^S(H_{m-1}, \eta_m \,|\eta_m = i) = D_k^S(H_{m-1}, i) = D_{ik}, \quad m = 0, 1, 2, \ldots,$$

thus we have obtained the Markov chains represented by (2.3.2), (2.3.4) and (2.3.5).

Let $B_m (m \geq 0)$ denote the net benefit obtained at time τ_m, and define its stationary conditional expectation as:

$$\mathbb{E}\{B_m \,|\eta_m = i \ , \Delta_m = k\} = b_{ik}$$

As assumed, an arriving customer decides whether to join the queue or not based on the observation of the queue length i observed upon arrival. If $i < c$, obviously the customer joins the system and spends there an expected total time of $\frac{1}{\mu}$. If $i \geq c$, there is an expected total time of $\frac{i-c+1}{c\mu} + \frac{1}{\mu} = \frac{i+1}{c\mu}$. We immediately have:

$$b_{i0} = -l, \quad \text{for all } i \geq 0, \tag{2.3.6}$$

$$b_{i1} = \begin{cases} g - \frac{C}{\mu}, & \text{if } i < c, \\ g - \frac{C}{c\mu}(i+1), & \text{if } i \geq c. \end{cases} \tag{2.3.7}$$

For customers who pursue individual optimization, they want to find a strategy composed of joining probabilities $D_1^S(H_{m-1}, i)(D_0^S(\cdot) = 1 - D_1^S(\cdot))$ such that for every history H_{m-1} and every i,

$$\mathbb{E}(B_m \,|\eta_m = i) = D_1^S(H_{m-1}, i)b_{i1} + [1 - D_1^S(H_{m-1}, i)]b_{i0},$$

is maximized. By (2.3.6) and (2.3.7) we have

$$\mathbb{E}(B_m \,|\eta_m = i) = \begin{cases} D_1^S(H_{m-1}, i)\Big(g - \frac{C}{\mu}\Big) - [1 - D_1^S(H_{m-1}, i)]l, & \text{if } i < c, \\ D_1^S(H_{m-1}, i)\Big[g - \frac{C}{c\mu}(i+1)\Big] - [1 - D_1^S(H_{m-1}, i)]l, & \text{if } i \geq c. \end{cases}$$

Since $b_{i1} = g - \frac{C}{c\mu}(i+1)$ is a decreasing function of i then, for $i > c$, there exists a smallest integer $n(c) \geq 0$, defined by $i = c + n(c)$, for which balking is better than joining. That is, $b_{c+n(c),1} < -l$ whereas $b_{c+n(c),1} \geq -l$. It follows then that $n(c)$ satisfies

$$g - \frac{C}{c\mu}(c + n(c) + 1) < -l \leq g - \frac{C}{c\mu}(c + n(c)). \tag{2.3.8}$$

It is therefore clear that for every history H_{m-1} and every i, the policy

$$S = \Big\{D_1^S(\cdot) \,\Big|\, D_1^S(H_{m-1}, i) = 1, i \leq c + n(s) - 1;\ D_1^S(H_{m-1}, i) = 0, i \geq c + n(s)\Big\} \tag{2.3.9}$$

is the one which maximized $\mathbb{E}(B_m \,|\eta_m = i)$. Note that if strict inequality holds in (2.3.8) then (2.3.9) is the unique optimal policy. From (2.3.8) we immediately obtain

$$\frac{c\mu}{C}\Big(b - \frac{C}{\mu}\Big) - 1 < n(s) \leq \frac{c\mu}{C}\Big(b - \frac{C}{\mu}\Big). \tag{2.3.10}$$

Since the policy S given by (2.3.9) is applied by all customers, the current $GI/M/c$ queueing process reduces to the $GI/M/c/n(c)$ queueing system with balking sequence

$$\{D_i \,|D_i = 1, i \leq c + n(c) - 1;\; D_i = 0, i \geq c + n(c)\}.$$

Such a sequence is called a *deterministic control limit rule* (see Yechiali 1971) with control limit being equal to $c + n(c)$. Therefore the optimal threshold joining strategy of customers is: if and only if the number of customers in the system is less than $c + n(c)$, the arriving customers will choose to join the queue.

2.3.3 Socially Optimal Strategy

For system administrators, the objective now is to find a rule or policy S which maximizes the long-ruin average expected net benefit of the entire customers' population. For any given policy S and an initial state $\eta_0 = j$ the expected net benefit at instant τ_m is:

$$\mathbb{E}_S(B_m \,|\eta_0 = j) = \sum_i \sum_{k=0}^{1} b_{ik}\mathbb{P}_S(\eta_m = i, \Delta_m = k \,|\eta_0 = j), \tag{2.3.11}$$

where $\mathbb{E}_S$ and $\mathbb{P}_S$ denote the expectation and probability under the policy S. Let

$$\phi_S(j) = \lim_{T\to\infty} \sup \left[\frac{1}{T+1}\sum_{m=0}^{T} \mathbb{E}_S(B_m \,|\eta_0 = j)\right], \tag{2.3.12}$$

be the long-run average expected net benefit.

For social optimization we seek the rule $S \in C_s$ that maximizes $\phi_S(j)$ for all j, where C_s is the class of all Markovian stationary policies $S = \{D_i \,|0 \leq D_i \leq 1, \text{for all } i\}$.

Consider now the class C_{DCL} of all deterministic control limit policies S_k^D, where we say that $S_k^D \in C_{DCL}$ if $S_k^D = \{D_i \,|D_i = 1, i \leq k;\; D_i = 0, i \geq k\}$ for some $k = 1, 2, \ldots$ and we let $S_\infty^D = \{D_i|D_i = 1, \text{for all } i = 0, 1, 2, \ldots\}$ Clearly, $C_{DCL} \subset C_s$ Following the procedure adopted in (1971), in searching for an optimal policy for social optimization in C_s, it is sufficient to consider only policies $S_k^D \in C_{DCL}$ where S_∞^D may be considered only when $\lambda < c\mu$. Moreover, since $b \geq \frac{G}{\mu}$ (i.e., $g - \frac{G}{\mu} \geq -l$) implies that the expected net benefit for an individual customer is always nonnegative if he joins the system when it is in state $i < c$, we may consider only policy S_k^D with $k \geq c$

Now, for any $k \geq c$ and $S_k = \{D_i \,|D_i > 0, i \leq k;\ D_i = 0, i \geq k\}$ applying Chung's well-known theorem Chung (1967) to expressions (2.3.11) and (2.3.12) yields, independently of the initial state $\eta_0 = j$,

$$\phi_{S_k} = \phi_{S_k}(j) = \sum_{i=0}^{k} \pi_i(S_k)\left[D_i b_{i1} + (1 - D_i) b_{i0}\right], \tag{2.3.13}$$

where the $\{\pi_i(S_k)\}$'s are the corresponding limiting probabilities of the process under S_k and they satisfy (2.3.3)

Using (2.3.7) for (2.3.13) yields:

$$\phi_{S_k} = \sum_{i=0}^{c-1} D_i \pi_i(S_k)\left(b - \frac{C}{\mu}\right) + \sum_{i=c}^{k} D_i \pi_i(S_k)\left[b - \frac{C}{c\mu}(i+1)\right] - l, \tag{2.3.14}$$

In particular, if we consider S_k^D instead of S_k we have $(D_k = 0)$:

$$\phi_{S_k} = \sum_{i=0}^{c-1} D_i \pi_i\left(S_k^D\right)\left(b - \frac{C}{\mu}\right) + \sum_{i=c}^{k-1} \pi_i\left(S_k^D\right)\left[b - \frac{C}{c\mu}(i+1)\right] - l. \tag{2.3.15}$$

In the sequel we will show that the optimal policy for social optimization is a control limit policy with finite control limit, $c + n(p)$, and that self-optimization need not bring upon social optimization. That is,

$$\phi_{S^D_{c+n(p)}} = \sup_{S \in C_s} \phi_S, \quad \phi_{S^D_{c+n(p)}} \geq \phi_{S^D_{c+n(s)}}.$$

For this purpose, the system is assumed physically limited to a waiting room of size $n > n(c)$ and by virtue of Howard's algorithm (1960) , we start the "value determination operation" with a policy

$$S^D_{c+n(s)} = \{D_i \,|D_i = 1, i < c + n(s);\ D_i = 0,\ c + n(s) \leq i \leq c + n\}.$$

We thus have to find $\phi, v_0, v_1, \ldots, v_{c+n}$ that satisfy

$$\begin{aligned}
\phi + v_i &= g - \tfrac{C}{\mu} + \sum_{j=0}^{i+1} q_{ij}(1)\, v_j, \quad 0 \leq i \leq c-1, \\
\phi + v_i &= g - \tfrac{C}{c\mu}(i+1) + \sum_{j=0}^{i+1} q_{ij}(1)\, v_j, \quad c \leq i \leq c + n(s) - 1, \\
\phi + v_i &= -l + \sum_{j=0}^{i} q_{ij}(0)\, v_j, \quad c + n(s) \leq i \leq c + n.
\end{aligned} \tag{2.3.16}$$

We need the following Lemma for solving (2.3.16).

Lemma 2.3.1 *For any solution of (2.3.16) we have*

$$v_0 \geq v_1 \geq \ldots \geq v_{c+n(s)-1} \geq v_{c+n(s)} \geq \ldots \geq v_{c+n}.$$

Proof We first subtract the $c + n(sc) + 1$st row of (2.3.16) from the $c + n(c)$th row to obtain

$$\begin{aligned} v_{c+n(c)-1} - v_{c+n(c)} &= g - \frac{C}{c\mu}(c + n(c)) + l + \sum_{j=0}^{c+n(c)} [q_{c+n(c)-1,j}(1) - q_{c+n(c),j}(0)]v_j \\ &= g - \frac{C}{c\mu}(c + n(c)) + l \geq 0. \end{aligned}$$

Similar to the proof given in Yechiali (1971), we use backward induction and forward induction respectively to show that $v_0 \geq v_1 \geq \ldots \geq v_{c+n(c)-1}$ and $v_{c+n(c)} \geq v_{c+n(c)+1} \ldots \geq v_{c+n}$. □

We can now prove the following:

Theorem 2.3.1 *There exists $n(p) \leq n(s)$ such that $\phi_{S^D_{c+n(p)}} = \sup_{S \in C_s} \phi_S$*

Proof Since only deterministic control limit rules need to be considered, it suffices to show that a rule S^D_{c+n} for some $n > n(c)$ is never an improvement on $S^D_{c+n(c)}$. To show this we define for every i,

$$g_i(k) = b_{ik} + \sum_j q_{ij}(k)\, v_j, \quad k = 0, 1.$$

and therefore, it needs to prove that for all $n(s) \leq i \leq n$ we have that $g_{s+i}0 > g_{s+i}(1)$ In virtue of (2.3.16) and the fact that $q_{c+i,j}(1) = q_{c+i+1,j}(0)$, we get

$$\begin{aligned} & g_{c+i}(1) - g_{c+i}(0) \\ &= g - \frac{C}{c\mu}(c + i + 1) + \sum_{j=0}^{c+i+1} q_{c+i+1,j}(0)\, v_j - \left[-l + \sum_{j=0}^{c+i} q_{c+i,j}(0)\, v_j \right] \\ &= g - \frac{C}{c\mu}(c + i + 1) + (\phi + v_{c+i+1} + l) - \left[-l + (\phi + v_{c+i}) \right] \\ &= g - \frac{C}{c\mu}(c + i + 1) + l + v_{c+i+1} - v_{c+i}. \end{aligned}$$

From (2.3.8) it follows that for $i \geq n(s)$, $g - \frac{G}{c\mu}(c + i + 1) + l < 0$. From Lemma 2.3.1 it follows that $v_{c+i+1} - v_{c+i} \leq 0$ Therefore, $g_{c+i}(0) > g_{c+i}(1)$ hold for any $n(c) \leq i < n$ and therefore, $S^D_{c+n(c)}$ is better than S^D_{c+n}. That is, $\phi_{S^D_{c+n(s)}} > \phi_{S^D_{c+n}}$ for all $n > n(c)$. This implies immediately that $\phi_{S^D_{c+n(s)}} > \phi_{S^D_\infty}$. Hence there exists $n(p) \leq n(c)$ such that

$$\phi_{S^D_{c+n(p)}} = \sup_{S \in C_{DCL}} \phi_S = \sup_{S \in C_s} \phi_S.$$

□

And $n(p)$ could now be calculated by solving a linear program as follows.

$$\text{Maximize} \left\{ \sum_{i=0}^{c-1} \chi_{i1}\left(b - \frac{G}{\mu}\right) + \sum_{i=c}^{n(s)} \chi_{i1}\left[b - \frac{G}{c\mu}(i+1)\right] \right\}$$

$$\text{Subject to} \sum_{k=0}^{1} \chi_{jk} - \sum_{i=0}^{n(s)} \sum_{k=0}^{1} \chi_{ik} q_{ij}(k) = 0, \quad j = 0, 1, \ldots, n(s),$$

$$\sum_{i=0}^{n(s)} \sum_{k=0}^{1} \chi_{ik}, \quad \chi_{ik} \geq 0, \quad i = 0, 1, \ldots, n(s); k = 0, 1,$$

where

$$\chi_{ik} = \pi_i(S^D_{c+n(s)}) D_{ik}, \quad i = 0, 1, \ldots, n(s); k = 0, 1,$$

i.e.,

$$\chi_{i1} = \pi_i(S^D_{c+n(s)}) D_i, \quad \chi_{i0} = \pi(S^D_{c+n(s)})(1 - D_i).$$

Since there exist at most $n(c) + 1$ variables χ_{ik} will assume positive values in the optimal solution, and since $\chi_{i1} = 0 \Rightarrow \pi_i = 0$, $n(p)$ is readily obtained from:

$$n(p) = \max\{i \,|\chi_{i1} > 0\}.$$

□

References

Altman, Eitan, and Refael Hassin, 2002. Non-threshold equilibrium for customers joining an $M/G/1$ queue. In *ISDG2002*, vol. I, II, 56–64. St. Petersburg: Saint Petersburg State University Institute of Chemistry.

Borgs, C., J. T. Chayes, S. Doroudi, M. Harchol-Baltera, and K. Xu. 2014. The optimal admission threshold in observable queues with state dependent pricing. *Probability in the Engineering and Informational Sciences* 28: 101–119.

Chen, Hong, and Murray Frank. 2004. Monopoly pricing when customers queue. *IIE Transactions* 36: 569–581.

Chung, K. L. 1967. *Markov Chains with Stationary Transition Probabilities*, 2nd ed. New York: Springer.

De Vany, A. 1976. Uncertainty, waiting time, and capacity utilization: A stochastic theory of product quality. *Journal of Political Economy* 84: 523–541.

Debo, L. G., and Senthil Veeraraghavan. 2013. Equilibrium in queues under unknown service rates and service value. *Operations Research* 62: 38–57.
Guo, Pengfei, and Zhe George Zhang. 2013. Strategic queueing behavior and its impact on system performance in service systems with the congestion-based staffing policy. *Manufacturing & Service Operations Management* 15: 118–131.
Hassin, Refael. 1985. On the optimality of first come last served queues. *Econometrica* 53: 201–202.
Hassin, Refael, and M. Henig. 1986. Control of arrivals and departures in a state-dependent input-output system. *Operations Research Letters* 5: 33–36.
Howard, R. A. 1960. *Dynamic Programming and Markov Processes*. New York: Wiley.
Kerner, Y. 2008. The conditional distribution of the residual service time in the $M_n/G/1$ queue. *Stochastic Models* 24: 364–375.
Kerner, Y. 2011. Equilibrium joining probabilities for an M/G/1 queue. *Games and Economic Behavior* 71: 521–526.
Larsen, C. 1998. Investigating sensitivity and the impact of information on pricing decisions in an M/M/1/∞ queueing model. *International Journal of Production Economics* 56 (57): 365–377.
Mandelbaum, A., and U. Yechiali. 1983. Optimal entering rules for a customer with wait option at an M/G/1 queue. *Management Science* 29: 174–187.
Miller, B. L., and A. G. Buckman. 1987. Cost allocation and opportunity costs. *Management Science* 33: 626–639.
Nalebuff, B. 1989. The arbitrage mirage, wait watchers, and more. *Journal of Economic Perspectives* 3: 165–174.
Naor, Pinhas. 1969. The regulation of queue size by levying tolls. *Econometrica* 37: 15–24.
Schroeter, R. 1982. The costs of concealing the customer queue, In *Working Paper*. Department of Economics, Iowa State University.
Stidham, S., Jr. 1978. Socially and individually optimal control of arrivals to a GI/M/1 queue. *Management Science* 24: 1598–1610.
Whitt, Ward. 1999. Improving service by informing customers about anticipated delays. *Management Science* 45: 192–207.
Yechiali, U. 1971. On optimal balking rules and toll charges in the GI/M/1 queuing process. *Operations Research* 19: 349–370.
Yechiali, U. 1972. Customers' optimal joining rules for the GI/M/s queue. *Management Science* 18: 434–443.
Zhu, Sheng, Jinting Wang, and Bin Liu. 2020. Equilibrium joining strategies in the $M_n/G/1$ queue with server breakdowns and repairs. *Operational Research: An International Journal* 20: 2163–2187.

Chapter 3
Unobservable Queueing Systems

For the unobservable queueing system, customers do not know the system queue length information before making behavior decisions, which also increases the uncertainty of the game between customers. Edelson and Hildebrand (1975) consider the Naor's model under the assumption that an arriving customer does not know the number of customers in the system, i.e., the FU case mentioned in Sect. 1.3. Littlechild (1974) studies the model of Edelson and Hildebrand (1975) by assuming the service reward is a function of arrival rate of customers. Chen and Frank (2004) discuss the robustness of the main results in Edelson and Hildebrand (1975), and point out that when the customer's preference is no longer linear, the service provider will choose the socially optimal admission cost. Balachandran (1991) considers the unobservable M/G/1 queueing model in which the service works with a fixed cost, and the cost does not depend on the number of customers and service time. He proves that the equilibrium arrival rate of customers is unique, and discussed the relationship between the arrival rate and the cost of service operation. Guo et al. (2011) study the customer equilibrium and socially optimal joining strategy in a queueing system with partially observable service time information. Guha et al. (2014) investigate the GI/M/C queueing system with multiple servers and impatient customers, and obtain the unobservable customer equilibrium joining strategy under the linear cost-reward structure.

In the queueing-game scenarios, the disclosure of information such as queue length may help or harm the benefits of customers and service providers. Hassin (1986) points out that for a service provider seeking to maximize benefits, depending on the setting of parameters, disclosing queue length information to customers may be beneficial or unfavorable to social benefits. Guo and Zipkin (2007) assume that the waiting cost of customers follows a uniform distribution on [0,1], and considered three delay information levels: no information, queue length and exact waiting time. They gave a method to obtain the performance indexes of these three systems. It is proved that when more accurate delay information is given, the revenue of customers and service providers will not necessarily increase. Allon and Bassamboo

J. Wang, *Fundamentals of Queueing-Game Models*,
https://doi.org/10.1007/978-981-95-0261-5_3

(2011) show the impact of delaying the provision of delay information on the service provider's ability to affect customer behavior. Armony et al. (2009) study the performance impact of disclosing delay information to newly arrived customers in a multi-server queueing system that allows customers to leave. In the model considered by Allon et al. (2011), service providers disclose non-verified dynamic delay information to customers, so a game is formed between service providers and customers. Hassin (1996) concludes that when service providers are in a competitive relationship, the service providers who disclose the queue length information have more advantages than those who do not. Whitt (1999) assumes that the increase of the amount of information will reduce the possibility of customers stopping, and shows how to improve the service by providing customers with delay information. Arnott et al. (1996; 1999) analyze that in a congested system, when the service capacity and demand fluctuate, the impact of information disclosure on customer participation and time use decisions. Shone et al. (2013) indicate that based on the different settings of arrival rate parameters and whether customers are selfish, whether service providers disclose queue length information may or may not affect customers' joining rate.

This chapter is mainly based on the Hassin and Haviv (2003). For the purpose of completeness we recapitulate these results on unobservable queueing game models with minor modifications.

3.1 The M/M/1 Queueing System

3.1.1 Model Description

Consider a first-come first-served single server queueing system, the arrival of customers follows the Poisson flow with the parameter Λ, and the service time follows the exponential distribution with the parameter μ. Assuming that customers can not observe the system queue length when they arrive, however, the choice of whether to enter the queue must be made. Each customer will receive a reward of R. And the unit waiting fee during their stay in the system (including waiting in line and being served) is C. In addition, each customer entering the queue needs to pay an admission fee of p. Assuming that customers are risk neutral, and their decisions are irrevocable. As usual, suppose $R\mu \geq C$ to ensure that customers who find that the system is empty will choose to join the system.

3.1.2 Nash Equilibrium Strategy

Let us start by studying the customers' behavior in equilibrium when an admission fee of size p is imposed and the potential arrival rate is Λ. Evidently, for an arriving customer, there are two pure strategies available: to join or not to join the queue. A pure or mixed strategy can be described by a fraction q $(0 \leq q \leq 1)$, which represents the probability of entering the system. Given that the admission fee is p, we denote the equilibrium probability of joining by $q_e(p)$ and $\lambda(p)$ the effective arrival rate. Obviously, λ needs to be satisfied under equilibrium conditions $\lambda_e(p) = \Lambda q_e(p) < \mu$. It is well known that when the effective arrival rate is $\lambda < \mu$, the expected waiting time of customers is $W(\lambda) = \frac{1}{\mu-\lambda}$; Otherwise, if $\lambda \geq \mu$, define $W(\lambda) = \infty$. This function is a continuous and monotone increasing function of the arrival rate. As a result, the expected net benefit for a customer who chooses to join the queue is $R - [p + CW(\lambda)]$. We distinguish three cases for a tagged customer who arrives and joins the queue.

(1) If $p + CW(0) \geq R$. In this case, even if no other customer joins, the net benefit of a customer who joins is non-positive. Therefore, the strategy of joining with probability $q_e(p) = 0$ is an equilibrium strategy and no other equilibrium is possible. Moreover, in this case, not joining is a dominant strategy.
(2) $p + CW(\Lambda) \leq R$. In this case, even if all potential customers join, they all enjoy a non-negative benefit. Therefore, the strategy of joining with probability $q_e(p) = 1$ is an equilibrium strategy and no other equilibrium is possible. Moreover, in this case, joining is a dominant strategy.
(3) $p + CW(\Lambda) \leq R \leq p + CW(\Lambda)$. In this case, if $q_e(p) = 1$ then a customer who joins suffers a negative benefit. Hence, this cannot be an equilibrium strategy. Likewise, if $q_e(p) = 0$, a customer who joins gets a positive benefit, more than by balking. Hence, this too cannot be an equilibrium. Therefore, there exists a unique equilibrium strategy $q_e = \frac{\lambda_e(p)}{\Lambda}$, where $\lambda_e(p)$ solves $CW(\lambda_e(p)) = R - p$.

It should be noted that, suppose that the probability of joining is q: if $q < q_e(p)$ then the unique best response is 1, if $q = q_e(p)$ then any strategy between 0 and 1 is a best response, and if $q > q_e(p)$ then the unique best response is 0. Since the best response is a monotone non-increasing function of the strategy, the model is of the ATC type and therefore, there is a unique equilibrium solution.

3.1.3 *Social Optimization*

We now turn our attention to social optimization. Let the socially optimal joining probability be q_{soc}, and let the socially optimal joining rate be λ_{soc}, where $\lambda_{soc} = q_{soc}\Lambda$. Then the optimal arrive rate λ^*_{soc} is the maximum arrival rate satisfies the objective function $\lambda[R - CW(\lambda)]$. Since $W(\lambda)$ is strictly convex, the function to be maximized is strictly concave and has a unique maximum. Substituting $W(\lambda) = \frac{1}{\mu-\lambda}$ we get that the solution $\mu - \sqrt{\frac{C\mu}{R}} = \arg \max\limits_{0 \leq \lambda < \mu} \{\lambda R - C\lambda\frac{1}{\mu-\lambda}\}$ is optimal as long as it is in $[0\Lambda]$. The fact that the solution is nonnegative follows from the assumption that $R\mu \geq C$. Thus, if $\Lambda \geq \mu - \sqrt{\frac{C\mu}{R}}$ then $\lambda^*_{soc} = \mu - \sqrt{\frac{C\mu}{R}}$; Otherwise, we have that $\lambda^*_{soc} = \Lambda$.

3.1.4 Profit Maximization Strategy

We consider now a monopolistic server that sets a profit-maximizing admission fee p_m, A monopoly does not leave a positive customer surplus, since in such a case the admission fee can be increased without reducing the arrival rate. Therefore, we have $p_m + CW(\lambda) = R$. The monopoly's problem is to maximize λp_m subject to $0 \leq \lambda \leq \Lambda$ and $p_m = R - CW(\lambda)$. It can be found that the objective function is consistent with the socially-optimal objective function. Therefore, when $\lambda^*_{soc} = \mu - \sqrt{\frac{C\mu}{R}}$, $p_m = p^*_{soc} = R - CW(\lambda^*_{soc}) = R - \sqrt{\frac{CR}{\mu}}$ When $\lambda^*_{soc} = \Lambda$, the profit maximizer chooses the maximum fee which induces this rate, i.e., $p_m = R - \frac{C}{\mu - \Lambda}$ A social planner would choose this fee, or any smaller fee, since any such choice induces the same optimal arrival rate, Λ.

3.2 M/G/1 Queues

3.2.1 Model Description

We consider a single server queueing system in which customers arrive according to a Poisson process with rate Λ. Arriving customers form a single waiting line based on the order of their arrivals and are served according to the first-come, first-served (FCFS) service discipline. The service times of customers are independent, identically distributed with common general distribution function $B(\cdot)$ and the corresponding first moment, second moment and variance are $\bar{s}_1$, $\bar{s}_2$ and σ_s^2, respectively. The customer cannot observe the system queue length after arrival, and must make the choice of whether to join the queue, and once the decision to join the queue or to balk is made, it is irrevocable and reneging is not allowed when customers waiting in the queue. A customer's benefit from completed service is R. The cost to a customer for staying in the system, either while waiting or while being served, is C per unit of time. Suppose $R \geq C\bar{s}_1$ to ensure that a customer finding the system empty to join.

3.2.2 Nash Equilibrium Strategy

Suppose arriving customers join the queue with probability $q(0 \leq 1)$, i.e., the effective arrival rate is $\lambda = \Lambda q$ The expected sojourn time of a customer in the M/G/1 queueing system is given by

$$\mathbb{E}[S] = \bar{s}_1 + \frac{\lambda \bar{s}_2}{2(1-\rho)}, \tag{3.2.1}$$

where $\rho = \lambda \bar{s}_1$ The expected payoff of customers after joining the queue is

$$U(q) = R - C\mathbb{E}[S(q)] = R - C\bar{s}_1 - C\frac{\lambda \bar{s}_2}{2(1-\rho)}.$$

Solving the equation $U(q) = 0$, we get

$$q_e = \frac{2(R - C\bar{s}_1)}{\Lambda[2R\bar{s}_1 + C(\bar{s}_2 - 2\bar{s}_1^2)]}. \tag{3.2.2}$$

If

$$\frac{2(R - C\bar{s}_1)}{[2R\bar{s}_1 + C(\bar{s}_2 - 2\bar{s}_1^2)]} \leq \Lambda,$$

we can obtain the equilibrium joining probability as given by (3.2.2). Otherwise, the equilibrium joining probability is $q_e = 1$

3.2.3 Socially Optimal Strategy

Assuming that all arriving customers join the queue with a probability of q, the social benefit per unit time is

$$SW(q) = \Lambda q\Big(R - C\Big(\bar{s}_1 + \frac{\Lambda q(\bar{s}_1^2 + \sigma_s^2)}{2(1 - \Lambda q\bar{s}_1)}\Big)\Big).$$

It is not difficult to prove that $SW(q)$ is a strictly convex function about q, so it can be obtained from $SW'(q)$ that

$$q_{soc}^* = \frac{2R\bar{s}_1 + C(\sigma_s^2 - \bar{s}_1^2) - \sqrt{C(\bar{s}_1^2 + \sigma_s^2)(2R\bar{s}_1 + C(\sigma_s^2 - \bar{s}_1^2))}}{\bar{s}_1\Lambda(2R\bar{s}_1 + C(\sigma_s^2 - \bar{s}_1^2))}. \tag{3.2.3}$$

By assumption of $R \geq C\bar{s}_1$ we know that q_{soc}^* is nonnegative. Therefore, if

$$\Lambda \geq \frac{2R\bar{s}_1 + C(\sigma_s^2 - \bar{s}_1^2) - \sqrt{C(\bar{s}_1^2 + \sigma_s^2)(2R\bar{s}_1 + C(\sigma_s^2 - \bar{s}_1^2))}}{\bar{s}_1\Lambda(2R\bar{s}_1 + C(\sigma_s^2 - \bar{s}_1^2))},$$

the socially optimal effective arrival rate is

$$\lambda_{soc}^* = \Lambda q_{soc}^* = \frac{2R\bar{s}_1 + C(\sigma_s^2 - \bar{s}_1^2) - \sqrt{C(\bar{s}_1^2 + \sigma_s^2)(2R\bar{s}_1 + C(\sigma_s^2 - \bar{s}_1^2))}}{\bar{s}_1\Lambda(2R\bar{s}_1 + C(\sigma_s^2 - \bar{s}_1^2))}.$$

Otherwise, we have $\lambda_{soc}^* = \Lambda \; q_{soc}^* = 1$

3.2.4 Revenue Maximization Strategy

For any given admission fee of p, the expected payoff of a customer joining the queue (with a probability of q) satisfies the following equation.

$$R - p = C\mathbb{E}[S(q)]. \tag{3.2.4}$$

Substituting (3.2.1) into (3.2.4) we have

$$q(p) = \frac{2(R - p - C\bar{s}_1)}{\Lambda[2\bar{s}_1(R - p) + C(\sigma_s^2 - \bar{s}_1^2)]}.$$

Therefore, the expected revenue of the service provider is

$$\Pi(p) = \Lambda q(p)p = \frac{2p(R - p - C\bar{s}_1)}{[2\bar{s}_1(R - p) + C(\sigma_s^2 - \bar{s}_1^2)]}.$$

By solving the equation $\Pi'(p) = 0$, the admission fee that maximizes the admission revenue can be obtained as follows

$$p^*_{rev} = \frac{2R\bar{s}_1 + C(\sigma_s^2 - \bar{s}_1^2) - \sqrt{C(\bar{s}_1^2 + \sigma_s^2)(2R\bar{s}_1 + C(\sigma_s^2 - \bar{s}_1^2))}}{2\bar{s}_1}.$$

Then the maximum revenue of the service provider per unit time is $\Pi^* = \Lambda q(p^*_{rev})p^*_{rev}$.

3.3 GI/M/c Queueing System

3.3.1 Model Description

Consider the unobservable GI/M/c queueing system, that is, the queue length of the system cannot be observed when customers arrive. The arrival times of customers are independent and identically distributed random variables with common general distribution function $A(u)(u \geq 0)$, probability density function $a(u)(u \geq 0)$, Laplace-Stieljes transform $a^*(s)$, and mean $\frac{1}{\lambda}$. The service rule is first-come, first-served (FCFS), and the service time follows the exponential distribution with parameter μ.

Take the time of customer arrival as the embedding point, and denote $t_0, t_1^-, t_2^-, \cdots$ as the time of successive arrival at the system. Inter-arrival times $T_{n+1} = t_{n+1}^- - t_n^-, n = 0, 1, 2, \cdots$ are independent and identically distributed random variables with common general distribution function $A(x)$. Define $\{N_s(t_i^-)\}$ as the system state at instant t_i^-, where $N_s(t_i^-)$ denotes the number of customers in the system. Stochastic process $\{n_s(t_i^-)\}$ is an embedded Markov chain, and its state space is $\Theta = \{(k), k \geq 0\}$. Let $\pi_n^- = \lim_{i\to\infty} \Pr(N_s(t_i^-) = n), n \geq 0$, where π_n^- is the probability of n customers in the system when a new customer arrives at the system.

Suppose that customers can choose to join the queue or to balk when they arrive. The reward of each customer after receiving the service is R, and the stay fee per unit time in the system is C. Suppose that all customers are risk neutral and want to maximize their own interests. Suppose that the probability of customers joining the queue after arrival is f, and once they make a choice, the decision is irrevocable and reneging is not allowed. A customer who balks the system is not allowed to retry to come back to the system.

3.3.2 *Queue Length Distribution Derivation Using Arrival Times*

With d_k represents the probability that k customers leave the system during an arrival interval given that c servers in the system are busy. Let $a_{k+1,j}$ be the probability that a newly arrived customer sees $k(k \leq c-1)$ customers in the system and the next arriving customer sees $j\,(0 \leq j \leq k)$ customers in the system. Then, $k+1-j$ customers left during this arrival interval. Similarly, define $b_{k+1,j}$ as the probability that a newly arrived customer sees $k(k \geq c)$ customers in the system and the next arriving customer sees $j\,(0 \leq j \leq c-1)$ customers in the system. Therefore, for all $k \geq 0$, we have

$$\begin{aligned}
a_{k,j} &= \Pr(N_s(t_i^-) = j | N_s(t_{i-1}^-) = k-1) = \int_0^\infty \binom{k}{j} e^{-\mu j t}(1-e^{-\mu t})^{k-j} dA(t),\\
&\quad 0 \leq k-1 \leq c-1, 1 \leq j \leq k,\\
b_{k,j} &= \Pr(N_s(t_i^-) = j | N_s(t_{i-1}^-) = k-1)\\
&= \int_0^\infty \int_0^t \frac{(\mu c)^{k-c} u^{k-c-1} e^{-c\mu u}}{(k-c-1)!} \binom{c}{j} e^{-\mu j(t-u)}(1-e^{-\mu(t-u)})^{c-j} du dA(t),\\
&\quad k-1 \geq c, 1 \leq j \leq c-1,\\
d_k &= \int_0^\infty \frac{(c\mu t)^k}{k!} e^{-c\mu t} dA(t),\\
a_{k,0} &= 1 - \sum_{j=1}^{k} a_{k,j}, \quad b_{k,0} = 1 - \sum_{j=1}^{c-1} b_{k,j} - \sum_{j=0}^{k-c} d_j.
\end{aligned}$$

Simplify $a_{k,j}$ and $b_{k,j}$ to

$$
\begin{aligned}
a_{k,j} &= \int_0^\infty \binom{k}{j} e^{-\mu j t}(1-e^{-\mu t})^{k-j} dA(t) \\
&= \int_0^\infty \binom{k}{j} e^{-\mu j t}(-1)^{k-j} \sum_{l=0}^{k-j} \binom{k-j}{l} e^{-\mu t(k-j-l)}(-1)^l dA(t) \\
&= \binom{k}{j} \sum_{l=0}^{k-j} (-1)^{k-j+l} \binom{k-j}{l} \int_0^\infty e^{-\mu t(k-l)} dA(t) \\
&= \binom{k}{j} \sum_{l=0}^{k-j} (-1)^{k-j+l} \binom{k-j}{l} a^*(\mu(k-l)),
\end{aligned} \tag{3.3.1}
$$

$$
\begin{aligned}
b_{k,j} &= \int_0^\infty \int_0^t \frac{(\mu c)^{k-c} u^{k-c-1} e^{-c\mu u}}{(k-c-1)!} \binom{c}{j} e^{-\mu j(t-u)}(1-e^{-\mu(t-u)})^{c-j} du dA(t) \\
&= \frac{(\mu c)^{k-c}}{(k-c-1)!} \binom{c}{j} \int_0^\infty \int_0^t e^{-c\mu u} u^{k-c-1} e^{-\mu j(t-u)}(1-e^{-\mu(t-u)})^{c-j} du dA(t) \\
&= \frac{(\mu c)^{k-c}}{(k-c-1)!} \binom{c}{j} \int_0^\infty \int_0^t g(u) h(t-u) du dA(t),
\end{aligned} \tag{3.3.2}
$$

where $g(u) = e^{-c\mu u} u^{k-c-1}$, $h(t-u) = e^{-\mu j(t-u)}(1-e^{-\mu(t-u)})^{c-j}$ Therefore, the quadratic integration is the convolution of $g(u)$ and $h(t-u)$, so the whole integration result is the Laplace-Stieltjes transformation of the convolution of the two functions. The Laplace-Stieltjes transforms of $g(t)$ and $h(t)$ are given by

$$
\int_0^\infty e^{-st} e^{-c\mu t} t^{k-c-1} dt = \frac{(k-c-1)!}{(s+c\mu)^{k-c}}, \tag{3.3.3}
$$

$$
\begin{aligned}
\int_0^\infty e^{-st} h(t) dt &= \int_0^\infty e^{-st} e^{-\mu j t}(1-e^{-\mu t})^{c-j} dt \\
&= \frac{\int_0^1 e^{-\mu t(s/\mu+j-1)} z^{c-j} dz}{\mu} \text{(Letting } z = 1-e^{-\mu t})
\end{aligned}
$$

$$= \frac{\int_0^1 (1-z)^{-\mu t(s/\mu+j-1)} z^{c-j} dz}{\mu}$$

$$= \frac{\Gamma(j+s/\mu)\Gamma(c-j+1)}{\mu\Gamma(c+s/\mu+1)}. \tag{3.3.4}$$

Consider those distributions of Laplace-Stieltjes transformation with proportional form such as $\upsilon(s) = P(s)/Q(s)$, where the dimension of $Q(s)$is n, and the dimension of $p(s)$is at most n. Thus, the convolution of the two functions (3.3.3) and (3.3.4) can be rewritten as

$$\frac{(k-c-1)!}{(s+c\mu)^{k-c}} \frac{\Gamma(j+s/\mu)\Gamma(c-j+1)}{\mu\Gamma(c+s/\mu+1)} \simeq \frac{(k-c-1)!(c-j)!}{\mu} \frac{P(s)}{Q(s)},$$

where $\frac{P(s)}{Q(s)} \simeq \frac{\Gamma(j+s/\mu)}{(s+c\mu)^{k-c}\Gamma(c+s/\mu+1)}$. In the sequel we consider a case that the interarrival times of customers follow a phase-type (PH) distribution with parameters $(\boldsymbol{\alpha}, \mathbf{B})$, the density function is $a(t) = \boldsymbol{\alpha} e^{\mathbf{B}t}\mathbf{B}^0$ Substituting $a(t) = \boldsymbol{\alpha} e^{\mathbf{B}t}\mathbf{B}^0$ into (3.3.2), we can write $b_{k,j}$ as follows.

$$\begin{aligned} b_{k,j} &= \frac{(\mu c)^{k-c}}{(k-c-1)!}\binom{c}{j} \int_0^\infty \int_0^t e^{-c\mu u} u^{k-c-1} e^{-\mu j(t-u)} (1-e^{-\mu(t-u)})^{c-j} du dA(t) \\ &= \frac{(\mu c)^{k-c}}{(k-c-1)!}\binom{c}{j} \int_0^\infty \boldsymbol{\alpha} e^{\mathbf{B}t} \int_0^t g(u)h(t-u) du dt \mathbf{B}^0 \\ &\simeq \frac{(\mu c)^{k-c} c!}{j!\mu} \boldsymbol{\alpha} \left[\frac{P(s)}{Q(s)}\bigg|_{s=-\mathbf{B}} \right] \mathbf{B}^0. \end{aligned} \tag{3.3.5}$$

The probability generating function of d_k is

$$\bar{D}(z) = \sum_{k=0}^{\infty} d_k z^k = a^*(c\mu - c\mu z). \tag{3.3.6}$$

By observing the state of the system at two consecutive arrival times, the following difference equation can be obtained:

$$\begin{aligned} \pi_0^- &= f\left(\sum_{k=0}^{c-1} \pi_k^- a_{k+1,0} + \sum_{k=c}^{\infty} \pi_k^- b_{k+1,0} \right) \\ &\quad + (1-f)\left(\pi_0^- + \sum_{k=1}^{c-1} \pi_k^- a_{k,0} + \sum_{k=c}^{\infty} \pi_k^- b_{k,0} \right), \end{aligned} \tag{3.3.7}$$

$$\pi_i^- = f\left(\sum_{k=i-1}^{c-1} \pi_k^- a_{k+1,i} + \sum_{k=c}^{\infty} \pi_k^- b_{k+1,i} \right)$$

$$+(1-f)\left(\sum_{k=i}^{c-1}\pi_k^- a_{k,i}+\sum_{k=c}^{\infty}\pi_k^- b_{k,i}\right),\quad 1\le i\le c-1, \tag{3.3.8}$$

$$\pi_i^- = f\left(\sum_{k=i-1}^{\infty}\pi_k^- d_{k+1-i}+(1-f)\sum_{k=i}^{\infty}\pi_k^- d_{k-i}\right),\quad i\ge c. \tag{3.3.9}$$

Define the probability generating function of $\pi_i^-(i \ge c)$ as $\pi_c^{-*}(z) = \sum_{k=c}^{\infty}\pi_i^- z^{i-c}$ Multiplying both sides of (3.3.9) by z^{i-c} and adding them to obtain

$$\pi_c^{-*}(z)=\frac{\pi_c^- - f\sum_{i=1}^{\infty}\frac{d_i}{z^{i-1}}\sum_{j=0}^{i-1}\pi_{c+j}^- z^j-(1-f)\sum_{i=0}^{\infty}\pi_{c+i}^- z^j\sum_{k=i}^{\infty}\frac{d_k}{z^k}}{1-(1-f+fz)\bar{D}(\frac{1}{z})}. \tag{3.3.10}$$

We observe that (3.3.10) is analytic and convergent under the condition of $|z| \le 1$. Since equation $1-(1-f+\frac{f}{z})\bar{D}(z)=0$ has a unique solution ω in the region of unit circle $|z|<1$ and $\pi_c^{-*}(z)$ is analytic in the region $|z| \le 1$, so we have

$$\pi_c^{-*}(z)=\frac{K_1}{1-\omega z}, \tag{3.3.11}$$

where K_1 is a constant to be determined. By comparing the coefficients decomposed into z^{i-c} from (3.3.11), it can be obtained that the probability of the system state at the time of arrival is

$$\pi_i^- = K_1\omega^{i-c}, i\ge c. \tag{3.3.12}$$

Combining (3.3.7) and (3.3.8), and normalizing by the condition $\sum_{i=0}^{\infty}\pi_i^- = 1$, we can get the solution of K_1 and the probability of all system states at the time of arrival $\pi_i^-, i \ge 0$.

3.3.3 Equilibrium Balking Strategy

Consider a newly arrived customer. If he chooses to join the queue, his expected stay in the system is

$$\mathbb{E}[W]=\sum_{j=0}^{c-1}\frac{\pi_j^-}{\mu}+\sum_{j=c}^{\infty}\pi_j^-\frac{j+1-c}{c\mu}, \tag{3.3.13}$$

where π_j^- is given by (3.3.7)-(3.3.9). Subsequently, the expected payoff of the customer joining the queue is $\Delta = R - C\mathbb{E}[W]$. In order to ensure that arriving customers will choose to join the queue when the system is empty, it is assumed that

$$R > \frac{C}{\mu}.$$

If $\Delta \geq 0$ when $f = 1$, then the choice of "joining the queue" is the customer's Nash equilibrium strategy. Otherwise, if $\Delta < 0$ when $f = 1$, then $f_e \in (0, 1)$ which satisfies $\Delta = R - C\mathbb{E}[W] = 0$ is the customer's equilibrium joining probability.

References

Allon, Gad, and Achal Bassamboo. 2011. The impact of delaying the delay announcements. *Operations Research* 59: 1198–1210.

Allon, Gad, Achal Bassamboo, and Itai Gurvich. 2011. We will be right with you: managing customer expectations with vague promises and cheap talk. *Operations Research* 59: 1382–1394.

Armony, Mor, Nahum Shimkin, and Ward Whitt. 2009. The impact of delay announcements in many-server queues with abandonment. *Operations Research* 57 (1): 66–81.

Arnott, Richard, André De Palma, and Robin Lindsey. 1996. Information and usage of free-access congestible facilities with stochastic capacity and demand. *International Economic Review* 37: 181–203.

Arnott, Richard, André De Palma, and Robin Lindsey. 1999. Information and time-of-usage decisions in the bottleneck model with stochastic capacity and demand. *European Economic Review* 43: 525–548.

Balachandran, K. R. 1991. Incentive and regulation in queues. *Lecture Notes in Economics and Mathematical Systems* 370: 162–176.

Chen, Hong, and Murray Frank. 2004. Monopoly pricing when customers queue. *IIE Transactions* 36: 569–581.

Edelson, Noel M., and David K. Hildebrand. 1975. Congestion tolls for Poisson queueing processes. *Econometrica* 43: 81–92.

Guha, D., A.D. Banik, V. Goswami, and S. Ghosh. 2014. Equilibrium balking strategy in an unobservable $GI/M/c$ queue with customers' impatience. *10th International Conference on Distributed Computing and Internet Technology, ICDCIT 2014, Proceedings*, 188–199.

Guo, Pengfei, and Paul Zipkin. 2007. Analysis and comparison of queues with different levels of delay information. *Management Science* 53: 962–970.

Guo, Pengfei, Wei Sun, and Yulan Wang. 2011. Equilibrium and optimal strategies to join a queue with partial information on service times. *European Journal of Operational Research* 214: 284–297.

Hassin, Refael. 1986. Consumer information in markets with random product quality: the case of queues and balking. *Econometrica* 54: 1185–1195.

Hassin, Refael. 1996. On the advantage of being the last server. *Management Science* 42: 618–623.

Hassin, Refael, and Moshe Haviv. 2003. *To Queue or Not to Queue: Equilibrium Behavior in Queueing Systems*. Boston: Kluwer Academic Publishers.

Littlechild, S. C. 1974. Optimal arrival rate in a simple queueing system. *International Journal of Production Research* 12: 391–397.

Shone, R., V. Knight, and J. Williams. 2013. Comparisons between observable and unobservable $M/M/1$ queues with respect to optimal customer behavior. *European Journal of Operational Research* 227: 133–141.

Whitt, Ward. 1999. Improving service by informing customers about anticipated delays. *Management Science* 45: 192–207.

Chapter 4
Optimal Disclosure of Information

The literature on the impact of queue length disclosure on consumer welfare and service provider's revenue dates back to Naor (1969). In Naor's model, homogeneous customers observe the real-time queue length upon their arrival to the system and make decisions to either join the queue or balk. In other words, the service provider discloses full queue length information, and such setting has since been referred to as an *observable* queue in the literature. Edelson et al. (1975) study the *unobservable* version of Naor's model where customers do not have any queue length information upon arriving at the service system. They decide to join the queue or balk based on anticipations of other customers' strategies.

Subsequently, both observable and unobservable queues were studied in the context of heterogeneous service values (e.g., Larsen (1998)), heterogenous time costs (e.g., Afèche et al. (2004)) and many others. The interested readers are referred to Hassin and Haviv (2003) and Hassin (2016) for comprehensive reviews of this literature. In particular, Hassin (1986), Chen and Frank (2004), Chambers and Kouvelis (2006) and Shone et al. (2013) compare the observable and unobservable queueing settings, and identified instances when one is superior than the other.

In this chapter, we study state-dependent queue length disclosure policies which include the classical observable and unobservable settings as two extreme cases. In practice, consider customers calling a customer-support phone number or engaged in some service live chat.[1] The service provider can announce the current queue length information to customers who arrive at certain states but not at other states (hereinafter, referred to as the "informed" and "uninformed" states/customers, respectively). Customers then make decisions to join the queue or balk with the state-dependent information provided by the service provider.

---

[1] Readers interested in the call center literature are referred to surveys by Gans et al. (2003) and Akşin et al. (2007).

J. Wang, *Fundamentals of Queueing-Game Models*,
https://doi.org/10.1007/978-981-95-0261-5_4

Let's suppose the server receives a fixed income from each customer and the goal of the server is to maximize revenue/throughput. To achieve this goal, the service provider informs customers of the queue length when the queue is short (that is, it is less than some threshold D) and hide the queue length information when the queue is long. In this way customers would be informed and encouraged to enter the system when the queue is short, and would not get informed or scared away when the queue is long. In the literature, Simhon et al. (2016) prove that such a policy is never optimal for the server, and Kim and Kim (2017) further show that the optimal throughput-maximizing policy is to announce the queue length information if and only if the queue is long.

It is noted that in the model of Simhon et al. (2016), even though a customer who arrives at an uninformed state is not informed of the current queue length, she is assumed to be aware of the threshold disclosure policy of the server and the value of the threshold D (e.g., she is an experienced customer who uses the service frequently). In other words, when an arriving customer is not informed of the current queue length, she can infer that the current queue length must be equal to or exceed the threshold D and uses such information to make a decision to join the queue or balk. Kim and Kim (2017) study a general queue length disclosure policy (which not necessarily needs to be of a threshold type), but continued to assume that customers are familiar with the queue length disclosure policy of the server.

In this chapter we study a different setting, that is, we consider unexperienced customers who are not aware of the information disclosure policy used by the service provider. Customers act according to the classical unobservable queueing game unless they receive (extra) queue length information from the server. This setting applies to situations where customers are either first-time or occasional service users who do not possess the knowledge of the service provider's information disclosure decisions. For example, individuals may need to call many different customer service lines (merchants, credit card companies, utility firms, etc.) from time to time, and they often cannot remember any specific service provider's queue length disclosure policy given that the time between two calls made to the same vendor can be long.

In addition, while Simhon et al. (2016) and Kim and Kim (2017) focus on throughput-maximizing threshold and general state-dependent policies, respectively, for their setting, we consider both throughput-maximizing and socially optimal (general state-dependent) policies for ours. We show that a threshold policy that announces the current queue length to customers when the queue is sufficiently short and does not announce otherwise indeed leads to the highest throughput for the server. Nevertheless, a socially optimal policy, which maximizes consumer welfare, contains a "hole," i.e., the service provider should announce the queue length to customers when the queue is sufficiently short or sufficiently long but not announce if the queue length is moderate.

This work belongs to the large volume of literature that examines the impact of information disclosure (including queue length, real-time waiting time, service rate, service quality) on arriving customers.[2] For example, Guo and Zipkim (2007) study

[2] Readers interested in a more complete review are referred to Hassin (2016).

M/M/1 queues with three levels of information: no information, partial information (the queue length), and full information (the exact waiting time). Econumou et al. (2008) and Guo et al. (2009) consider situations where some partitioned queue information (such as range of queue lengths) is available to the customers when they make decisions. Armony et al. (2009) consider a setting where the server announces the delay information of the last customer to have entered service, and benchmarked it to the case when long-run average delay information is provided instead. Kremer and Debo (2015) use lab experiments to study the impact of waiting time on customers' purchasing behavior when the service quality is known to some customers but unknown to others. Cui et al. (2016) study the impact of service rate revelation when customers do not possess the information. Recently, Hassin et al. (2017) consider a scenario where arriving customers to an unobservable queue can purchase the queue length information for a fee, and Hu et al. (2018) study a model where a portion of the customer population are aware of the queue length information while others are not. In both of these papers, similar to our setting, the system is observable to some customers while unobservable to others. Wang and Wang (2021) explore the effect of information on queue-scalping service systems. Nevertheless, in all of these works mentioned above, the information disclosure policy is not state-dependent.

In addition, Allon et al. (2011) study a "cheap talk" game when non-verifiable queue length information is provided to the customers under the unobservable queueing setting. Hassin and Koshman (2017) introduce high-low pricing to achieve the highest possible server profit in Naor's model when arriving customers are only informed of whether the queue congestion is high or low. Yu et al. (2018) extend the work of Allon et al. (2011) to include multiple customer classes, and considered both admission and scheduling policies of the server. This chapter is mainly adapted from Li, Cui and Wang (2021) which focus on investigation about optimal disclosure of queue length information.

4.1 Preliminaries

We consider an M/M/1 queue first. Later in the chapter, we will establish robustness of our results numerically with M/G/1 queues. Potential customers arrive to the system according to a Poisson process with rate λ, and make decisions upon arrival to either join the queue or balk (We assume that reneging is not allowed.) Joining customers are served according to the FCFS discipline (first come first served), and their service times are independent and exponentially distributed with rate μ. We assume that λ and μ are common knowledge, as is typically done in the literature. All joining customers incur a linear waiting cost of C per unit of time while they are in the system. Upon completion of the service, each customer receives a service reward R which is net of service price. We do not consider the server's pricing decision in the paper, but instead focus on its queue length disclosure policy. We assume without loss of generality that $R > \frac{C}{\mu}$ because otherwise no customers would consider the service in the first place, and that R is not a multiple of $\frac{C}{\mu}$ to avoid discussing trivial mixed strategies.

We use $i \geq 0$ to denote the state of the system that corresponds to the total number of customers currently in the system (including the customer at the server if any). The server can implement a state-dependent queue length disclosure policy. Such a policy can be denoted as $X = (x_0, x_1, \ldots, x_i, \ldots)$, where

$$x_i = \begin{cases} 1 & \text{the server reveals the current queue length to customers arriving at state } i; \\ 0 & \text{the server conceals the current queue length from customers arriving at state } i. \end{cases}$$

For instance, a policy that reveals only odd states can be represented as $X^{odd} \triangleq (0, 1, 0, 1, 0, 1, 0, 1, \ldots)$.

As mentioned, Simhon et al. (2016) considered the set of threshold policies in their paper.[3] We can write such a policy for an integer threshold $D \in \{0, 1, 2, \ldots, \infty\}$ as X^D where

$$x_i = \begin{cases} 1 & \text{if } 0 \leq i < D; \\ 0 & \text{if } i \geq D. \end{cases} \tag{4.1.1}$$

X^D specifies that the server should reveal the current queue length up to state D and conceal it otherwise.

The server's information disclosure policy in an observable M/M/1 queue, introduced by Naor (1969), is always revealing, that is, $\forall i, x_i = 1$. Thus, it can be denoted as X^∞ according to the definition in Eq. (4.1.1). In contrast, the server's information disclosure policy in an unobservable queue, introduced by Edelson et al. (1975), is never revealing, that is, $\forall i, x_i = 0$, and we can denote it as X^0. In other words, X^0 and X^∞ are extreme cases of a threshold policy given by (4.1.1) that correspond to the unobservable and observable queueing settings in the literature, respectively.

For notational convenience, we also define a policy X^{D_1, D_2} for $D_1, D_2 \in \{0, 1, 2, \ldots, \infty\}$ and $D_1 < D_2$ as the following

$$x_i = \begin{cases} 1 & \text{if } i \leq D_1 \text{ or } i \geq D_2; \\ 0 & \text{if } D_1 < i < D_2. \end{cases} \tag{4.1.2}$$

A policy that satisfies the definition in (4.1.2) contains two thresholds D_1 and D_2, and it specifies that the server should disclose the queue length information to arriving customers only when the queue is either short ($\leq D_1$) or long ($\geq D_2$), but not inbetween—that is, the information disclosure policy contains a "hole."

We now detail the impact of the server's information disclosure policy on customers' decisions. Our study focuses on settings where customers are occasional users of the service so that they do not have the information of or cannot recall the

---

[3] In defining the threshold policies, D in our paper is actually equivalent to $D - 1$ in Simhon et al. (2016), as they defined X^D to be $x_i = 1$ if and only if $0 \leq i \leq D$ (not $i < D$). Simhon et al. (2016) also used different notations for X^∞ and X^0.

server's information disclosure policy through past experience. Therefore, when the server does not provide any queue length information (for any state), customers in our model will face the traditional unobservable queueing setting as described in Edelson et al. (1975).

Given a policy $X = (x_0, x_1, \ldots, x_i, \ldots)$, if $x_i = 1$ for some $i \geq 0$ meaning that the server will be providing the queue length information to customers arriving at state i, then those customers will use the additional information received to make their joining versus balking decisions. In particular, upon learning the queue length, a customer will join the system if and only if the net service payout exceeds the expected wait or $R > \frac{(i+1)C}{\mu}$, regardless of other customers' actions. This is the same decision as one would make under the observable queueing setting described by Naor (1969). Let N denote Naor's balking threshold, that is, $N \triangleq \lfloor \frac{R\mu}{C} \rfloor \geq 1$. Then customers' decision upon arrival at state i when $x_i = 1$ can be characterized as

$$\begin{cases} i < N & \text{the customer joins the queue;} \\ i \geq N & \text{the customer balks.} \end{cases}$$

Because customers in our model have no information of the server's information disclosure policy, those who have not received additional (queue length) information from the server upon arrival will not know that some other customers could be receiving the queue length information. In other words, a customer who does not observe the queue length (those who arrive at state $i \geq 0$ where $x_i = 0$) believe that neither of other customers observe the queue length information. As a consequence, customers who arrive at state i with $x_i = 0$ will behave according to the equilibrium strategy of an unobservable queue. It is well-known that a customer's strategic decision in equilibrium is to join the system with probability $q = \min\{\frac{1}{\lambda}\left(\mu - \frac{C}{R}\right), 1\}$ and balk with probability $(1 - q)$ under the unobservable queueing setting (see, e.g., Sect. 3.1 of Hassin and Haviv (2003)), which in turn characterizes customers' decision in our model when they arrive at a state i with $x_i = 0$.

4.2 Throughout-Maximizing Policy

In this section, we set out to determine the throughput-maximizing policy. If the server receives a fixed income from each customer, then it is also a revenue-maximizing policy. Let $\{\pi_0^X, \pi_1^X, \ldots, \pi_i^X, \ldots\}$ denote the steady-state probabilities for the system under policy X. The system throughput is equal to $\mu(1 - \pi_0^X)$ (see, e.g., Larsen (1998) and Cui et al. (2016)). Because μ is treated as an exogenous parameter in our model, the throughout-maximizing policy must minimize the idle probability π_0.

In the following Theorem 4.2.1, we show that to maximize throughput, it is optimal for the server to disclose the queue length information to customers when the queue is shorter than Naor's balking threshold N, but conceal the information when the queue length is longer than N, namely it is optimal to use the X^N policy as defined by (4.1.1). Our proof method is to represent the idle probability π_0 as a function of x_i's of policy X and show that the idle probability is minimized when $x_i = 1$ for $i \in \{0, 1, \ldots, N-1\}$ and $x_i = 0$ otherwise.

Theorem 4.2.1 *X^N is a throughout-maximizing queue length disclosure policy.*

Proof We start the proof by picking any arbitrary policy $X = (x_0, x_1, \ldots, x_i, \ldots)$. When $i < N$, the effective arrival rate at state i of the underlying queue is λ if $x_i = 1$, and λq if $x_i = 0$. So the effective arrival rate at any state $i \in \{0, 1, \ldots, N-1\}$ can be written as $\lambda_i \triangleq \lambda\,[x_i + q\,(1 - x_i)]$.

Define $k \triangleq \min\{i \mid i \geq N, x_i = 1\}$. Then no state higher than k will be ever reached. Moreover, the effective arrival rate at states $\{N, N+1, \ldots, k-1\}$ is λq. It follows that the effective arrival rate, denoted by λ_e, is given by

$$\lambda_e \triangleq \begin{cases} \lambda_i & i < N; \\ \lambda q & N \leq i < k. \end{cases} \tag{4.2.1}$$

Given that the transition rate form state i to state $i-1$ is μ for all i, we can calculate the steady-state probabilities $\{\pi_i^X\}$. The balance equations are give by

$$\begin{cases} \lambda_i \pi_i^X = \mu \pi_{i+1}^X & 0 \leq i < N; \\ \lambda q \pi_i^X = \mu \pi_{i+1}^X & N \leq i < k. \end{cases}$$

Then the steady-state probabilities can be derived as follows.

$$\pi_j^X = \begin{cases} \left(\prod_{i=0}^{j-1} \frac{\lambda_i}{\mu}\right) \pi_0^X & 1 \leq j \leq N; \\ \left(\prod_{i=0}^{N-1} \frac{\lambda_i}{\mu}\right) \left(\frac{\lambda q}{\mu}\right)^{j-N} \pi_0^X & N < j \leq k. \end{cases} \tag{4.2.2}$$

By the condition $\sum_{j=0}^{k} \pi_j^X = 1$, we have $\pi_0^X + \sum_{j=1}^{N} \pi_j^X + \sum_{j=N+1}^{k} \pi_j^X = 1$. And thus,

$$\begin{aligned} \pi_0^X &= \left[1 + \sum_{j=1}^{N} \left(\prod_{i=0}^{j-1} \frac{\lambda_i}{\mu}\right) + \sum_{j=N+1}^{k} \left(\prod_{i=0}^{N-1} \frac{\lambda_i}{\mu}\right) \left(\frac{\lambda q}{\mu}\right)^{j-N}\right]^{-1} \\ &= \left[1 + \sum_{j=1}^{N} \left(\frac{\lambda}{\mu}\right)^j \prod_{i=0}^{j-1} [(1-q)x_i + q] + \left(\frac{\lambda}{\mu}\right)^N \left\{\prod_{i=0}^{N-1} [(1-q)x_i + q]\right\} \frac{\frac{\lambda q}{\mu} - \left(\frac{\lambda q}{\mu}\right)^{k-N}}{1 - \frac{\lambda q}{\mu}}\right]^{-1} \\ &= [1 + g(X)]^{-1}, \end{aligned} \tag{4.2.3}$$

where

$$g(X) \triangleq \sum_{j=1}^{N} \left(\frac{\lambda}{\mu}\right)^j \prod_{i=0}^{j-1} [(1-q)x_i + q] + \left(\frac{\lambda}{\mu}\right)^N \left\{\prod_{i=0}^{N-1} [(1-q)x_i + q]\right\} \frac{\frac{\lambda q}{\mu} - \left(\frac{\lambda q}{\mu}\right)^{k-N}}{1 - \frac{\lambda q}{\mu}}.$$

To search for a policy X that minimizes π_0^X, it is equivalent to find one that maximizes $g(X)$. We proceed with two different cases.

Case (1): $\mu < \lambda + \frac{C}{R}$. We have $q = \frac{1}{\lambda}\left(\mu - \frac{C}{R}\right) \in (0, 1)$. In this case, $g(X)$ can be simplified into

$$g(X) = \sum_{j=1}^{N} \left(\frac{\lambda}{\mu}\right)^j \prod_{i=0}^{j-1} [(1-q)x_i + q] + \left(\frac{\lambda}{\mu}\right)^N \left\{\prod_{i=0}^{N-1} [(1-q)x_i + q]\right\} \frac{R\mu}{C} \left[1 - \frac{C}{R\mu} - \left(1 - \frac{C}{R\mu}\right)^{k-N}\right]. \tag{4.2.3}$$

The first term $\sum_{j=1}^{N} \left(\frac{\lambda}{\mu}\right)^j \prod_{i=0}^{j-1}[(1-q)x_i + q]$ reaches maximum when $x_i = 1$ because $q < 1$. And $\left(1 - \frac{C}{R\mu}\right)^{k-N}$ is decreasing in k because $0 < 1 - \frac{C}{R\mu} < 1$ due to the fact that $R > \frac{C}{\mu}$. Therefore, $g(X)$ is maximized when $x_i = 1$ for $i \in \{0, 1, \ldots, N-1\}$ and $k \to \infty$. That is, X^N corresponds to the optimal policy.

Case (2): $\mu \geq \lambda + \frac{C}{R}$. In this case, we have $q = 1$ and $g(X)$ is reduced to

$$g(X) = \sum_{j=1}^{N} \left(\frac{\lambda}{\mu}\right)^j + \frac{\left(\frac{\lambda}{\mu}\right)^{N+1} - \left(\frac{\lambda}{\mu}\right)^k}{1 - \frac{\lambda}{\mu}}.$$

The maximum value of $g(X)$ is obtained as long as $k \to \infty$, which means it is optimal to have $x_i = 0$ when $i \geq N$. Since x_i can be either 0 or 1 when $i < N$, there are 2^N many different optimal policies although all of them lead to the same queueing dynamics. Clearly, X^N is one of the optimal policies. □

The above Theorem 4.2.1 reveals that to maximize throughput, the server should announce the queue length to customers that arrive when the queue is shorter than Naor's balking threshold N, but hide the information from them when the queue is longer than N. Moreover, it is intuitive to understand why such a policy, X^N, can maximize system throughput in our setup: When the queue length is less than the balking threshold N, it is wise for the server to reveal the queue length information to ensure that any arriving customer will join the queue. On the other hand, when the queue length is N or greater, any arriving customer will not join the queue if they knew it, thus it is in the server's best interest to conceal the information. Last but not least, we want to point out that X^N will not be optimal under the setting studied in Simhon et al. (2016) when customers are aware that X^N is the server's information disclosure policy, because any customer who does not receive the queue information can infer that the queue length must be greater or equal to N. These customers will refuse to join the queue, reducing the system throughput.

4.3 Socially Optimal Policy

Next, we turn to determine the socially optimal policy that maximizes overall consumer welfare. (We will consider an alternative objective of the social welfare that includes both consumer welfare and server revenue in Sect. 4.4.1.) We first introduce the following technical lemma which will be needed to prove the main result.

Lemma 4.3.1 $w(x) \triangleq \ln x + 1 - x \leq 0$ *and* $l(x) \triangleq \ln x + \frac{1}{x} - 1 \geq 0$ *for* $x > 0$.

Proof Since $\frac{dw(x)}{dx} = \frac{1}{x} - 1 = \frac{1-x}{x}$ and $\frac{dl(x)}{dx} = \frac{1}{x} - \frac{1}{x^2} = \frac{x-1}{x^2}$, it is clear that $w(x)$ increases in x for $x \in (0, 1]$ and decreases for $x \in (1, \infty)$ whereas $l(x)$ decreases in x for $x \in (0, 1]$ and increases for $x \in (1, \infty)$. Therefore, $w(x) \leq w(1) = 0$ and $l(x) \geq l(1) = 0$. □

Lemma 4.3.2 $y(n) \triangleq n(\rho q)^n \left[\rho\left(1 - (\rho q)^N\right) + 1 - \rho q\right] - N(\rho q)^N$ $\left[\rho\left(1 - (\rho q)^n\right) + 1 - \rho q\right] \geq 0$ *for* $n \in \{2, \ldots, N - 1\}$.

Proof Note that

$$
\begin{aligned}
&y(n) - y(n-1)\\
&= (\rho q)^{n-1}\left\{[n(\rho q - 1) + 1]\left[\rho\left(1 - (\rho q)^N\right) + 1 - \rho q\right] + N(\rho q)^N(\rho q - 1)\right\}\\
&= (\rho q)^{n-1}\left\{-n(1-\rho q)\left[\rho\left(1 - (\rho q)^N\right) + 1 - \rho q\right] + \rho\left(1 - (\rho q)^N\right) + 1 - \rho q + N(\rho q)^N(\rho q - 1)\right\}\\
&= (\rho q)^{n-1} z(n),
\end{aligned}
$$

where $z(n) \triangleq \left\{-n(1-\rho q)\left[\rho\left(1 - (\rho q)^N\right) + 1 - \rho q\right] + \rho\left(1 - (\rho q)^N\right) + 1 - \rho q + N(\rho q)^N(\rho q - 1)\right\}$.

Since $\rho q = 1 - \frac{C}{R\mu} < 1$, $(\rho q)^{n-1}$ is positive and $z(n)$ is decreasing in n. There are three possible cases:

- $z(n) \geq 0$ for all n, then $y(n) - y(n-1) \geq 0$ for all n, and $y(n)$ is increasing for $n \in \{2, \ldots, N-1\}$, that is, minimum value is $y(2)$,
- $z(n) \leq 0$ for all n, then $y(n) - y(n-1) \leq 0$ for all n, so $y(n)$ is decreasing for $n \in \{2, \ldots, N-1\}$, that is, minimum value is $y(N-1)$,
- $z(n) \geq 0$ for all $n \leq n^*$ for some n^* and $z(n) < 0$ for all $n > n^*$ (since $z(n)$ decreases in n), then $y(n)$ is increasing in $n \in \{2, \ldots, n^*\}$ and decreasing in $n \in \{n^* + 1, \ldots, N-1\}$, that is, minimum value is either $y(2)$ or $y(N-1)$.

It then suffices to show that $y(2) \geq 0$ and $y(N-1) \geq 0$.
First, we will prove that $y(N-1) \geq 0$. We have

$$
\begin{aligned}
y(N-1) &= (N-1)(\rho q)^{N-1}\left[\rho\left(1 - (\rho q)^N\right) + 1 - \rho q\right] - N(\rho q)^N\left[\rho\left(1 - (\rho q)^{N-1}\right) + 1 - \rho q\right]\\
&= (\rho q)^{N-1}\left\{[N(1-\rho q) - 1](\rho + 1 - \rho q) + \rho(\rho q)^N\right\}\\
&= (\rho q)^{N-1}\left\{\left(N\frac{C}{R\mu} - 1\right)\left(\rho + \frac{C}{R\mu}\right) + \rho\left(1 - \frac{C}{R\mu}\right)^N\right\}\\
&= (\rho q)^{N-1}\left\{\rho\left[N\frac{C}{R\mu} - 1 + \left(1 - \frac{C}{R\mu}\right)^N\right] + \frac{C}{R\mu}\left(N\frac{C}{R\mu} - 1\right)\right\}.
\end{aligned}
\tag{4.3.1}
$$

We first show that the coefficient for ρ in (4.3.1), namely $\left[N\frac{C}{R\mu}-1+\left(1-\frac{C}{R\mu}\right)^{N}\right]$, is greater than 0.

For a given N, we have $N \le \frac{C}{R\mu} < N+1$ by definition, which means $\frac{C}{R\mu} \in \left(\frac{1}{N+1}, \frac{1}{N}\right]$. Since

$$\frac{\partial\left[N\frac{C}{R\mu}-1+\left(1-\frac{C}{R\mu}\right)^{N}\right]}{\partial(\frac{C}{R\mu})} = N\left[1-\left(1-\frac{C}{R\mu}\right)^{N-1}\right] > 0,$$

it is readily seen that $\left[N\frac{C}{R\mu}-1+\left(1-\frac{C}{R\mu}\right)^{N}\right]$ increases in $\frac{C}{R\mu}$. Thus,

$$\left[N\frac{C}{R\mu}-1+\left(1-\frac{C}{R\mu}\right)^{N}\right] > \left(\frac{N}{N+1}\right)^{N} - \frac{1}{N+1},$$

which is positive because $\left(\frac{N}{N+1}\right)^{N}$ decreases in N and $\lim_{N\to\infty}\left(\frac{N}{N+1}\right)^{N} = \frac{1}{e} > \frac{1}{4} \ge \frac{1}{N+1}$ for $N \ge 3$.

With $\left[N\frac{C}{R\mu}-1+\left(1-\frac{C}{R\mu}\right)^{N}\right] > 0$ and $\rho > \rho q = 1-\frac{C}{R\mu}$, we have

$$\begin{aligned} y(N-1) &= (\rho q)^{N-1}\left\{\rho\left[N\frac{C}{R\mu}-1+\left(1-\frac{C}{R\mu}\right)^{N}\right]+\frac{C}{R\mu}\left(N\frac{C}{R\mu}-1\right)\right\} \\ &> (\rho q)^{N-1}\left[N\frac{C}{R\mu}-1+\left(1-\frac{C}{R\mu}\right)^{N+1}\right], \end{aligned}$$

where $\left[N\frac{C}{R\mu}-1+\left(1-\frac{C}{R\mu}\right)^{N+1}\right]$ increases in $\frac{C}{R\mu}$ because

$$\begin{aligned} \frac{\partial\left[N\frac{C}{R\mu}-1+\left(1-\frac{C}{R\mu}\right)^{N+1}\right]}{\partial(\frac{C}{R\mu})} &= N-(N+1)\left(1-\frac{C}{R\mu}\right)^{N} \\ &> N-(N+1)\left(\frac{N}{N+1}\right)^{N} = N\left[1-\left(\frac{N}{N+1}\right)^{N-1}\right] > 0. \end{aligned}$$

Plugging in the lower end $\frac{C}{R\mu} \to \frac{1}{N+1}$, we have

$$y(N-1) > (\rho q)^{N-1}\left[N\frac{C}{R\mu}-1+\left(1-\frac{C}{R\mu}\right)^{N+1}\right] > (\rho q)^{N-1}\left[\left(\frac{N}{N+1}\right)^{N+1}-\frac{1}{N+1}\right].$$

To show $y(N-1) > 0$, it now suffices to show that $\left[\left(\frac{N}{N+1}\right)^{N+1}-\frac{1}{N+1}\right] > 0$. Note that by Lemma 4.3.1

$$\frac{\partial\left[\left(\frac{N}{N+1}\right)^{N+1}-\frac{1}{N+1}\right]}{\partial N} = \frac{1}{(N+1)^2}+\left(\frac{N}{N+1}\right)^{N+1}\left[\ln\frac{N}{N+1}+\frac{N+1}{N}-1\right] > 0.$$

Thus, $\left[\left(\frac{N}{N+1}\right)^{N+1} - \frac{1}{N+1}\right] > \left(\frac{3}{3+1}\right)^{3+1} - \frac{1}{3+1} = \frac{17}{256} > 0$ because $N \geq 3$.

Next, we turn to prove that $y(2) \geq 0$. We have

$$\begin{aligned} y(2) &= 2(\rho q)^2\left[\rho\left(1-(\rho q)^N\right)+1-\rho q\right] - N(\rho q)^N\left[\rho\left(1-(\rho q)^2\right)+1-\rho q\right] \\ &= (\rho q)^2\left\{\left[2-N(\rho q)^{N-2}\right](1-\rho q+\rho)+(N-2)\,\rho(\rho q)^N\right\} \\ &= (\rho q)^2\left\{\left[2-N\left(1-\frac{C}{R\mu}\right)^{N-2}\right]\left(\frac{C}{R\mu}+\rho\right)+(N-2)\,\rho\left(1-\frac{C}{R\mu}\right)^N\right\} \\ &= (\rho q)^2\left\{\rho\left[2-N\left(1-\frac{C}{R\mu}\right)^{N-2}+(N-2)\left(1-\frac{C}{R\mu}\right)^N\right]+\frac{C}{R\mu}\left[2-N\left(1-\frac{C}{R\mu}\right)^{N-2}\right]\right\} \end{aligned} \quad (4.3.2)$$

Similar to the proof that shows $y(N-1) > 0$, we first verify that the coefficient for ρ in (4.3.2), that is,

$$\left[2-N\left(1-\frac{C}{R\mu}\right)^{N-2}+(N-2)\left(1-\frac{C}{R\mu}\right)^N\right],$$

is greater than 0. Since

$$\frac{\partial\left[2-N\left(1-\frac{C}{R\mu}\right)^{N-2}+(N-2)\left(1-\frac{C}{R\mu}\right)^N\right]}{\partial\left(\frac{C}{R\mu}\right)} = N(N-2)\left[\left(1-\frac{C}{R\mu}\right)^{N-3}-\left(1-\frac{C}{R\mu}\right)^{N-1}\right] > 0,$$

it readily seen that $\left[2-N\left(1-\frac{C}{R\mu}\right)^{N-2}+(N-2)\left(1-\frac{C}{R\mu}\right)^N\right]$ increases in $\frac{C}{R\mu}$.

Plugging in the lower end $\frac{C}{R\mu} \to \frac{1}{N+1}$, we have

$$\begin{aligned} \left[2-N\left(1-\frac{C}{R\mu}\right)^{N-2}+(N-2)\left(1-\frac{C}{R\mu}\right)^N\right] &> 2-N\left(\frac{N}{N+1}\right)^{N-2}+(N-2)\left(\frac{N}{N+1}\right)^N \\ &= \left[2-\left(\frac{N}{N+1}\right)^N\left(4+\frac{1}{N}\right)\right]. \end{aligned}$$

Since

$$\frac{\partial\left[2-\left(\frac{N}{N+1}\right)^N\left(4+\frac{1}{N}\right)\right]}{\partial N} = \left(\frac{N}{N+1}\right)^N\left[\frac{1}{N^2}-\left(4+\frac{1}{N}\right)\left(\ln\frac{N}{N+1}+1-\frac{N}{N+1}\right)\right] > 0,$$

by Lemma 4.3.1, we know $\left[2-\left(\frac{N}{N+1}\right)^N\left(4+\frac{1}{N}\right)\right]$ increases in N. Thus, $\left[2-\left(\frac{N}{N+1}\right)^N\left(4+\frac{1}{N}\right)\right] \geq 2-\left(\frac{3}{3+1}\right)^3\left(4+\frac{1}{3}\right) = \frac{11}{64} > 0$ because $N \geq 3$.

Now that the coefficient for ρ, $\left[2-N\left(1-\frac{C}{R\mu}\right)^{N-2}+(N-2)\left(1-\frac{C}{R\mu}\right)^N\right]$, is greater than 0, we have

$$
\begin{aligned}
y(2) &= (\rho q)^2 \left\{ \rho \left[2 - N\left(1-\frac{C}{R\mu}\right)^{N-2} + (N-2)\left(1-\frac{C}{R\mu}\right)^{N} \right] + \frac{C}{R\mu}\left[2 - N\left(1-\frac{C}{R\mu}\right)^{N-2}\right] \right\} \\
&= (\rho q)^2 \left\{ \left[2 - N\left(1-\frac{C}{R\mu}\right)^{N-2} \right]\left(\frac{C}{R\mu}+\rho\right) + (N-2)\,\rho\left(1-\frac{C}{R\mu}\right)^{N} \right\} \\
&> (\rho q)^2 \left[2 - N\left(1-\frac{C}{R\mu}\right)^{N-2} + (N-2)\left(1-\frac{C}{R\mu}\right)^{N+1} \right].
\end{aligned}
$$

Since

$$
\begin{aligned}
&\frac{\partial\left[2 - N\left(1-\frac{C}{R\mu}\right)^{N-2} + (N-2)\left(1-\frac{C}{R\mu}\right)^{N+1}\right]}{\partial(\frac{C}{R\mu})} \\
&= (N-2)(N+1)\left[\left(\frac{N}{N+1}\right)\left(1-\frac{C}{R\mu}\right)^{N-3} - \left(1-\frac{C}{R\mu}\right)^{N}\right] \\
&> N(N-2)\left[\left(1-\frac{C}{R\mu}\right)^{N-3} - \left(1-\frac{C}{R\mu}\right)^{N-1}\right] > 0,
\end{aligned}
$$

we know $\left[2 - N\left(1-\frac{C}{R\mu}\right)^{N-2} + (N-2)\left(1-\frac{C}{R\mu}\right)^{N+1}\right]$ is increasing in $\frac{C}{R\mu}$. Plugging in the lower end $\frac{C}{R\mu} \to \frac{1}{N+1}$, we have

$$
\begin{aligned}
y(2) &> (\rho q)^2 \left[2 - N\left(\frac{N}{N+1}\right)^{N-2} + (N-2)\left(\frac{N}{N+1}\right)^{N+1} \right] \\
&= (\rho q)^2 \left[2 - \left(\frac{N}{N+1}\right)^{N+1}\left(5 + \frac{3}{N} + \frac{1}{N^2}\right) \right].
\end{aligned}
$$

Since $\left(\frac{N}{N+1}\right)^{N+1}$ increases in N and is bounded above by $\frac{1}{e}$, and $\left(5 + \frac{3}{N} + \frac{1}{N^2}\right)$ decreases in N. We have for $N \geq 8$ that

$$
2 - \left(\frac{N}{N+1}\right)^{N+1}\left(5 + \frac{3}{N} + \frac{1}{N^2}\right) \geq 2 - \frac{1}{e}\left(5 + \frac{3}{N} + \frac{1}{N^2}\right) \geq 2 - \frac{1}{e}\left(5 + \frac{3}{8} + \frac{1}{8^2}\right) = 0.0169 > 0.
$$

For $N = 3, 4, 5, 6, 7$, it follows that $\left[2 - \left(\frac{N}{N+1}\right)^{N+1}\left(5 + \frac{3}{N} + \frac{1}{N^2}\right)\right] = \frac{17}{256}, \frac{298}{3125}, \frac{1729}{15552}, \frac{99662}{823543}, \frac{2142149}{16777216}$, respectively, all strictly positive. Therefore, we can conclude that $y(2) > 0$ for all $N \geq 3$.

We have successfully shown so far that $y(2) > 0$ and $y(N-1) > 0$, which implies that $y(n) > 0$ for $n \in \{2, \ldots, N-1\}$. □

Lemma 4.3.3

$$h(q_1, q_2 \ldots, q_{n-1}) \triangleq \left[\left(R - \frac{C}{\mu}\right) + \sum_{i=1}^{n-1} \rho^i \prod_{j=1}^{i} q_j \left(R - C\frac{i+1}{\mu}\right)\right]\left(\sum_{i=n+1}^{N} \rho^i q^{i-n-1}\right)$$
$$- \left[\sum_{i=n}^{N-1} \rho^i q^{i-n} \left(R - C\frac{i+1}{\mu}\right)\right]\left(1 + \rho + \sum_{i=2}^{n} \rho^i \prod_{j=1}^{i-1} q_j\right) \geq 0.$$

Proof Note that

$$\begin{aligned}
&h(q_1, q_2, \ldots, q_{n-1}) \\
&= \left(R - \frac{C}{\mu}\right)\left(\sum_{i=n+1}^{N} \rho^i q^{i-n-1}\right) + \sum_{i=1}^{n-1} \rho^i \prod_{j=1}^{i} q_j \left(R - C\frac{i+1}{\mu}\right)\left(\sum_{i=n+1}^{N} \rho^i q^{i-n-1}\right) \\
&\quad - \left[\sum_{i=n}^{N-1} \rho^i q^{i-n} \left(R - C\frac{i+1}{\mu}\right)\right](1+\rho) - \left[\sum_{i=n}^{N-1} \rho^i q^{i-n} \left(R - C\frac{i+1}{\mu}\right)\right]\left(\sum_{i=2}^{n} \rho^i \prod_{j=1}^{i-1} q_j\right) \\
&= \left(R - \frac{C}{\mu}\right)\left(\sum_{i=n+1}^{N} \rho^i q^{i-n-1}\right) + \sum_{i=1}^{n-1} \rho^i \prod_{j=1}^{i} q_j \left(R - C\frac{i+1}{\mu}\right)\left(\sum_{i=n+1}^{N} \rho^i q^{i-n-1}\right) \\
&\quad - \left[\sum_{i=n+1}^{N} \rho^{i-1} q^{i-n-1} \left(R - C\frac{i}{\mu}\right)\right](1+\rho) - \left[\sum_{i=n+1}^{N} \rho^{i-1} q^{i-n-1} \left(R - C\frac{i}{\mu}\right)\right]\left(\sum_{i=1}^{n-1} \rho^{i+1} \prod_{j=1}^{i} q_j\right) \\
&= \sum_{i=n+1}^{N} \rho^i q^{i-n-1} \left\{\rho\left(R - \frac{C}{\mu}\right) - \left(R - C\frac{i}{\mu}\right)(1+\rho)\right\} \\
&\quad + \sum_{i=1}^{n-1} \rho^i \left(\prod_{j=1}^{i} q_j\right) \sum_{m=n+1}^{N} \rho^m q^{m-n-1} \left\{\left(R - C\frac{i+1}{\mu}\right) - \left(R - C\frac{m}{\mu}\right)\right\} \\
&= \sum_{i=n+1}^{N} \rho^i q^{i-n-1} \left\{\rho\left(R - \frac{C}{\mu}\right) - \left(R - C\frac{i}{\mu}\right)(1+\rho)\right\} + \sum_{l=1}^{n-1} \rho^l \left(\prod_{j=1}^{l} q_j\right) \sum_{m=n+1}^{N} \rho^m q^{m-l-1} \frac{C}{\mu}(m-l-1) \\
&\geq \left(\sum_{i=n+1}^{N} \rho^i q^{i-n-1}\right) \left\{\rho\left(R - \frac{C}{\mu}\right) - \left(R - C\frac{i}{\mu}\right)(1+\rho)\right\} + \sum_{l=1}^{n-1} (\rho q)^l \sum_{m=n+1}^{N} \rho^m q^{m-l-1} \frac{C}{\mu}(m-l-1) \\
&= q^{1-n} \cdot y(n) \geq 0,
\end{aligned} \tag{4.3.3}$$

where the first inequality holds because $m \in \{n+1, n+2, \ldots, N\}$, $l \in \{1, 2, \ldots, n-1\}$, so $m > l + 1$ and the second inequality holds due to Lemma 4.3.2.

Thus, we have $h(q_1, q_2, \ldots, q_{n-1}) \geq 0$. □

Now, let S^X_{soc} denote the average individual consumer welfare under policy X. We show in the following Theorem 4.3.1 that the socially optimal policy is either $X^{0,N}$ or $X^{1,N}$ as defined by (4.2.2). And here is the outline of the otherwise lengthy proof: First, we show that it is socially optimal for the server to reveal the queue length when it is N (or above). Then we proceed to discuss two situations when the equilibrium joining probability for the uninformed customers, q, is either less than 1 or equal to 1. When $q < 1$, we establish at optimality that $x_0 = x_N = 1$ and $x_2 = x_3 = \ldots = x_{N-1} = 0$ must be satisfied, and derive condition for the optimal value for $x_1 \in \{0, 1\}$. Finally, we show that $X^{0,N}$ or $X^{1,N}$ is also optimal when $q = 1$.

Theorem 4.3.1 *The socially optimal general policy is given by* $X^{0,N}$ *(resp.,* $X^{1,N}$*) as defined in* (4.2.2) *if* $\rho q\left[1-\rho q+\rho(1-(\rho q)^N)\right]-N(\rho q)^N(1-\rho q)(1+\rho)\geq 0$ *(resp.,* <0*).*

Proof We continue to use $k=\min\{i\mid i\geq N, x_i=1\}$, and no states higher than k will be ever reached. Given any policy X, if k exceeds N (that is, $x_N=0$), the average individual social welfare is given by

$$S_{soc}^X=\sum_{i=0}^{N-1}\pi_i q_i\left(R-C\frac{i+1}{\mu}\right)+\sum_{i=N}^{k}\pi_i q\left(R-C\frac{i+1}{\mu}\right)\text{ where } q_i=q(1-x_i)+x_i. \quad (4.3.4)$$

The average individual consumer welfare can be improved simply by setting $x_N=1$ (that is, $k=N$) without changing the values of $\{x_0,x_1,\ldots,x_{N-1}\}$. Therefore, for an optimal policy, we must have $k=N$ and no state higher than N will be reached, that is, the optimal strategy can have arbitrary values for $\{x_{N+1},x_{N+2},\ldots\}$. Expression (4.3.4) is thus reduced to

$$S_{soc}^X=\sum_{i=0}^{N-1}\pi_i q_i\left(R-C\frac{i+1}{\mu}\right). \quad (4.3.5)$$

The effective arrival rate, λ_e, given in (4.3.1), becomes

$$\lambda_e=\begin{cases}\lambda_i & i<N;\\ 0 & i\geq N.\end{cases}$$

Plugging (4.3.4) and (4.2.3) into (4.3.5), we have

$$S_{soc}^X=\frac{\sum_{i=0}^{N-1}\rho^i\prod_{j=0}^{i}q_j\left(R-C\frac{i+1}{\mu}\right)}{1+\sum_{i=1}^{N}\rho^i\prod_{j=0}^{i-1}q_j}. \quad (4.3.6)$$

The goal is to identify values of $\{x_0,x_1,\ldots,x_{N-1}\}$ that would maximize (4.3.6). We proceed with two cases, when the equilibrium joining probability for the uninformed customers, q, is either less than 1 or equal to 1.

Case (1): $q<1\Leftrightarrow\mu<\lambda+\frac{C}{R}\Leftrightarrow\rho>1-\frac{C}{R\mu}$. Let $S^{(x_0,x_1,\ldots,x_{N-1})}$ denote the average consumer welfare under the policy $(x_0,x_1,\ldots,x_{N-1},x_N=1)$. First, we determine the optimal value of x_0 given $x_1,x_1,\ldots,x_{N-1}$. Since x_0 can only be 0 or 1, we can compare the social welfare under the policies $(x_0=0,x_1,\ldots,x_{N-1},x_N=1)$ and $(x_0=1,x_1,\ldots,x_{N-1},x_N=1)$. Since

$$
\begin{aligned}
& S^{(1,x_1,\ldots,x_{N-1})} - S^{(0,x_1,\ldots,x_{N-1})} \\
&= \frac{R - \frac{C}{\mu} + \sum_{i=1}^{N-1} \rho^i \prod_{j=1}^{i} q_j \left(R - C\frac{i+1}{\mu}\right)}{1 + \rho + \sum_{i=1}^{N} \rho^i \prod_{j=2}^{i-1} q_j} - \frac{q\left(R - \frac{C}{\mu}\right) + q\sum_{i=1}^{N-1} \rho^i \prod_{j=1}^{i} q_j \left(R - C\frac{i+1}{\mu}\right)}{1 + \rho q + q\sum_{i=2}^{N} \rho^i \prod_{j=2}^{i-1} q_j} \\
&= \frac{\left[R - \frac{C}{\mu} + \sum_{i=1}^{N-1} \rho^i \prod_{j=1}^{i} q_j \left(R - C\frac{i+1}{\mu}\right)\right]\left\{1 + \rho q + q\sum_{i=2}^{N} \rho^i \prod_{j=2}^{i-1} q_j - q\left[1 + \rho + \sum_{i=1}^{N} \rho^i \prod_{j=2}^{i-1} q_j\right]\right\}}{\left[1 + \rho + \sum_{i=1}^{N} \rho^i \prod_{j=2}^{i-1} q_j\right]\left[1 + \rho q + q\sum_{i=2}^{N} \rho^i \prod_{j=2}^{i-1} q_j\right]} \\
&= \frac{\left[R - \frac{C}{\mu} + \sum_{i=1}^{N-1} \rho^i \prod_{j=1}^{i} q_j \left(R - C\frac{i+1}{\mu}\right)\right](1-q)}{1 + \rho q + q\sum_{i=2}^{N} \rho^i \prod_{j=2}^{i-1} q_j} > 0,
\end{aligned}
$$

It is optimal to have $x_0 = 1$.

In the case $N = 1$, the optimal policy is $(x_0 = 1, x_1 = x_N = 1)$ which can be shown to satisfy the condition in the statement of Theorem 4.3.1 when $N = 1$.

In the case $N = 2$, we have

$$
\begin{aligned}
S^{(x_0=1,x_1=0)} - S^{(x_0=1,x_1=1)} &= \frac{R - \frac{C}{\mu} + \rho q\left(R - \frac{2C}{\mu}\right)}{1 + \rho + \rho^2 q} - \frac{R - \frac{C}{\mu} + \rho\left(R - \frac{2C}{\mu}\right)}{1 + \rho + \rho^2} \\
&= \frac{R\rho(1-q)\left\{\rho\left(1 - \frac{C}{R\mu}\right) - (1+\rho)\left(1 - \frac{2C}{R\mu}\right)\right\}}{\left[1 + \rho + \rho^2\right]\left[1 + \rho + \rho^2 q\right]} \\
&= \frac{R\rho(1-q)\left\{(\rho+2)\frac{C}{R\mu} - 1\right\}}{\left[1 + \rho + \rho^2\right]\left[1 + \rho + \rho^2 q\right]}. \qquad (4.3.7)
\end{aligned}
$$

Thus if $\rho \geq \frac{R\mu}{C} - 2$, (4.3.7) is positive and $(x_0 = 1, x_1 = 0, x_2 = 1)$ is the optimal policy. Otherwise, $(x_0 = 1, x_1 = 1, x_2 = 1)$ is the optimal policy. It can be shown that these conditions are also consistent with the condition given in the statement of Theorem 4.3.1 when $N = 2$.

We now consider the case $N \geq 3$. We have already shown that $x_0 = x_N = 1$ are optimal, and will establish next that $x_n = 0$ is optimal for all $n \in \{2, 3, \ldots, N-1\}$ by induction. The inductive step is to show that $x_n = 0$ is optimal if $x_i = 0$ for $2 \leq n < i < N$ (the base case is to show that $x_{N-1} = 0$ is optimal). Under the policy $(x_0 = 1, x_1, \ldots, x_{n-1}, x_n, 0, \ldots, 0, x_N = 1)$, the average consumer welfare is given by

$$
\begin{aligned}
& S^{(1,x_1,x_2,\ldots,x_{n-1},x_n,0,\ldots,0)} \\
&= \frac{\left(R - \frac{C}{\mu}\right) + \sum_{i=1}^{n-1} \rho^i \prod_{j=1}^{i} q_j \left(R - C\frac{i+1}{\mu}\right) + \left(\prod_{j=1}^{n} q_j\right)\sum_{i=n}^{N-1} \rho^i q^{i-n}\left(R - C\frac{i+1}{\mu}\right)}{1 + \rho + \sum_{i=2}^{n} \rho^i \prod_{j=1}^{i-1} q_j + \left(\prod_{j=1}^{n} q_j\right)\sum_{i=n+1}^{N} \rho^i q^{i-n-1}}.
\end{aligned}
$$

It is then sufficient to show that $S^{(1,x_1,x_2,\ldots,x_{n-1},x_n=0,0,\ldots,0)} - S^{(1,x_1,x_2,\ldots,x_{n-1},x_n=1,0,\ldots,0)} > 0$ for $n \in \{N-1, N-2, \ldots, 4, 3, 2\}$. Note that $n = N - 1$ establishes the base case. We have

$$
\begin{aligned}
&S^{(1,x_1,x_2,\ldots,x_{n-1},x_n=0,0,\ldots,0)} - S^{(1,x_1,x_2,\ldots,x_{n-1},x_n=1,0,\ldots,0)}\\
&= \frac{\left(R-\frac{C}{\mu}\right)+\sum_{i=1}^{n-1}\rho^i\prod_{j=1}^{i}q_j\left(R-C\frac{i+1}{\mu}\right)+q\left(\prod_{j=1}^{n-1}q_j\right)\sum_{i=n}^{N-1}\rho^i q^{i-n}\left(R-C\frac{i+1}{\mu}\right)}{1+\rho+\sum_{i=2}^{n}\rho^i\prod_{j=1}^{i-1}q_j+q\left(\prod_{j=1}^{n-1}q_j\right)\sum_{i=n+1}^{N}\rho^i q^{i-n-1}}\\
&- \frac{\left(R-\frac{C}{\mu}\right)+\sum_{i=1}^{n-1}\rho^i\prod_{j=1}^{i}q_j\left(R-C\frac{i+1}{\mu}\right)+\left(\prod_{j=1}^{n-1}q_j\right)\sum_{i=n}^{N-1}\rho^i q^{i-n}\left(R-C\frac{i+1}{\mu}\right)}{1+\rho+\sum_{i=2}^{n}\rho^i\prod_{j=1}^{i-1}q_j+\left(\prod_{j=1}^{n-1}q_j\right)\sum_{i=n+1}^{N}\rho^i q^{i-n-1}}\\
&= \frac{(1-q)\left(\prod_{j=1}^{n-1}q_j\right)\left[\left(R-\frac{C}{\mu}\right)+\sum_{i=1}^{n-1}\rho^i\prod_{j=1}^{i}q_j\left(R-C\frac{i+1}{\mu}\right)\right]\left(\sum_{i=n+1}^{N}\rho^i q^{i-n-1}\right)}{\left[1+\rho+\sum_{i=2}^{n}\rho^i\prod_{j=1}^{i-1}q_j+\left(\prod_{j=1}^{n-1}q_j\right)\sum_{i=n+1}^{N}\rho^i q^{i-n-1}\right]\left[1+\rho+\sum_{i=2}^{n}\rho^i\prod_{j=1}^{i-1}q_j+q\left(\prod_{j=1}^{n-1}q_j\right)\sum_{i=n+1}^{N}\rho^i q^{i-n-1}\right]}-\\
&\frac{(1-q)\left(\prod_{j=1}^{n-1}q_j\right)\left(\sum_{i=n}^{N-1}\rho^i q^{i-n}\left(R-C\frac{i+1}{\mu}\right)\right)\left[1+\rho+\sum_{i=2}^{n}\rho^i\prod_{j=1}^{i-1}q_j\right]}{\left[1+\rho+\sum_{i=2}^{n}\rho^i\prod_{j=1}^{i-1}q_j+\left(\prod_{j=1}^{n-1}q_j\right)\sum_{i=n+1}^{N}\rho^i q^{i-n-1}\right]\left[1+\rho+\sum_{i=2}^{n}\rho^i\prod_{j=1}^{i-1}q_j+q\left(\prod_{j=1}^{n-1}q_j\right)\sum_{i=n+1}^{N}\rho^i q^{i-n-1}\right]} \qquad (4.3.8)\\
&= \frac{(1-q)\left(\prod_{j=1}^{n-1}q_j\right)h(q_1,q_2\ldots,q_{n-1})}{\left[1+\rho+\sum_{i=2}^{n}\rho^i\prod_{j=1}^{i-1}q_j+\left(\prod_{j=1}^{n-1}q_j\right)\sum_{i=n+1}^{N}\rho^i q^{i-n-1}\right]\left[1+\rho+\sum_{i=2}^{n}\rho^i\prod_{j=1}^{i-1}q_j+q\left(\prod_{j=1}^{n-1}q_j\right)\sum_{i=n+1}^{N}\rho^i q^{i-n-1}\right]}\geq 0,
\end{aligned}
$$

where the last inequality holds due to Lemma 4.3.3.

Thus, $S^{(1,x_1,x_2,\ldots,x_{n-1},x_n=0,0,\ldots,0)} - S^{(1,x_1,x_2,\ldots,x_{n-1},x_n=1,0,\ldots,0)} > 0$.

It follows from previous analysis that (i) $x_{N-1}=0$ is optimal, and (ii) $x_n=0$ is optimal if $x_i=0$ is optimal for all $i: 2\leq n<i<N$.

Knowing that $x_0=x_N=1$ and $x_2=x_3=\ldots=x_{N-1}=0$ are optimal, what is left to consider is the optimal value of $x_1\in\{0,1\}$. From (4.3.8), we have

$$
S^{(1,0,0,\ldots,0)} - S^{(1,1,0,\ldots,0)} = \frac{(1-q)\left\{\left(R-\frac{C}{\mu}\right)\left(\sum_{i=2}^{N}\rho^i q^{i-2}\right)-\sum_{i=1}^{N-1}\rho^i q^{i-1}\left(R-C\frac{i+1}{\mu}\right)[1+\rho]\right\}}{\left[1+\rho\sum_{i=2}^{N}\rho^i q^{i-2}\right]\left[1+\rho\sum_{i=2}^{N}\rho^i q^{i-n-1}\right]}. \tag{4.3.9}
$$

Since $q<1$, we have $1-q>0$. The denominator is also positive. Note that

$$
\begin{aligned}
&\left(R-\frac{C}{\mu}\right)\left(\sum_{i=2}^{N}\rho^i q^{i-2}\right)-\sum_{i=1}^{N-1}\rho^i q^{i-1}\left(R-C\frac{i+1}{\mu}\right)(1+\rho) \qquad (4.3.10)\\
&=\frac{Cq^{-1}}{\mu}\left\{(\rho q)\rho\left[(\rho q)-(\rho q)^N\right]-\left[N(\rho q)^N-(\rho q)\right][1-\rho q+\rho(1-(\rho q))]\right\}\\
&=\frac{Cq^{-1}}{\mu}\left\{\rho q\left[1-\rho q+\rho(1-(\rho q)^N)\right]-N(\rho q)^N(1-\rho q)(1+\rho)\right\}.
\end{aligned}
$$

Therefore, if $\rho q\left[1-\rho q+\rho(1-(\rho q)^N)\right]-N(\rho q)^N(1-\rho q)(1+\rho)\geq 0$, then (4.3.10) is positive, making (4.3.9) positive in which case $x_1=0$ is optimal; otherwise, $x_1=1$ is optimal. This also concludes our proof that the socially optimal general policy is either $D^{0,N}$ or $D^{1,N}$ when $\mu<\lambda+\frac{C}{R}$.

Case (2): $q=1\Leftrightarrow\mu\geq\lambda+\frac{C}{R}$. In this case, all customers will join the queue at any state $i<N$ regardless of the server's queue length information disclosure policy

(that is, regardless of the value of x_i for $i < N$), thus any policy that contains $x_N = 1$ is optimal. □

Recall the $X^{0,N}$ (resp., $X^{1,N}$) policy specifies that the server should announce the queue length when there are no customers in the system (resp., in the queue) or there are more than N customers in the system, and should not announce the queue length otherwise. Therefore, Theorem 4.3.1 demonstrates that the socially optimal policy contains a "hole," that is, the service provider should announce the queue length to customers if and only if it is sufficiently short or sufficiently long. Such non-threshold equilibrium is also identified in Debo et al. (2012) and Hassin et al. (2017) for other settings in queues.

To explain the result, recall that the socially optimal policy maximizes consumer welfare. It is intuitive that the server should reveal any queue length $\geq N$ to prevent customers from joining because these customers will receive a negative expected payoff. On the other hand, when the queue is sufficiently small (e.g., when the server is idle or when the line is empty), customers should be encouraged to join the system as they impose little negative externalities to other customers. When the queue length is moderate, the negative externalities become significant enough that the socially optimal outcome is achieved by hiding the queue length information from customers arriving at these states.

4.4 Robustness of Findings

We have studied queue length disclosure policies in this paper when customers are not familiar with them. This applies to service settings that are infrequently needed by customers (e.g., customers may contact a particular customer service hot-line only once in a while). Contrary to Simhon et al. (2016) and Kim et al. (2017), we find that the intuitive policy, which reveals queue length information if the queue is short and conceals the information if the queue is long, indeed maximizes the server's throughput. Interestingly, the socially optimal policy, which maximizes consumer welfare, contains a "hole."

In this section, we first present numerical experiments to support our findings and then discuss robustness issues. In Tables 4.1 and 4.2, we set $N = 5$ and $N = 6$, respectively, and calculate the idle probability and social welfare for a variety of information disclosure policies. Note that these are just some example policies and they do not cover all the cases. However, we ran numerical comparison for all possible policies, and can show that any policies not listed here are suboptimal.

A policy that corresponds to the lowest idle probability (resp., the highest social welfare) is the throughput-maximizing (resp., socially optimal) policy. Consistent with our analytical findings, the numerical examples suggest that the throughput-maximizing policy is X^N whereas the socially optimal policy can be either $X^{0,N}$ or $X^{1,N}$.

Table 4.1 $R = 5, C = 1, \mu = 1, \Lambda = 2$ and $N = 5$

Policy	Idle probability	Social welfare
X^0	0.2	0
X^∞	0.0159	0.4127
X^5	**0.0052**	−1.5393
$X^{0,5}$	0.1295	**1.0607**
$X^{1,5}$	0.0675	0.9978
$(1, 0, 1, 0, 0, 1, 1, 1, \ldots)$	0.0806	0.8769
$(1, 1, 1, 0, 0, 1, 1, 1, \ldots)$	0.0337	0.7994
$(1, 1, 1, 0, 1, 1, 1, \ldots)$	0.0292	0.6199
$(1, 1, 1, 1, 0, 1, 1, 1, \ldots)$	0.0228	0.5936
$(1, 0, 0, 0, 0, 0, 1, 1, 1, \ldots)$	0.1194	0.9386
$(1, 1, 0, 0, 0, 0, 1, 1, 1, \ldots)$	0.0608	0.8306
$(1, 0, 1, 1, 0, 0, 1, 1, 1, \ldots)$	0.0427	0.4592
$(1, 1, 0, 1, 1, 0, 1, 1, 1, \ldots)$	0.0252	0.2846
$(1, 0, 0, 0, 0, 0, 0, 1, 1, 1, \ldots)$	0.1123	0.8245
$(1, 1, 0, 0, 0, 0, 0, 1, 1, 1, \ldots)$	0.0563	0.6955
$(1, 0, 1, 0, 0, 0, 0, 1, 1, 1, \ldots)$	0.0651	0.5698
$(1, 0, 1, 1, 0, 0, 0, 1, 1, 1, \ldots)$	0.0375	0.2800

Table 4.2 $R = 6, C = 1, \mu = 1, \Lambda = 0.88$ and $N = 6$

Policy	Idle probability	Social welfare
X^0	0.1667	0
X^∞	0.2029	2.5987
X^6	**0.1379**	−0.4174
$X^{0,6}$	0.2216	2.6722
$X^{1,6}$	0.2146	**2.6725**
$(1, 0, 1, 0, 0, 0, 1, 1, 1, \ldots)$	0.2165	2.6588
$(1, 1, 1, 0, 0, 0, 1, 1, 1, \ldots)$	0.2096	2.6558
$(1, 1, 1, 0, 1, 0, 1, 1, 1, \ldots)$	0.2075	2.6352
$(1, 1, 1, 1, 0, 0, 1, 1, 1, \ldots)$	0.2061	2.6329
$(1, 0, 0, 0, 0, 0, 0, 1, 1, 1, \ldots)$	0.2081	2.4387
$(1, 1, 0, 0, 0, 0, 0, 1, 1, 1, \ldots)$	0.2012	2.4340
$(1, 0, 1, 1, 0, 0, 0, 1, 1, 1, \ldots)$	0.1991	2.3868
$(1, 1, 1, 1, 0, 1, 0, 1, 1, 1, \ldots)$	0.1908	2.3576
$(1, 0, 0, 0, 0, 0, 0, 0, 1, 1, 1, \ldots)$	0.1979	2.2097
$(1, 1, 0, 0, 0, 0, 0, 0, 1, 1, 1, \ldots)$	0.1912	2.2006
$(1, 0, 1, 0, 0, 0, 0, 0, 1, 1, 1, \ldots)$	0.1927	2.1820
$(1, 0, 1, 1, 0, 0, 0, 0, 1, 1, 1, \ldots)$	0.1888	2.1459

Next, we establish the robustness of our findings in a few important ways. In each of these, we shall relax one assumption or change one perspective of the base model while keeping the remaining parts unchanged.

4.4.1 Alternative Objective of Social Welfare

So far in the paper, the socially optimal policy is defined as one that maximizes consumer welfare (mainly because we do not consider the service provider's pricing decision in our study). If the service provider can choose an optimal service price to maximize revenue, and the social welfare includes both consumer welfare and server revenue, our findings will continue to hold, that is, the throughput-maximizing policy is still X^N whereas the socially optimal policy either $X^{0,N}$ or $X^{1,N}$. This is because i) our current results are valid for any given price, including the price that would maximize server revenue, and ii) the objective functions will remain the same under the alternative objective of social welfare. To be exact about the second point, the system throughout will still equal to $\mu(1-\pi_0)$ and the social welfare expression remain the same as Eq. (4.3.4):

$$S_{soc} = \sum_{i=0}^{N-1} \pi_i q_i \left(R - C\frac{i+1}{\mu} \right) + \sum_{i=N}^{k} \pi_i q \left(R - C\frac{i+1}{\mu} \right) \text{ where } q_i = q(1-x_i) + x_i$$

as long as we interpret R as the actual service value without subtracting price (instead of the previously net service payout which is the service value less the given price)—when social welfare includes both consumer welfare and server revenue, the service price becomes a transfer price between the customers and the server.

4.4.2 Other Service Time Distributions

The derivation of our analytical results in the base model are based on explicit calculations of the M/M/1 model due to analytical tractability. A natural question that arises is how the optimal policies (identified in the M/M/1 model) perform under more general queueing settings. To this end, we ran extensive numerical experiments with M/G/1 queues when the service times follow an Erlang or a uniform distribution, and found that the throughput-maximizing policy is always X^N whereas the socially optimal policy either $X^{0,N}$ or $X^{1,N}$. We present two specific examples in Tables 4.3 and 4.4, where a policy that corresponds to the lowest idle probability (resp., the highest social welfare) is the throughput-maximizing (resp., socially optimal) policy. Similar to Tables 4.1 and 4.2, Tables 4.3 and 4.4 are just showing some example policies and they do not cover all the cases. However, we ran numerical comparison for all possible policies, and can show that any policies not listed here are suboptimal.

Table 4.3 $R = 3, C = 1, \Lambda = 1, N = 5$, and M/G/1 with Erlang (3, 3.1) for the service time distribution

Policy	Idle probability	Social welfare
X^0	0.2438	0.0356
X^∞	0.1992	0.7097
X^5	**0.1266**	−0.7632
$X^{0,5}$	0.2750	**0.9097**
$X^{1,5}$	0.2042	0.8655
$(1, 0, 1, 0, 0, 1, 1, 1\ldots)$	0.2302	0.5971
$(1, 1, 1, 0, 0, 1, 1, 1\ldots)$	0.1959	0.5432
$(1, 1, 1, 0, 1, 1, 1, \ldots)$	0.2099	0.7664
$(1, 1, 1, 1, 0, 1, 1, 1, \ldots)$	0.2021	0.7186
$(1, 0, 0, 0, 0, 0, 1, 1, 1, \ldots)$	0.2530	0.5085
$(1, 1, 0, 0, 0, 0, 1, 1, 1, \ldots)$	0.2167	0.8306
$(1, 0, 1, 1, 0, 0, 1, 1, 1, \ldots)$	0.1992	0.2517
$(1, 1, 0, 1, 1, 0, 1, 1, 1, \ldots)$	0.1875	0.1968
$(1, 0, 0, 0, 0, 0, 0, 1, 1, 1, \ldots)$	0.2451	0.3369
$(1, 1, 0, 0, 0, 0, 0, 1, 1, 1, \ldots)$	0.2102	0.2806
$(1, 0, 1, 0, 0, 0, 0, 1, 1, 1, \ldots)$	0.2120	0.2115
$(1, 0, 1, 1, 0, 0, 0, 1, 1, 1, \ldots)$	0.1887	0.0418

Table 4.4 $R = 5, C = 1, \Lambda = 0.9, N = 6$, and M/G/1 with Uniform (0.6, 1.4) for service time distribution

Policy	Idle probability	Social welfare
X^0	0.1234	0.1756
X^∞	0.1818	2.5307
X^6	**0.1127**	−0.0149
$X^{0,6}$	0.1887	2.5553
$X^{1,6}$	0.1859	**2.5573**
$(1, 0, 1, 0, 0, 0, 1, 1, 1, \ldots)$	0.1869	2.5580
$(1, 1, 1, 0, 0, 0, 1, 1, 1, \ldots)$	0.1844	2.5458
$(1, 1, 1, 0, 1, 0, 1, 1, 1, \ldots)$	0.1828	2.5390
$(1, 1, 1, 1, 0, 0, 1, 1, 1, \ldots)$	0.1817	2.5371
$(1, 0, 0, 0, 0, 0, 0, 1, 1, 1, \ldots)$	0.1724	2.2778
$(1, 1, 0, 0, 0, 0, 0, 1, 1, 1, \ldots)$	0.1700	2.2719
$(1, 0, 1, 1, 0, 0, 0, 1, 1, 1, \ldots)$	0.1677	2.2582
$(1, 1, 1, 1, 0, 1, 0, 1, 1, 1, \ldots)$	0.1633	2.2510
$(1, 0, 0, 0, 0, 0, 0, 0, 1, 1, 1, \ldots)$	0.1601	2.0198
$(1, 1, 0, 0, 0, 0, 0, 0, 1, 1, 1, \ldots)$	0.1583	2.0143
$(1, 0, 1, 0, 0, 0, 0, 0, 1, 1, 1, \ldots)$	0.1578	2.0126
$(1, 0, 1, 1, 0, 0, 0, 0, 1, 1, 1, \ldots)$	0.1555	1.9906

4.4.3 Unknown Arrival Rate and Service Rate Information

The base model assumes that customers know the arrival rate and service rate information. In practice, however, it is plausible that sometimes customers do not know the information. We study how that would impact our main findings in this subsection.

When the arrival rate and service rate information is readily known to the customers (as in the base model), they can compute the expected waiting time in equilibrium when the underlying state they arrive at is uninformed. Without the arrival rate and service rate information, customers can no longer compute the expected waiting time. As a result, we assume that (uninformed) customers' beliefs of the expected waiting time follow some distribution characterized by cumulative distribution function F, i.e., each (uninformed) customer has a belief of her expected waiting time upon arriving to the system and these beliefs form a distribution F because customers arrive randomly. Different factors can affect customers' expected waiting time beliefs, e.g., customers believing the expected wait to be long (or short) may think that the arrival rate is high (or low) and/or the service rate is low (or high). We do not model how individual customers' beliefs on their expected wait are formed by treating the waiting time belief distribution F as given.

When a customer with expected waiting time belief of w arrives at an uninformed state, she will join the system if and only if the net service payout exceeds the expected wait, i.e., $R > C \cdot w$ or $w < R/C$. This translate into that the joining probability among customers who arrive at an uninformed state is $F(R/c)$. Assuming the stability condition $\lambda F(R/c) < \mu$ is satisfied, we have the following analytic result.

Theorem 4.4.1 *X^N is a throughout-maximizing queue length disclosure policy.*

Proof Using the same argument from the proof of Theorem 4.2.1, for any arbitrary policy $X = (x_0, x_1, \ldots, x_{N-1})$ and $k(> N)$, we can derive the steady-state probabilities as follows.

$$\pi_0^X = [1 + g_1(X)]^{-1} \tag{4.4.1}$$

where

$$g_1(X) \triangleq \sum_{j=1}^{N} \left(\frac{\lambda}{\mu}\right)^j \prod_{i=0}^{j-1} [(1 - F(R/c))x_i + F(R/c)]$$
$$+ \left(\frac{\lambda}{\mu}\right)^N \left\{\prod_{i=0}^{N-1} [(1 - F(R/c))x_i + F(R/c)]\right\} \frac{\frac{\lambda F(R/c)}{\mu} - \left(\frac{\lambda F(R/c)}{\mu}\right)^{k-N}}{1 - \frac{\lambda F(R/c)}{\mu}}. \tag{4.4.2}$$

To search for a policy X that minimizes π_0^X, it is equivalent to find one that maximizes $g(X)$. We proceed with the following three cases.

Case (1): $F(R/c) \in (0, 1)$. In this case, the first term of $g_1(X)$ reaches maximum when $x_i = 1$ because $\lambda F(R/c) < \mu$. The second term is increasing in k, so it is optimal to have $k \to \infty$. Therefore, X^N is the optimal policy.

Case (2): $F(R/c) = 0$. We can derive that

$$g_1(X) \triangleq \sum_{j=1}^{N} \left(\frac{\lambda}{\mu}\right)^j \prod_{i=0}^{j-1} x_i.$$

Clearly, to maximize $g_1(X)$, we only need to set $x_i = 1$ for $i \in \{0, 1, \ldots, N-1\}$, and k can be of any value in $\{N, N+1, N+2, \ldots\}$. Thus, X^N is one of the optimal policies.

Case (3): $F(R/c) = 1$. In this case, $g_1(X)$ is reduced to

$$g_1(X) = \sum_{j=1}^{N} \left(\frac{\lambda}{\mu}\right)^j + \frac{\left(\frac{\lambda}{\mu}\right)^{N+1} - \left(\frac{\lambda}{\mu}\right)^k}{1 - \frac{\lambda}{\mu}}.$$

Because $g_1(X)$ increases with k, we have $k \to \infty$. For the valve of x_i, they can be either 0 or 1. Thus, X^N is one of the optimal policies. □

Note that the result of Theorem 4.4.1 is free of the specific waiting time belief distribution F. It says that when customers do not know the arrival rate and service rate information (and follow any waiting time belief distribution), X^N–which specifies that the server should disclose the queue length information to customers when the queue is shorter than Naor's balking threshold N, but conceal the information when the queue length is longer than N–continues to maximize system throughput. Therefore, Theorem 4.4.1 establishes robustness of our first main result presented in Theorem 4.2.1. Next, we explore to what extent the other main result namely Theorem 4.3.1 holds under the assumption that customers do not possess the arrival rate and service rate information, by investigating how $X^{0,N}$ and $X^{1,N}$ perform in terms of maximizing social welfare.

Through extensive numerical experiments, we find that although the socially optimal policy can deviate from $X^{0,N}$ and $X^{1,N}$ depending on the underlying waiting time belief distribution (e.g., numerically we found a waiting time belief distribution for which $X^{3,N}$ is socially optimal), both $X^{0,N}$ and $X^{1,N}$ continue to perform remarkably well in terms of achieving ideal social welfare, for all belief distributions that we have tested. In what follows, we show the results of a specific example when the waiting time belief distribution F is uniform with lower limit equal to 0 and upper limit equal to the smallest value that guarantees the stability condition $\lambda F(R/c) < \mu$.

Table 4.5 depicts the tested parameter values. In particular, we choose Λ and μ to each include $\{0.1, 0.2, 0.3, \ldots, 0.8, 0.9\}$ such that both stable systems ($\mu > \lambda$) and overloaded systems ($\mu < \lambda$) are well represented, and $\frac{C}{R\mu}$ to include $\{0.1, 0.2, 0.3, \ldots, 0.8, 0.9\}$ because $\frac{C}{R\mu} \in (0, 1)$ (see the discussion in the first paragraph of Sect. 4.1). We exclude cases of $\mu = \lambda$ in our simulations. Therefore, all combinations of the tested parameter values constitute $9 \times 8 \times 9 = 648$ scenarios.

We check the performance, in terms of social welfare, of $X^{0,N}$ and $X^{1,N}$ against that of the socially optimal policy in ratio form for each of the 648 scenarios, and

Table 4.5 Tested parameter values for Sect. 4.4.3

Parameters	Included values
Λ	$\{0.1, 0.2, 0.3, 0.4, 0.5, 0.6, 0.7, 0.8, 0.9\}$
μ	$\{0.1, 0.2, 0.3, 0.4, 0.5, 0.6, 0.7, 0.8, 0.9\}$
$\frac{C}{R\mu}$	$\{0.1, 0.2, 0.3, 0.4, 0.5, 0.6, 0.7, 0.8, 0.9\}$

Note All combinations of these values constitute 648 scenarios

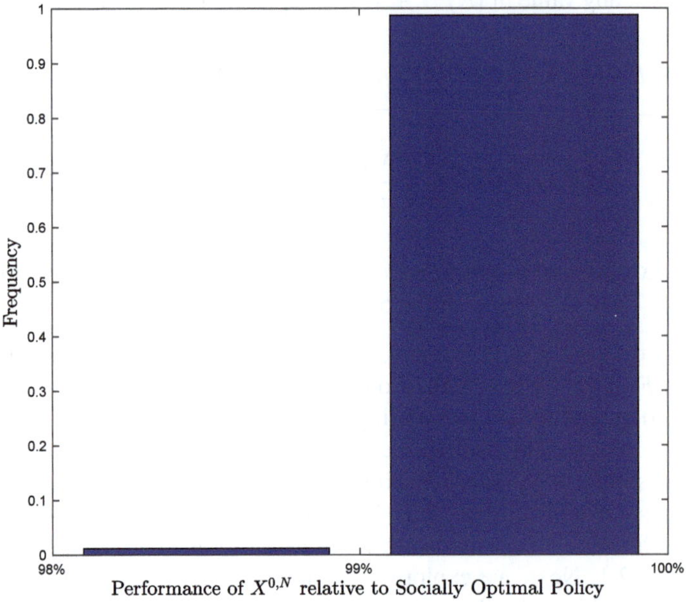

Fig. 4.1 Social welfare performance of the $X^{0,N}$ policy when F is a uniform distribution

present the result in Figs. 4.1 and 4.2. The figures suggest that the $X^{0,N}$ (resp., $X^{1,N}$) policy–which specifies that the server should announce the queue length when there are no customers in the system (resp., in the queue) or there are more than N customers in the system, and not announce the queue length otherwise—works well in most cases to produce high social welfare. In particular, $X^{0,N}$ (resp., $X^{1,N}$) achieves social welfare within 5% of that of the socially optimal policy (that is, the performance of $X^{0,N}$ (resp., $X^{1,N}$) is greater than 95% of the optimal value) for 100% (resp., 98.3%) of all the tested scenarios.

The analytical and numerical results presented in this subsection demonstrates strongly that even when customers do not know the arrival rate and service rate information, our main findings of the base model will largely continue to hold, i.e., the X^N policy continues to maximize system throughput whereas the $X^{0,N}$ or $X^{1,N}$ policy can provide near-optimal, if not optimal social welfare.

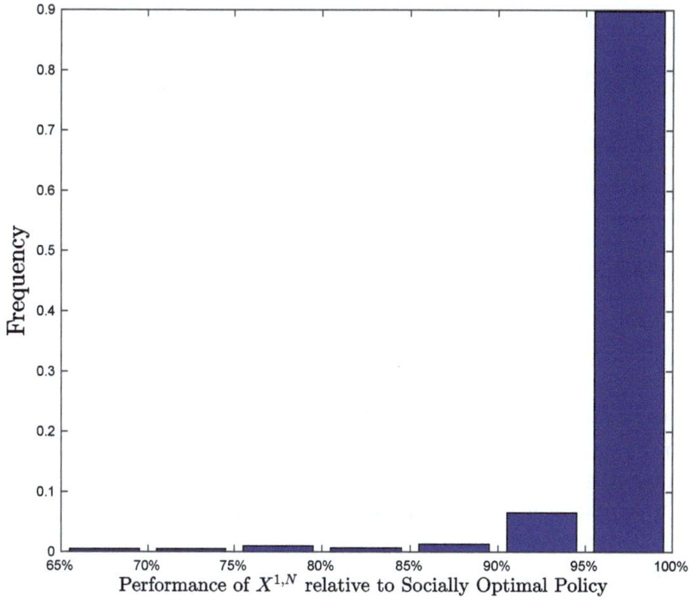

Fig. 4.2 Social welfare performance of the $X^{1,N}$ policy when F is a uniform distribution

4.4.4 Customer Heterogeneity with Different Service Rewards

In practice, customers may have different service values. We now extend the base model to consider two classes of customers whose actual service reward are given be R_1 and R_2, respectively. Suppose α is the fraction of customers who have service reward R_1, and $(1 - \alpha)$ the fraction of customers who have service reward R_2.

If we were to conduct an equilibrium analysis as in the base model, we need have additional assumptions on how customers believe what others' service rewards are and what their belief of α are. In particular, when a customer, say with service reward R_1, arrives at an uninformed state, she will join the system if and only if $R_1 - C \cdot w > 0$ where w is her belief of the expected waiting time. However, the expected waiting time calculation for the customer (i.e., value of w) will be dependent on how she believes what other customers' service rewards are which dictates how the customer believes what other customers' joining decisions are. That is, w will be different if the customer believes that other customers all have service reward R_1, versus the case she believes that 50% of the other customers have service reward R_1 and 50% R_2, versus the case 70% R_1 and 30% R_2, versus the case that she believes some other customers have a service reward other than R_1 and R_2.

Without getting into too much of customers' belief about each other's service reward, we adopt the approach from Sect. 4.4.3 by assuming that customers' beliefs on waiting time follow some distribution F, independent of anything else. In what follows, we check how well X^N (resp., $X^{0,N}$ or $X^{1,N}$) continues to perform in terms

Table 4.6 Tested parameter values for Sect. 4.4.4

Parameters	Included values
α	$\{0, 0.1, 0.2, 0.3, 0.4, 0.5, 0.6, 0.7, 0.8, 0.9, 1\}$
Λ	$\{0.1, 0.2, 0.3, 0.4, 0.5, 0.6, 0.7, 0.8, 0.9\}$
μ	$\{0.1, 0.2, 0.3, 0.4, 0.5, 0.6, 0.7, 0.8, 0.9\}$
$\frac{C}{R_1\mu}$	$\{0.1, 0.2, 0.3, 0.4, 0.5, 0.6, 0.7, 0.8, 0.9\}$
$\frac{C}{R_2\mu}$	$\{0.1, 0.2, 0.3, 0.4, 0.5, 0.6, 0.7, 0.8, 0.9\}$

Note All combinations of these values constitute 64152 scenarios

of achieving ideal throughput (resp. social welfare) for the service system when F is a uniform distribution with lower limit equal to 0 and upper limit equal to the smallest value that guarantees the stability condition $\alpha\lambda F(R_1/c) + (1-\alpha)\lambda F(R_2/c) < \mu$.

Table 4.6 depicts the tested parameter values. We include $\alpha \in \{0, 0.1, 0.2, \ldots, 0.9, 1\}$ because $\alpha \in [0, 1]$, and $\frac{C}{R_1\mu}, \frac{C}{R_2\mu} \in \{0.1, 0.2, 0.3, \ldots, 0.8, 0.9\}$ because they fall in $(0, 1)$. Λ and μ are chosen as before to include $\{0.1, 0.2, 0.3, \ldots, 0.8, 0.9\}$ such that both stable systems ($\mu > \lambda$) and overloaded systems ($\mu < \lambda$) are well represented. We exclude cases of $\mu = \lambda$ in our simulations. Therefore, all combinations of the tested parameter values constitute $11 \times 9 \times 8 \times 9 \times 9 = 64152$ scenarios.

We check the performance, in terms of throughput (resp., social welfare), of X^N (resp., $X^{0,N}$ and $X^{1,N}$) against that of the throughput-maximizing (resp., socially optimal) policy in ratio form for each of the 64152 scenarios, and present the result in Fig. 4.3 (resp., Figs. 4.4 and 4.5). Figure 4.3 shows that the X^N policy works well in most cases to produce high system throughput. For example, it achieves throughput within 5% of that of the throughput-maximizing policy (that is, the performance of X^N is greater than 95% of the optimal value) for 85.38% of all the tested scenarios. Figure 4.7 (resp., Fig. 4.8) shows that the $X^{0,N}$ (resp., $X^{1,N}$) policy works well in most cases to produce high social welfare. In particular, $X^{0,N}$ (resp., $X^{1,N}$) achieves social welfare within 5% of that of the socially optimal policy (that is, the performance of $X^{0,N}$ (resp., $X^{1,N}$) is greater than 95% of the optimal value) for 73.53% (resp., 75.85%) of all the tested scenarios.

4.4.5 Presence of Experienced Customers

Finally, in our base model we assumed that customers are not familiar with the server's information disclosure policy. It is plausible, however, in practice that not all of the customers are unfamiliar with a server's information disclosure policy. For example, customers who frequently shop a service may be able to infer the server's information disclosure policy based on their experiences. As such, we now consider a mixture of two types of customers, experienced and unexperienced, who are familiar and not familiar with the server's information disclosure policy.

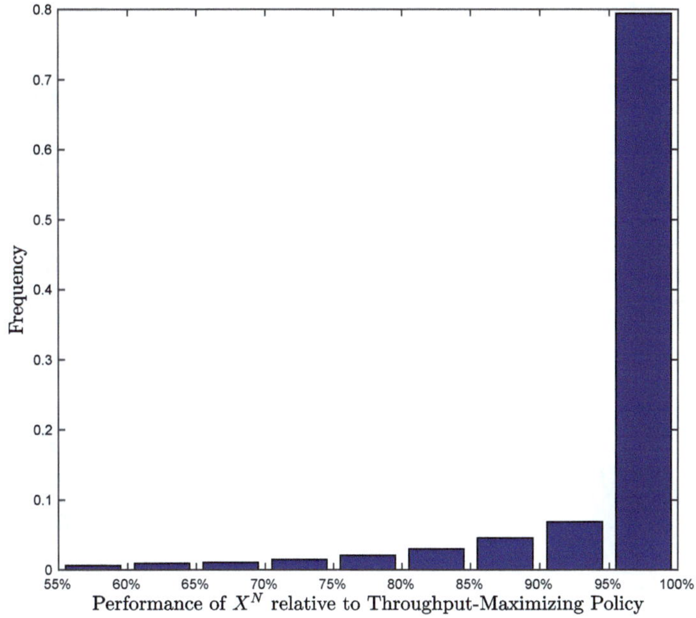

Fig. 4.3 Throughput performance of the X^N policy with two types of service rewards

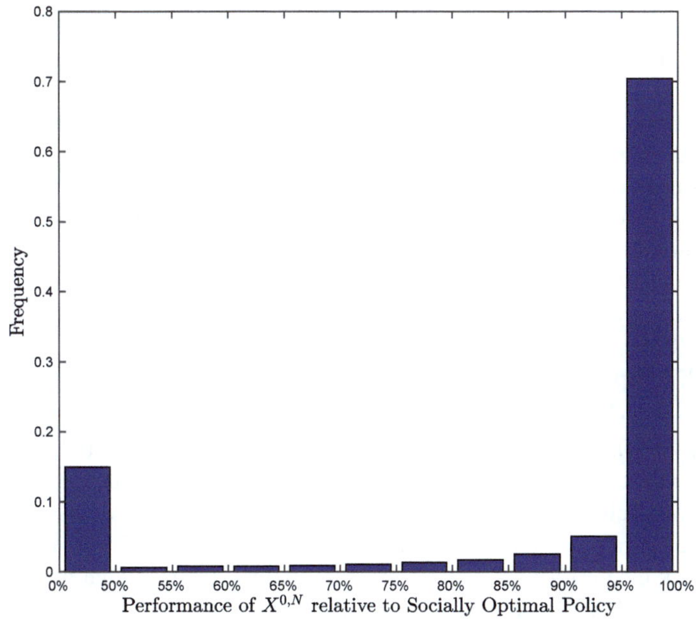

Fig. 4.4 Social welfare performance of the $X^{0,N}$ policy with two types of service rewards

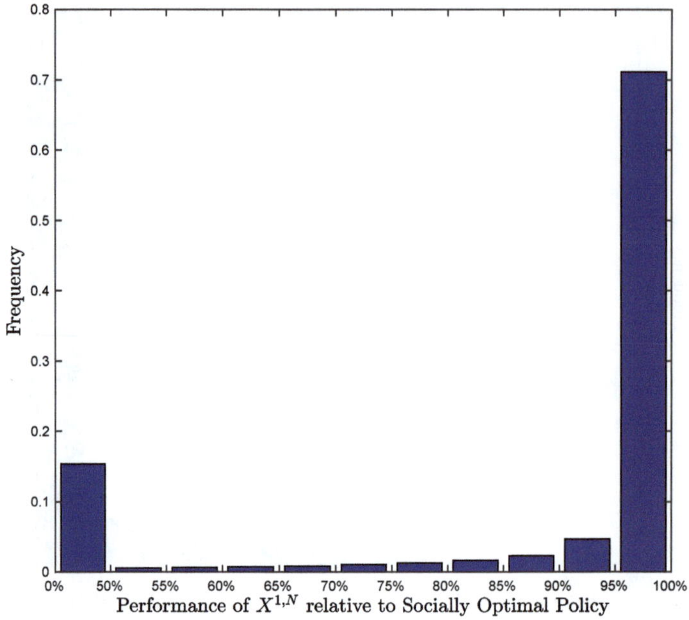

Fig. 4.5 Social welfare performance of the $X^{1,N}$ policy with two types of service rewards

We assume a fraction $\gamma \in [0, 1]$ among the total arrivals Λ are now customers as in Simhon et al. (2016) who are familiar with the server's information disclosure policy. The rest, a fraction $(1 - \gamma)$ of the customers, are "our" customers who do not know the server's information disclosure policy. It is clear that when $\gamma = 0$, the model reduces to the base model for which we have shown that the throughput-maximizing policy is X^N whereas the socially optimal policy either $X^{0,N}$ or $X^{1,N}$. In what follows, our goal is to demonstrate numerically that X^N (resp., $X^{0,N}$ or $X^{1,N}$) is generally a great policy in terms of achieving ideal throughput (resp. social welfare) for the service system even with the presence of experienced customers, that is, even when $\gamma > 0$.

Table 4.7 depicts the tested parameter values. In particular, we choose $\gamma \in \{0, 0.1, 0.2, \ldots, 0.9, 1\}$ because $\gamma \in [0, 1]$, and $\frac{C}{R\mu} \in \{0.1, 0.2, 0.3, \ldots, 0.8, 0.9\}$ because $\frac{C}{R\mu} \in (0, 1)$. Λ and μ are still chosen to include $\{0.1, 0.2, 0.3, \ldots, 0.8, 0.9\}$ such that both stable systems ($\mu > \lambda$) and overloaded systems ($\mu < \lambda$) are well represented. We exclude cases of $\mu = \lambda$ in our simulations. Therefore, all combinations of the tested parameter values constitute $11 \times 9 \times 8 \times 9 = 7128$ scenarios.

We check the performance, in terms of throughput (resp., social welfare), of X^N (resp., $X^{0,N}$ and $X^{1,N}$) against that of the throughput-maximizing (resp., socially optimal) policy in ratio form for each of the 7128 scenarios, and present the result in Fig. 4.6 (resp., Figs. 4.7 and 4.8). Figure 4.6 shows that the X^N policy works well in most cases to produce high system throughput. For example, it achieves throughput

Table 4.7 Tested parameter values for Sect. 4.4.5

Parameters	Included values
γ	{0, 0.1, 0.2, 0.3, 0.4, 0.5, 0.6, 0.7, 0.8, 0.9, 1}
Λ	{0.1, 0.2, 0.3, 0.4, 0.5, 0.6, 0.7, 0.8, 0.9}
μ	{0.1, 0.2, 0.3, 0.4, 0.5, 0.6, 0.7, 0.8, 0.9}
$\frac{C}{R\mu}$	{0.1, 0.2, 0.3, 0.4, 0.5, 0.6, 0.7, 0.8, 0.9}

Note All combinations of these values constitute 7128 scenarios

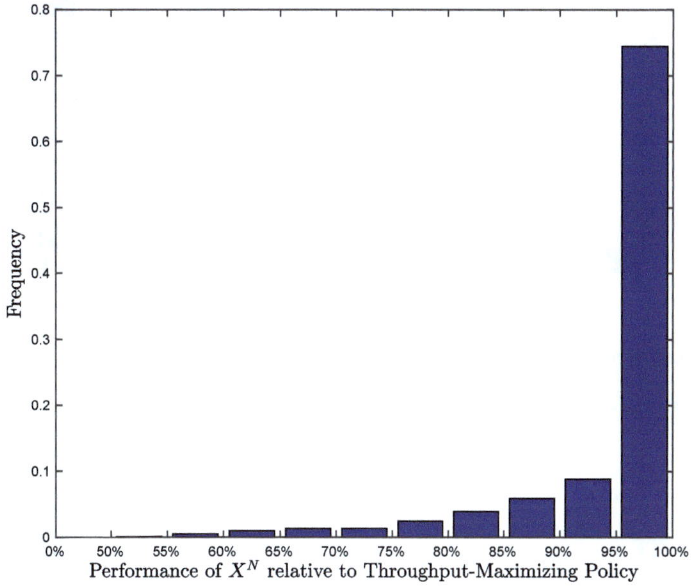

Fig. 4.6 Throughput performance of the X^N policy with presence of experienced customers

within 5% of that of the throughput-maximizing policy (that is, the performance of X^N is greater than 95% of the optimal value) for 74.42% of all the tested scenarios. Figure 4.7 (resp., Fig. 4.8) shows that the $X^{0,N}$ (resp., $X^{1,N}$) policy works well in most cases to produce high social welfare. In particular, $X^{0,N}$ (resp., $X^{1,N}$) achieves social welfare within 5% of that of the socially optimal policy (that is, the performance of $X^{0,N}$ (resp., $X^{1,N}$) is greater than 95% of the optimal value) for 83.36% (resp., 74.96%) of all the tested scenarios.

It should be noted that all of the extensions considered in this section provide strong evidence that X^N (resp., $X^{0,N}$ or $X^{1,N}$) can be a very good and simple universal information disclosure policy for a service provider who wants to maximize throughput (resp., social welfare), which is the key takeaway of this chapter.

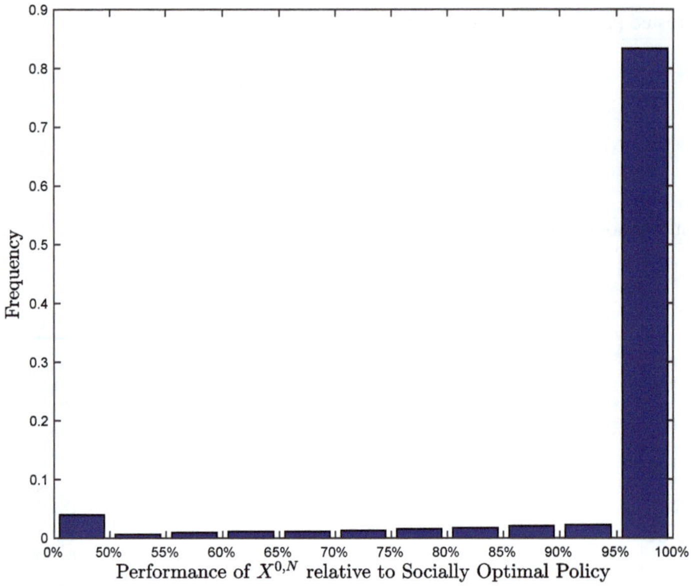

Fig. 4.7 Social welfare performance of the $X^{0,N}$ policy with presence of experienced customers

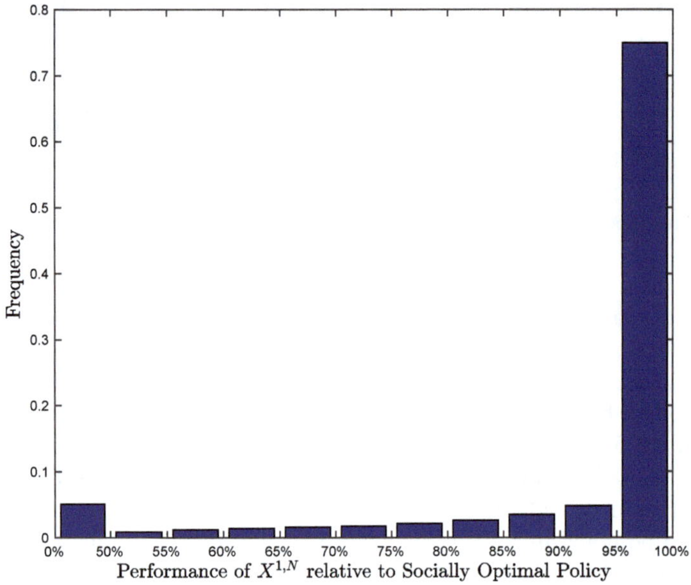

Fig. 4.8 Social welfare performance of the $X^{1,N}$ policy with presence of experienced customers

References

Afèche, Philipp, and Haim Mendelson. 2004. Pricing and priority auctions in queueing systems with a generalized delay cost structure. *Management Science* 50 (7): 869–882.

Akşin, Zeynep, Mor Armony, and Vijay Mehrotra. 2007. The modern call center: A multi-disciplinary perspective on operations management research. *Production and Operations Management* 16 (6): 665–688.

Allon, Gad, Achal Bassamboo, and Itai Gurvich. 2011. We will be right with you: Managing customer expectations with vague promises and cheap talk. *Operations Research* 59: 1382–1394.

Armony, Mor, Nahum Shimkin, and Ward Whitt. 2009. The impact of delay announcements in many-server queues with abandonment. *Operations Research* 57 (1): 66–81.

Chambers, Chester, and Panagiotis Kouvelis. 2006. Modeling and managing the percentage of satisfied customers in hidden and revealed waiting line systems. *Production and Operations Management* 15 (1): 103–116.

Chen, Hong, and Murray Frank. 2004. Monopoly pricing when customers queue. *IIE Transactions* 36: 569–581.

Cui, Shiliang, and Senthil Veeraraghavan. 2016. Blind queues: The impact of consumer beliefs on revenues and congestion. *Management Science* 62 (12): 3656–3672.

Debo, L. G., C. Parlour, and U. Rajan. 2012. Signaling quality via queues. *Management Science* 58: 876–891.

Economou, Antonis, and Spyridoula Kanta. 2008. Optimal balking strategies and pricing for the single server Markovian queue with compartmented waiting space. *Queueing Systems* 59 (3–4): 237–269.

Edelson, Noel M., and David K. Hildebrand. 1975. Congestion tolls for Poisson queueing processes. *Econometrica* 43: 81–92.

Gans, Noah, Ger Koole, and Avishai Mandelbaum. 2003. Telephone call centers: Tutorial, review, and research prospects. *Manufacturing & Service Operations Management* 5 (2): 79–141.

Guo, Pengfei, and Paul Zipkin. 2007. Analysis and comparison of queues with different levels of delay information. *Management Science* 53: 962–970.

Guo, Pengfei, and Paul Zipkin. 2009. The effect of availability of waiting-time information on a balking queue. *European Journal of Operational Research* 198: 199–209.

Hassin, Refael. 2016. *Rational Queueing*. Chapman and Hall/CRC Press.

Hassin, Refael. 1986. Consumer information in markets with random product quality: The case of queues and balking. *Econometrica* 54: 1185–1195.

Hassin, Refael, and Moshe Haviv. 2003. *To Queue or Not to Queue: Equilibrium Behavior in Queueing Systems*. Boston: Kluwer Academic Publishers.

Hassin, Refael, and Alexandra Koshman. 2017. Profit maximization in the M/M/1 queue. *Operations Research Letters* 45 (5): 436–441.

Hassin, Refael, and Ricky Roet-Green. 2017. The impact of inspection cost on equilibrium, revenue, and social welfare in a single-server queue. *Operations Research* 65 (3): 804–820.

Hu, Ming, Yang Li, and Jianfu Wang. 2018. Efficient ignorance: Information heterogeneity in a queue. *Management Science* 64 (6): 2650–2671.

Kim, Bara, and Jeongsim Kim. 2017. Optimal information disclosure policies in a strategic queueing model. *Operations Research Letters* 45 (2): 181–186.

Kremer, Mirko, and Laurens Debo. 2015. Inferring quality from wait time. *Management Science* 62 (10): 3023–3038.

Larsen, C. 1998. Investigating sensitivity and the impact of information on pricing decisions in an $M/M/1/\infty$ queueing model. *International Journal of Production Economics* 56 (57): 365–377.

Li, Kaili, Shiliang Cui, and Jinting Wang. 2021. On the optimal disclosure of queue length information. *Naval Research Logistics* 68 (5): 615–630.

Naor, Pinhas. 1969. The regulation of queue size by levying tolls. *Econometrica* 37: 15–24.

Shone, R., V. Knight, and J. Williams. 2013. Comparisons between observable and unobservable $M/M/1$ queues with respect to optimal customer behavior. *European Journal of Operational Research* 227: 133–141.

Simhon, Eran, Yezekael Hayel, David Starobinski, and Quanyan Zhu. 2016. Optimal information disclosure policies in strategic queueing games. *Operations Research Letters* 44 (1): 109–113.

Wang, Zhongbin, and Jinting Wang. 2021. The effect of information on queue-scalping service systems. *Operations Research Letters* 49 (4): 485–491.

Yu, Qiuping, Gad Allon, Achal Bassamboo, and Seyed Iravani. 2018. Managing customer expectations and priorities in service systems. *Management Science* 64 (8): 3942–3970.

Chapter 5
Risk-Sensitive Queueing Game

In most of the past studies in the queueing-game literature, customers are assumed to be risk-neutral, making their decisions based on the expected monetary benefit of joining the queue. Such an assumption ignores the fact that customers may be risk-averse or even risk-seeking when making decisions under uncertain wait times. Chen and Frank (2004) study the pricing problem of a monopolist in a queueing setting in both risk-neutral and risk-sensitive cases. They find that while equilibrium pricing and capacity strategy is socially efficient for the risk-neutral customer case, it is usually not socially efficient for the risk-sensitive case. Guo and Zipkin (2007) utilize a customers' delay-risk sensitivity parameter to model the customers' attitude towards the wait time.

However, very few work in the literature addressed this risk factor in game-theoretic queueing analysis. Guo et al. (2011) consider a quadratic cost function of the wait time in a queueing system with risk-sensitive customers. With limited information about the service times such as the range and/or first or second moments, they apply the entropy-maximization principle to estimate the service time distribution and related waiting cost. Such information is provided to customers making their joining or balking decisions. They considered both linear and quadratic cost functions of wait time or $c(w) = cw$ and $c(w) = cw^2$ in an M/G/1 queue to model risk-neutral and risk-averse customer behaviors, respectively. However, a quadratic *cost* function of wait time lacks flexibility for modeling different types of customer attitudes or the mean-variance trade-off of wait time. Afèche et al. (2013) study a service system that charges up-front fees but reimburses customers for delivery delay. For multiple delay-sensitive customer types, they investigated revenue-maximizing tariffs which depend on realized lead times for a provider serving customers of either risk averse or risk-neutral type with respect to payoff uncertainty. Taking the customers as the buyers and the server as the supplier of the market, Afèche et al. (2013) focus on the supplier's strategy but not the buyers'. To analyze customers' strategy, we utilize a quadratic *utility* function of the wait time to model customer risk attitudes in a queueing system. This type of function is similar to the mean-variance utility

J. Wang, *Fundamentals of Queueing-Game Models*,
https://doi.org/10.1007/978-981-95-0261-5_5

function for portfolio selection problems in finance literature (see Gerber and Pafum 1998, Maccheroni et al. 2009) and is more flexible in modeling different types of risk attitudes including risk-seeking and risk-averse types. This flexibility is important because customer risk attitude can be one of the most critical factors affecting the performance of a stochastic service system. Unfortunately, this important factor has been ignored in most of the past studies on queueing models with strategic customers.

In a risk-ubiquitous world, risk-averse customers are concerned more with the variability than with the mean of wait time. This is because that customers may make use of the relatively stable wait time to do other things (a long mean wait time with low variability may even preferable in some situations). An example is the "fast-pass" line for some popular attractions in Disneyland theme parks. The "fast-pass" customers will return after a certain number of hours (long but almost deterministic wait time) to get service without any further delay. In this example, those "fast-pass" customers are highly risk-averse about wait time variability (they dislike the high variability of wait time more than the long mean of wait time) and prefer to visit other less congested attractions during the certain or less variable wait time. On the other hand, risk-seeking customers enjoy the uncertainty and hope that they may be lucky enough to wait less than the average waiting time. Thus they may join a congested queue, or follow the crowd. For example, the wait time of the emergency room (ER) in a busy hospital is highly uncertain for non-life threatening illnesses of different types, ranging from almost no wait to several hours. Some patients with non-life threatening conditions may still choose to go to the ER if they have the health insurance to cover the cost. These patients are considered to be risk-seeking as they feel that it is worth going to the ER (or joining the queue) if there is a chance of getting the medical attention quickly due to the high variability of wait time. In other words, these patients prefer visiting the ER with highly uncertain wait time to making a scheduled appointment with a family doctor with almost certain (but long) wait time.

The risk attitude of queueing customers for service is more dynamic than that of investors in finance and may depend on many factors. Usually, queueing customers are less risk-averse or even risk loving if they have no scheduled or urgent tasks for them to do right after the service. For example, regarding wait time uncertainty, banking customers are less risk-averse on weekends or holidays than weekdays; and retired people are less risk-averse than working people about waiting for medical service in a walk-in clinic. de Palma and Picard (2005) study travel route choice behaviors for travelers in the Paris area based on survey data. In their empirical study, traveler choice depends both on expected travel time (like mean system time in a queueing system) and travel time variability. They found that while 66% of the sample are risk-averse or risk-neutral, the remaining 33% of sample are risk-seeking. Specifically, they observed that transit users, blue collar workers, and people with business appointments are more risk-averse about delay variability because they have to be punctual for their work or business. These examples illustrate that in a variety of real service systems it is important to take customer risk attitude into account when addressing the congestion issue in these systems. Queueing systems with risk-averse customers have been treated in several studies as mentioned earlier. However, there is no systematic analysis on queueing systems with risk-seeking customers despite the existence of such a risk attitude in reality.

To cater to emerging need of risk analysis with regard to queueing game application, this chapter is adapted from Wang and Zhang (2018) that makes a contribution by developing a model that can be applied for both risk-averse and risk-seeking customers. As mentioned earlier, risk-seeking attitude is not commonly assumed in the queueing literature although it does exist in some service systems. Therefore, it is legitimate to consider its implications under certain reasonable constraints.

5.1 Preliminaries

The basic quadratic utility function for an investment portfolio in finance literature is

$$U(r_p) = \mathbb{E}(r_p) - cA\mathbb{V}\text{ar}(r_p), \tag{5.1.1}$$

where $\mathbb{E}(r_p)$ and $\mathbb{V}\text{ar}(r_p)$ are the expected return and variance of the portfolio, respectively; c is a model specific constant; and A is the Arrow-Pratt measure of absolute risk aversion of the investor (see Maccheroni et al. 2009, Chopra and Ziemba 1993, Carlsson et al. 2002, Bodnar et al. 2013). We model the mean-variance trade-off for the service utility in a queueing system using the expected utility function of the same form as Eq. (5.1.1). That is

$$U(X) = \mathbb{E}X - \frac{1}{2}A\mathbb{V}\text{ar}(X), \tag{5.1.2}$$

where X is the net benefit of a customer joining the queue and A is the absolute risk aversion coefficient of customers. Note that here our utility function is an expected value function of a random variable of X. Clearly $A > 0, = 0$, and < 0 represent the risk averse, risk neutral, and risk seeking cases, respectively. It is worth noting that such a mean-variance quadratic utility function lacks the monotonicity of customers' strategies in the portfolio selection problem where a investor's asset allocation can be significantly affected by the investor's risk tolerance, as mentioned in Maccheroni et al. (2009). However, this is not an issue in modeling service utility for a queueing customer. Thus we utilize this quadratic utility function to study the impact of customer risk attitude on the stochastic service system performance.

Another important issue for queueing systems with strategic customers is to determine whether the queue length information is beneficial for customers or the server. In other words, we need to find the performance difference by comparing the observable and unobservable queues. The information level may be classified into three categories: no information (unobservable queue), which is corresponding to the fully unobservable (FU) case defined in Chap. 1; partial information (almost observable (AO) and almost unobservable (AU) cases); and full information (fully observable

(FO) case; exact waiting time—not only queue length but also each customer service time is known). Guo and Zipkin (2007) investigate the effect of these three levels of delay information on system performance. Although they mentioned the risk-averse and risk-seeking cases, they did not explicitly examine the different customer joining strategies for individual self-interests, social welfare maximization, and server's profit maximization. In this chapter, since the customer risk attitude is the main factor to consider and the unobservable queue case involves more uncertainty (more risk) in wait time, we will focus on the unobservable queue to study customer joining strategies for different objectives. An important implication of our results is that some well-known optimal joining strategies in classical risk-neutral models may not be applicable to situations where a customer's risk attitude plays an important role in customer choice. In addition, we also investigate the effect of the queue length information on the system performance.

5.2 Model and Analysis

Consider an M/M/1 queueing system with arrival rate λ and the service rate μ. A joining customer receives a constant service reward R and incurs a waiting cost of C per time unit in the system. If the actual waiting time is denoted by W, the net benefit of the customers is $X = R - CW$, which is random due to the random W. Customers make their joining or baking decisions independently at their arrival instants. Specifically, the customer's decision is based on the mean-variance quadratic utility function of Eq. (5.1.1). Therefore, the utility function can be written as

$$U(W) = R - C\mathbb{E}W - \frac{1}{2}AC^2\mathbb{V}\mathrm{ar}(W), \tag{5.2.1}$$

in which the risk aversion coefficient A plays an critical role. As long as $U(W)$ is non-negative, the customer joins the queue. Compared with the risk-neutral utility function which depends only on the mean waiting time, the quadratic utility function incorporates the second moment or the variability of the waiting time into the utility determination and can model different types of risk attitudes. We assume that A is constant for all customers. In a practical situation, A is most likely a random variable and follows a certain distribution. However, as long as the distribution is symmetric, we can use its mean as the constant A in our model. Such an assumption will simplify our analysis and provide a good approximation to the randomly distributed A case (verified by extensive simulations). Obviously, a negative A represents that the customer prefers the variability and is risk-seeking type; a positive A represents that the customer dislikes the variability and is risk averse; and the $A = 0$ represents the risk-neutral case. We will investigate the system performance of the M/M/1 queue with different types of customer risk attitudes.

Assume that customers do not know the number of customers in the system at their arrival or decision instants (also called unobservable queue case) and join the queue with probability q. Thus the effective arrival rate is λq for the M/M/1 queue. Let $N(t)$ denote the number of customers in the system at time t and $\pi_n (n = 0, 1, 2, ...)$ denote the stationary distribution. According to the classical results given in Bocharov et al. (2004, pp. 89–90), the stationary distribution is given by

$$\pi_n = \lim_{t\to\infty} P(N(t) = n) = (1-\rho)\rho^n, n = 0, 1, 2, ..., \tag{5.2.2}$$

where $\rho = \frac{\lambda q}{\mu} < 1$, and the stationary wait time, denoted by W, has the probability density function as

$$f(w) = (\mu - \lambda q)e^{-(\mu-\lambda q)w}, w > 0. \tag{5.2.3}$$

Note that the stationary wait time can be written as $W(q)$, a function of q. In this section, we evaluate the customers' strategies under three scenarios of the individual interest maximization (self-interest equilibrium), social welfare maximization, and server profit maximization over the feasible risk aversion and service reward ranges on an A-R parameter plane. Such an evaluation reflects our focus on the risk and return tradeoff in customer's choice behavior. The following theorem determines the equilibrium probability of joining the queue under self-interest maximization over the feasible A-R region.

Theorem 5.2.1 *For the M/M/1 queue with no queue length information for customers, when $\lambda < \mu$, there exists the equilibrium strategy of "joining the queueing system with probability q_e". The equilibrium probability q_e is given according to the combination of A and R. These combinations are classified into different regions on the A-R plane formed by three boundary lines, denoted by $M_1(A)$, $M_2(A)$, and $R = -1/(2A)$.*

(I). If $A \le \frac{-\mu}{C}$,

$$q_e = \begin{cases} 0, & \textit{if } R < M_2(A); \\ 0, \textit{ or } q_2, \textit{ or } 1, & \textit{if } M_2(A) \le R < M_1(A); \\ 1, & \textit{if } R \ge M_1(A), \end{cases} \tag{5.2.4}$$

(II). If $-\frac{\mu}{C} < A < \frac{\lambda-\mu}{C}$,

$$q_e = \begin{cases} 0, & \textit{if } R \le \min\{M_1(A), M_2(A)\}; \\ 0, \textit{ or } q_2, \textit{ or } 1, & \textit{if } M_2(A) \le R < M_1(A); \\ q_1, & \textit{if } M_1(A) \le R < M_2(A); \\ q_1, \textit{ or } q_2, \textit{ or } 1, & \textit{if } \max\{M_1(A), M_2(A)\} \le R < -\frac{1}{2A}; \\ q^*, \textit{ or } 1, & \textit{if } R = -\frac{1}{2A}; \\ 1, & \textit{if } R > -\frac{1}{2A}, \end{cases} \tag{5.2.5}$$

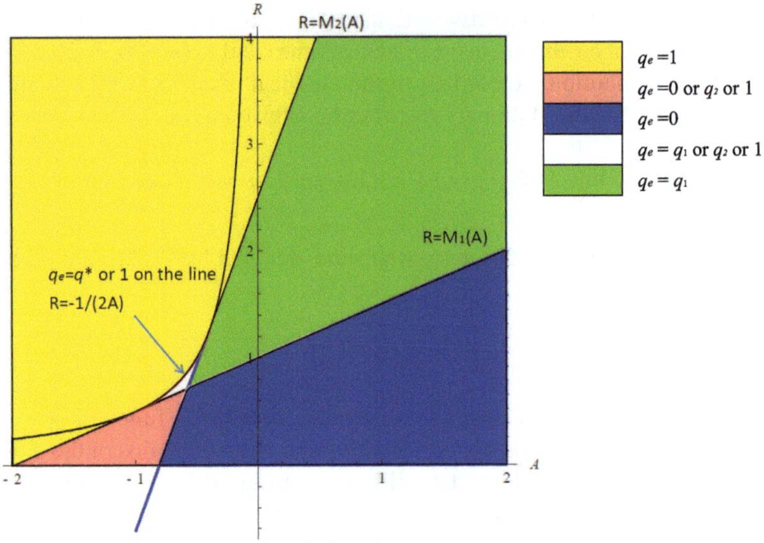

Fig. 5.1 Individual equilibrium joining probability

(III). If $A \geq \frac{\lambda-\mu}{C}$,

$$q_e = \begin{cases} 0, & \textit{if}\, R \leq M_1(A); \\ q_1, & \textit{if}\, M_1(A) < R < M_2(A); \\ 1, & \textit{if}\, R \geq M_2(A), \end{cases} \tag{5.2.6}$$

where $q_1 = \{2R\mu - C - C\sqrt{1+2AR}\}/\{2R\lambda\}$, $q_2 = \{2R\mu - C + C\sqrt{1+2AR}\}/\{2R\lambda\}$, $q^* = (\mu + AC)/\lambda$, *and*

$$M_1(A) = \frac{C}{\mu} + \frac{AC^2}{2\mu^2}, \tag{5.2.7}$$

$$M_2(A) = \frac{C}{\mu - \lambda} + \frac{AC^2}{2(\mu-\lambda)^2}. \tag{5.2.8}$$

The pure equilibria 0, 1, and mixed equilibrium q_1 *are evolutionarily stable strategies (ESS) (see Chap. 1 or the monograph of Hassin and Haviv (2003, p. 5). Mixed equilibria* q^* *and* q_2 *are not ESS. For simplifying the notations in the following proofs, we henceforth suppress the argument A in the boundary functions and write* $M_i = M_i(A), i = 1, 2$.

Note that for all cases of q_e, Fig. 5.1 shows all corresponding regions in different colors formed by the three R function boundary lines and the three ranges of $A \leq -\frac{\mu}{C}$, $-\frac{\mu}{C} < A < \frac{\lambda-\mu}{C}$, and $A \geq \frac{\lambda-\mu}{C}$. To make it more clear, we produce three separate Figs. 5.1, 5.2, and 5.3 to show the three cases of A value range, respectively.)

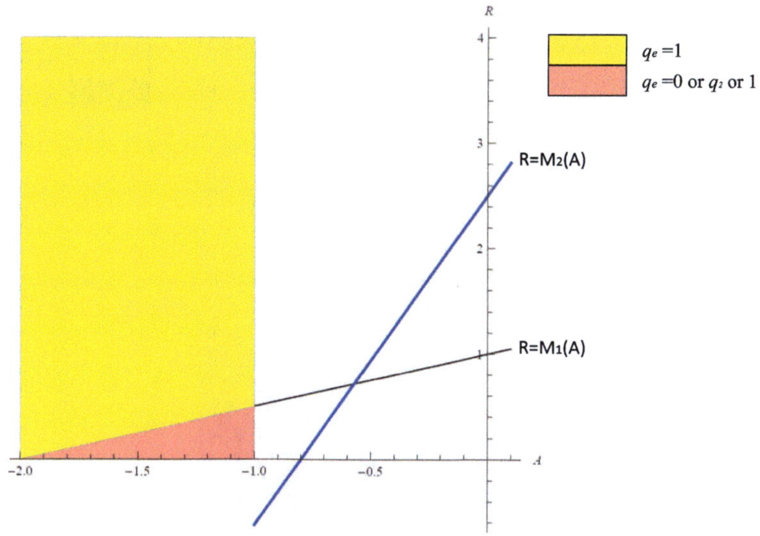

Fig. 5.2 Individual equilibrium joining probability for the case of $A \leq -\frac{C}{\mu}$

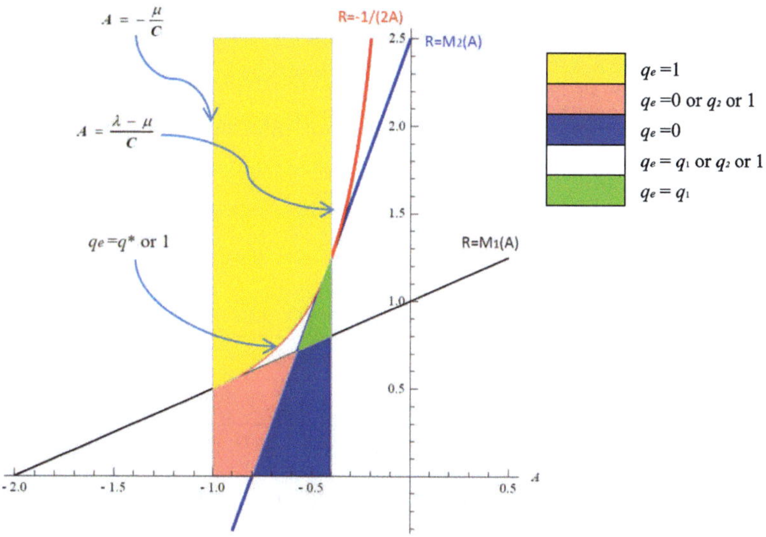

Fig. 5.3 Individual equilibrium joining probability for the case of $-\frac{\mu}{C} < A < -\frac{\mu-\lambda}{C}$

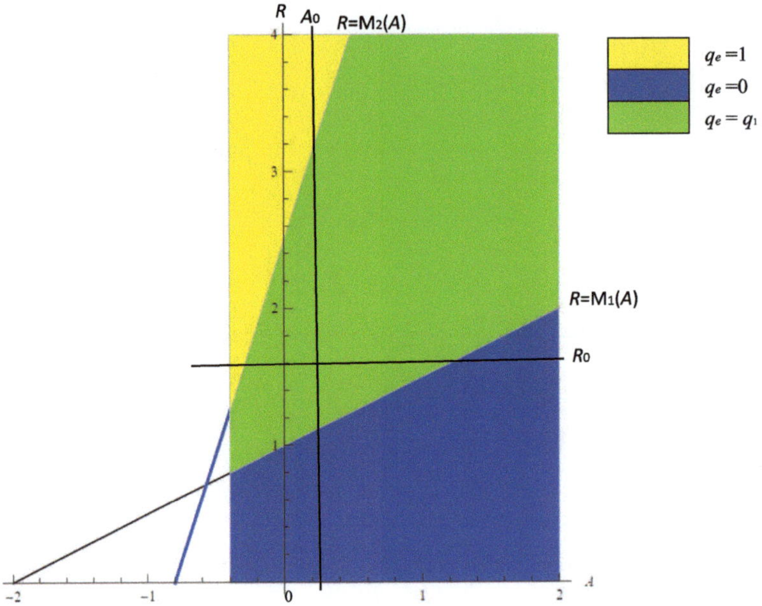

Fig. 5.4 Individual equilibrium joining probability for the case of $A \geq -\frac{\mu-\lambda}{C}$

Proof Suppose all customers without congestion information follow the same joining strategy $q \in [0, 1]$. From (5.2.3) we get

$$\mathbb{E}(W(q)) = \frac{1}{\mu - \lambda q}, \ \mathbb{V}\text{ar}(W(q)) = \frac{1}{(\mu - \lambda q)^2}.$$

The expected utility function (5.2.1) then can be written as a function of q due to the dependence of W on q:

$$U(q) = R - C\frac{1}{\mu - \lambda q} - \frac{1}{2}AC^2\frac{1}{(\mu - \lambda q)^2}. \tag{5.2.9}$$

When $2AR \geq -1 (A \neq 0)$, the above function has two real roots:

$$q_1 = \frac{2R\mu - C - C\sqrt{1 + 2AR}}{2R\lambda}, \tag{5.2.10}$$

$$q_2 = \frac{2R\mu - C + C\sqrt{1 + 2AR}}{2R\lambda}, \tag{5.2.11}$$

otherwise, a pair of complex roots exist. Note that q_2 is meaningful and considered only in $A < 0$ case so that $\lambda q_2 < \mu$ is satisfied to ensure the stability of the queue. It is obvious that $\lambda q_1 < \mu$ always holds as long as the real roots exist.

Differentiating $U(q)$ with respect to q, we obtain

$$U^{'}(q) = -C\lambda \left\{ \frac{\mu - \lambda q + AC}{(\mu - \lambda q)^3} \right\}. \tag{5.2.12}$$

Note that the monotonicity of function $U(q)$ depends on customer parameter A. Therefore, we consider the following cases separately:
(I). When $A \le -\frac{\mu}{C}$, $U(q)$ is increasing in $q \in [0, 1]$. Hence, a customer joins the system with probability $q_e = 1$ if $U(0) \ge 0$, which is equivalent to $R \ge M_1$. This pure joining strategy is dominant. A customer joins the system with probability 0, q_2 or 1 if $U(0) < 0$ and $U(1) \ge 0$, which is equivalent to $M_2 \le R < M_1$, depending on the strategy chosen by others.

Finally, a customer balks if $U(1) < 0$, which is equivalent to $R < M_2$. The pure strategy $q_e = 0$ is dominant. Clearly, this is a case where an individual's best response is a monotone increasing function of the strategy selected by the other customers, therefore we have the "follow the crowd" (FTC) situation where multiple equilibria exist.

The pure strategies are $q_e = 0$ and $q_e = 1$ are ESS, while the mixed strategy $q_e = q_2$ is not an ESS.
(II). When $-\frac{\mu}{C} < A < \frac{\lambda-\mu}{C} \Leftrightarrow$ (which is equivalent to) $U^{'}(0) < 0$ and $U^{'}(1) > 0$, so there exists a unique $q^* \in (0, 1)$ that minimizes $U(q)$. It follows from $U^{'}(q^*) = 0$ that $q^* = (\mu + AC)/\lambda$ and the corresponding minimum $U(q^*)$ is $R + \frac{1}{2A}$. Now we distinguish three sub-cases:

(II)-1. If $R + \frac{1}{2A} > 0$, we have $1 + 2AR < 0$ (note that $A < 0$). This means that there are no real roots to (5.2.9). Customers certainly join the queue or $q_e = 1$ is a dominant strategy.

(II)-2. If $R + \frac{1}{2A} = 0$ (or $q_1 = q_2 = q^* \in (0, 1)$), it follows from $U(0) > 0$ and $U(1) > 0$ that there are two equilibria of $q_e = q^*$ and 1. While the pure equilibrium is an ESS, the mixed strategy is not an ESS. For $q < q^*$, it is an "avoid the crowd" (ATC) and for $q \ge q^*$, it becomes an FTC.

(II)-3. If $R + \frac{1}{2A} < 0$, we have $1 + 2AR > 0$ and there exist two real roots of q_1 and q_2 for $U(q) = 0$. This is a more complex case and we need to consider four separate situations.

(II)-3-a. If $R \le \min\{M_1, M_2\}$, then we have $q_2 \ge 1$ and $q_1 \le 0$ and $U(q) \le 0$ for $0 \le q \le 1$. Thus the equilibrium strategy is $q_e = 0$ or nobody joins the queue.

(II)-3-b. If $A \le (-2\mu(\mu - \lambda))/(2C\mu - \lambda C)$, then $M_2 \le M_1$. When $M_2 \le R < M_1$, then we have $0 < q_2 \le 1$ and $q_1 < 0$ and $U(q) \le 0$ for $q \le q_2$. Thus there are three equilibria of $q_e = 0, q_2,$ and 1. For $q < q_2$, $U(q) < 0$ and nobody joins the queue; for $q > q_2$, $U(q) > 0$ and joining the queue is a dominant strategy; and for $q = q_2$, it is not an ESS.

(II)-3-c. If $A > (-2\mu(\mu-\lambda))/(2C\mu-\lambda C)$, then $M_2 > M_1$. When $M_1 \le R < M_2$, then we have $0 < q_1 \le 1$ and $q_2 > 1$. Note that $U(q) \ge 0$ for $q \le q_1$ and $U(q) < 0$ for $q > q_1$. This is an ATC case and the unique ESS equilibrium is $q_e = q_1$.

(II)-3-d. If $R \ge \max\{M_1, M_2\}$, then there exist two real roots of q_1 and q_2 in [0,1]. For $0 \le q \le q^*$, $U(q)$ is a decreasing function and q_1 is an ESS equilibrium due to the ATC case. For $q^* < q \le 1$, there are two equilibria of $q_e = q_2$ and 1. $q_e = 1$ and q_1 are ESS while $q_e = q_2$ is not ESS. This is a mixture of ATC and FTC cases.

When $-\frac{\mu}{C} < A < \frac{\lambda-\mu}{C}$, $R = -\frac{1}{2A}$ is an increasing function for $A \in (-\frac{\mu}{C}, \frac{\lambda-\mu}{C})$ and $-\frac{1}{2A} = M_1$ (when $A = -\frac{\mu}{C}$), $-\frac{1}{2A} = M_2$ (when $A = \frac{\lambda-\mu}{C}$), respectively. Furthermore, we have $-\frac{1}{2A} > M_1 \ge M_2$ (when $A \le \frac{2\mu(\lambda-\mu)}{C(2\mu-A)}$) and $-\frac{1}{2A} > M_2 > M_1$ (when $A > \frac{2\mu(\lambda-\mu)}{C(2\mu-A)}$). So we have $R = -\frac{1}{2A} > \max\{M_1, M_2\}$ for any $A \in (-\frac{\mu}{C}, \frac{\lambda-\mu}{C})$. Therefore, the conclusion can be rearranged as separate cases of (II) in Theorem 5.2.1.

(III). When $A \ge \frac{\lambda-\mu}{C}$, $U(q)$ is a decreasing function of $q \in [0, 1]$ and $M_1 < M_2$. The unique maximum and minimum utility points are

$$U(0) = R - \frac{C}{\mu} - \frac{1}{2}\frac{AC^2}{\mu^2},$$

$$U(1) = R - \frac{C}{\mu-\lambda} - \frac{1}{2}\frac{AC^2}{(\mu-\lambda)^2},$$

respectively, over [0,1]. Then we distinguish three sub-cases:

(III)-1. If $U(0) \le 0 \Leftrightarrow R \le M_1$, $U(q)$ is always nonpositive, so the customer equilibrium strategy is balking.

(III)-2. If $U(0) > 0$ and $U(1) < 0 \Leftrightarrow M_1 < R < M_2$, there exists a unique solution of $q_e^* = q_1$ such that $U(q_e^*) = 0$. Thus the customer equilibrium strategy is joining the queue with probability q_e^*. Note that in this sub-case, due to the monotonicity of $U(q)$, the best response is 1 whenever the joining probability q_e is smaller than q_e^*; and is 0 whenever $q_e > q_e^*$. Furthermore if $q_e = q_e^*$, customers are indifferent between joining and balking. This shows that an individual's best response is a decreasing function of the strategy selected by the other customers, therefore, we have the ATC situation and $q_e^* = q_1$ is an ESS.

(III)-3. If $U(1) \ge 0 \Leftrightarrow R \ge M_2$, then the best strategy is joining the queue. □

Remark 5.2.1 In the risk-neutral situation, the classical condition $R > \frac{C}{\mu}$ should be satisfied to ensure that the queueing system is not empty. In the risk-averse situation ($A > 0$), a stronger condition of $R > \frac{C}{\mu} + \frac{AC^2}{2\mu^2}$ than the risk-neutral case should be satisfied to avoid an empty queue. Here the term $\frac{AC^2}{2\mu^2}$ is the required compensation for taking the risk. In the risk-seeking situation ($A < 0$), a weaker condition of $R > 0$ than the risk-neutral case is sufficient. For highly risk-loving customers, the expected utility can be positive even if the expected net monetary reward is negative (or $R < \frac{C}{\mu}$). However, it is not practically meaningful to consider the case where customer utility is still positive even for the $R = 0$ case. To avoid such a case, we need $A > -\frac{2\mu}{C}$ from the requirement that the utility of joining an empty queue with

$R = 0$ must be negative. In the classical risk-neutral customer case, we have the "ATC" situation in the M/M/1 queueing system. However, when customers are not risk-neutral, both "ATC" and "FTC" situations can occur. Theorem 5.2.1 reveals that customers with negative risk aversion coefficient (risk-seeking) may still prefer to join the queue when others choose to join. The risk-seeking customers are like speculators and they bet that the actual wait time will be shorter than the average wait time when the wait time variability is high. In contrast, risk-averse customers behave more conservatively and join the queue only when the expected reward is large enough to compensate the risk of uncertain outcomes and only "ATC" situation occurs. The implication of Theorem 5.2.1 is that in an unobservable queue, the customer's behavior is influenced significantly by both his risk aversion coefficient A and service reward R. □

In Theorem 5.2.1 we assume that $\lambda < \mu$. For the $\lambda \geq \mu$ case (or called overloaded queue), using the similar analysis we can obtain the equilibrium joining strategy q_e under the stability condition of $\lambda q < \mu$ (or the joining probability is subject to a constraint of $q < \lambda/\mu$) as follows:

Corollary 5.2.1 *For the M/M/1 queue with no information for customers, when $\lambda \geq \mu$, the equilibrium joining probability q_e is given by*

(I). When $A \leq -\frac{\mu}{C}$,

$$q_e = 0, or\ q_2,\ \ if\ R < M_1(A), \tag{5.2.13}$$

otherwise, if $R \geq M_1$, no equilibrium strategy exists.

(II). When $-\frac{\mu}{C} < A < 0$,

$$q_e = q_1, or\ q_2, if\ R \leq -\frac{1}{2A}, \tag{5.2.14}$$

otherwise, if $R > -\frac{1}{2A}$, no equilibrium strategy exists.

(III). When $A \geq 0$,

$$q_e = \begin{cases} 0, & if\ R < M_1(A); \\ q_1, & if\ R \geq M_1(A). \end{cases} \tag{5.2.15}$$

The pure equilibrium 0 and mixed equilibrium q_1 are ESS, and q_2 is not an ESS.

The proof of Corollary 5.2.1 is similar to Theorem 5.2.1 and therefore is omitted here.

Remark 5.2.2 Corollary 5.2.1 indicates that in an overloaded queueing system the ESS equilibrium q_1 can be reached for the risk-seeking customer case only with very restrictive combination of the risk aversion and reward of $-\frac{\mu}{C} < A < 0$ and $R \leq -\frac{1}{2A}$. This implies that when customers are not extremely risk-seeking or the negative A is lower bounded, then the system can reach an equilibrium only if the service reward is not too large or R is upper bounded. Otherwise, we cannot perform the equilibrium or stationary analysis on the $A < 0$ cases. On the other hand, the ESS equilibrium can always be reached in risk-averse cases with $R > M_1(A)$. Thus while the stationary analysis can be applied to the overloaded queues with risk-averse customer choice, it may not be applicable to the cases with risk-seeking customers. □

It is well known that an individual makes the joining decision based on his or her own utility gain and ignores the negative externality imposed on others. Thus for the risk-neutral case, the individual equilibrium joining probability is greater than the joining probability for maximizing the social welfare which is defined as the sum of all customers' utilities in this paper. However, for the non-risk-neutral case, the social welfare optimal solution becomes more complicated and the relation between different strategies for the risk-neutral case may not hold or fail to be determined . The optimal social-maximization strategies are given as follows.

Theorem 5.2.2 *For the M/M/1 queue with no information for customers, when $\lambda < \mu$, there exists a unique optimal strategy of "joining the queueing system with probability q_{soc}" that maximizes the social welfare. The social welfare optimal probability q_{soc} is given by*

(I). If $A \le -\frac{\mu}{C}$, (high risk-seeking case)

$$q_{soc} = \begin{cases} 0, & \text{if } R \le M_2(A); \\ 1, & \text{if } R > M_2(A). \end{cases} \tag{5.2.16}$$

(II). If $-\frac{\mu}{C} < A < \frac{2\mu(\lambda-\mu)}{C(\lambda+2\mu)}$, (medium risk-seeking case)

$$q_{soc} = \begin{cases} 0, & \text{if } R \le \min\{M_1(A), M_3(A)\}, \text{ or } M_3(A) < R < \min\{M_1(A), M_2(A)\}; \\ q^*_{soc}, & \text{if } M_1(A) < R < M_3(A); \\ q^*_{soc1}, & \text{if } \max\{M_1(A), M_3(A)\} \le R < M_4(A) \text{and } U_{soc}(1) < U_{soc}(q^*_{soc1}); \\ 1, & \text{if } R \ge M_4(A), \text{ or } M_2(A) < R < M_1(A), \text{ or } \max\{M_1(A), M_3(A)\} \le R < M_4(A) \\ & \text{and } U_{soc}(1) \ge U_{soc}(q^*_{soc1}). \end{cases} \tag{5.2.17}$$

(III). If $A \ge \frac{2\mu(\lambda-\mu)}{C(\lambda+2\mu)}$, (from low risk-seeking to risk-averse cases)

$$q_{soc} = \begin{cases} 0, & \text{if } R \le M_1(A); \\ q^*_{soc}, & \text{if } M_1(A) < R < M_3(A); \\ 1, & \text{if } R \ge M_3(A), \end{cases} \tag{5.2.18}$$

where q^*_{soc} is the root of the equation

$$\frac{2C\mu}{(\mu-\lambda q)^2} + AC^2\frac{\mu+\lambda q}{(\mu-\lambda q)^3} = 2R, \tag{5.2.19}$$

if the above equation has only one positive root in $q \in (0, 1)$*. If the above equation has two roots in* $q \in (0, 1)$*, we denote them by* q^*_{soc1} , q^*_{soc2}*, where* $q^*_{soc1} < q^*_{soc2}$.

$$M_3(A) = \frac{C\mu}{(\mu-\lambda)^2} + \frac{1}{2}\frac{AC^2(\mu+\lambda)}{(\mu-\lambda)^3}, \tag{5.2.20}$$

$$M_4(A) = \frac{(2\mu - AC)^2(10\mu + AC)}{54\,A^2C\mu^2}, \tag{5.2.21}$$

and $M_1(A)$ *and* $M_2(A)$ *are defined in Theorem 5.2.1. The corresponding optimal social welfare probability is illustrated in Fig. 5.1, and the three cases of* $A \le -\frac{\mu}{C}$, $-\frac{\mu}{C} < A < \frac{2\mu(\lambda-\mu)}{C(\lambda+2\mu)}$, $A \ge \frac{2\mu(\lambda-\mu)}{C(\lambda+2\mu)}$*, are shown in Figs. 5.2, 5.3, and 5.4, respectively.*

Proof Consider that all customers follow the same joining strategy $q \in [0, 1]$, then the utility of an individual joining customer is $U(q)$. Thus the social welfare is

$$U_{soc}(q) = \lambda q U(q). \tag{5.2.22}$$

Differentiating $U_{soc}(q)$ with respect to q, we obtain

$$U^{'}_{soc}(q) = \lambda\left\{R - \left[\frac{C\mu}{(\mu-\lambda q)^2} + \frac{1}{2}AC^2\frac{\mu+\lambda q}{(\mu-\lambda q)^3}\right]\right\}. \tag{5.2.23}$$

And the second derivative is

$$U^{''}_{soc}(q) = \frac{C\lambda^2 f(q)}{(\mu-\lambda q)^4}, \tag{5.2.24}$$

where $f(q) = (2\mu - AC)\lambda q - 2\mu^2 - 2AC\mu$. Based on the sign of $U^{''}_{soc}(q)$, we consider the following two cases separately, labeled as **(A)** and **(B)**:

(A). When $A \ge \frac{2\mu}{C}$, $f(q)$ is negative and decreasing in $q \in [0, 1]$. Thus $U^{''}_{soc}(q) < 0$ or $U^{'}_{soc}(q)$ is decreasing in $q \in [0, 1]$. We distinguish three sub-cases according to the boundary conditions of $U^{'}_{soc}(q)$:

(A)-1. If $U^{'}_{soc}(0) \le 0 \Leftrightarrow R \le M_1$, then $U_{soc}(q)$ is decreasing in $q \in [0, 1]$. It follows from $U_{soc}(0) = 0$ that $U_{soc}(q) \le 0$ for $q \in [0, 1]$. Thus, customers balk.

(A)-2. If $U^{'}_{soc}(0) > 0$ and $U^{'}_{soc}(1) < 0 \Leftrightarrow M_1 < R < M_3$, function $U_{soc}(q)$ will first increase and then decrease. Hence, there exists a q^*_{soc} that maximizes $U_{soc}(q)$ and q^*_{soc} is the solution to $U^{'}_{soc}(q) = 0$.

(A)-3. If $U^{'}_{soc}(1) \ge 0 \Leftrightarrow R \ge M_3$, $U_{soc}(q)$ is increasing in $q \in [0, 1]$. Thus, customers join the system for sure.

(B). When $A < \frac{2\mu}{C}$, $f(q)$ increases in $q \in [0, 1]$. The solution to $f(q) = 0$ is easily obtained as $q^{'} = \frac{2\mu^2+2AC\mu}{(2\mu-AC)\lambda}$, the inflection point of $U_{soc}(q)$. We again distinguish three sub-cases according to the location of the inflection point:

(B)-1. If $q^{'} \geq 1 \Leftrightarrow A \geq \frac{2\mu(\lambda-\mu)}{C(\lambda+2\mu)}$, then we have $U^{''}_{soc}(q) \leq 0$ for $q \in [0, 1]$, which implies that $U^{'}_{soc}(q)$ is non-increasing in $q \in [0, 1]$. Therefore, customers will: (i) with probability 1 join the system if $U^{'}_{soc}(1) \geq 0 \Leftrightarrow R \geq M_3$; (ii) with probability q^*_{soc} join the system if $U^{'}_{soc}(0) > 0$ and $U^{'}_{soc}(1) < 0 \Leftrightarrow M_1 < R < M_3$; (iii) balk if $U^{'}_{soc}(0) \leq 0 \Leftrightarrow R \leq M_1$. (Note: (*III*) of the theorem is proved by combining Case (A) and sub-case (B)-1.)

(B)-2. If $q^{'} \leq 0 \Leftrightarrow A \leq -\frac{\mu}{C}$, then we have $U^{''}_{soc}(q) \geq 0$ for $q \in [0, 1]$ which implies that $U^{'}_{soc}(q)$ is non-decreasing in $q \in [0, 1]$. Customers will join for sure if $U^{'}_{soc}(0) \geq 0 \Leftrightarrow R \geq M_1$; If $U^{'}_{soc}(0) < 0$ and $U^{'}_{soc}(1) > 0 \Leftrightarrow M_3 < R < M_1$, customers' optimal join strategy depends on the value of the social welfare $U_{soc}(1)$. In particular, customers will join for sure when $U_{soc}(1) > 0$ and balk when $U_{soc}(1) \leq 0$; Finally, customers will balk when $U^{'}_{soc}(1) \leq 0 \Leftrightarrow R \leq M_3$. Therefore, customers will join the system with $q_{soc} = 1$ when $R \geq M_1$ or max $\{M_2, M_3\} < R < M_1$; and with $q_{soc} = 0$ when $M_3 < R < \min\{M_1, M_2\}$ or $R \leq M_3$. Note that when $A \leq -\frac{\mu}{C}$, max $\{M_2, M_3\} = M_2$ and min $\{M_1, M_2\} = M_2$. Hence, customers' optimal joining strategy is: (i) to join the system with $q_{soc} = 1$ when $R > M_2$; (ii) to balk when $R \leq M_2$. (Note: (*I*) of the theorem is proved in this sub-case).

(B)-3. If $0 < q^{'} < 1 \Leftrightarrow -\frac{\mu}{C} < A < \frac{2\mu(\lambda-\mu)}{C(\lambda+2\mu)}$, then $U^{'}_{soc}(q)$ is decreasing in q for $q \in [0, q^{'}]$ and increasing in q for $q \in [q^{'}, 1]$. Hence, $U^{'}_{soc}(q)$ will attain the minimum at $q = q^{'}$. Now we consider the following two situations according to the sign of the minimum of $U^{'}_{soc}(q')$:

(B)-3-a. When $U^{'}_{soc}(q^{'}) \geq 0 \Leftrightarrow R \geq M_4$, $U_{soc}(q)$ is increasing in $q \in [0, 1]$. Thus all customers join the system or $q_{soc} = 1$.

(B)-3-b. When $U^{'}_{soc}(q^{'}) < 0 \Leftrightarrow R < M_4$, and we have $M_4 > \max\{M_1, M_2, M_3\}$ when $A \in (-\frac{\mu}{C}, \frac{2\mu(\lambda-\mu)}{C(\lambda+2\mu)})$, so we need to consider the following four sub-situations according to the boundary conditions of $U^{'}_{soc}(q)$:

(B)-3-b1. If $U^{'}_{soc}(0) \leq 0$ and $U^{'}_{soc}(1) \leq 0 \Leftrightarrow R \leq \min\{M_1, M_3\}$, then $U_{soc}(q)$ is decreasing in $q \in [0, 1]$, hence all customers balk.

(B)-3-b2. If $U^{'}_{soc}(0) < 0$ and $U^{'}_{soc}(1) > 0 \Leftrightarrow M_3 < R < M_1$ (when $A \in (-\frac{\mu}{C}, \frac{2\mu(\mu-\lambda)}{C(\lambda-2\mu)})$), then $U_{soc}(q)$ will first decrease and then increase. Then customers join the system for sure when $U_{soc}(1) > 0 \Leftrightarrow R > M_2$ and balk when $U_{soc}(1) \leq 0 \Leftrightarrow R \leq M_2$. Thus customers will join when $\max\{M_2, M_3\} < R < M_1$; and balk when $M_3 < R < \min\{M_1, M_2\}$. Note again that $\max\{M_2, M_3\} = M_2$ when $-\frac{\mu}{C} < A < \frac{2\mu(\mu-\lambda)}{C(\lambda-2\mu)}$. So customers' optimal joining strategy follows: (i) $q_{soc} = 1$ when $M_2 < R < M_1$; (ii) $q_{soc} = 0$ when $M_3 < R < \min\{M_1, M_2\}$.

(B)-3-b3. If $U^{'}_{soc}(0) > 0$ and $U^{'}_{soc}(1) < 0 \Leftrightarrow M_1 < R < M_3$, then $U_{soc}(q)$ will first increase and then decrease. $U_{soc}(q)$ will attain the maximum at q^*_{soc}, where q^*_{soc} is the unique root of the equation $U^{'}_{soc}(q) = 0$ in $q \in (0, 1)$. Hence the optimal strategy is $q_{soc} = q^*_{soc}$.

(B)-3-b4. If $U^{'}_{soc}(0) > 0$ and $U^{'}_{soc}(1) > 0 \Leftrightarrow R > \max\{M_1, M_3\}$, then $U_{soc}(q)$ will first increase, then decrease, and finally increase again. We have $U^{'}_{soc}(0)U^{'}_{soc}(q^{'}) < 0$, so there is a unique root q^*_{soc1} of the equation (3.18) in $(0, q^{'})$ and $U_{soc}(q^*_{soc1})$ is a local maximum point of the function $U_{soc}(q)$. So we have to consider the

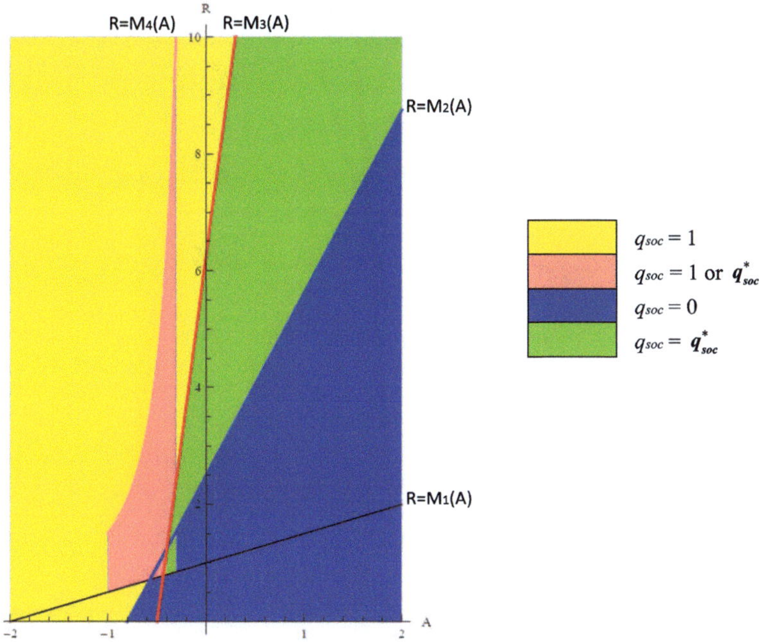

Fig. 5.5 Social optimal joining probability

social welfare at $q = 1$. Therefore, customers will join the system: (i) with probability $q_{soc} = q^*_{soc1}$ when $U_{soc}(1) < U_{soc}(q^*_{soc1})$; (ii) with probability $q_{soc} = 1$ when $U_{soc}(1) \geq U_{soc}(q^*_{soc1})$. (Note: *(II)* of the theorem is proved in the sub-case (B)-3).

As a summary, **(I)** is proved in **(B)-2**; **(II)** is proved in **(B)-3**; and **(III)** is proved by combining **(A)** and **(B)-1**. □

Theorem 5.2.2 is for $\lambda < \mu$ case. The $\lambda \geq \mu$ case can be treated similarly and its social welfare optimal strategy q_{soc} is determined as follows:

Corollary 5.2.2 *For the M/M/1 queue with no information for customers, when $\lambda \geq \mu$, the social welfare optimal joining probability q_{soc} is given by (Figs. 5.5, 5.6 and 5.7)*

(I). When $A < 0$, we have

$$q_{soc} = \left(\frac{\mu}{\lambda}\right)^{-}, \tag{5.2.25}$$

where $(.)^{-}$ denotes the left limit of a real number sequence.

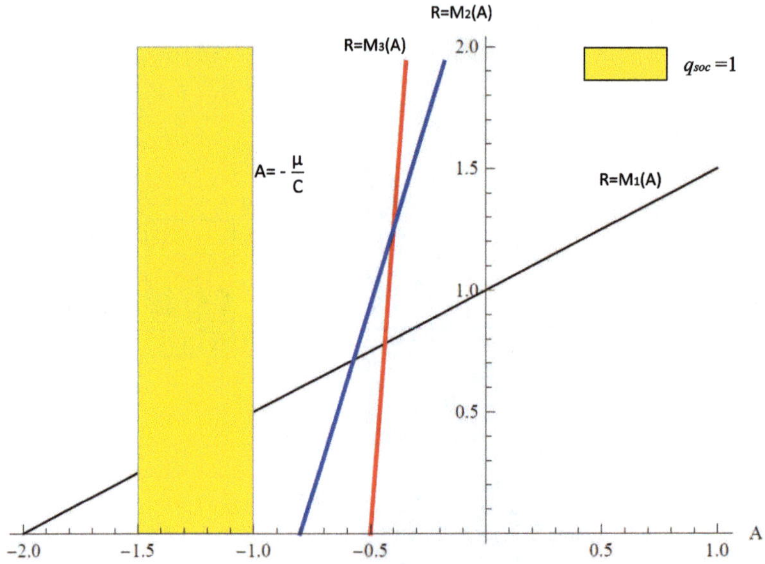

Fig. 5.6 Social optimal joining probability for the case of $A \le -\frac{\mu}{C}$

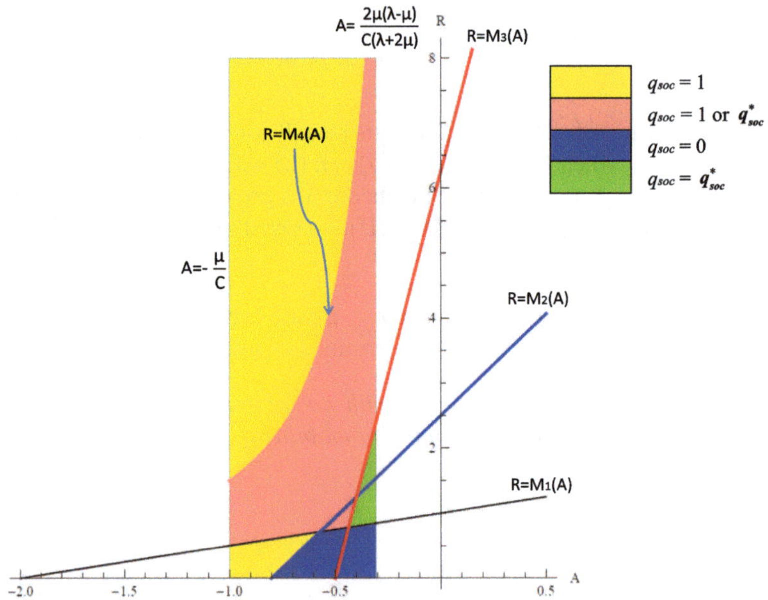

Fig. 5.7 Social optimal joining probability for the case of $-\frac{\mu}{C} < A < \frac{2\mu(\lambda-\mu)}{C(\lambda+2\mu)}$

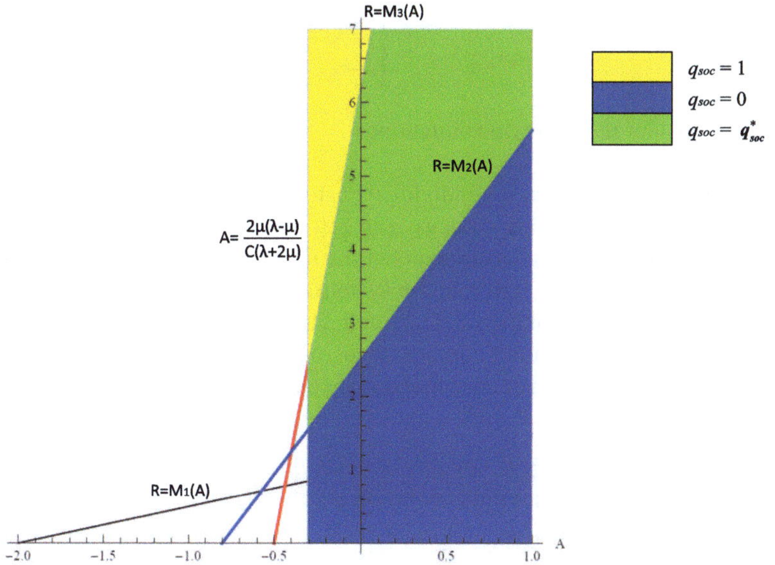

Fig. 5.8 Social optimal joining probability for the case of $A \geq \frac{2\mu(\lambda-\mu)}{C(\lambda+2\mu)}$

(II). When $A \geq 0$, we have

$$q_{soc} = \begin{cases} 0, & if\, R \leq M_1; \\ q^*_{soc}, & R > M_1, \end{cases} \tag{5.2.26}$$

where q^*_{soc} is defined in Theorem 5.2.2.

Proof In this case, when all the customers follow the same joining strategy $q \in [0, \frac{\mu}{\lambda})$ such that the system is stable, then the expected social utility (social welfare) $U_{soc}(q)$ and its first and second derivatives are given by

$$U_{soc}(q) = \lambda q U(q), \tag{5.2.27}$$

$$U'_{soc}(q) = \lambda \left\{ R - \left[\frac{C\mu}{(\mu - \lambda q)^2} + \frac{1}{2} AC^2 \frac{\mu + \lambda q}{(\mu - \lambda q)^3} \right] \right\}, \tag{5.2.28}$$

$$U''_{soc}(q) = \frac{C\lambda^2 f(q)}{(\mu - \lambda q)^4}, \tag{5.2.29}$$

where $f(q) = (2\mu - AC)\lambda q - 2\mu^2 - 2AC\mu$. Note that

$$U_{soc}((\frac{\mu}{\lambda})^-) = \begin{cases} +\infty, & A < 0; \\ -\infty, & A \geq 0, \end{cases} \tag{5.2.30}$$

and

$$U'_{soc}((\frac{\mu}{\lambda})^-) = \begin{cases} +\infty, & A < 0; \\ -\infty, & A \geq 0. \end{cases} \quad (5.2.31)$$

Hence we consider the following situations:
(I): $A < 0$.
In this situation, the risk-seeking individual customer's expected utility may change dramatically when the joining strategy q is getting close to $\frac{\mu}{\lambda}$ from left and attain its optimal value at $(\frac{\mu}{\lambda})^-$. As customers bet the uncertainty of the wait time as well as the expected utility, they will join the system with probability $q_{soc} = (\frac{\mu}{\lambda})^-$ or as closer to $\frac{\mu}{\lambda}$ as possible from below. The implication is that a social welfare optimal policy is to serve the maximum number of customers with the constrained capacity (not sufficient for serving the entire market) if customers are risk-seeking.

(II): $A \geq 0$.
In this situation, $U''_{soc}(q) \leq 0$ for any $q \in [0, \frac{\mu}{\lambda})$ and $U'_{soc}(q)$ is decreasing in $q \in [0, \frac{\mu}{\lambda})$. so there exists a unique solution solution $q^*_{soc} \in (0, \frac{\mu}{\lambda})$ that maximizes the social welfare $U_{soc}(q)$ if $U'_{soc}(0) > 0 \Leftrightarrow R > M_1$, otherwise customers always balk. □

Remark 5.2.3 Here a special left limit strategy of social welfare maximization is introduced for the risk-seeking customer case in the overloaded case. Note that compared with the individual equilibrium existed in the overloaded situation (ignore the non-equilibrium cases in Corollary 5.2.1), such a left limit strategy is greater than 0, or q_1 in the risk-seeking ($A < 0$) case. However, without any incentive, such an equilibrium will not be reached by the customer's self-interest choice. In addition, as shown later (after presenting the profit optimal joining probability), this kind of equilibrium, although existed in theory, may not be practically meaningful. This is because that for a service system the service reward is a constant while the payment of a gamble varies greatly. Therefore customers in the service system usually do not have the high risk-seeking behavior of gamblers in casinos. Recall Corollary 5.2.1 reveals that the case of reaching an ESS self-interest equilibrium (meeting the condition of reaching ESS q_1) is not likely either for the risk-seeking customers in an overloaded queue. Therefore, the risk-seeking case of $A < 0$ in an overloaded queue is also not practically meaningful and can be ignored (i.e. although customers are risk-seeking, it does not make much sense to consider them to keep joining an overloaded queue with an increasing variance of wait time). In contrast, for the risk-averse customer case, a well defined joining probability for optimizing the social welfare exists if the service reward R is greater than a threshold. □

Now we consider the server's objective of maximizing the profit if he can charge a fee for the service. For a monopolistic server, he will not leave a positive customer surplus, since in such a case a higher profit may be attained by increasing the fee. Denote the joining probability by q_{prof} for the case of "charging a fee". We can obtain the following result.

Theorem 5.2.3 *For the M/M/1 queue with no queue length information for customers, if the server wants to maximize his profit by charging a fee, then the profit optimal strategy q_{prof} coincides with the social welfare optimal strategy q_{soc}, i.e.*

$$q_{prof} = q_{soc}, \tag{5.2.32}$$

and the optimal entrance fee is

$$p^* = R - \frac{C}{\mu - \lambda q_{prof}} - \frac{AC^2}{2(\mu - \lambda q_{prof})^2}, \tag{5.2.33}$$

for $q_{prof} > 0$. But for the the situation $q_{prof} = 0$, the optimal entrance fee is undefined.

Proof Suppose that customers join the system with the same probability of q at equilibrium. This joining probability and the fee charged by the server are interdependent. Thus we can write the fee as a function of the joining probability and denote it by $p(q)$. The server's objective is to maximize its profit $U_{prof}(q)$, where

$$U_{prof}(q) = \lambda q p(q). \tag{5.2.34}$$

The admission fee is based on the customers' surplus value or the surplus utility $U(q)$. So if the customers' surplus utility $U(q) \leq 0$, the server cannot charge a fee. When the customers' surplus utility $U(q) > 0$, then the server's maximum profit can be attained by leaving a zero customer surplus which means that $p(q)$ satisfies

$$R - p(q) - \frac{C}{\mu - \lambda q} - \frac{AC^2}{2(\mu - \lambda q)^2} = 0, \tag{5.2.35}$$

which is equivalent to

$$p(q) = R - C\frac{1}{\mu - \lambda q} - \frac{1}{2}AC^2\frac{1}{(\mu - \lambda q)^2}. \tag{5.2.36}$$

Substituting (5.2.36) into (5.2.34) yields $U_{prof}(q) = U_{soc}(q)$. This implies that the profit maximization and social welfare maximization have the same solution or $q_{prof} = q_{soc}$ and for the $q_{prof} > 0$ case the optimal entrance fee is $p^* = p(q_{prof}) = p(q_{soc})$. When the entrance probability $q_{prof} = 0$, customers will always balk and it's meaningless to charge a fee. □

Remark 5.2.4 In a classical M/M/1 queueing system with risk-neutral customers, inequality $q_e \geq q_{soc}$ always holds (the individual equilibrium joining probability is no less than the social welfare optimal joining probability). Such a relation remains true for risk-averse customer cases. However, for risk-seeking cases, this relation is either reversed or cannot be determined. For $\lambda < \mu$, by comparing Theorem 5.2.1 and Theorem 5.2.2, we can verify that $q_{soc} \geq q_e$ when $A \leq -\frac{\mu}{C}$ and $q_{soc} \leq q_e$ when

$A \geq 0$. However, we are not sure about the relation between q_{soc} and q_e for $-\frac{\mu}{C} < A < 0$. The reason for such a phenomenon is that the expected quadratic utility function is based on both the mean and variance of the wait time and may either decrease or increase with the effective arrival rate depending on the customer's risk attitude. The implication is that if the social planner (or the server) wants to achieve social optimization (or profit maximization), it may either discourage customers (risk-averse case or $A \geq 0$) or encourage customers (risk-seeking case or $A \leq -\frac{\mu}{C}$) to join the system by charging a fee. Charging a fee to encourage customers to join in a risk-seeking case sounds counter-intuitive. However, it can be verified that under the condition of $A \leq -\frac{\mu}{C}$ (a quite risk-loving case), charging a fee will increase the variability of the wait time which in turn increases the joining probability of highly risk-seeking customers. Such a case is similar to the situation where a gambler is willing to pay for an unfavorable game with a negative expected return in a casino but a high variance. □

In a summary, the practical range of A should be from minor or moderate risk-seeking to risk averse in a waiting line service system. Therefore, Figs. 5.4 and 5.8 show the joining behaviors of customers in most realistic waiting line systems. An important implication generated by our model is that the predicted performance of a queueing system by the classical model with risk-neutral customers can be far away from the true system performance when customers are not risk-neutral. Consider a given service reward level R_0 (horizontal line) in Fig. 5.4, for the risk-neutral customer case, the joining probability is determined by the intersection of the R_0 line and the R-axis (vertical axis). Note that the joining probability can vary from 0 (blue zone) to 1 (yellow zone) when A is decreased from positive value (risk averse) to negative value (minor or moderate risk-seeking). Thus, for a case with $\rho = \lambda/\mu$ closer to 1, even for the same level of reward R_0, the system performance can vary from almost no queue to highly congested queue, depending on the customer's risk attitudes or joining probability. In a practical service system such as healthcare, the service reward R (treating an illness) is more likely to be stable or constant over time. In contrast, the customer's risk attitude A may vary with time or day. For example, customers for a walk-in clinic can be less risk averse or even somewhat risk-seeking (willing to take a chance of getting served quickly at the risk of waiting a long time) on weekends than weekdays. Thus, our model provides a more flexible and reliable tool to predict the performance of the walk-in clinic on different days. Another example is for an urban highway network where the traffic congestion pattern may depend more on the travelers risk attitude in selecting the major routes than the mean travel times for these major routes. The other advantage of our model is to allow us to see that the sensitivity of the customer's joining probability on the service reward depends on the risk attitude significantly. For example, as shown in Fig. 5.4, at a fixed A_0 level (vertical line), the joining probability can vary from 0 to 1 as R increases. It is easy to see that when customers are less risk-averse or slightly risk-seeking (as A decreases), the equilibrium joining probability zone (green) is shrinking, indicating that the customer joining probability is more sensitive to the service reward than in a more risk-averse case. This implies that when customers are less risk-averse or somewhat risk-seeking, the service reward is more critical in affecting the customer joining behavior. A small improvement in the service value (increase in R) may change the proportion of customer joining system from 0 to 100%!

5.3 Effects of Queue Length Information

In this section, we investigate the effects of queue length information on the system performance. Assume that customers can observe the queue at their arrival instants (also called observable queue case) and make their joining or balking decisions based on the expected quadratic utility assessed by real-time queue length. Of course, with the queue length information, the wait time uncertainty is significantly reduced and the chances of customers being either more risk-seeking or more risk-averse are slim. Hence, the practical range of A should be narrower when the queue length information is available for customers. To investigate the information effects on the server's profit and social welfare, we need to determine the optimal strategies of customers for the observable queue case.

5.3.1 Customer Joining Strategies with Queue Length Information

Consider a tagged arriving customer joining the queue at the nth position. For the M/M/1 queue, the wait time of the tagged customer follows an Erlang distribution with parameters n and μ. We first establish the equilibrium strategy due to the self-interest choice. Note that now the customer strategies become the threshold type for the queue length.

Theorem 5.3.1 *For the M/M/1 queue with the queue length information for customers, if $A > -\frac{2\mu}{C}$, there exists a unique threshold strategy of "joining the system if there are at most $n_e - 1$ customers in the system", where the threshold n_e is given by*

$$n_e = \left\lfloor \frac{2R\mu^2}{2C\mu + AC^2} \right\rfloor. \tag{5.3.1}$$

If $A \leq -\frac{2\mu}{C}$, there is no threshold joining strategy (Nash equilibrium).

Proof For an arriving customer finding $n - 1$ customers in the system, his joining puts him in the nth position of the system. Thus his wait time follows an Erlang distribution with parameters n and μ. The mean and variance of his wait time are given by

$$\mathbb{E}W(n) = \tfrac{n}{\mu}, \tag{5.3.2}$$

$$\mathbb{V}\text{ar}W(n) = \tfrac{n}{\mu^2}. \tag{5.3.3}$$

Denote his utility by $U(n)$ and recall the expected quadratic utility function

$$U(n) = R - C\mathbb{E}W(n) - \frac{1}{2}AC^2\mathbb{V}\text{ar}[W(n)]. \tag{5.3.4}$$

Using (5.3.2), (5.3.3), and (5.3.4), we get

$$U(n) = R - \frac{Cn}{\mu} - \frac{1}{2}\frac{AC^2n}{\mu^2}. \tag{5.3.5}$$

It follows from (5.3.5) that $U(n)$ is increasing in n if $A \leq -\frac{2\mu}{C}$ hence the threshold strategy never exists. If $A > -\frac{2\mu}{C}$, $U(n)$ is decreasing in n, so customers will join the system if and only if $U(n) \geq 0$ which yields the threshold. Condition $A > -\frac{2\mu}{C}$ ensures that threshold n_e is always nonnegative. □

Note that the requirement of $A > -2\mu/C$ is consistent with the no information case discussed in Remark 1. We can also show that there exists a server's profit maximization threshold policy, denoted by n_{prof}.

Theorem 5.3.2 *For the M/M/1 queue with the queue length information for customers, if $A > -\frac{2\mu}{C}$, there exists a threshold strategy of "joining the queue if no more than $n_{prof} - 1$ customers in the system" that maximizes the server's profit. Specifically, if there exists a positive real number x_{prof} that satisfies the first order condition of the profit function or $U^{'}_{prof}(x) = 0$, the optimal threshold n_{prof} is given by*

$$n_{prof} = arg\ max\{U_{prof}(\lfloor x_{prof}\rfloor), U_{prof}(\lfloor x_{prof}\rfloor + 1)\}, \tag{5.3.6}$$

where $U_{prof}(n)$ is

$$U_{prof}(n) = \lambda(\frac{1-\rho^n}{1-\rho^{n+1}})(R - C\frac{n}{\mu} - \frac{AC^2n}{2\mu^2}),$$

and the profit maximization fee is given by

$$p(n_{prof}) = R - \frac{Cn_{prof}}{\mu} - \frac{AC^2n_{prof}}{\mu^2}. \tag{5.3.7}$$

Otherwise, customers enter the system if and only if the system is empty (or $n_{prof} = 1$); if $A \leq -\frac{2\mu}{C}$, there is no threshold strategy for maximizing the server's profit.

Proof By imposing an admission fee p, a customer enters the system if and only if $U(n) = R - p(n) - \frac{Cn}{\mu} - \frac{AC^2n}{2\mu^2} \geq 0$ or $p(n) \leq R - \frac{Cn}{\mu} - \frac{AC^2n}{2\mu^2}$. Hence, to maximize the profit, the server will set the admission fee to be $p^*(n) = R - \frac{Cn}{\mu} - \frac{AC^2n}{2\mu^2}$ such that the customers' surplus is reduced to zero, i.e., $U(n) = 0$. The server's profit is

$$U_{prof}(n) = \lambda(1 - \pi_n(n))p^*(n). \tag{5.3.8}$$

If $A \leq -\frac{2\mu}{C}$, $U_{prof}(n)$ is strictly increasing in n as both $\lambda(1 - \pi_n(n))$ and $p^*(n)$ are nonnegative and increasing in n. Such a case is not practically meaningful. Hence, we only consider the case of $A > -\frac{2\mu}{C}$ (this requirement is again consistent with the previous sections). Substituting $p^*(n)$ into (5.3.8) yields

$$U_{prof}(n) = \lambda(1 - \pi_n(n))(R - \frac{Cn}{\mu} - \frac{AC^2n}{2\mu^2}). \tag{5.3.9}$$

Considering the following function (when $\rho \neq 1$)

$$U_{prof}(x) = \lambda\left(\frac{1-\rho^x}{1-\rho^{x+1}}\right)\left(R - C\frac{x}{\mu} - \frac{AC^2x}{2\mu^2}\right), \quad x \in \mathbf{R}^+, \tag{5.3.10}$$

we further denote $\lambda\frac{1-\rho^x}{1-\rho^{x+1}}$ and $R - C\frac{x}{\mu} - \frac{AC^2x}{2\mu^2}$ by $f(x)$ and $g(x)$, respectively. Then we have $g(x) > 0$, $g^{'}(x) < 0$, $g^{''}(x) = 0$ and

$$f^{'}(x) = \lambda\rho^x\frac{ln\rho(\rho-1)}{\left(1-\rho^{x+1}\right)^2}, \tag{5.3.11}$$

$$f^{''}(x) = \lambda(ln\rho)^2(\rho^x + \rho^{2x+1})\frac{(\rho-1)}{\left(1-\rho^{x+1}\right)^3}. \tag{5.3.12}$$

It's obvious that $f^{'}(x) > 0$, $f^{''}(x) < 0$, and $U^{''}_{prof}(x) = 2f^{'}(x)g^{'}(x) + f^{''}(x)g(x) + f(x)g^{''}(x) < 0$ or $U_{prof}(x)$ is concave. Thus, if there exists a real positive number $x_{prof} \in (0, \frac{2R\mu^2}{2C\mu+AC^2})$ such that $U^{'}_{prof}(x_{prof}) = 0$, then the threshold $\lfloor x_{prof} \rfloor$ or $\lfloor x_{prof} \rfloor + 1$ will maximize $U_{prof}(n)$; otherwise, a customer enters the system if and only if the system is empty.

In the situation of $\rho = 1$, we get $f(x) = \frac{x}{x+1}$ and $g(x) = R - C\frac{x}{\mu} - \frac{AC^2x}{2\mu^2}$. It's easy to see that $f(x)$ is concave with respect to x. This implies that $U_{prof}(x)$ is concave. Therefore a profit maximization threshold exists. □

Next we investigate the social welfare optimal solution. Due to the customer choice in an observable queue scenario, we assume that customers follow an $n-1$ threshold joining strategy, i.e. a customer enters the system if and only if at his or her arrival instant the number of customers in the system does not exceed $n-1$.

According to Bocharov et al. (2004, pp. 98–99), when customers follow the $n-1$ threshold strategy, the stationary queue length distribution function and queueing time probability density function, denoted by $\pi_k(n)$ and $f_{W_Q}(t)$, respectively, are given by

$$\pi_k(n) = \pi_0(n)\rho^k, \quad k = 0, 1, 2, ..., n, \tag{5.3.13}$$

$$f_{W_Q}(t) = \frac{\pi_0(n)}{1-\pi_n(n)}\delta(t) + \sum_{k=1}^{n-1} \frac{\pi_k(n)}{1-\pi_n(n)} \frac{\mu^k t^{k-1} e^{-\mu t}}{(k-1)!}, t \geq 0, \tag{5.3.14}$$

where $\pi_0(n) = \frac{1-\rho}{1-\rho^{n+1}}$, $\rho = \frac{\lambda}{\mu}$, $\delta(t)$ is 1, if $t = 0$, and is 0 otherwise. Hence using (5.3.13) and (5.3.14), after some tedious calculations (presented in the Appendix), we can obtain the first two moments of the wait time (when $n \geq 1$) as follows

$$\mathbb{E}W_Q(n) = \frac{\pi_0(n)\rho^2}{(1-\rho)^2\lambda(1-\pi_n(n))}(1 - n\rho^{n-1} + (n-1)\rho^n), \tag{5.3.15}$$

$$\mathbb{E}W_Q^2(n) = \frac{\pi_0(n)\rho}{(1-\pi_n(n))\mu^2} \frac{(n-n^2)\rho^{n+1} + (2n^2-2)\rho^n - (n(n+1))\rho^{n-1} + 2}{(1-\rho)^3}, \tag{5.3.16}$$

$$\mathbb{V}\mathrm{ar}[W_Q(n)] = \frac{1}{(\mu-\lambda)^2(1-\rho^n)^2} \times (V_1(n) + V_2(n)), \tag{5.3.17}$$

where

$$V_1(n) = (n-1)\rho^{2n+2} - 2(n-1)\rho^{2n+1} + n\rho^{2n} - (n-1)(n+2)\rho^{n+2}, \tag{5.3.18}$$

$$V_2(n) = 2(n-1)(n+2)\rho^{n+1} - n(n+1)\rho^n + 2\rho - \rho^2. \tag{5.3.19}$$

If $n = 0$, the first and the second moments of the wait time do not exist.

If customers follow a threshold $n-1$ joining strategy, then the rate of customer entering the system is $\lambda(1-\pi_n(n))$. Using $\mathbb{E}[W(n)] = \mathbb{E}W_Q(n) + \frac{1}{\mu}$ and $\mathbb{V}\mathrm{ar}[W(n)] = \mathbb{V}\mathrm{ar}[W_Q(n)] + \frac{1}{\mu^2}$, we have the social welfare as

$$U_{soc}(n) = \lambda(1-\pi_n(n))\left\{R - C(\mathbb{E}W_Q(n) + \frac{1}{\mu}) - \frac{AC^2}{2}(\mathbb{V}\mathrm{ar}[W_Q(n)] + \frac{1}{\mu^2})\right\}.$$

To analyze the property of $U_{soc}(n)$, we write $U_{soc}(n) = f(n) \times h(n)$, where

$$f(n) = \lambda(1-\pi_n(n)), \tag{5.3.20}$$

$$h(n) = R' - C\mathbb{E}W_Q(n) - \frac{AC^2}{2}\mathbb{V}\mathrm{ar}[W_Q(n)], \tag{5.3.21}$$

where $R' = R - C/\mu - AC^2/(2\mu^2)$. For any function $g(n)$, we define $\Delta g(n) = g(n) - g(n-1)$, then we have the following equivalent inequalities

$$\Delta U_{soc}(n) \geq 0 \Leftrightarrow f(n)h(n) - f(n-1)g(n-1) \geq 0$$
$$\Leftrightarrow \frac{h(n)}{h(n-1)} \geq f_1(n) \Leftrightarrow A \leq -B(n), \tag{5.3.22}$$

where

$$f_1(n) = \frac{(1-\rho^{n-1})(1-\rho^{n+1})}{(1-\rho^n)^2}, \tag{5.3.23}$$
$$B(n) = \frac{2(R-\frac{C}{\mu})(f_1(n)-1) + 2C[\mathbb{E}W_Q(n) - f_1(n)\mathbb{E}W_Q(n-1)]}{C^2\{\mathbb{V}\mathrm{ar}[W_Q(n)] - f_1(n)\mathbb{V}\mathrm{ar}[W_Q(n-1)] + \frac{1-f_1(n)}{\mu^2}\}}. \tag{5.3.24}$$

After some algebraic manipulations, we have the following limits:

$$\lim_{n\to\infty} B(n) = 0, \quad \text{when} \quad \rho < 1, \tag{5.3.25}$$
$$\lim_{n\to\infty} B(n) = \frac{2\mu}{C}, \quad \text{when} \quad \rho > 1. \tag{5.3.26}$$

In the situation of $\rho > 1$, we can show numerically in Fig. 5.9 that $B(n)$ is increasing in n with the limit as its upper bound or $B(n) \leq \frac{2\mu}{C}, \forall n \geq 0$. So we have the following cases with different A values: (i) $U_{soc}(n)$ is an increasing function when $A \leq -\frac{2\mu}{C}$ and no finite threshold strategy exists; (ii) $U_{soc}(n)$ is first increasing and then decreasing when $-\frac{2\mu}{C} < A < -B(1)$ and there exists a finite threshold strategy that maximizes the social welfare; and (iii) $U_{soc}(n)$ is a decreasing function when $A \geq -B(1)$, indicating that a customer enters the system if and only if the system is empty. In the situation of $\rho < 1$, we cannot obtain an explicit upper bound for $B(n)$ although we know it has a zero limit as n goes to infinity. The monotone property of $B(n)$ is observed in Fig. 5.10 as follows: as n increases, $B(n)$ first increases from negative value to positive value and then decreasingly approaches to zero as the limit. Therefore, we conclude that a threshold policy for optimizing the social welfare may exist (the situation of $\rho = 1$ can be analyzed in the same way).

Remark 5.3.1 Although Theorems 5.3.1 and 5.3.2 indicate that while the individual equilibrium and profit maximization thresholds exist in reasonably practical (stable) observable queue situations, the social welfare optimal threshold may or may not exist in these situations. Unlike the unobservable queue case where the relation between policy parameters $q_{soc} = q_{prof}$ can be shown, we cannot compare n_{soc} with n_{prof} in an observable queue case as n_{soc} may not even exist. Therefore, the queue length information does affect the equilibrium analysis in service systems with customer choice. Next, we perform numerical analysis to gain more managerial insights about the queue length information effects on the social welfare and server's profit when the corresponding threshold strategies exist.

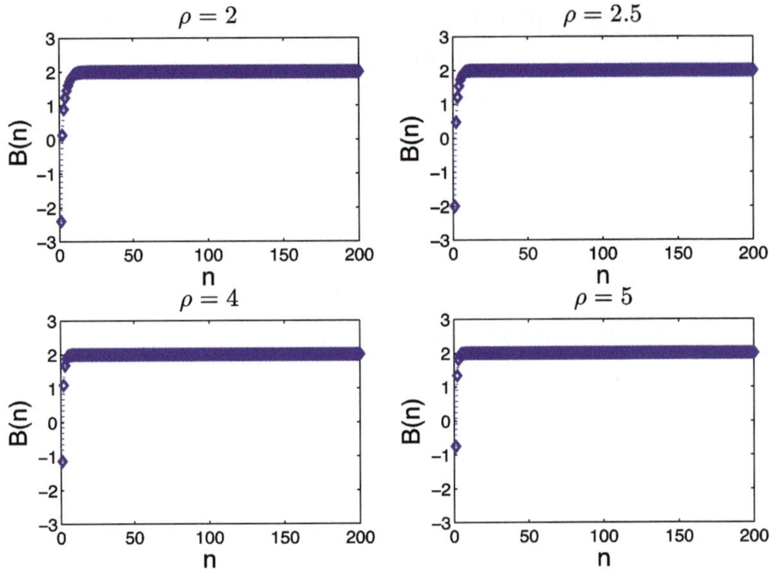

Fig. 5.9 $B(n)$ with an upper bound of 2, for the case $\rho > 1$, $C = 1$, $\mu = 1$

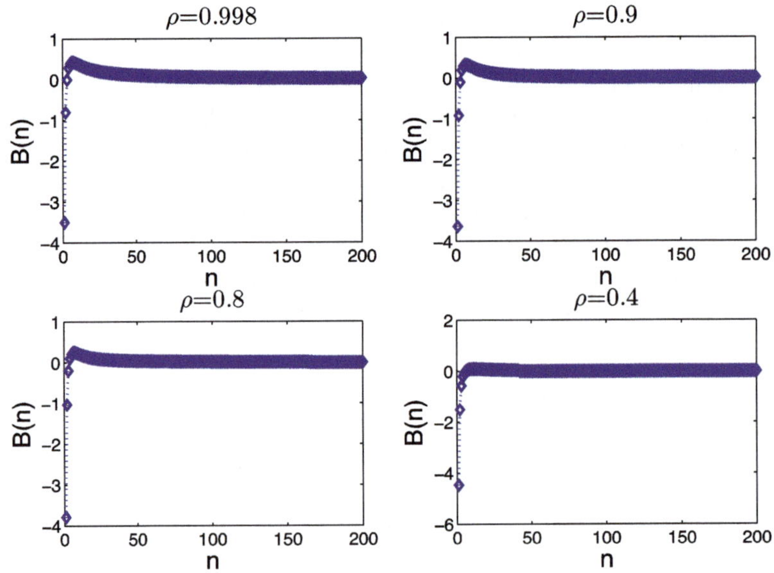

Fig. 5.10 The range of $B(n)$, for the case $\rho < 1$, $C = 1$, $\mu = 1$

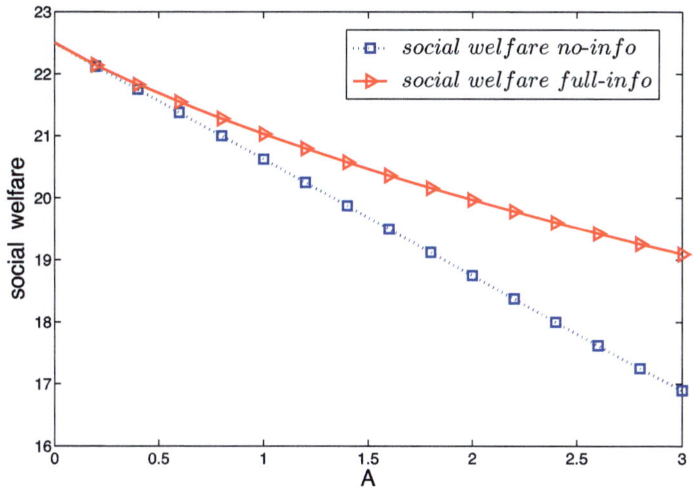

Fig. 5.11 Expected social welfare under two levels of information with respect to A for $\lambda = 0.6$, $\mu = 1$, $R = 40$, $C = 1$ (stable queue)

5.3.2 Numerical Illustrations

Based on the results obtained above, we examine the queue-length information effect on the server's profit and social welfare. Figure 5.11 reveals how social welfare will change with $A > 0$ under two information scenarios. It is observed that providing the queue length information will improve the social welfare. Such an improvement increases with A. This is very intuitive result. However, as shown in Fig. 5.12, depending on the degree of risk aversion, providing the queue length information may either hurt or improve the server's profit for the same case. For less risk-averse customers, hiding real-time information can improve the server's profit but for more risk-averse customers, disclosing real-time information can improve the server's profit. This finding is intriguing and somewhat counter-intuitive as it implies that concealing real-time information in a system serving risk-averse customers may still increase the server's profit. Therefore, a for-profit server must assess the degree of customer's risk aversion accurately before deciding if real-time information should be disclosed. Just knowing the risk-averse type attitude is not enough for making the information level decision. By comparing Fig. 5.11 with Fig. 5.12, we find the preference inconsistency about disclosing the real-time queue length information between the profit maximization and the social welfare maximization. When A is below the intersection of the two profit lines or $A < 2.352$ in Fig. 5.5, the server wants to conceal the information to maximize the profit but the social planner prefers to disclose the information to maximize the social welfare. When $A \geq 2.352$, the two goals are consistent and both the server and social planner want to disclose the queue length information to customers.

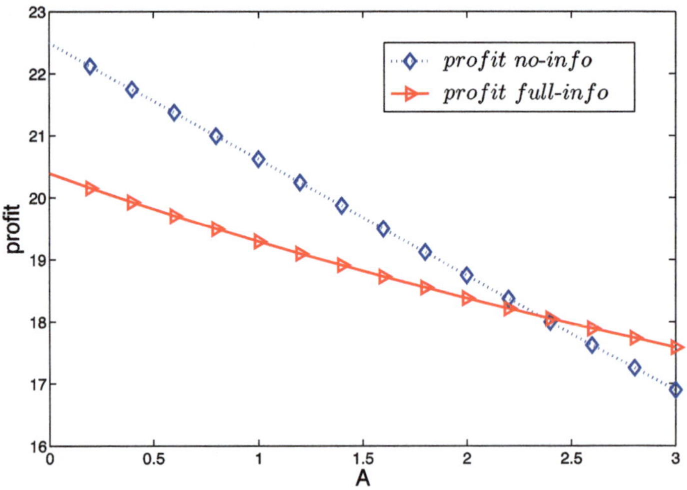

Fig. 5.12 Expected server's profit under two levels of information with respect to A for $\lambda = 0.6$, $\mu = 1$, $R = 40$, $C = 1$ (stable queue)

5.4 Concluding Remarks

We have analyzed the performance of an M/M/1 queueing system with strategic and risk-sensitive customers. An innovative feature of this study is to model the customer's risk attitude using a quadratic service utility function. Such a utility function allows us to study a variety of risk attitudes ranging from risk-seeking to risk-averse, with risk-neutral as a special case. In many practical stochastic service systems, customers make their joining or balking decisions based on the trade-off between return and risk. By taking the customer risk attitude into account, we are able to examine customer strategies in a congested service system under different goals. These goals are individual self-interest, social welfare maximization, and server's profit maximization, which have all been considered in past studies for classical queueing systems with risk-neutral customers. An important contribution of our study is to show that some well-known results for customer strategies under the risk-neutral customer assumption may not hold in more practical systems with risk-sensitive customers. For example, in an unobservable queue scenario, while the individual equilibrium joining probability is always greater than the social optimal joining probability in the risk-neutral case, this relation may be reversed or undetermined in a risk-seeking case. This is because a customer's joining action will impose both negative and positive externalities on other customers, hence, either ATC or FTC customer behaviors may occur. Although we obtain the intuitive results that risk-averse customers follow the ATC type joining strategies, the degree of the risk-aversion has significant effects on the optimal strategies and resulting system performance. Furthermore, we have investigated the impact of the queue length information on the system performance such as social welfare and server's profit. For a risk-averse customer case, while the queue length information benefits the social welfare, it may either hurt or improve the sever's profit, depending on the degree of customer risk aversion. An implication of this finding is that to improve the social welfare the social planner may need to encourage or force the server to provide the congestion information to customers in a less risk-averse situation. Therefore, assessing customer risk attitudes accurately becomes critical in predicting or evaluating customer strategies under different goals. Our results provide important insights for managers of real world service systems where customers are likely to be risk-sensitive.

Recently, Wang et al. (2024) investigate a fully unobservable single-server Markovian queue with strategic customers under the mean-variance criterion. It is reveals that the customers' non-risk-neutral attitudes can only be stimulated under particular circumstances, while social managers' various risk preference only become active when the overall queueing system's risk level is sufficiently high. Also, it is noted that in this chapter the risk aversion coefficient is assumed to be constant for all customers. Although this model can provide a good approximation to the case with symmetrically distributed risk-aversion, it cannot be applied to the case where the risk-aversion coefficient is highly skewed. Possibilities for future research include extending our analysis to cases with heterogeneous customers having different risk attitudes. Estimating the risk aversion coefficients for customers based on empirical data may also be an attractive area for future research.

References

Afèche, Philipp, O. Baron, and Y. Kerner. 2013. Pricing time-sensitive services based on realized performance. *Manufacturing & Service Operations Management* 15 (3): 492–506.

Bocharov, P., C. D'Apice, A. V. Pechinkin, and S. Salerno. 2004. *Modern Probability and Statistics: Queueing Theory*. Utrecht, Boston: VSP Publisher.

Bodnar, T., N. Parolya, and W. Schmid. 2013. On the equivalence of quadratic optimization problems commonly used in portfolio theory. *European Journal of Operational Research* 229: 637–644.

Carlsson, C., R. Fullér, and P. Majlender. 2002. A possibilistic approach to selecting portfolios with highest utility score. *Fuzzy Sets and Systems* 131: 13–21.

Chen, Hong, and Murray Frank. 2004. Monopoly pricing when customers queue. *IIE Transactions* 36: 569–581.

Chopra, W. K., and W. T. Ziemba. 1993. The effect of errors in means, variances, and covariances on optimal portfolio choice. *Journal of Portfolio Management* 19 (2): 6–11.

De Palma, A., and N. Picard. 2005. Route choice decision under travel time uncertainty. *Transportation Research: Part A* 39: 295–324.

Gerber, H. U., and G. Pafum. 1998. Utility functions: From risk theory to finance. *North American Actuarial Journal* 2 (3): 74–91.

Guo, Pengfei, and Paul Zipkin. 2007. Analysis and comparison of queues with different levels of delay information. *Management Science* 53: 962–970.

Guo, Pengfei, Wei Sun, and Yulan Wang. 2011. Equilibrium and optimal strategies to join a queue with partial information on service times. *European Journal of Operational Research* 214: 284–297.

Hassin, Refael, and Moshe Haviv. 2003. *To Queue or Not to Queue: Equilibrium Behavior in Queueing Systems*. Boston: Kluwer Academic Publishers.

Maccheroni, F., M. Marinacci, A. Rustichini, and M. Taboga. 2009. Portfolio selection with monotone mean–variance preferences. *Mathematical Finance* 19: 487–521.
Wang, Jinting, and Zhe George Zhang. 2018. Strategic joining in an M/M/1 queue with risk-sensitive customers. *Journal of the Operational Research Society* 69 (8): 1197–1214.
Wang, Jinting, Shouyang Wang, and Sheng Zhu. 2024. Risk analysis of queueing game in service systems under mean-variance criterion. *Scientia Sinica Mathematica In press*. https://doi.org/10.1360/SSM-2023-0204.

Chapter 6
Queueing Games with Priority

Following Naor (1969), strategic customer behavior in queues has been studied in a large volume of literature with multiple dimensions, such as heterogeneous customers, reneging or retrial behavior, and delay information announcements. We will primarily review papers with priorities below and interested readers are referred to the monographes of Hassin and Haviv (2003), Stidham (2009) and Hassin (2016) for comprehensive reviews on other variations to Naor's model.

In many systems, such as communication systems, electronic countermeasures systems, computer network systems, and many other service systems, some classes of "customers" receive priority service, and this kind of queueing systems is called priority queueing systems. In a priority queueing system, customers with higher priority are served ahead of those with lower priority, regardless of the order in which they enter the system. Priority queues are generally more difficult to analyze than non-priority ones.

In the literature, Balachandran (1972) first studies the equilibrium behavior of customers in the queueing system with priority. He considers an observable M/M/1 queuing model, in which each arriving customer could choose the number of possible payments from a discrete infinite set $\{b(0) < b(1) < \cdots\}$, customers' priority levels are based on how much they paid, and customers are not informed on the payments made by other customers. Tilt and Balachandran (1979) consider the model of customers with different time values, and the results showed that multiple pure equilibrium strategies may exist and that an equilibrium may result in various service orders including FCFS and LCFS. Haviv and van der Wal (1997) assume that customers are allowed to choose any non-negative number of purchasing relative priority according to their own situation. Dolan (1978) considers a model that giving customers the corresponding priority according to the time cost they provided. Hassin (1995) considers the overcrowding problem of equilibrium arrival in unobservable queuing system and provided a solution. Adiri and Yechiali (1974), Hassin and Haviv (1997) analyze the queueing system with two classes customers with different priorities. They assume that customers can choose whether to buy priority or not upon

J. Wang, *Fundamentals of Queueing-Game Models*,
https://doi.org/10.1007/978-981-95-0261-5_6

arrival and obtain the pure and mixed joining threshold Nash equilibrium strategies respectively. Lillo (2001) considers an M/G/1 queueing system of two classes of customers with different priorities. Based on the linear cost structure, they obtained the optimal control strategy with two thresholds. Jouini et al. (2009) establish a call center model with multiple types of customers with priority and impatience. They assumed that customers were informed of delay information upon arrival.

Wang et al. (2019) consider an M/M/1 queueing system with a pay-for-priority option, and study customers' joint decisions between joining/balking and pay-for-priority. The equilibrium strategies are two-dimensional. They fully characterize the equilibrium structure and identify the Pareto-dominant strategies of such a game analytically, under both the observable and unobservable settings. Interestingly, the equilibrium structure, the system throughput, and the service provider's optimal price for priority premium can all be non-monotone in the service reward, which departs from the existing models of priority queues without balking. In particular, they find that an increase in service reward can actually hurt the firm's revenue (everything else being equal). Second, they also compare the server's revenue between the observable and the unobservable settings. It is found that the service provider is better off with the observable setting when the system load is either low or high, but benefits more from the unobservable setting when the system load is medium. The fact that the optimal setting switches twice as the system load increases, is rather interesting. More recently, Cui et al. (2023) summarize innovative priority mechanisms in service operations from both theoretic and practical perspectives. See also Gilbert and Weng (1998), Guo et al. (2014), Wang and Wang (2019), Cui et al. (2020), Yang et al. (2021), Hu (2019) in the related research areas.

6.1 M/M/1 Queueing Systems with Priority

6.1.1 Model Description

Consider that two FCFS queues are formed in front of a single server, one line for priority customers and one for ordinary customers. When the customer with priority arrives, if ordinary customers are served, the service is immediately stopped and the service is diverted to serve the preferred customer. Preempted ordinary customers will resume service only when no priority customer in the system. Suppose that customers arrive as a Poisson process with parameter λ and service time is exponential distributed with parameter μ. Upon arrival, a customer must decide whether to purchase priority and join in the queue, and cannot buy priority once has decided to enter the general queue. All customers bear a cost of C per unit of waiting time. The price for priority is θ. For stability, we assume that $\rho = \frac{\lambda}{\mu} < 1$ Denote by $B \equiv T_{BT} \frac{1}{\mu - \lambda}$ the expected length of a busy period.

6.1.2 Equilibrium Strategy in Observable Queue

In the case of observable queue, upon arrival and after observing the length of the two queues, a customer decides whether to purchase priority. Denote $(i, j), i, j \geq 0$, a typical state of the system observed by an arrival, where i and j denote the number of priority and ordinary customers, respectively.

Assume that all customers follow the pure threshold strategy n, i.e., they buy priority if and only if the number of customers in the queue is at least a threshold n. If all use the threshold strategy n, then the only recurrent states are $(0, j), j = 0, \ldots, n$ and $(i, n), i = 1, 2, \ldots$, whereas the other states are transient.

Assume that all customers follow the mixed threshold strategy $x = n + p, n \in \mathbb{N}, 0 \leq p < 1$, i.e., a customer who observes a total of k customers in the system joins the ordinary queue if $k \leq n - 1$, does so with probability p and joins the priority queue with probability $1 - p$ if $k = n$. And purchase priority if $k > n$. Denote $\mathbb{E}(x)$ be the expected time in the system of the last customer in the ordinary queue when the state is $(0, \lceil x \rceil)$, where $\lceil x \rceil$ indicates the largest integer not great than x. Here is the algorithm to calculate $\mathbb{E}(x)$.

Let $H_{i,k}(x)$ be the expected waiting time for an ordinary customer at position i in his queue, when k ordinary customers are behind him and the priority queue is empty. When all customers follow the mixed threshold strategy $x = n + p, i + k \leq \lceil x \rceil, i \geq 1, k \geq 0$ holds. By definition, $\mathbb{E}(x)$ is computed by $\mathbb{E}(x) = H_{\lceil x \rceil,0}(x)$, we just need to calculate H ($\lceil x \rceil = n + 1$) (Set a time unit so that $\lambda + \mu = 1$):

$$
\begin{aligned}
H_{1,n}(x) &= B, \\
H_{1,n-1}(x) &= 1 + \lambda p H^{x}_{1,n} + \lambda(1 - p)(B + H_{1,n-1}(x)), \\
H_{1,k}(x) &= 1 + \lambda H_{1,k+1}(x), \quad k = 0, 1, \ldots, n - 2, \\
H_{i,k}(x) &= 1 + \lambda H_{i,k+1}(x) + \mu H_{i-1,k}(x), \quad i = 2, \ldots, n - 1, \\
&\qquad k = 0, \ldots, n - i - 1, \\
H_{i,n-i}(x) &= 1 + \lambda p H_{i,n-i+1}(x) + \lambda(1 - p)(B + H_{i,n-i}(x)) + \mu H_{i-1,n-i}(x), \\
&\qquad i = 2, \ldots, n, \\
H_{i,n-i+1}(x) &= 1 + \lambda(B + H_{i,n-i+1}(x)) + \mu H_{i-1,n-i+1}(x), \quad i = 2, \ldots, n + 1.
\end{aligned}
$$

From the above equation, the following algorithm can be obtained to calculate H (in the similar case of $p = 0$, it is omitted here):

---

1. First step:
If $p \neq 0$, let $H_{1,n}(x) = B$;
Let$H_{1,n-1}(x) = (1 + \lambda B)/(1 - \lambda(1 - p))$;
From $k = n - 2$ to $k = 0$, $H_{1,k}(x) = 1 + \lambda H_{1,k+1}(x)$.
2. Recursive step:
From $i = 2$ to $i = n$, $H_{i,n-i+1}(x) = (1 + \lambda B + \mu H_{i-1,n-i+1}(x))/\mu$;
$H_{i,n-i}(x) = (1 + \lambda p H_{i,n-i+1}(x) + \mu H_{i-1,n-i}(x) + \lambda(1 - p)B)/(1 - \lambda(1 -$

$p)$).

If $i \neq n$, from $k = n - i - 1$ to $k = 0$, $H_{i,k}(x) = 1 + \lambda H_{i,k+1}(x) + \mu H_{i-1,k}(x)$.

If $p \neq 0$, let $H_{n+1,0}(x) = (1 + \lambda B + \mu H_{n,0}(x))/\mu$.

Theorem 6.1.1 *The integer threshold strategy n, $n \geq 1$, specifies an equilibrium if and only if $\frac{C}{\mu} + \theta - CB \leq C\mathbb{E}(n) \leq \frac{C}{\mu} + \theta$. The threshold $n = 0$ specifies an equilibrium if and only if $\frac{C}{\mu} + \theta \leq CB$.*

Proof Assume that all customers use an integer-value threshold strategy $n (n \geq 1)$. To be a best response strategy for an individual, two conditions are necessary for n:

(1) First, if the arriving customer observes system state $(0, n-1)$, no buying priority is an optimal strategy, therefore we have $C\mathbb{E}(n) \leq \theta + \frac{C}{\mu}$.
(2) Second, if the arriving customer observes state $(0, n)$, his optimal strategy is to buy priority, and therefore one has $C[B + \mathbb{E}(n)] \geq \theta + \frac{C}{\mu}$.

Moreover, these conditions are also sufficient: if it is optimal to buy priority at state $(0, n)$, it is also optimal to do so at state $(i, n), n \leq 1$; In the same way, if it is optimal not to buy priority at state $(0, n-1)$, then it is also optimal not to do so at states $(0, j)$, $j \leq n-2$.

Finally, the fact that if $n = 0$ is a equilibrium strategy, the necessary and sufficient conditions is $\theta + \frac{C}{\mu} \leq CB$ is straightforward. □

We can see that there are FTC cases in this observable situation, so there may be multiple equilibrium threshold strategies. Now let's figure out the upper bound on the number of equilibrium solutions.

Theorem 6.1.2 *For $n \geq 1$,*

$$\frac{1}{\mu} \leq \mathbb{E}(n+1) - \mathbb{E}(n) \leq B,$$

and

$$\lim_{n \to \infty} [\mathbb{E}(n+1) - \mathbb{E}(n)] = \frac{1}{\mu}.$$

Therefore, there are at least 1 and at most $\lfloor \frac{1}{1-\rho} \rfloor$ pure threshold equilibria.

Proof Assume that the initial state of the system is $(0, n)$, and all use the threshold strategy n. Note the customer C_n is the last customer in the ordinary queue at initial state. When customer C_n left after receiving service, customer C_{n+1} has arrived at the head of the line and must go through at least one service cycle before leaving the system. So $\mathbb{E}(n+1) - \mathbb{E}(n) \geq \frac{1}{\mu}$ and $\mathbb{E}(n+1) - \mathbb{E}(n) \leq B$.

Consider two ordinary customers, one in position n and one in position $n+1$ where in both cases no priority customers are present. The former customer is in a system where all use the threshold strategy n and the latter where all use $n+1$.

Looking at the two under the same realization of events, the latter commences service for the first time as soon as the former leaves. When n is large it is unlikely that he will ever be preempted since the expected queue length behind him is bounded. Hence, his additional time in the system approaches $\frac{1}{\mu}$ when $n \to \infty$.

According to Theorem 6.1.1 and the first conclusion of this theorem, the number of pure threshold equilibria at is between 1 and $\lfloor \frac{B}{1/\mu} \rfloor = \lfloor \frac{1}{1-\rho} \rfloor$. The lower and upper bounds of 1 and $\lfloor \frac{1}{1-\rho} \rfloor$ on the number of pure equilibria are attainable. □

If mixed strategy $x = n + p, 0 < p < 1$ is an equilibrium strategy, a customer who observes state $(0, \lfloor x \rfloor)$ upon arrival is indifferent between buying priority and not. Similar to the proof of Theorem 6.1.1, the following theorem can be obtained.

Theorem 6.1.3 *The non-integer threshold x defines an equilibrium strategy if and only if* $C\mathbb{E}(n) = \frac{C}{\mu} + \theta$

6.1.3 Equilibrium Strategy in Unobservable Queue

In the case of unobservable queue, customers decide whether to purchase priority at the price of θ with a probability p without observing the queue. When all customers adopt strategy p to purchase priority, the arrival process of priority customers is Poisson with rate λp and therefore the expected waiting of a priority customer is $\frac{1}{\mu - \lambda p}$. Considering the equilibrium behavior of customers, the following conclusions can be drawn.

Theorem 6.1.4 *In an M/M/1 unobservable queue with priority, the equilibrium strategy of customers are listed in the following theorem:*

1. *The model is of the FTC type;*
2. *If $\theta \le \frac{C\rho}{\mu(1-\rho)}$ then $p = 1$ is a dominant strategy;*
3. *If $\theta \ge \frac{C\rho}{\mu(1-\rho)^2}$ then $p = 0$ is a dominant strategy;*
4. *If $\frac{C\rho}{\mu(1-\rho)} < \theta < \frac{C\rho}{\mu(1-\rho)^2}$ then there are three equilibria: $p = 0$, $p = 1$ and $p = \frac{1}{\rho} - \frac{C}{\theta\mu(1-\rho)}$.*
5. *The pure equilibria are ESS. Mixed equilibria are not.*

Proof Let W denote the expected waiting time of an ordinary customer. Then, the expected waiting time of a random customer can be computed in two ways. First, it equals the expected waiting time in a FCFS queue, which is $\frac{1}{\mu-\lambda}$. Second, it is $\frac{1}{\mu-\lambda p}$ with probability p and W with probability $1 - p$. Thus,

$$\frac{1}{\mu - \lambda} = \frac{p}{\mu - \lambda p} + (1 - p)W,$$

from which we conclude that $W = \frac{\mu-\lambda}{\mu-\lambda p}$.

The reduction, $f(p)$, in expected waiting costs if the customer choose to buy priority is

$$f(p) = \frac{\lambda C}{(\mu - \lambda)(\mu - \lambda p)} = \frac{\rho C}{\mu(1-\rho)(1-\lambda p)}$$

This function is monotone increasing with p, from which the FTC property follows.

1. When $\theta \leq f(0)$, it is uniquely optimal for a customer to purchase priority no matter what the others do. In other words, this is a dominant strategy.
2. When $\theta \geq f(1)$, it is uniquely optimal for a customer not to purchase priority no matter what the others do. Again, this is a dominant strategy.
3. When $f(0) < \theta < f(1)$, there exists a unique value $p_e(0 < p_e < 1)$, such that given that p_e is adopted by the other customers, a customer is indifferent between purchasing priority or not. Solving $f(p) = \theta$ leads to $p_e = \frac{1}{\rho} - \frac{C}{\theta\mu(1-\rho)}$. This equilibrium is not ESS: if p is larger than p_e, the unique best response is to buy priority, whereas if p is smaller than p_e, then the unique best response is not to buy. Therefore, in addition to the mixed equilibrium, there are two pure equilibria, where all buy priority and where none do. These pure equilibria are ESS. □

6.2 M/M/1 Queues with Priority and Processor Sharing

Processor sharing refers to situations where servers split service capacity among the customers presenting in the system. Consequently, given a fixed service capacity, the larger the number of customers present in the system, the slower the service rate they receive. If customers evenly share the capacity of a single server, it is referred to as an egalitarian processor sharing model (EPS); while when each customer in the system obtains a relative priority parameter and receives service proportionally to the associated parameter, this system is called discriminatory processor sharing model (DPS). We refer to Altman et al. (2011), Altman and Shimkin (1998), Haviv (2014), and the monograph of Hassin and Haviv (2003) for further reading.

6.2.1 *Model Description*

Consider a queueing system with priority and processor sharing. Suppose that customers arrive as a Poisson process with parameter λ and service time is exponential distributed with parameter μ. Define $\rho = \frac{\lambda}{\mu}$. Without loss of generality assume that $\lambda + \mu = 1$. Processor sharing assigns service capacity based on the prices paid by all customers who are present in the system, that is if the price paid by the i customer is x_i, $1 \le i \le n$, then the proportion of the service capacity obtained by him is $x_i / \sum_{j=1}^{n} x_j$. All customers bear a cost of C per unit of waiting time.

Each arriving customer knows neither his service time, nor the number of customers in the current system, and he should choose the optimal payment strategy to minimize his total cost when joining the queue.

6.2.2 Equilibrium Payment Strategy

Tag a customer who has arrived and just joined the queue. For any $n \ge 0$, let $f(n, x)$ be the expected residual time of a customer who paid x when all others pays 1 and that he is in service with n other customers. Observe that $f(n, 0)$ is the expected sum of $n + 1$ busy periods. And for $x = \infty$, the tagged customer has absolute priority, thus

$$f(n, 0) = (n + 1)\frac{1}{1 - \rho}\frac{1}{\mu}, \tag{6.1}$$

$$f(n, \infty) = \frac{1}{\mu}, \tag{6.2}$$

and for $0 < x < \infty$,

$$f(n, \infty) < f(n, x) < f(n, 0). \tag{6.3}$$

For $n \ge 0$ $f(n, x)$ satisfies the difference equation

$$f(n, x) = 1 + \lambda f(n + 1, x) + \mu \frac{n}{n + x} f(n - 1, x), \quad n \ge 0. \tag{6.4}$$

Lemma 6.2.1 *$f(n, x)$ is an affine function of n. In other words, for some functions $A(x)$ and $B(x)$,*

$$f(n, x) = A(x)n + B(x), \quad n \ge 0. \tag{6.5}$$

Proof At time 0, tag a customer who paid x while others in the system pay 1. In case 1, there are n customers in the system except tagged customer while there are m customers in the system except tagged customer in case 2. Suppose that an additional customer joins the system. Suppose that service is given in a weighted round-robin fashion, namely for some (small) quantum Δ, the server dedicates the service capacity for an amount of time Δ to each of the other customers and then serves the tagged customer for an amount of time $x\Delta$. This process is repeated until the tagged customer leave the queue. It can be concluded that in both cases, additional customers have an equal impact on the waiting time of tagged customers. In fact, consider the service times of the additional customer, those who arrive during his service, those who arrive during their periods of service,... , ad infinitum. The tagged customer may be present during some of these periods but the distribution of his added time in the system due to the extra customer is independent of how many others are present in the system at time 0. Thus, when Δ goes to 0, the expected waiting time of tagged customer inflicted by the added customer is $A(x)$. □

Lemma 6.2.2 *The functions* $A(x)$ *and* $B(x)$ *defined in Lemma 6.2.1 are*

$$A(x) = \frac{1}{1+x-\rho}\frac{1}{\mu}, \quad B(x) = \frac{1+x}{1+x-\rho}\frac{1}{\mu}. \tag{6.6}$$

Proof Substituting (6.5) into (6.4) will obtain the conclusion. □

Lemma 6.2.3 *Denote by* $g(x)$ *the expected time in the system for tagged customer who pays* x *when the other customers pay* 1. *Then,*

$$g(x) = \frac{1+x-\rho x}{1+x-\rho}\frac{1}{1-\rho}\frac{1}{\mu}. \tag{6.7}$$

Proof The probability that a customer who pays x observes n other customers in the system upon his arrival is $(1-\rho)\rho^n$. Thus,

$$g(x) = A(x)\sum_{n=0}^{\infty}(1-\rho)\rho^n n + B(x). \tag{6.8}$$

By calculating, we can obtain (6.7) and $g(x)$ is convex. □

The customer's equilibrium payment strategy is determined below.

Theorem 6.2.1 *In an M/M/1 queueing system with priority and processor sharing, there exists a unique equilibrium payment which is*

$$\frac{\rho}{\mu(1-\rho)(2-\rho)}C.$$

Proof Suppose that all customers pay the price of 1. In order for 1 to be the optimal price for a tagged customer, it is necessary that

$$\frac{d}{dx}(Cg(x)+x)\mid_{x=1} = 0, \tag{6.9}$$

where $g(x)$ is given in (6.7). Since $g(x)$ is convex, this condition is also sufficient. Solving (6.9), we obtain

$$C = \frac{\mu(1-\rho)(2-\rho)}{\rho}.$$

Therefore, customers' equilibrium payment strategy is

$$\frac{\rho}{\mu(1-\rho)(2-\rho)}C$$

□

6.3 M/M/1 Queueing Systems with Priority and Random Service

6.3.1 Model Description

Consider a queueing system with priority and random service. Suppose that customers arrive as a Poisson process with parameter λ and service time is exponential distributed with parameter μ. Define $\rho = \frac{\lambda}{\mu}$. Without loss of generality assume that $\lambda + \mu = 1$. The server serves the next customer only after serving one customer, and determines the probability of each customer being served based on the prices paid by all customers who are present in the system. In other words, if the price paid by the i customer is x_i, $1 \leq i \leq n$, then the probability that he will be served after the current customer is served is $x_i / \sum_{j=1}^{n} x_j$.

Each arriving customer knows neither his service probability, nor the number of customers in the current system, and he should choose the optimal payment strategy to minimize his total cost when joining the queue.

6.3.2 Equilibrium Payment Strategy

For any $n \geq 0$, let $h(n, x)$ be the expected residual time of a customer who paid x when all others pays 1 and that he is in service with n other customers. Same as the previous analysis in Sect. 6.2, it is easy to obtain that

$$(n, 0) = (n+1)\frac{1}{1-\rho}\frac{1}{\mu}, \tag{6.10}$$

$$(n, \infty) = \frac{1}{\mu}. \tag{6.11}$$

Further, for $0 < x < \infty$,

$$(n, \infty) < (n, x) < (n, 0). \tag{6.12}$$

For $n \geq 0$ $h(n, x)$ satisfies the difference equation

$$(n, x) = 1 + \lambda(n+1, x) + \mu \frac{n}{n+x}(n-1, x), \quad n \geq 0. \tag{6.13}$$

Lemma 6.3.1 *For $n \geq 0$, $h(n, x)$ is an affine function of n. In other words, for some functions $A^*(x)$ and $B^*(x)$,*

$$h(n, x) = A^*(x)n + B^*(x), \quad n \geq 0, \tag{6.14}$$

where $B^*(x) = (1+x)A^*(x)$

Proof Tag a customer who paid x, when other customers in the system and all other customers pay 1. let $B^*(x)$ indicate the expected residual time of tagged customer when the system is empty and busy when he arrives. And let $A^*(x)$ be the expected residual time added to the tagged customer when there is a waiting customer in the system. The waiting customer is called an additional customer, and $B^*(x) = (1+x)A^*(x)$. In fact, tagged customer have priority over additional customer in receiving service with probability $\frac{x}{1+x}$. In this case, the expected residual time of tagged customer is $B^*(x)$. On the other hand, tagged customer being served later than the additional customer with probability $\frac{1}{1+x}$, so his expected residual time is $2B^*(x)$. Due to total expectation formula, $A^*(x) = \frac{B^*(x)}{1+x}$ can be obtained. Similarly, when there are n other customers in the system, the expected residual time increased by each customer to the tagged customers is independent with n, and is $A^*(x)$. □

Substituting (6.14) into (6.13) will obtain:

$$A^*(x) = \frac{1}{1+x-\rho}\frac{1}{\mu}, \quad B^*(x) = \frac{1+x}{1+x-\rho}\frac{1}{\mu}. \tag{6.15}$$

Lemma 6.3.2 *Denote by $w(x)$ the expected time in the system for tagged customer who pays x when the other customers pay 1. Then,*

$$w(x) = \frac{1+x-\rho x}{1+x-\rho}\frac{\rho}{1-\rho}\frac{1}{\mu}, \tag{6.16}$$

and $w(x)$ is convex.

Proof When server is busy, the probability that a customer who pays x observes n other customers in the system upon his arrival is $(1-\rho)\rho^n$, thus

$$w(x) = \rho\Big[A^*(x)\sum_{n=0}^{\infty}(1-\rho)\rho^n n + B^*(x)\Big], \tag{6.17}$$

By calculating, we can obtain (6.16) □

Theorem 6.3.1 *In the $M/M/1$ queueing system with priority and random service, there exists a unique equilibrium payment which is*

$$\frac{\rho^2 C}{2-\rho}\frac{1}{1-\rho}\frac{1}{\mu},$$

Proof Suppose that all customers pay the price of 1. In order for 1 to be the optimal price for a tagged customer, it is necessary that

$$\frac{d}{dx}(Cw(x)+x)\mid_{x=1}=0, \tag{6.18}$$

where $w(x)$ is given in (6.16). Since $w(x)$ is convex, this condition is also sufficient. Solving (6.18), and obtain

$$C\rho=\frac{\mu(1-\rho)(2-\rho)}{\rho}.$$

Therefore, customers' equilibrium payment strategy is

$$\frac{\rho^2 C}{2-\rho}\frac{1}{1-\rho}\frac{1}{\mu}$$

□

6.4 M/M/1 Queueing Systems with Priority and Balking

This section is adapted from Wang et al. (2019) and we excerpt some of the main results for the sake of conciseness, interested readers can refer to Wang et al. (2019) for more details. In service industries, charging for priority access is an efficient and prevalent business scheme for managing congestion and revenue. Afèche and Sarhangian (2015) investigate the rational abandonment from priority queues and discuss equilibrium strategy and pricing implications. In principle, both the consumer welfare and the service provider's revenue can be improved with the optimal design of an appropriate priority system (Gavirneni and Kulkarni 2016). And, on the other hand, the concept of pay-for-priority is natural for customers to understand—one can obtain priority access to the service by paying a premium in addition to the basic service fee, or remain a non-priority customer who is responsible only for the basic service fee but may be overtaken by future customers who opt in for priority.

From the perspective of the revenue-maximizing service provider, both service fee and priority premium contribute to revenue. Correspondingly, when the service provider sets an optimal price for priority access, it should take into consideration two possible market-force consequences: (i) The price of priority premium can affect the total customer throughput, and therefore the total amount collected from the basic service charges; and (ii) for customers that decide to use the service, priority pricing also determines the number of priority purchases and thus the portion of revenue that derives from priority premium. At a more granular level, because customers can make both joining/balking and priority/non-priority decisions upon arrival to the service system, the trade-off identified in the point (i) above can be related to customers joining/balking decision while point (ii) is based on their priority-purchasing decision.

In the literature of Operations Management, pay-for-priority decisions have been well-studied with strategic customers, however, joining/balking decisions are typically not studied concurrently. Such a treatment (which ignores balking), could be sometimes unrealistic especially when customers have alternative service seeking options. Motivated by this research gap, our work studies customers' joint decisions on joining/balking and pay-for-priority. Consider the following priority-queue examples. (1) (Tourism) To obtain a UK tourist visa, there is an optional service where the applicant "can pay an extra fee to fast-track the application and have the visa application processed ahead of the non-priority applications".[1] However due to the high price and otherwise lengthy visa process, some customers choose to tour a different country (*The Daily Telegraph*, 2013). (2) (Recreation & Entertainment) Visitors of the Universal Orlando Resort can pay an extra $85 to get in the priority line. Similarly, the Park City Mountain Resort in Utah offers a $25/day Fast Track lift-ticket upgrade that allows patrons to use express lanes at popular lifts. Alternative balking options exist for customers in parks (e.g., Walt Disney World) and ski resorts (e.g., Canyons) nearby. (3) (Restaurants) Papa John's moves a pizza order to the front of the line "so that it is made faster" if customers pay a $3 "Papa Priority" fee[2]; RAAD, a start-up mobile app, allows its users to skip the long lines at bars and restaurants by paying a certain fee set by the vendors. In both settings, customers can freely decide to join, purchase priority, or balk (using a different restaurant). Later in this chapter, we will use the Papa John's Pizza example again to illustrate some of our findings (Sect. 6.4.4.3). (4) (Manufacturing) Tesla customers can pay $1000 to reserve a Model 3 car which will be delivered according to the order when the reservation is received. Customers who do not reserve will not get the car until all the pre-orders are fulfilled. The money is a deposit, so the priority premium in this example is the opportunity cost of the capital. Customers also have the option of purchasing a car from a different company.

All of the above-mentioned priority examples include customer's balking options. Studying balking altogether with priority decisions is however not only relevant to real-word practices, but also brings about technical merits. Particularly, there are three modeling advantages. First, without balking only a system with workload less than one can be studied due to the stability requirement. Second, if customers do not have a balking option, they are assumed to join the queue even when the expected utility under the service provider's optimal price is negative, causing inconsistency with

[1] https://www.gov.uk/government/publications/usa-apply-for-a-uk-visa/apply-for-a-uk-visa-in-the-usa.

[2] http://www.papajohns.com/papa-priority/.

behavior of strategic consumers.[3] Last but not least, revenue maximizing is purely equal to priority premium maximizing when balking is not allowed (i.e., throughput is fixed) because revenue collected from the basic service fee is constant. This totally ignores the impact of priority pricing on system throughput and customers' joining/balking decisions, and can lead to suboptimal decisions and revenues for the firm.

To this end, this chapter incorporates customers' joining/balking decisions into a pay-for-priority system. The problem is nontrivial due to the two-dimensional decision scope. The following research questions arise: (1) How do strategic customers make simultaneous joining and priority decisions? (2) What is the service provider's optimal price for priority premium?

Furthermore, observing the real-life examples mentioned earlier can be broadly categorized into observable and unobservable settings and this section aims to study question 1 and question 2 mentioned above for both the observable and unobservable cases. Doing so enables us to understand the impact of queue length information on the server's revenue. In particular, we develop one more research question: (3) Comparing the server's revenues between the observable and unobservable setting, which setting is best for the service provider?

To address the research questions, in this section we consider a single-server queue with a pay-for-priority option. Homogeneous customers arrive according to a Poisson process with potential arrival rate Λ. The service times are independent and exponentially distributed with rate μ. We assume that the values of Λ and μ are common knowledge among the customers. Upon arrival, a new customer decides whether to join the queue or balk (value of outside option is normalized to zero), with or without seeing the state of the system. A joining customer also decides whether or not to purchase priority for a premium P set by the service provider. A priority customer obtains non-preemptive priority for service over the non-priority customers (i.e. a non-priority customer who has started service will not be cut in line by a priority customer), and the queue disciplines within the priority and within the non-priority lines are both first-come, first-serve (FCFS). We note that the non-preemptive priority discipline is technically more challenging than the preemptive case,[4] and also more consistent with real-world examples. Customers are risk neutral and act to maximize their expected utilities. The options they have upon arrival to the system can be distilled into (1) balk; (2) join the queue and purchase priority; and (3) join the queue without purchasing priority. We assume that any decision made is irrevocable, and reneging or purchasing priority after joining the non-priority line is not allowed.

---

[3] For example, the optimal price for the priority premium is $\frac{\rho^2 C}{\mu-\Lambda}$ for an observable queue without balking that contains two priority classes, and customers' expected utility is negative if value of the service is less than $\frac{\rho^2 C}{\mu-\Lambda}$.

[4] Under the preemptive scheme, as long as there are priority customers in the system, the customer at the server (currently being served) is always a priority customer. But this is not the same for the non-preemptive scheme.

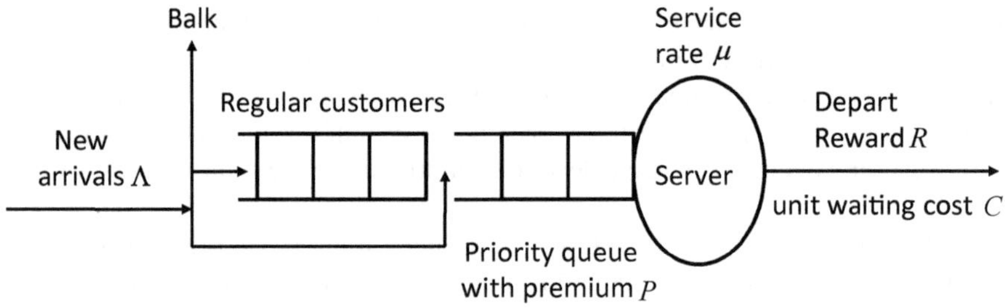

Fig. 6.1 Schematic representation of the $M/M/1$ priority queueing system with balking

A typical cost-reward structure in the literature is adopted, as in Naor (1969): Each customer receives a reward R, which is equal to the service value V less the basic service fee F, from completion of the service, but incurs a waiting cost C per unit of time while in the system. The fee for priority and non-priority customers are therefore given by $(P + F)$ and F, respectively. For notational convenience, we denote the ordinary/non-priority/regular and the priority customers as type-o and type-p customers for the rest of the paper. We assume $R > C/\mu$, because otherwise customers would not be interested in getting the service. Figure 6.1 provides a schematic representation of the system.

In the observable case, customers make rational decisions based on the state of the system observed upon arrival, including the number of customers in the priority queue and in the entire system. While they do not know the state of the system in the unobservable case, customers make decisions based on evaluating the expected waiting times from the system parameters. As all the customers are ex-ante symmetric in our model, we will only consider symmetric equilibria under which all of the customers adopt the same strategy.

The main results of this section are as follows: (1) The equilibrium structure of the model is fully characterized and the necessary analytical tools to identify the Pareto-dominant strategy is developed. In particular, for the observable case, we prove that pure equilibrium can only be a threshold strategy with two thresholds. For the unobservable case, we prove that a mixed strategy can be an ESS.[5] Furthermore, we show that *avoid the crowd* (ATC) behavior can exist in the presence of multiple equilibria in a two-dimensional decision problem. (2) It is demonstrated that the equilibrium structure, system throughput and the service provider's optimal price for the priority premium can all be non-monotone in the service reward R. Interestingly, with customers making both joining/balking and priority decisions, increasing service value without increasing service fees can actually lead to reduced firm revenue. (3) Comparing the server's optimal revenues under the observable and unobservable

[5] Whereas in a priority queue without balking, it is well known that mixed equilibria is not ESS (e.g., see Theorem 4.5 in Hassin and Haviv 2003).

setting, it is found that the service provider is better off with the observable setting when the system load is either low or high, but with the unobservable setting when the system load is medium.

6.4.1 Related Literatures

This section studies equilibrium strategies of strategic customers in a priority queueing system with balking and the subsequent revenue management problem for the server. We study both the observable and the unobservable settings by allowing the customers to make both joining/balking and pay-for-priority decisions. Hassin (1986), Chen and Frank (2004), Chambers and Kouvelis (2006), Shone et al. (2013) and Hu et al. (2018), all compare the outcomes between an observable queue and an unobservable queue with joining/balking customers, however, to the extent of our knowledge, the work in this chapter seems to be the first one that compares observable and unobservable priority queueing systems.

There are different regimes studied in the priority literature. Inspired by the examples listed above, the model studied in this section allows for balking and focuses on non-preemptive priority. The unobservable part of the model can be seen as an extension to the M/M/1 model discussed in Sect. 4.2 of Hassin and Haviv (2003). In their model, customers decide whether to purchase priority for a given price or to remain a non-priority customer for free. Their model is extended by allowing the customers to balk when their expected payoff from joining the system is negative.

Gavirneni and Kulkarni (2016) study a similar unobservable setting where customers balk when their expected payoff is lower than that in a first in, first out (FIFO) system without balking or priorities. In their paper, customers have heterogeneous waiting costs (modeled as a Burr distribution), so the ones with high/median/low waiting costs choose to become priority/non-priority/balking customers, respectively. In contrast, customers in our model are ex-ante symmetric, and our work solves a two-dimensional equilibrium with simultaneous joining/balking and priority purchasing decisions under the unobservable setting which has not been done before.

The observable setting of the current model is built on the models in Adiri and Yechiali (1974) and Hassin and Haviv (1997). Adiri and Yechiali (1974) analyze an M/M/1 model with two priority classes when the priority discipline is either preemptive or non-preemptive. They identified a single pure equilibrium of threshold-type that specifies that arriving customers will purchase priority, if and only if the number of customers in the non-priority queue exceeds a certain threshold. Hassin and Haviv (1997) extend the set of equilibrium strategies to include mixed strategies under the preemptive regime. The authors add to Adiri and Yechiali (1974) by showing that multiple equilibria exist in this case and they also give an example of a non-threshold (pure) equilibrium.

Compared to Adiri and Yechiali (1974), it is shown that multiple threshold equilibria exist under the non-preemptive regime, and both necessary and sufficient equilibrium conditions are obtained. The single equilibrium (threshold) that Adiri and

Yechiali (1974) identified in equation (20), on page 1057 of their paper, coincides with the biggest possible equilibrium threshold in the result obtained in this section. It is also proved analytically that a pure equilibrium strategy under the non-preemptive regime must be of a threshold type. This is a nice contrast to the preemptive result studied in Hassin and Haviv (1997). The non-threshold pure equilibrium example in their paper no longer survives as an equilibrium in our case.

For both the unobservable and observable cases, we are able to identify the full equilibrium structure and pin down the Pareto-dominant equilibrium by analyzing the structure of the consumer welfare function. We then study the pricing problem for the server when it can adjust the priority premium. Alperstein (1988) considers a similar revenue maximization problem in the preemptive regime. Mendelson and Whang (1990) show that maximum social welfare can be induced using the $C\mu$-rule when the customer classes are exogenously segmented. Afèche and Mendelson (2004) extend the results to a generalized delay cost structure.

Rather than paying the priority premium determined by the service provider, another priority pricing scheme known as "bribery" occurs when customers choose the amount of money they wish to pay/bribe in exchange for a level of priority which is in turn determined by the payment amount. Typically, customers who pay a larger bribe are placed in the queue before those who pay less. We refer interested readers to Sect. 4.5 in Hassin and Haviv (2003) for an excellent review of the bribery literature.

As mentioned in previous chapters, the strategy of customers can be defined as either *avoid the crowd* (ATC) or *follow the crowd* (FTC). Typically, a unique equilibrium can be identified because of the ATC behavior, while FTC allows for multiple equilibria (e.g., see Afimeimounga et al. 2005; Burnetas and Economou 2007; Guo and Hassin 2011; Guo and Zhang 2013). We show, however, that ATC behavior can exist in the presence of multiple equilibria in a two-dimensional decision problem like ours. There seems little prior research that studies two-dimensional equilibrium strategies of strategic customers. An exception is Hassin and Roet-Green (2017), in whose setting customers first decide whether to join an unobservable queue, inspect the queue length or balk and then make a second join-or-balk decision if inspecting.

6.4.2 The Unobservable Model

In our unobservable model, we assume customers do not observe the state of the system. They make decisions based on the information of the system including Λ (potential arrival rate) and μ (service rate) as well as the cost/reward parameters C (unit waiting cost), P (price of the priority premium) and $R \triangleq V - F$ (service reward $\triangleq$ service value—service fee). To consider a symmetric equilibrium, suppose that customers will join the system with probability $q \in [0, 1]$ (and balk with probability $1 - q$) upon arrival, and any joining customer will purchase priority with probability $p \in [0, 1]$. The decision process is shown in Fig. 6.2.

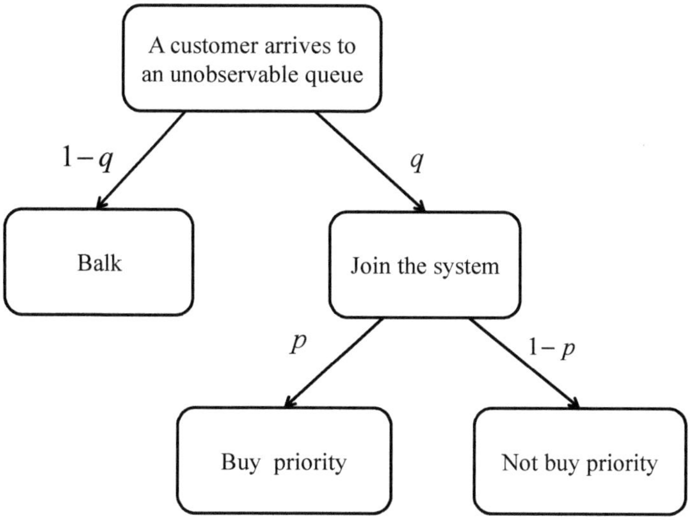

Fig. 6.2 The decision process under the unobservable setting

The strategy space of the customers can be described as a pair $(p, q) \in [0, 1] \times [0, 1]$ in the two-dimensional space. Under (p, q), the effective arrival rate to the system is $\lambda \triangleq \Lambda q$, where the effective arrival rate for type-o and type-p customers are $\lambda(1 - p)$ and λp, respectively. The three pure strategies, balking, joining without priority and joining with priority, can be characterized by $(0, 0)$, $(0, 1)$ and $(1, 1)$, respectively. Otherwise, (p, q) defines a mixed strategy. Our goal is to identify all the equilibrium strategies. Because of the one-to-one correspondence between q and λ (i.e., $\lambda = \Lambda q$), we will use (p, λ) for notational convenience to signify a strategy for the customers.

In this section, we first derive the equilibrium strategy for the customers under some given priority premium P, and then find the optimal revenue for the service provider when it can adjust P. To simplify notations, we will use $\nu_R \triangleq R\mu/C$ and $\nu_P \triangleq P\mu/C$ which are the service reward and priority premium relative to the average waiting cost for a service cycle.

6.4.2.1 Equilibrium Strategies

The expected sojourn times for the two types of customers are functions of effective arrival rate λ and the probability to purchase priority p, which are given as follows (see Chap. 33 of Harchol-Balter (2013)):

$$w_o(p, \lambda) \triangleq \frac{\lambda}{(\mu - \lambda)(\mu - \lambda p)} + \frac{1}{\mu}, \tag{6.19}$$

$$w_p(p,\lambda) \triangleq \frac{\lambda}{\mu(\mu-\lambda p)} + \frac{1}{\mu}, \tag{6.20}$$

where $\lambda < \mu$. That is, $w_o(p,\lambda)$ and $w_p(p,\lambda)$ are the expected sojourn times for a type-o customer and a type-p customer, respectively, when all the other customers follow the strategy (p,λ). We can characterize the reduction in waiting costs due to purchasing priority as follows:

$$D(p,\lambda) \triangleq [w_o(p,\lambda) - w_p(p,\lambda)]C = \left[\frac{\lambda^2}{\mu(\mu-\lambda)(\mu-\lambda p)}\right]C. \tag{6.21}$$

Remark 6.4.1 $D(p,\lambda)$ is an important criteria for customers' decision making. By purchasing priority, a customer can lower expected waiting cost by $D(p,\lambda)$. Thus, if the priority premium is less than $D(p,\lambda)$, it is wise for the customer to purchase priority. Otherwise, it is not. The following lemma describes an important property of this reduction function.

Lemma 6.4.1 *When $\lambda < \mu$ and $0 \le p \le 1$, $D(p,\lambda)$ is monotone increasing in p and λ.*

Proof Result follows directly from Eq. (6.21). □

Lemma 6.4.1 shows that the reduction in waiting cost with purchasing the priority increases with the total effective arrival rate to the system, as well as the probability of the priority purchase among the joining customers. Recall that $D(p,\lambda) = [w_o(p,\lambda) - w_p(p,\lambda)]C$. Intuitively, as more customers choose to enter the system or purchase priority, a non-priority customer will on average be more likely to be overtaken by a greater number of priority customers who arrive in the future. Therefore, this customer will be more incentivized to purchase priority. This is an FTC behavior described by Hassin and Haviv (2003).

The strategy of the customers depends on both the parameters of the system and the announced price P for the priority premium. When all customers follow the strategy (p,λ) (hereinafter referred to as "under" the strategy (p,λ)), we define the expected total costs for the two types of customers as $C_o(p,\lambda) \triangleq w_o(p,\lambda)C$ and $C_p(p,\lambda) \triangleq w_p(p,\lambda)C + P$, respectively. These costs are shown in Fig. 6.3 as functions of λ. In particular, four curves, $C_o(0,\lambda)$, $C_o(1,\lambda)$, $C_p(1,\lambda)$ and $C_p(1,\lambda)$, are plotted.

Note that $D(p,\lambda) = P$ holds if and only if $C_o(p,\lambda) = C_p(p,\lambda)$, so the possible (p,λ) that satisfy $C_o(p,\lambda) = C_p(p,\lambda)$ are given by the shaded areas in Fig. 6.3a, b. To derive the equilibrium strategy of the customers, however, we need to introduce a lemma first.

Lemma 6.4.2 *If (p,λ) is an internal equilibrium, i.e., $p \in (0,1)$, then $w_o(p(\lambda),\lambda)$ decreases in λ.*

Proof When (p,λ) is an internal equilibrium, we must have (i) $D(p(\lambda),\lambda) = P$ (customers are indifferent between purchasing priority and not) and (ii) $\lambda < \mu$.

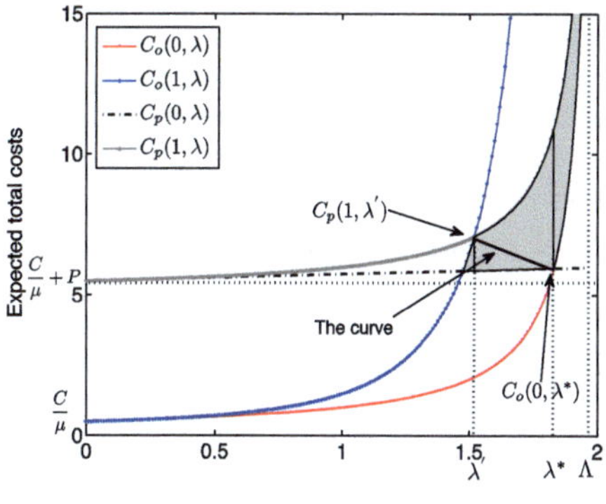

(a) when $0 < \nu_P \leq \frac{\rho^2}{1-\rho}$

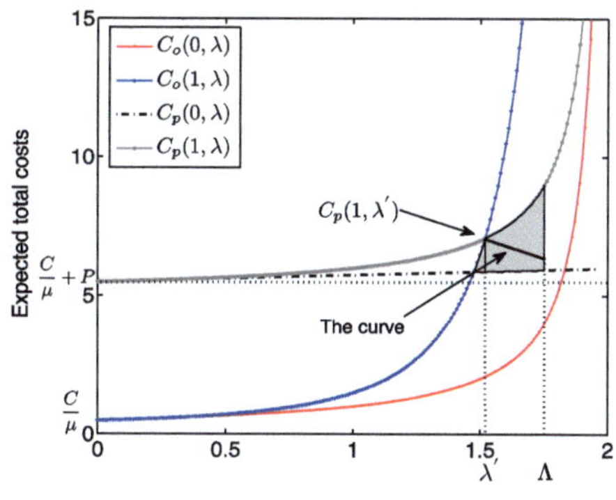

(b) when $\frac{\rho^2}{1-\rho} < \nu_P \leq \frac{\rho^2}{(1-\rho)^2}$

Fig. 6.3 Expected total costs for the customers versus effective arrival rate

By Eqs. (6.19) and (6.20), we have

$$\left[w_o(p(\lambda), \lambda) - \frac{1}{\mu}\right] \cdot (\mu - \lambda) = \left(w_p(p(\lambda), \lambda) - \frac{1}{\mu}\right) \cdot \mu. \tag{6.22}$$

Using (i), we get

$$P = (w_o(p(\lambda), \lambda) - w_p(p(\lambda), \lambda)) \cdot C. \tag{6.23}$$

Combining Eqs. (6.22)–(6.23) leads to

$$\left[w_o(p(\lambda), \lambda) - \frac{1}{\mu}\right] \cdot (\mu - \lambda) = w_o(p(\lambda), \lambda)\mu - \nu_P - 1. \tag{6.24}$$

After algebraic manipulations, we can obtain $w_o(p(\lambda), \lambda) = \frac{\nu_P}{\lambda} + \frac{1}{\mu}$ from which it is straightforward to see that $w_o(p(\lambda), \lambda)$ decreases with λ, which completes the proof. □

Lemma 6.4.2 establishes the monotonicity of the expected waiting time for non-priority customers under strategy (p, λ) when $p \in (0, 1)$. In Fig. 6.3a, b, we bold the points in black that satisfy $C_o(p(\lambda), \lambda) = C_p(p(\lambda), \lambda)$ when $0 < \nu_P \leq \frac{\rho^2}{1-\rho}$ and when $\frac{\rho^2}{1-\rho} < \nu_P \leq \frac{\rho^2}{(1-\rho)^2}$. Consistent with the lemma, they are both decreasing in λ. Define $\rho \triangleq \Lambda/\mu$. We can now fully characterize the equilibrium strategies of the customers when $\rho < 1$ and when $\rho \geq 1$.

Theorem 6.4.1 *When $\rho < 1$, for given cost-reward values C, P, R, the unique Pareto-dominant equilibrium strategy for the customers, denoted by (p_e, λ_e), is given as follows:*

(1) When $\nu_P = 0$, $p_e = 1$ is the dominant strategy and the system reduces to an FCFS-based $M/M/1$ queue where

$$(p_e, \lambda_e)=\begin{cases}(1, \mu - \frac{\mu}{\nu_R}), & \text{if } 1 < \nu_R \leq \frac{1}{1-\rho};\\ (1, \Lambda), & \text{if } \frac{1}{1-\rho} < \nu_R.\end{cases}$$

(2) When $0 < \nu_P \leq \frac{\rho^2}{1-\rho}$,

$$(p_e, \lambda_e)=\begin{cases}(0, \mu - \frac{\mu}{\nu_R}), & \text{if } 1 < \nu_R \leq \rho^* + 1 + \nu_P;\\ (\frac{\nu_R-1}{\nu_P} - \frac{1}{\nu_R-\nu_P-1}, \frac{\nu_P\mu}{\nu_R-1}), & \text{if } \rho^* + 1 + \nu_P < \nu_R \leq \frac{1}{1-\rho'} + \nu_P;\\ (1, \mu - \frac{\mu}{\nu_R-\nu_P}), & \text{if } \frac{1}{1-\rho'} + \nu_P < \nu_R \leq \frac{1}{1-\rho} + \nu_P;\\ (1, \Lambda), & \text{if } \frac{1}{1-\rho} + \nu_P < \nu_R.\end{cases}$$

(3) When $\nu_P > \frac{\rho^2}{1-\rho}$,

$$(p_e, \lambda_e)=\begin{cases}(0, \mu - \frac{\mu}{\nu_R}), & \text{if } 1 < \nu_R \leq \frac{1}{1-\rho};\\ (0, \Lambda), & \text{if } \frac{1}{1-\rho} < \nu_R.\end{cases}$$

where $\rho^* = \frac{\sqrt{\nu_P^2+4\nu_P}-\nu_P}{2}$ *and* $\rho' = \frac{\sqrt{\nu_P}}{\sqrt{\nu_P}+1}$, $\rho' < \rho^*$ *but both are increasing with P. Note that in Case (3) there exist multiple equilibria when* $\nu_R > \frac{1}{1-\rho}$, *and we have only presented the Pareto-dominant equilibrium in the theorem.*

Remark 6.4.2 For the case $\rho \geq 1$, analogous results can be derived through a similar proof, and the equilibrium strategies of the customers are as follows:

(1) When $\nu_P = 0$, $(p_e, \lambda_e) = (1, \mu - \frac{\mu}{\nu_R})$.
(2) When $\nu_P > 0$,

$$(p_e, \lambda_e) = \begin{cases}(0, \mu - \frac{\mu}{\nu_R}), & \text{if } 1 < \nu_R \leq \rho^* + 1 + \nu_P;\\ (\frac{\nu_R-1}{\nu_P} - \frac{1}{\nu_R-\nu_P-1}, \frac{\nu_P\mu}{\nu_R-1}), & \text{if } \rho^* + 1 + \nu_P < \nu_R \leq \frac{1}{1-\rho'} + \nu_P;\\ (1, \mu - \frac{\mu}{\nu_R-\nu_P}), & \text{if } \frac{1}{1-\rho'} + \nu_P < \nu_R.\end{cases}$$

We have presented the full equilibrium structure for the unobservable setting in Theorem 6.4.1. Even though multiple equilibria exist in some cases, we cannot assert that they represent the FTC situation (as opposed to ATC), since we are studying a two-dimensional decision problem. For example, when $p_e = 0$, the expected payoff for the customers is $R - Cw_o(0, \Lambda q_e)$, i.e. no one purchases priority. Since $w_o(0, \Lambda q)$ is increasing in q, when all the other customers choose to join the system with higher probability $q > q_e$, it reduces the expected payoff for the tagged individual, and their best response will be to balk. That is, an individual's best response for the joining probability is a decreasing function of the same decision made by other customers when $p_e = 0$. Therefore, it is an ATC situation in terms of the joining probability/decision q.

It is widely known in literature that mixed equilibria cannot be ESS in an unobservable priority queue without balking (e.g., see Theorem 4.5 in Hassin and Haviv 2003). This is consistent with our Lemma 6.4.1 which observes FTC customer behavior with respect to the priority decision. However, due to customers' ATC behavior in q, we can actually identify mixed equilibrium that are unique and ESS, e.g., the one

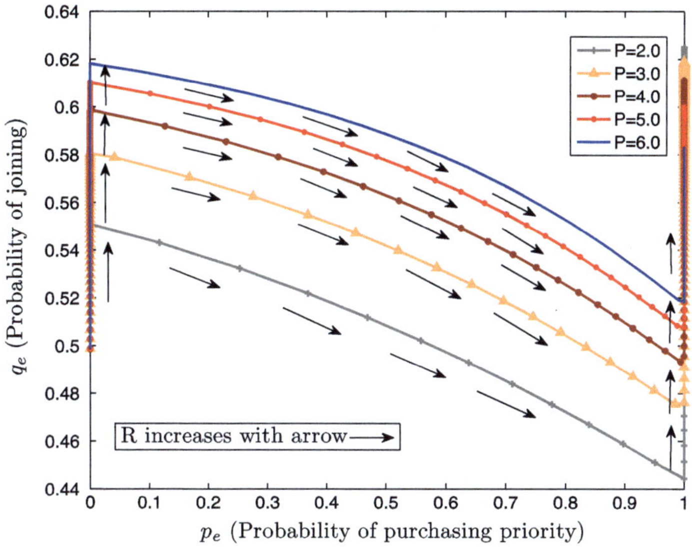

Fig. 6.4 The equilibrium strategy versus R and P when $\Lambda = 3$, $\mu = 2$ and $C = 1$

in Case (2) of Theorem 6.4.1 is proven to be so in Appendix A of Wang et al. (2019), which is omitted here for the sake of compactness.

Figure 6.4 illustrates the (Pareto-dominant) equilibrium strategy of the customers, (p_e, q_e), when $\Lambda = 3$, $\mu = 2$, $C = 1$, and both the service reward R and the priority premium P vary. The arrows indicate increasing values in R, while the different curves from bottom to top correspond to increasing values in P. As expected, as the service becomes more rewarding ($R \uparrow$), customers tend to purchase more priority at the equilibrium ($p_e \uparrow$). Surprisingly, the equilibrium joining probability q_e is not monotone in R. In particular, when R is sufficiently small (c.f. $p_e = 0$ in Fig. 6.4) or sufficiently large (c.f. $p_e = 1$ in Fig. 6.4), q_e is increasing in R. Otherwise (c.f. $0 < p_e < 1$ in Fig. 6.4), q_e is decreasing in R. This is because when R is sufficiently small or large, a marginal increase in R does not change the equilibrium priority probability (i.e., p_e remains 0 or 1). It prompts more customers to enter the system. Yet when R has an intermediate value, a marginal increase drives more priority activities ($p_e \uparrow$). This leads to higher expected waiting time for non-priority customers, and the equilibrium joining probability decreases ($q_e \downarrow$).

The firm's revenue is given by the summation of the basic service fee plus the priority premium, i.e., $\Lambda q_e F + \Lambda q_e p_e P$. In this case, for given service/priority fees F and P, increasing the service value V (thus $R \uparrow$) may actually reduce the firm's revenue (because $\Lambda q_e F \downarrow$), especially when the amount of the basic service fee F is significant. This is a counter-intuitive result: increasing a customer's service value without charging them more could lead to the firm's decrease in revenue. This is due to the intervention of balking and priority decisions explained earlier, and does not occur in $M/M/1$ queues only with priority (no balking) or only with balking (no priority).

6.4.2.2 Server's Revenue

In Sect. 6.4.2.1, we identified the equilibrium behavior of the customers given the price of priority premium P. In this section, we endogenize P to study the service provider's revenue maximization problem. We assume the basic service fee F is fixed in this paper to focus on the pricing strategy of the priority premium. Since customers respond to P, we model the interaction between the server and the customers as a Stackelberg game, where the server is the leader and the customers are the followers. The goal of the server is to maximize its revenue while anticipating customers' equilibrium strategies. The following lemma delivers an expression for the server's revenue.

Lemma 6.4.3 *When $\rho < 1$, the server's expected revenue (rate) under the unobservable case, denoted by S_{un}, is given as follows:*

(1) When $\nu_P = 0$ or $\nu_P > \frac{\rho^2}{1-\rho}$

$$S_{un}=\begin{cases}(\mu - \frac{\mu}{\nu_R})F, & \text{if } 1 < \nu_R \le \frac{1}{1-\rho};\\ \Lambda F, & \text{if } \frac{1}{1-\rho} < \nu_R;\end{cases}$$

(2) When $0 < \nu_P \le \frac{\rho^2}{1-\rho}$,

$$S_{un}=\begin{cases}\Lambda(P+F), & \text{if } 0 < P < P_0;\\ (\mu - \frac{\mu}{\nu_R-\nu_P})(P+F), & \text{if } P_0 \le P < P_1;\\ (\mu - \frac{\nu_P\mu}{(\nu_R-1)(\nu_R-\nu_P-1)})P + \frac{\nu_P\mu}{\nu_R-1}F, & \text{if } P_1 \le P < P_2;\\ (\mu - \frac{\mu}{\nu_R})F, & \text{if } P_2 \le P.\end{cases}$$

where $P_0 = (R - \frac{C}{\mu-\Lambda}, 0)^+$, $P_1 = \frac{C}{\mu}\frac{(\nu_R-1)^2}{2\nu_R-1}$ and $P_2 = \frac{C}{\mu}\frac{(\nu_R-1)^2}{\nu_R}$.

Remark 6.4.3 For $\rho \ge 1$, we have

$$S_{un}=\begin{cases}(\mu - \frac{\mu}{\nu_R-\nu_P})(P+F), & \text{if } 0 < P < P_1;\\ (\mu - \frac{\nu_P\mu}{(\nu_R-1)(\nu_R-\nu_P-1)})P + \frac{\nu_P\mu}{\nu_R-1}F, & \text{if } P_1 \le P < P_2;\\ (\mu - \frac{\mu}{\nu_R})F, & \text{if } P_2 \le P.\end{cases}$$

Note that in both cases, $\rho < 1$ and $\rho \ge 1$, the server's revenue function S_{un} is continuous and non-negative in P. Continuity can be easily verified by comparing the left and the right limits of the piecewise function S_{un} evaluated at the points P_0, P_1 and P_2. We present the optimal revenue for the server under the unobservable case in the following theorem:

Theorem 6.4.2 *When $\rho < 1$, the optimal revenue for server is given by $S^*_{un} = \max\{S_1, S_2, S_3, S_4\}$:*

$$S_1=\begin{cases}(f_1(P_1), 0)^+ \cdot \mathbb{I}_{\{P_1\in\Omega_1\}}, & \text{if } \overline{y}_1 < P_1;\\ (f_1(\overline{y}_1), 0)^+ \cdot \mathbb{I}_{\{\overline{y}_1\in\Omega_1\}}, & \text{if } P_1 < \overline{y}_1 \le P_2.\end{cases}$$

$$S_2=\begin{cases}f_2(P_0) \cdot \mathbb{I}_{\{P_0\in\Omega_1\}}, & \text{if } \overline{x} \le P_0 \text{ and } F < (\mu R/C - 1)R;\\ f_2(\overline{x}) \cdot \mathbb{I}_{\{\overline{x}\in\Omega_1\}}, & \text{if } P_0 < \overline{x} \le P_1 \text{ and } F < (\mu R/C - 1)R;\\ f_2(P_1) \cdot \mathbb{I}_{\{P_1\in\Omega_1\}}, & \text{if } P_1 < \overline{x}_1 \text{ and } F \ge (\mu R/C - 1)R.\end{cases}$$

$$S_3 = \Lambda(P_0 + F) \cdot \mathbb{I}_{\{P_0\in\Omega_1\}}.$$

$$S_4 = \begin{cases} f_1(\overline{P}), & \textit{if } \Omega_1 \subset (0, P_0); \\ f_2(\overline{P}), & \textit{if } \Omega_1 \subset (P_0, P_1); \\ \Lambda(\overline{P} + F), & \textit{if } \Omega_1 \subset (P_1, P_2). \end{cases}$$

where $\overline{P} = \frac{\rho^2 C}{\mu - \Lambda}$, $\overline{x} = R - \sqrt{(R+F)C/\mu}$, $\overline{y}_1$ *is the smaller solution of* $f_1'(x) = 0$, $A = \frac{\mu}{\nu_R - 1}$, $f_1(x) = (\mu - \frac{x\mu A}{((R-x)\mu - C)})x + Ax\mu F/C$, $f_2(x) = (\mu - \frac{C}{R-x})(x+F)$, $\Omega_1 = \{P | 0 < \nu_P \le \frac{\rho^2}{1-\rho}\}$, *and* $\mathbb{I}_{\{\cdot\}}$ *is the indicator function.*

Remark 6.4.4 When $\rho \ge 1$, the optimal revenue for the server is given by $S^*_{un} = \max\{S_1', S_2'\}$, where

$$S_1' = \begin{cases} (f_1(P_1), 0)^+, & \text{if } \overline{y}_1 < P_1; \\ (f_1(\overline{y}_1), 0)^+, & \text{if } P_1 < \overline{y}_1 \le P_2. \end{cases}$$

$$S_2' = \begin{cases} f_2(\overline{x}), & \text{if } \overline{x} \le P_1; \\ f_2(P_1), & \text{if } \overline{x} > P_1. \end{cases}$$

We now compare our findings to those in the Gavirneni and Kulkarni (2016) which also studies a priority queueing system with balking customers. In our paper, customers balk when their expected utility is negative. In particular, we model the service reward R as an explicit parameter. As R increases, it affects customers' utility and can lead to more or less joining behavior in equilibrium. However, in the setting of Gavirneni and Kulkarni (2016), customers balk when their expected utility is lower (or equivalently cost is higher) compared to that in a first in, first out (FIFO) system (no balking or priorities) with the same service reward—customers' joining/balking decisions are thus not affected by the absolute value of the service reward. As a result, even though in both papers an increasing service reward leads to a higher probability of priority purchase among the joining customers, it always increases a firm's revenue in the case of Gavirneni and Kulkarni (2016), but not necessarily in our setting (because there can be less joining customers overall).

The fundamental difference between the two papers is thus driven by the modeling trade-off where the work in this chapter focuses on revenue and pricing and theirs on cost and social welfare. The next subsection will study observable systems. Then through comparing the unobservable and observable settings, it provides managerial insights as to when the service provider is beneficial from revealing the queue length information.

6.4.3 The Observable Model

In the observable setting, all the arriving customers can observe the state of the system including the number of customers at the server, in the priority queue and in the regular (non-priority) queue. Since $R > C/\mu$, an arriving customer who finds the server idle will immediately obtain service without purchasing priority. Otherwise, this customer decides whether or not to join the system, and if they join, whether or not to purchase priority, in order to maximize expected utility. Clearly, these decisions are state-dependent, which is different from the unobservable model.

6.4.3.1 Equilibrium Strategies

A customer strategy is a mapping from the system states to decisions. For simplicity, we will only consider pure and symmetric equilibria for the observable case. Although all the customers adopt the same strategy in equilibrium, when they arrive at different states of the system they can end up making different joining and priority purchasing decisions. Furthermore, unlike a FCFS queue, customers' decision making is affected by future arrivals. For example, the expected waiting time of type-o customers is not only affected by other regular customers queued in front of them, but also by type-p customers who are either in the system or will arrive in the future to overtake them.

Let us denote the state of the server by I, i.e., $I = 0$ indicates that the server is idle and $I = 1$ is occupied. Let N_o and N_p denote the number of customers in the regular queue and in the priority queue, respectively (not including the person at the server if any). Then, $N \triangleq I + N_o + N_p$ corresponds to the total number of customers in the system.

Let $\phi_{\mathcal{A}}(1, j+1, k)$ denote the expected waiting time (excluding service time) of a customer who finds the system in state $I = 1$, $N_o = j$ and $N_p = k$ upon arrival and chooses to be an ordinary customer, when strategy $\mathcal{A}$ is adopted by all other customers. This customer's expected utility without and with purchasing the priority are given by

$$S_o(1, j, k) \triangleq R - [\phi_{\mathcal{A}}(1, j+1, k) + \frac{1}{\mu}]C;$$

$$S_p(1, j, k) \triangleq R - [\frac{(k+2)C}{\mu} + P].$$

Naturally, this customer, if joining, will purchase priority if and only if $S_o(1, j, k) \leq S_p(1, j, k)$, or equivalently $\nu_P \leq \mu\phi_{\mathcal{A}}(1, j+1, k) - k - 1$. The next lemma helps pin down the possible form of strategy $\mathcal{A}$ if it were to sustain an equilibrium.

Lemma 6.4.4 *For any strategy $\mathcal{A}$ and $j, k \geq 0$, if $\nu_P \leq \mu\phi_{\mathcal{A}}(1, j+1, k) - k - 1$, we must have $\nu_P \leq \mu\phi_{\mathcal{A}}(1, j+1, k+1) - k - 2$.*

Leaving out the balking option for the moment, Lemma 6.4.4 states that when strategy $\mathcal{A}$ is adopted by all other customers and given any fixed number of customers in the regular queue (j), if an arrival prefers to purchase priority when there are k customers in the priority queue, then the same customer would also prefer to purchase priority when there are more than k customers in the priority queue. If strategy $\mathcal{A}$ were an equilibrium, it must be a best response to itself. Thus, there would be some threshold queue length of the priority queue, beyond which strategy $\mathcal{A}$ would specify the customers to either join the priority queue or balk (since becoming a type-o customer is dominated by becoming a type-p customer). With Lemma 6.4.4, we are able to prove the unique form of an equilibrium strategy, which is surprisingly clean, in the following theorem.

Theorem 6.4.3 *Any pure symmetric equilibrium strategy of the game can be described by two thresholds $m, n \geq 0$ where the strategy, written as $[m, n]$, specifies a customer to*

$$\begin{cases} \textit{join the regular queue} & \textit{if } 0 \leq N \leq m; \\ \textit{join the priority queue} & \textit{if } m < N \leq m + n; \\ \textit{balk} & \textit{if } N = m + n + 1. \end{cases} \tag{6.25}$$

Theorem 6.4.3 provides a few important insights. (i) If the system is not busy ($0 \leq N \leq m$), customers join the system and do not need to purchase priority. As the system gets busier ($m < N \leq m + n$), customers would still join but it would be wiser to purchase priority to reduce the waiting cost. Finally, as the system gets too crowded ($N > m + n$), customers do not join any more. (ii) A pure equilibrium strategy must be of the type $[m, n]$ as described in the theorem, i.e. other pure equilibrium forms do not exist. However, for a given set of system parameters $\{\Lambda, \mu, C, P, R\}$, the equilibrium may not be unique, i.e. m may not be unique (see Theorem 6.4.4). (iii) Hassin and Haviv (1997) show by a counter-example that a threshold strategy is not the only pure equilibrium type when the priority discipline is preemptive, while in our case non-preemptive priority is studied and we have a different result. It is easy to verify that the example in their paper is not an equilibrium strategy in our model. To further elaborate on why non-threshold equilibrium can exist in preemptive queues but not in non-preemptive queues, let us consider a particular customer at the server when it is busy. Under the non-preemptive scheme, the decisions of future arrivals do not depend on whether the status of this customer is a priority or an ordinary customer. In contrast, under the preemptive scheme, an arrival benefits more from a priority purchase if the customer currently being served is an ordinary customer (the new arrival can preempt) as opposed to a priority customer (the new arrival cannot preempt). As a consequence, customer decisions not only depend on the queue length but also on the status of the customer at the server, leading to possible non-threshold equilibria.

Remark 6.4.5 Under an equilibrium strategy $[m, n]$, whenever there is any customer waiting in the priority queue (i.e., $N_p > 0$), the server must be busy (i.e., $I = 1$) and there must be m customers in the non-priority queue (i.e., $N_o = m$), so we have $N = I + N_o + N_p > m + 1$. It implies that $N_p = 0$ whenever $N \leq m + 1$. Therefore, when $m, n > 0$, Eq. (6.4.5) is equivalent to

$$\begin{cases} \text{join the regular queue} & \text{if } N_o < m; \\ \text{join the priority queue} & \text{if } N_o = m \text{ and } N_p < n; \\ \text{balk} & \text{if } N_p = n. \end{cases}$$

The decision process of the system under the observable setting is illustrated in Fig. 6.5, according to Theorem 6.4.3. Note that the expected waiting time of a type-p customer is not affected by type-o customers and its expression is straightforward. The challenge is to find the expected waiting time of a type-o customer who might be overtaken by future arrivals.

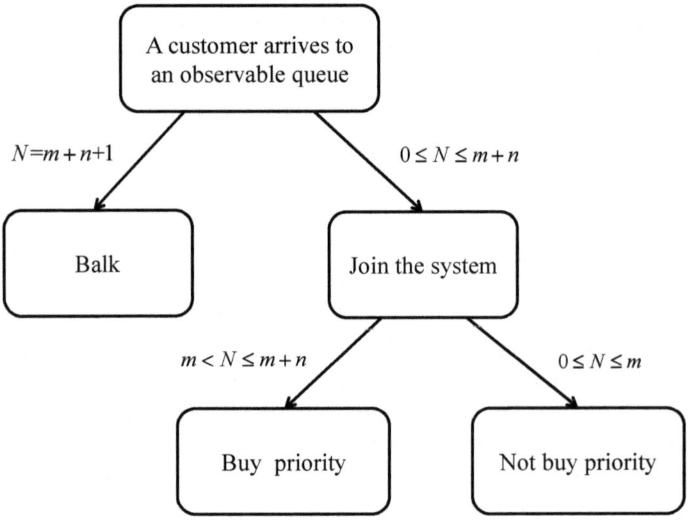

Fig. 6.5 The decision process under the observable setting

Under strategy $[m, n]$, the stochastic process $\{(I(t), N_o(t), N_p(t)), t \geq 0\}$ is a three-dimensional continuous-time Markov chain where $N_p(t) = 0$ whenever $N_o(t) < m$. Let A_k be the expected waiting time for the mth customer in the regular queue when there are k priority customers, $k \leq n$. Define $B_k \triangleq A_k - A_0$ where B_k is the expected time it takes the system to reach state $(1, m, 0)$ from state $(1, m, k)$ for the first time. A_k satisfies the following equations:

$$A_k = \frac{1}{\mu + \Lambda} + \frac{\Lambda}{\mu + \Lambda} A_{k+1} + \frac{\mu}{\mu + \Lambda} A_{k-1} \quad \text{for } k = 1, 2, ..., n-1; \tag{6.26}$$

$$A_n = A_{n-1} + \frac{1}{\mu}. \tag{6.27}$$

By solving (6.26) and (6.27), we can obtain the following lemma.

Lemma 6.4.5 *When all customers adopt strategy* $[m, n]$,

$$B_k = \frac{k}{\mu - \Lambda} + (\frac{1}{\mu} - \frac{1}{\mu - \Lambda}) \frac{\rho^{n-k} - \rho^n}{1 - \rho} \quad \textit{for } k = 1, 2, ..., n. \tag{6.28}$$

Under the strategy $[m, n]$, we define $H_{i,j}^{[m,n]}$ as the expected waiting time for an ordinary customer who is the ith in the regular queue with j other ordinary customers behind them, when the server is occupied and the priority queue is empty. We have $i \geq 0$, $j \geq 0$ and $i + j \leq m$.

When there are k customers in the priority queue, the expected residual waiting time for the ith customer in the regular queue would be $H_{i,m-i}^{[m,n]} + B_k$. To investigate the expected waiting time for the customers, it is then sufficient to derive $H_{i,j}^{[m,n]}$ for $i \geq 0$, $j \geq 0$ and $i + j \leq m$. Similar to Hassin and Haviv (1997), we have the following set of relations:

$$H_{i,m-i}^{[m,n]} = \frac{1}{\mu + \Lambda} + \frac{\Lambda}{\mu + \Lambda}(B_1 + H_{i,m-i}^{[m,n]}) + \frac{\mu}{\mu + \Lambda} H_{i-1,m-i}^{[m,n]}, \tag{6.29}$$

$$H_{1,a}^{[m,n]} = \frac{1}{\mu + \Lambda} + \frac{\Lambda}{\mu + \Lambda} H_{1,a+1}^{[m,n]}, \tag{6.30}$$

$$H_{j,k}^{[m,n]} = \frac{1}{\mu + \Lambda} + \frac{\Lambda}{\mu + \Lambda} H_{j,k+1}^{[m,n]} + \frac{\mu}{\mu + \lambda} H_{j-1,k}^{[m,n]}. \tag{6.31}$$

where $i = 1, 2, ..., m$; $a = 0, 1, ..., m - 2$; $j = 1, 2, ..., m - k - 1$ and $k = 0, 1, ..., m - j - 1$. $H_{i,m-i}^{[m,n]}$ can be calculated recursively using the following algorithm.

When $m = 1$, *we have* $H_{1,0}^{[1,n]} = \frac{1+\Lambda B_1}{\mu}$ *by noticing that* $H_{0,m}^{[m,n]} = 0$ *in* (6.29).

When $m > 1$, *we can first derive the initializations* $H_{1,a}^{[m,n]}$ *for* $a = 0, 1, ..., m - 2$ *from* (6.30), *then derive* $H_{i,j}^{[m,n]}$ *recursively as follows:*

For $i = 2$, *calculate* $H_{i,m-i}^{[m,n]} = \frac{1+\Lambda B_1}{\mu} + H_{i-1,m-i}^{[m,n]}$ *recursively until* $i = m - 1$.

For $k = m - i - 1$, *calculate* $H_{i,k}^{[m,n]} = \frac{1}{\mu+\Lambda} + \frac{\Lambda}{\mu+\Lambda} H_{i,k+1}^{[m,n]} + \frac{\mu}{\mu+\Lambda} H_{i-1,k}^{[m,n]}$ *recursively until* $k = 0$.

Calculate $H_{m,0}^{[m,n]} = \frac{1+\Lambda B_1}{\mu} + H_{m-1,0}^{[m,n]}$.

A priority customer incurs an expected sojourn time of at least $2/\mu$ units of time. Otherwise, the server would have been idle when this customer arrived, and the customer should have not purchased the priority. Thus when $P \geq R - \frac{2C}{\mu} \Leftrightarrow \nu_P \geq \nu_R - 2$, it is too expensive for a customer to purchase priority. In this case, our system is reduced to the model in Naor (1969). The equilibrium strategy $[m, n]$ also degenerates into the single-threshold balking policy in Naor's model where $m_e = \lfloor \nu_R \rfloor - 1$ (and $n_e = 0$).

Now consider $\nu_P < \nu_R - 2$. An equilibrium strategy $[m, n]$ given in Eq. (6.4.5) specifies that being the n_e^{th} customer in the priority queue has a positive expected payoff, but not being the $(n_e + 1)^{th}$ customer. For a type-p customer, their expected waiting time is purely determined by the other priority customers queued in front of them and the person at the server, but not any type-o customers. Thus, we must have $\frac{(n_e+1)C}{\mu} + P \leq R < \frac{(n_e+2)C}{\mu} + P$. It follows that $n_e = \lfloor \nu_R - \nu_P \rfloor - 1$, which is deterministic for any given cost-reward parameters (R, C, P) and independent of the other threshold m. To find m_e, we need to separate the cases $N_p(t) = 0$ and $N_p(t) > 0$.

Case (1): $N_p(t) = 0$.

(i) If $\mu H_{j,0}^{[m,n_e]} > \nu_P + 1$ for some $j \in [0, m]$ and all the other customers use strategy $[m, n_e]$, then for a new arrival who observes the state $(1, j - 1, 0)$, purchasing priority would be the best response. Thus, strategy $[m, n_e]$ is not a best response to itself and not a symmetric equilibrium.

(ii) Suppose $\mu H_{j,0}^{[m,n_e]} \leq \nu_P + 1$ for all $j \in [0, m]$ and all the other customers use strategy $[m, n_e]$. Consider a new arrival. If this customer finds less than m customers in the regular queue upon arrival, joining the regular queue would be the best response. On the other hand, if $N_o = m$, the best response would be purchasing priority if and only if $\mu H_{m+1,0}^{[m,n_e]} > \nu_P + 1$. Therefore, $[m, n_e]$ is a best response to itself and an equilibrium if and only if $\mu H_{m+1,0}^{[m,n_e]} > \nu_P + 1$ (and $\mu H_{j,0}^{[m,n_e]} \leq \nu_P + 1$ for all $j \in [0, m]$).

Case (2): $N_p(t) > 0$.

If $[m, n_e]$ is an equilibrium, we must have $\mu H_{m+1,k}^{[m,n_e]} > \nu_P + k + 1$ such that $[m, n_e]$ is a best response to itself. However, the condition identified in Case (1)/(ii), i.e., $\mu H_{m+1,0}^{[m,n_e]} > \nu_P + 1$, already implies that $\mu H_{m+1,k}^{[m,n_e]} > \nu_P + k + 1$, because $H_{m+1,k}^{[m,n_e]} = H_{m+1,0}^{[m,n_e]} + B_k > H_{m+1,0}^{[m,n_e]} + k/\mu$ and it leads to $\mu H_{m+1,k}^{[m,n_e]} \geq \mu H_{m+1,0}^{[m,n_e]} + k > \nu_P + k + 1$. Therefore, $[m, n_e]$ is an equilibrium strategy if and only if $\mu H_{j,0}^{[m,n_e]} \leq \nu_P + 1$ for all $j \in [0, m]$ and $\mu H_{m+1,0}^{[m,n_e]} > \nu_P + 1$.

Lemma 6.4.6 *The expected waiting time of ordinary customers $H_{i,j}^{[m,n_e]}$ has the following properties:*

(i) Under $[m, n_e]$, $H_{i,0}^{[m,n_e]}$ is strictly increasing in i for $i \geq 0$. In particular, we have $H_{i+1,0}^{[m,n_e]} - H_{i,0}^{[m,n_e]} > \frac{1}{\mu}$.
(ii) $H_{m,0}^{[m,n_e]}$ is strictly increasing in m for $m \geq 1$. In particular, we have $H_{m+1,0}^{[m+1,n_e]} - H_{m,0}^{[m,n_e]} > \frac{1}{\mu}$.
(iii) $H_{m+1,0}^{[m+1,n_e]} - H_{m,0}^{[m,n_e]} < \frac{1+\Lambda B_1}{\mu}$.

Using the fact that $H_{1,0}^{[1]} = \frac{1+\Lambda B_1}{\mu}$ from (6.29), we have the following main result of the paper that establishes the full equilibrium structure of the system under the observable setting.

Theorem 6.4.4 *When $\nu_P \geq \overline{\nu_P} \triangleq \nu_R - 2$, $[m_e, n_e] = [\lfloor \nu_R \rfloor - 1, 0]$ is the unique equilibrium. When $\nu_P < \underline{\nu_P} \triangleq \Lambda B_1$, $[m_e, n_e] = [0, \lfloor \nu_R - \nu_P \rfloor - 1]$ is the unique equilibrium. Otherwise, when $\underline{\nu_P} < \nu_P < \overline{\nu_P}$, any m that satisfies*

$$\nu_P - (1 + \Lambda B_1) \leq \mu H_{m,0}^{[m,n_e]} - 1 < \nu_P \tag{6.32}$$

for $n_e = \lfloor \nu_R - \nu_P \rfloor - 1$ gives an equilibrium strategy $[m, n_e]$. Furthermore, the number of equilibrium strategies in this case is at least one and at most $\lfloor \frac{1-\rho^{\lfloor \nu_R - \nu_P \rfloor}}{1-\rho} \rfloor$.

Table 6.1 summarizes the equilibrium types given by Theorem 6.4.4. (i) When the price of the priority premium is too high ($\nu_P \geq \overline{\nu_P}$), all customers avoid buying priority, and the system and the equilibrium strategy coincide with those in Naor (1969). (ii) When the price of the priority premium is low ($\nu_P < \underline{\nu_P}$), all customers

Table 6.1 Equilibrium type versus $\nu_P = \frac{P\mu}{C}$

	$\nu_P \leq \underline{\nu_P}$	$\underline{\nu_P} < \nu_P < \overline{\nu_P}$	$\nu_P \geq \overline{\nu_P}$
Equilibrium type	One threshold $[0, n]$	Two thresholds $[m, n]$	One threshold $[m, 0]$

choose to be priority customers (except those who arrive at an idle server). The equilibrium again degenerates to a single-threshold strategy.

(iii) When $\underline{\nu_P} < \nu_P < \overline{\nu_P}$, multiple equilibria in the form of two-threshold type $[m, n]$ may exist. See Fig. 6.6a, c for an illustration. Define $\mathscr{M} \triangleq \{m \text{ s.t. } [m, n_e] \text{ is an equilibrium}\}$. It can be seen that $\mathscr{M}$ contains a continuous sequence. That is, if there are k elements in $\mathscr{M}$ denoted by $m_e^{\min} \triangleq m_e^{(1)} < m_e^{(2)} < \cdots < m_e^{(k)} \triangleq m_e^{\max}$, then we have $m_e^{(i+1)} = m_e^{(1)} + i$ for $i = 1, 2, \ldots, k-1$. Furthermore, $m_e^{\max}$ satisfies $\mu H_{m_e,0}^{[m_e,n_e]} < \nu_P + 1 < \mu H_{m_e+1,0}^{[m_e+1,n_e]}$, and $m_e^{\min}$ satisfies $\mu H_{m_e-1,0}^{[m_e-1]} < \nu_P + 1 - (1 + \Lambda B_1) < \mu H_{m_e,0}^{[m_e,n_e]}$. The uniqueness of $m_e^{\max}$ and $m_e^{\min}$ can be verified through (i) and (ii) of Lemma 6.4.6.

We note that (6.4.4) is a necessary and sufficient condition for an equilibrium threshold m_e. Adiri and Yechiali (1974) identified a single equilibrium in Eq. (20) of their paper, which is only a sufficient condition. Their equilibrium coincides with $m_e^{\max}$ in this chapter.

In the case of multiple equilibria, we seek to identify the Pareto-dominant equilibrium, denoted by $[m_e^*, n_e]$, just as in the unobservable case. Under strategy $[m, n_e]$, the expected individual utility of the customers is given by

$$V(m) = (1 - P_{balk})(R - \frac{C}{\mu}) - p_{1,0,0}\left(\sum_{i=1}^{m} \rho^{i-1} C H_{i,0}^{[m,n_e]} + \rho^m \sum_{i=1}^{n_e} \rho^{i-1}(P + \frac{iC}{\mu})\right),$$

$$= \frac{1-\rho^{m+n_e+1}}{1-\rho^{m+n_e+2}}(R - \frac{C}{\mu}) - \frac{1-\rho}{1-\rho^{m+n_e+2}}\left(\sum_{i=1}^{m} \rho^{i} C H_{i,0}^{[m,n_e]} + \rho^m \sum_{i=1}^{n_e} \rho^{i}(P + \frac{iC}{\mu})\right),$$

where P_{balk} and $p_{1,0,0}$ will be defined in the later Corollary 6.4.1 and Theorem 6.4.7 respectively. Since the Pareto-dominant equilibrium $[m_e^*, n_e]$ maximizes consumer welfare, we have $m_e^* = \arg\max_{m \in \mathscr{M}} V(m)$. To show that m_e^* is unique, we will establish the unimodality of $V(\cdot)$ in Theorem 6.4.5 based on the following lemma.

Lemma 6.4.7 *For $m \geq 1$, we have the following inequality*

$$\sum_{i=1}^{m+2} \rho^i H_{i,0}^{[m+2,n_e]} - \sum_{i=1}^{m+1,n_e} \rho^i H_{i,0}^{[m+1,n_e]} > \rho\left(\sum_{i=1}^{m+1} \rho^i H_{i,0}^{[m+1,n_e]} - \sum_{i=1}^{m} \rho^i H_{i,0}^{[m,n_e]}\right) \tag{6.33}$$

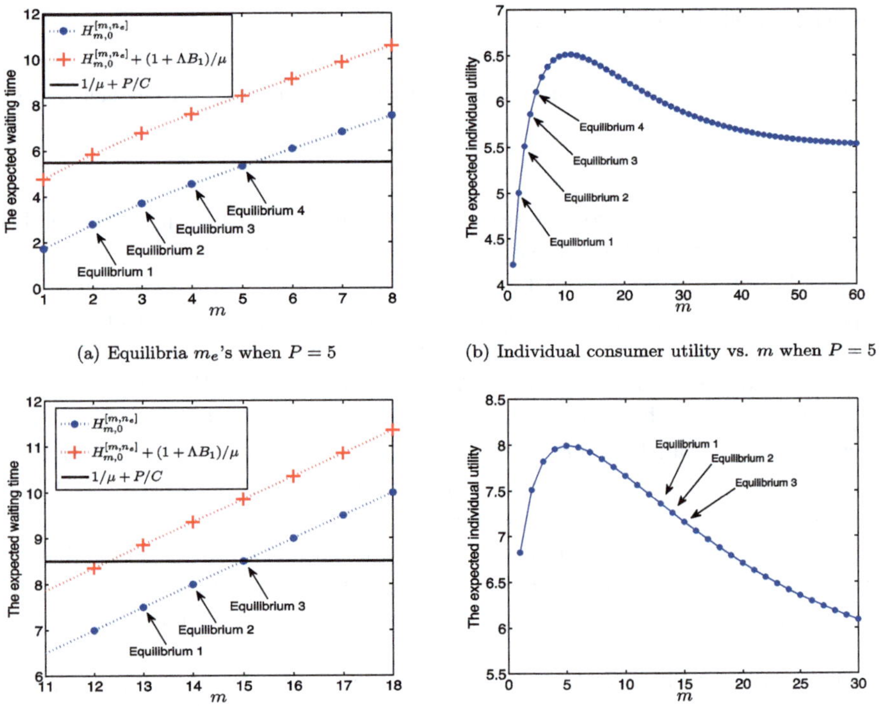

(a) Equilibria m_e's when $P = 5$

(b) Individual consumer utility vs. m when $P = 5$

(c) Equilibria m_e's when $P = 8$

(d) Individual consumer utility vs. m when $P = 8$

Fig. 6.6 Illustration of multiple equilibria ($R = 10$, $C = 1$, $\mu = 2$, $\Lambda = 1.8$)

Theorem 6.4.5 *Among* $\{[m_e, n_e]\}_{m_e \in \mathscr{M}}$, *there exists a unique Pareto-dominant equilibrium. In particular,* $V(m_e)$ *is unimodal in* m_e.

When multiple equilibria exist under the $\underline{\nu_P} < \nu_P < \overline{\nu_P}$ case, Theorem 6.4.5 indicates that we can always find a unique Pareto-dominant equilibrium. Figure 6.6 provides a couple of illustrations. When $P = 5$ (c.f. (a) and (b)), the equilibrium set is $\mathscr{M} = \{2, 3, 4, 5\}$ and we have $V(2) < V(3) < V(4) < V(5)$, so the Pareto-dominant equilibrium threshold for m is $m_e^* = m_e^{max} = 5$. Similarly, when $P = 8$ (c.f. (c) and (d)), $\mathscr{M} = \{13, 14, 15\}$ and $m_e^* = m_e^{min} = 13$. In general, we can use the following algorithm to identify the unique Pareto-dominant equilibrium $[m_e^*, n_e]$ given system parameters Λ, μ, C, P, R.

Step 1. $n_e = \lfloor \nu_R - \nu_P \rfloor - 1$.
Step 2. For $m = 1$ do $m = m + 1$
  if $\mu H_{m,0}^{[m,n_e]} \leq \nu_P - \Lambda B_1$,
  Else, $m_e^{min} = m$, *come to Step 3.*
Step 3. For $m = m_e^{min}$ do $m = m + 1$
  if $\mu H_{m+1,0}^{[m+1,n_e]} \leq \nu_P + 1$,
  Else, $m_e^{max} = m$, *come to Step 4.*
Step 4. If $V(m_e^{min} + 1) - V(m_e^{min}) < 0$,
  $m_e^* = m_e^{min}$, *end.*
  If $V(m_e^{max}) - V(m_e^{max} - 1) > 0$,
  $m_e^* = m_e^{max}$, *end.*
  Else, come to Step 5.
Step 5. For $m = m_e^{min} + 1$, *do* $m = m + 1$
  If $V(m + 1) > V(m)$,
  Else $m_e^* = m$, *end.*

In a special case covered in the following theorem, we always have $m_e^* = m_e^{max}$, which was the equilibrium threshold identified in Adiri and Yechiali (1974).

Theorem 6.4.6 *When* $\rho < 1$, *there exists a threshold* $\overline{R} > 0$ *s.t. when* $R > \overline{R}$, m_e^{max} *is the Pareto-dominant equilibrium.*

Under Theorem 6.4.6's condition, i.e., when $\rho \downarrow$ (so waiting time for priority customers is relatively short) and the service value $R \uparrow$, new arrivals can afford to purchase priority to overtake customers waiting in the regular queue. As a consequence, the equilibrium regular-queue thresholds m_e's all shrink so the additional value of overtaking is neutralized. When $R > \overline{R}$, all the m_e's are now small enough to be located to the left of the peak of the unimodally-distributed consumer welfare function $V(m)$, and m_e^{max} becomes the Pareto-dominant equilibrium. In this case, the equilibrium threshold on the priority queue, $n_e = \lfloor \nu_R - \nu_P \rfloor - 1$, increases in μ or R as expected.

Similarly, we can investigate the impact of Λ and P on the Pareto-dominant equilibrium, m_e^*. In Fig. 6.7, we fix $R = 10$, $C = 1$ and $\mu = 2$ while varying Λ (or equivalently ρ) and P, and the created plane can be divided into four regions: (i) $m_e^* =$ *unique* m_e (region I); (ii) $m_e^* = m_e^{max}$ (region II); (iii) $m_e^* = m_e^{min}$ (region III); (iv) $m_e^* \in (m_e^{min}, m_e^{max})$ (region IV).

When multiple equilibria exist, we observe that m_e^* shifts towards m_e^{min} as $\rho \uparrow$ or $P \uparrow$, and towards m_e^{max} as $\rho \downarrow$ or $P \downarrow$. When $\rho \uparrow$ (or $P \uparrow$), the regular queue is more congested (or longer), so joining customers generate more negative externalities. As a consequence, customers over-join the regular queue in equilibrium compared to the socially optimal outcome. The Pareto-dominant equilibrium threshold, which maximizes customers' social welfare, shifts towards m_e^{min}. The same phenomenon was identified by Naor (1969) in a FCFS queue. On the other hand, the impact of $\rho \downarrow$ or $P \downarrow$ is similar to that of $R \uparrow$ as discussed earlier.

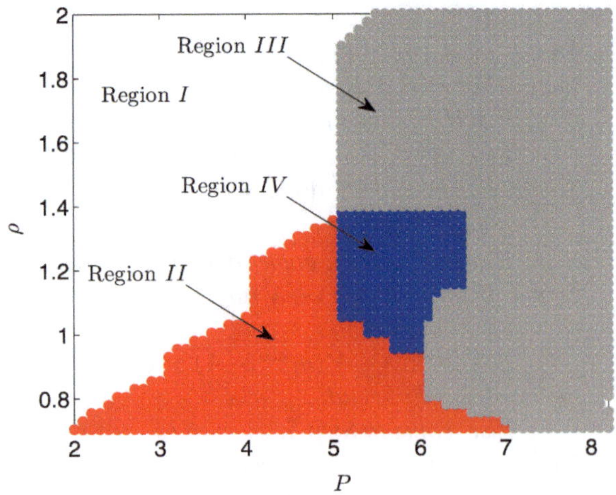

Fig. 6.7 Forms of Pareto-dominant equilibrium

6.4.3.2 Server's Revenue

Like in the unobservable setting, we model a Stackelberg game between the server and the customers, where the server is able to maximize its expected revenue by optimizing over price P for the priority premium. The basic service fee F is fixed. Let $p_{i,j,k}$ denote the steady-state probability of state (i, j, k) for the underlying queueing process $\{(I(t), N_o(t), N_p(t)), t \geq 0\}$. Recall that $N_p(t) = 0$ whenever $N_o(t) < m$. The stationary distribution of the system can be derived as follows.

Theorem 6.4.7 *Under strategy* $[m, n]$*, we have*

$$p_{0,0,0} = \frac{1-\rho}{1-\rho^{m+n+2}},$$

$$p_{1,j,0} = \frac{\rho^{j+1}-\rho^{j+2}}{1-\rho^{m+n+2}} \quad \textit{for } j = 0, 1, 2, ..., m-1,$$

$$p_{1,m,k} = \frac{\rho^{m+k+1}-\rho^{m+k+2}}{1-\rho^{m+n+2}} \quad \textit{for } k = 0, 1, 2, ..., n.$$

We can then derive the unconditional probability of an arrival choosing to balk, join as a regular customer and join as a priority customer, denoted by P_{balk}, P_o and P_p, respectively. Clearly, $P_{balk} + P_o + P_p = 1$.

Corollary 6.4.1 *Under strategy* $[m, n]$*, we have*

$$P_{balk} = \frac{\rho^{m+n+1}-\rho^{m+n+2}}{1-\rho^{m+n+2}}; \quad P_o = \frac{1-\rho^{m+1}}{1-\rho^{m+n+2}}; \quad P_p = \frac{\rho^{m+1}-\rho^{m+n+1}}{1-\rho^{m+n+2}}.$$

Furthermore, P_{balk} *and* P_p *decrease and* P_o *increases in* m*, while* P_{balk} *and* P_o *decrease and* P_p *increases in* n*.*

Corollary 6.4.1 is intuitive, as the balking probability is expected to decrease in m or n (recall that balking happens when $N > m + n$). Similarly, the probability of becoming an ordinary or a priority customer increases with the corresponding threshold value, m or n.

Under the Pareto-dominant equilibrium $[m_e^*, n_e]$, the percentage of arrivals who purchase priority is $P_p = \frac{\rho^{m_e^*+1} - \rho^{m_e^*+n_e+1}}{1-\rho^{m_e^*+n_e+2}}$, and the server's revenue (rate) under the observable case, denoted by S_{ob}, is given as:

$$S_{ob} = \Lambda \left[\frac{\rho^{m_e^*+1} - \rho^{m_e^*+\lfloor \nu_R - \nu_P \rfloor - 1}}{1 - \rho^{m_e^*+\lfloor \nu_R - \nu_P \rfloor}} \right] P + \Lambda \left[\frac{1 - \rho^{m_e^*+\lfloor \nu_R - \nu_P \rfloor - 1}}{1 - \rho^{m_e^*+\lfloor \nu_R - \nu_P \rfloor}} \right] F. \tag{6.34}$$

Unfortunately, it is too complicated to derive the explicit expression for the server's maximum revenue when the service provider can adjust P. However, we will examine it numerically in Sect. 6.4.4 and compare the performance to that in the unobservable case.

6.4.4 Numerical Experiments

In this section, we present numerical results using graphs and tables for both the unobservable and observable cases. In particular, we first examine the impact of the model's parameters (including Λ, C, P, R) on the equilibrium behavior of the customers (Section §6.4.4.1). We focus our discussion on the Pareto-dominant equilibrium, denoted by $[m_e, n_e]$ with a slight abuse of notation (instead of $[m_e^*, n_e]$). Then, we study and compare revenue maximization of the server under the two information cases (Section §6.4.4.2). Unless otherwise noted, the service rate is $\mu = 2$, unit waiting time is $C = 1$, and the basic service fee is $F = 1$ for all of the examples provided in §6.4.4.1 and §6.4.4.2. Finally, we apply our results to a mini case study (Papa John's Pizza) where parameters for the model are extrapolated from real-world information that is publicly available (Sect. 6.4.4.3).

6.4.4.1 On the Equilibrium Strategies

Fig. 6.8 illustrates the impact of $\nu_R = R\mu/C$ on the effective arrival rate λ_e under the unobservable case with $\Lambda = 2.5$ and $\nu_P = 6$, where $\lambda^* = \rho^*\mu$ and $\lambda' = \rho'\mu$ are as defined in Theorem 6.4.1. In the graph, the light grey area corresponds to the joining rate at the equilibrium of the regular type-o customers, while the dark grey area corresponds to the priority type-p customers.

It is clear that the proportion of type-p customers increases as $\nu_R \uparrow$. When the reward R is sufficiently big (i.e., $\nu_R = R\mu/C \geq 1/(1-\rho') + \nu_P$), all joining customers end up purchasing priority. It is interesting to note that the joining rate of the regular customers first increases then decreases. This is consistent with Case (3) of

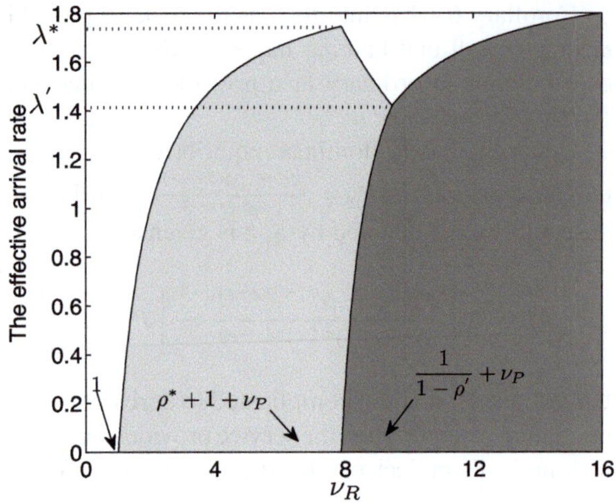

Fig. 6.8 Effective arrival rate versus ν_R when $\Lambda = 2.5$ and $\nu_P = 6$ under the unobservable case

Theorem 6.4.1 which predicts that it increases in ν_R over $1 < \nu_R \leq \rho^* + 1 + \nu_P$ and decreases in ν_R over $\rho^* + 1 + \nu_P < \nu_R \leq 1/(1 - \rho') + \nu_P$. As a result, the effective arrival rate (combining both the type-o and type-p customers) at equilibrium is not monotone in ν_R or R, which is different from the results in FCFS queueing systems such as Naor (1969) and Edelsen and Hildebrand (1975); the effective arrival rate always increases in ν_R in those papers. The difference is driven by the fact that customers are self-interested. As $\nu_R \uparrow$, an arriving customer has more incentives to purchase priority, which explains the monotonicity of the joining rate of the priority customers in ν_R. However, this imposes negative externalities to the customers in the regular queue, and the proportion of these type-o customers decreases and eventually all of them vanish.

Next, we discuss the impact of system parameters on the equilibrium strategy $[m_e, n_e]$ for the observable case in Table 6.2. When P is sufficiently small (such as $P = 3$ in this example), customers will always prefer to purchase priority because of the low premium (unless the server is idle upon arrival), and the equilibrium becomes a single threshold strategy with n. These equilibria are highlighted in red in the table.

Table 6.2 Equilibrium strategy $[m_e, n_e]$ under the observable case with $\Lambda = 1.8$ and different R, P

$[m_e, n_e]$	P = 3	P = 4	P = 5	P = 6	P = 7	P = 8	P = 9	P = 10
R = 10	[0, 12]	[3, 10]	[5, 8]	[8, 6]	[9, 4]	[15, 2]	[18, 0]	[18, 0]
R = 11	[0, 14]	[2, 12]	[4, 10]	[6, 8]	[9, 6]	[11, 4]	[17, 2]	[20, 0]
R = 12	[0, 16]	[2, 14]	[3, 12]	[5, 10]	[7, 8]	[11, 6]	[12, 4]	[19, 2]
R = 13	[0, 18]	[2, 16]	[3, 14]	[4, 12]	[6, 10]	[9, 8]	[12, 6]	[14, 4]
R = 14	[0, 20]	[0, 18]	[2, 16]	[4, 14]	[5, 12]	[8, 10]	[10, 8]	[14, 6]

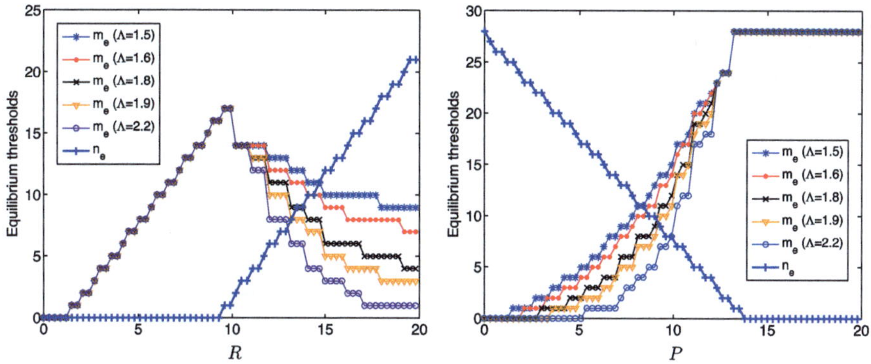

(a) m_e, n_e vs. R when $P = 8$ with different Λ (b) m_e, n_e vs. P when $R = 15$ with different Λ

Fig. 6.9 Equilibrium thresholds under the observable case versus R and P with different Λ

On the other hand, when P is sufficiently big, it will not be worthwhile for customers to purchase priority because of the high premium. Thus, the system is reduced to a FCFS queue without priorities as in Naor (1969), whose equilibrium is a single threshold strategy with m. The equilibria highlighted in blue in the table correspond to this type of a situation. In other cases, the equilibrium strategy follows the two-threshold form $[m, n]$ as predicted by Theorem 6.4.4. Additionally, the equilibrium thresholds m_e and n_e seem to be monotone with respect to R and P. This prompts us to examine their relations with greater details in Fig. 6.9.

In Fig. 6.9a, we study the impact of reward R on m_e and n_e when facing different Λ levels. It is interesting that while n_e (which is the same for all Λ levels according to Theorem 6.4.4) is monotone in R, m_e is not. This is because when $R < P + 2C/\mu$, no customers purchase priority at equilibrium. Thus, $R \uparrow$ drives the increase in customers joining the regular queue ($m_e^* \uparrow$). However, when $R \geq P + 2C/\mu$, $R \uparrow$ induces more and more priority customers, which worsens the expected waiting time for a regular customer. Thus, fewer customers enter the regular queue and $m_e^* \downarrow$. This is the same intuition we discussed for the unobservable case in Fig. 6.8. On the other hand, in Fig. 6.9/(b) we observe that both the equilibrium thresholds m_e and n_e are monotone in the premium P. As $P \uparrow$, customers have less incentives to purchase priority, as opposed to joining the regular queue. As a consequence, $m_e \uparrow$ and $n_e \downarrow$.

6.4.4.2 On the Server's Optimal Revenue

Figure 6.10 provides a comparison of the server's optimal revenue between the observable and unobservable settings as the values of the service reward R and the system load ρ vary. Interestingly, the service provider has an incentive to reveal the state of the system to the customers when the system load is either low or high in order to maximize revenue (corresponding to Region A and Region C of Fig. 6.10 respectively), but it has an incentive to hide it when the load is at a medium level

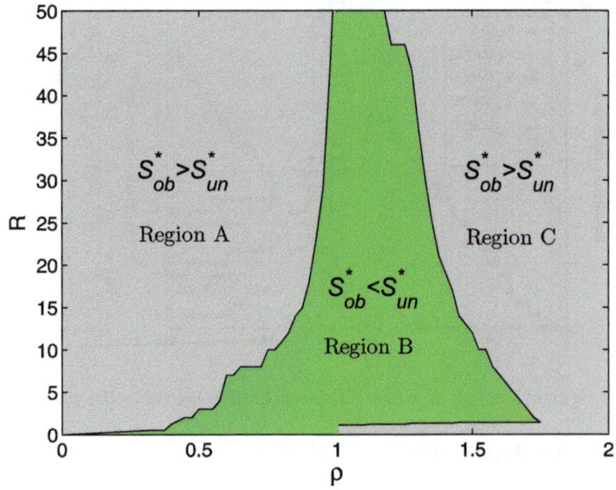

Fig. 6.10 Comparison of optimal revenues under unobservable and observable settings

(Region B in Fig. 6.10). This observation is different from the results presented in Chen and Frank (2004) and Hu et al. (2018) where there are no priority customers. To elaborate on the intuitions, we zoom into each region and plot the server's revenue against price for both the observable and unobservable cases. In particular, we set $R = 6, 8, 10$ and use $\rho = 0.6$, $\rho = 1.2$ and $\rho = 1.8$ to study Region A, Region B and Region C in Fig. 6.11a–c correspondingly.

First, when the system load is sufficiently low (i.e., $\rho = 0.6$ in Fig. 6.11a), customers anticipate a short expected waiting time under the unobservable case, so the server needs to charge a sufficiently low premium to induce priority purchases ($P^*_{un} = 0.45$ in this particular example). When $\nu_P = \frac{P\mu}{C} > \rho^2/(1-\rho)$ (or $P > 0.45$ in this example), no one deems purchasing priority as worthwhile and thus the server receives zero revenue (consistent with Case (1) of Lemma 6.4.3). However, if the state of the system is observable, customers make state-dependent decisions. Even though the system load is on average low, there are times when the system reaches higher states due to the stochastic nature and customers who arrive during those times are willing to purchase priority for a higher price. The optimal revenue for the server would also be higher under the observable case accordingly.

Second, when the value of the system load falls into an intermediate range (i.e., $\rho = 1.2$ in Fig. 6.11b), customers anticipate an increased expected waiting time compared to the case $\rho = 0.6$ under the unobservable case, and are comfortable with purchasing priority at a higher premium. Under the observable case, even though the system load is higher compared to the case $\rho = 0.6$, the increase is not significant enough compared to the case $\rho = 1.8$ and there are times when the system will still be in low states. At equilibrium, there will be less customers who purchase priorities in the observable setting compared to the unobservable setting, leading to an overall lower revenue.

Finally, consider a system load that is sufficiently high (i.e., $\rho = 1.8$ in Fig. 6.11c). In this case, the relationship between the optimal revenues are again flipped (the

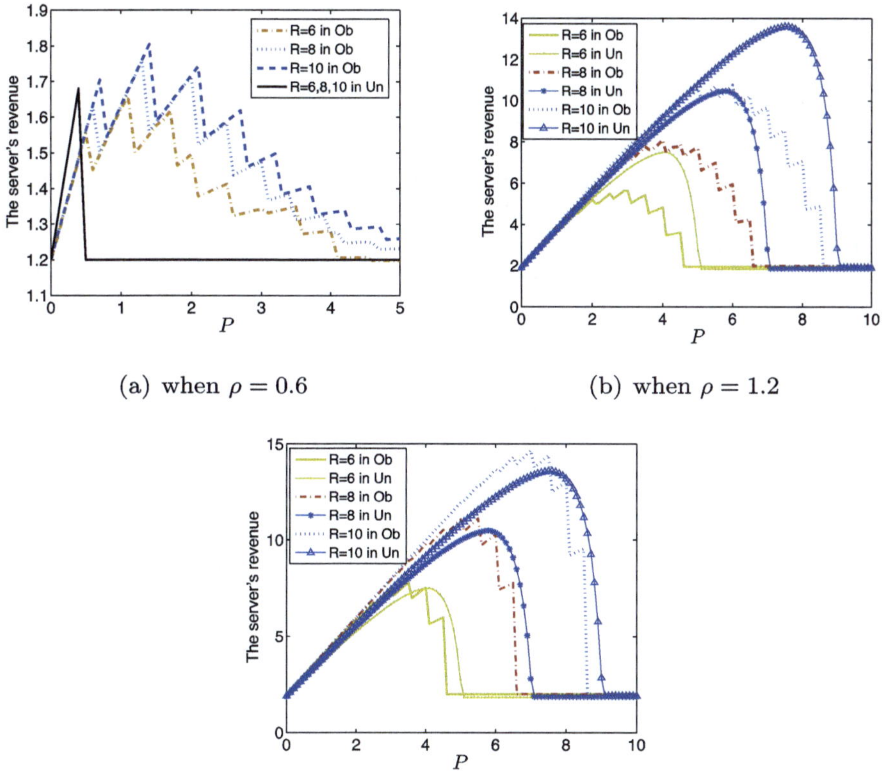

(a) when $\rho = 0.6$

(b) when $\rho = 1.2$

(c) when $\rho = 1.8$

Fig. 6.11 Server's revenues in observable [Ob] and unobservable [Un] cases versus P

server is better off with the observable setting). This happens firstly because the server's revenue under the unobservable case remains unchanged in the system's workload ρ as long as the potential arrival rate is sufficiently large (see Lemma 6.4.3 for the $\rho > 1$ case). Therefore, the server's revenue under the unobservable case are identical in Fig. 6.11b, c. Secondly, for the observable case, we can observe that revenue continues to increase in ρ, so eventually the revenue under the observable case surpasses that under the unobservable case.

The fact that the optimal setting switches twice between the observable and unobservable case, as the system load increases, is rather interesting. To the best of our knowledge, this is a new result in the literature. It also raises the need for practitioners to have a deeper understanding of their services and industries before making relevant adjustments. In what follows, we provide a concrete example to demonstrate the implications of this finding.

6.4.4.3 A Mini Case Study

In this section, we conduct a case study on Papa John's Pizza, the third largest pizza restaurant chain in the United States. In the spring of the year 2017, Papa John's introduced a $3 Papa Priority fee that allows customers to bump their pizza orders to the front of the line.[6] Our goal is to study this priority practice under our model's framework, derive the optimal price for the priority premium and present some managerial insights.

We first proceed to extrapolate the model's parameters using real Papa John's data that is publicly available, including information provided in the company's 10-K report. Papa John's received a total of 16.61 million orders in the United States within a 30-day span in spring 2017.[7] In the spring, the company operated 3,428 locations in the U.S. (including both company-owned and franchised stores).[8] The average number of orders that a U.S. location receives in a day can be calculated as

$$[(16.61\text{M orders})/(30 \text{ days})/(3428 \text{ stores})] = 161.5 \text{ orders}/(\text{day} \cdot \text{store}).$$

Papa John's receives customer orders through its digital channel including online/mobile orders (approx. 60%), over the phone (approx. 20%) and walk-in customers (approx. 20%).[9] We will focus on orders that require delivery because the waiting time for in-store pick-up or dining has small variability. Typically, orders received from the digital and phone channels require delivery (80% of the total orders). For simplicity, we assume that it is always the case. Furthermore, a Papa John's location is normally open Monday through Sunday from 10am till 11pm (13 h in total).[10] Therefore, we can estimate the effective arrival rate λ to a U.S. store as

$$[(161.5 \text{ orders/day}) \times 80\%/(13 \text{ hrs})] = 9.94 \text{ orders/hr} \approx 10 \text{ orders/hr}.$$

To estimate the average store capacity under our model's framework ($M/M/1$), we set the mean sojourn time (including pizza baking and delivery), $1/(\mu - \lambda)$, equal to 45 min (or 0.75 hr), the typical waiting time for pizza orderings in the U.S. This leads to $\mu = 11.3$ orders/hr.

Next, we estimate the basic service fee F for the model which in this case corresponds to the U.S. average customer order dollar amount. From the company's 2016 10-K report,[11] Papa John's annual revenue was $1.71 billion, including sales from company-owned restaurants and royalties/fees from franchised restaurants in its U.S. and international locations. We identify the average order dollar amount in a U.S.

---

[6] https://www.usatoday.com/story/money/2017/02/28/papa-johns-testing-fee-priority-orders/98544092/#.

[7] https://www.statista.com/statistics/230978/people-who-visited-papa-johns-pizza-usa/.

[8] http://ir.papajohns.com/releasedetail.cfm?releaseid=1024108.

[9] http://www.businessinsider.com/papa-johns-digital-ordering-2017-8.

[10] http://hours.ninja/papa-johns-operating-hours/

[11] Available at http://eproxymaterials.com/interactive/pzza2016/.

Table 6.3 P_{un}^* as a function of R and Λ under the unobservable case

P_{un}^*	$\Lambda = 7$/hr	$\Lambda = 8$/hr	$\Lambda = 9$/hr	$\Lambda =$ 10/hr	$\Lambda =$ 11/hr	$\Lambda =$ 12/hr	$\Lambda =$ 13/hr	$\Lambda =$ 14/hr
$R = \$6$	\$0.89	\$1.20	\$1.20	\$1.20	\$1.20	\$1.20	\$1.20	\$1.20
$R = \$8$	\$0.89	\$1.51	\$2.75	\$3.02	\$3.02	\$3.02	\$3.02	\$3.02
$R = \$10$	\$0.89	\$1.51	\$2.75	\$4.86	\$4.86	\$4.86	\$4.86	\$4.86
$R = \$12$	\$0.89	\$1.51	\$2.75	\$6.02	\$6.68	\$6.68	\$6.68	\$6.68
$R = \$14$	\$0.89	\$1.51	\$2.75	\$6.02	\$11.00	\$11.00	\$11.00	\$11.00

Table 6.4 P_{ob}^* as a function of R and Λ under the observable case

P_{ob}^*	$\Lambda = 7$/hr	$\Lambda = 8$/hr	$\Lambda = 9$/hr	$\Lambda =$ 10/hr	$\Lambda =$ 11/hr	$\Lambda =$ 12/hr	$\Lambda =$ 13/hr	$\Lambda =$ 14/hr
$R = \$6$	\$0.69	\$0.69	\$0.69	\$0.69	\$1.57	\$1.57	\$1.57	\$1.57
$R = \$8$	\$2.69	\$2.69	\$2.69	\$2.69	\$2.69	\$2.69	\$3.57	\$3.57
$R = \$10$	\$1.80	\$3.80	\$3.80	\$3.80	\$3.80	\$4.69	\$4.69	\$4.69
$R = \$12$	\$1.84	\$1.84	\$3.38	\$4.92	\$4.92	\$5.80	\$5.80	\$6.69
$R = \$14$	\$1.84	\$2.49	\$3.09	\$3.38	\$6.92	\$6.92	\$7.80	\$7.80

location by only considering the U.S. company-owned location since sales information for the franchised stores is not available, and more importantly, because the legal structure of a store (company-owned vs. franchised) should have little impact on customers' order sizes. According to the annual report, sales of U.S. company-owned restaurants amounted to 47.6% of total revenue, contributed by 702 locations. Thus F can be estimated as

$$(\$1.71\text{B}/\text{year}) \times 47.6\%/(702 \text{ stores})/(365 \text{ days})/(161.5 \text{ orders/day}) = \$19.67/\text{order} \approx \$20/\text{order}$$

Now we proceed to study the optimal priority pricing strategy for the server (an average Papa John's store) under different service rewards R and potential arrival rates Λ. In practice, customers will not be willing to wait more than an hour and half for their pizzas, so we control the reward-to-waiting-cost ratio R/C to be under 1.5 hrs. To do that, we normalize the unit waiting cost C to be \$10/hr, i.e., customers' utility of the food decreases \$10 an hour by the delay of the delivery (when the average order is \$20), and vary $R < \$15$. On the other hand, because the effective arrival rate λ is 10 orders/hr, we investigate scenarios where the potential arrival rate Λ is in its neighborhood (from 7 orders/hr to 14 orders/hr).

Tables 6.3 and 6.4 present the optimal prices for Papa Priority predicted using our theoretical unobservable and observable models studied in Sects. 6.4.3 and 6.4.2, respectively. First of all, the optimal price P^* for the priority premium always increases with Λ in both cases (for fixed R), as the server can charge a higher premium with a greater number of potential customers. Second, P^* increases monotonically

Table 6.5 Optimal revenue as Λ increases ($R = \$8$)

Optimal revenue	$\Lambda =$ 7/hr	$\Lambda =$ 8/hr	$\Lambda =$ 9/hr	$\Lambda =$ 10/hr	$\Lambda =$ 11/hr	$\Lambda =$ 12/hr	$\Lambda =$ 13/hr	$\Lambda =$ 14/hr
Unobservable case	<u>$146.23</u>	<u>$172.08</u>	<u>$204.75</u>	<u>$213.90</u>	$213.90	$213.90	$213.90	$213.90
Observable case	$140.18	$160.28	$179.95	$198.25	<u>$213.92</u>	<u>$226.16</u>	<u>$235.24</u>	<u>$242.24</u>

Table 6.6 Optimal price of priority premium as Λ increases ($R = \$8$)

Optimal price	$\Lambda =$ 7/hr	$\Lambda =$ 8/hr	$\Lambda =$ 9/hr	$\Lambda =$ 10/hr	$\Lambda =$ 11/hr	$\Lambda =$ 12/hr	$\Lambda =$ 13/hr	$\Lambda =$ 14/hr
Unobservable case	<u>$0.89</u>	<u>$1.51</u>	<u>$2.75</u>	<u>$3.02</u>	$3.02	$3.02	$3.02	$3.02
Observable case	$2.69	$2.69	$2.69	$2.69	<u>$2.69</u>	<u>$2.69</u>	<u>$3.57</u>	<u>$3.57</u>

in R under the unobservable case, but not under the observable case (see the underlined values in Table 6.4). This is because the server's revenue is a piecewise function under the observable case (see Fig. 6.11 for an example). Furthermore, the current practice of Papa John's reflects an unobservable setting. Assuming customers' waiting tolerance is 48 min (i.e., utility of having a pizza becomes negative after 48 min of waiting or $R/C = 0.8$ hr), our model's predicted optimal price ($3.02) is very close to Papa Priority's actual charge of $3 for $\Lambda = 10, 11, 12, 13, 14$ orders/hr (which is consistent with λ of 10 orders/hr).

Next, we investigate whether it is beneficial for Papa John's to release real-time queue length information (e.g., they can show wait information on their existing app). To do that, we fix $R/C = 0.8$ hr. Table 6.5 presents the optimal revenues for the unobservable and observable cases as the potential arrival rate Λ increases. We observe that $\Lambda = 11$ orders/hr is the turning point from the unobservable to the observable setting (see Table 6.5), i.e., when $7 \leq \Lambda < 11$ orders/hr, Papa John's earns more revenue under the unobservable setting, but when $\Lambda \geq 11$ orders/hr, it has a higher revenue from the observable setting (the more profitable case is underlined accordingly). This also corresponds to the shift from Region B to Region C described in Fig. 6.10.

Given that their effective arrival rate is estimated to be $\lambda = 10$ orders/hr, it is unknown to us whether the potential arrival rate Λ is between 10 orders/hr and 11 orders/hr, or greater than 11 orders/hr. If it is already greater than 11 orders/hr, or if at some point in the future it surpasses 11 orders/hr, our analysis would suggest that the company can benefit from revealing the number of waiting orders to the customers by operating under an observable model. However, there are also pricing implications. In particular, the company should decrease the Papa Priority fee by about a quarter of a dollar at the time of the switch (from $3.02 to $2.69, see the underlined values in Table 6.6). Eventually, the Papa Priority fee can be increased again as demand continues to grow.

In practice, Papa John's potential arrival rate must vary by the time of the day, the day of the week, and the year's season. Our results suggest that the company may be interested in implementing a dynamic priority pricing strategy with respect to the potential arrival rate to further maximize their revenue. However, there are two caveats: (1) Priority premium should increase with demand (i.e., potential arrival rate Λ) *within* the observable or unobservable case, but may not be the case when switched from one setting to another. (2) As a company with the motto "Better Ingredient, Better Pizza", Papa John's priority premium may need to be reduced in some cases when its pizza quality/taste improves (because the optimal price for the priority premium is non-monotone in R).

6.4.5 Concluding Remarks

In this section, a single-server queueing system with non-preemptive pay-for-priority options in the presence of strategic customers has been studied. Under both the unobservable and observable settings, the equilibrium strategies of the customers for given priority premiums is first derived, and then the server's revenue maximization problem is studied. The primary contribution of this work to the literature is two-fold: (1) The first contribution is to study customers' simultaneous joining/balking and pay-for-priority decisions. Including balking in priority queueing systems not only is more realistic to real-world practices, but also increases the complexity of the problem due to the two-dimensional decision scope. For games with one decision, it is well-known that ATC typically leads to a unique symmetric equilibrium, while FTC leads to multiple equilibria. However, in our two dimensional settings we are able to identify ATC activity in the presence of multiple equilibria. We refer interested readers to the discussion in Sect. 1.8.2. of Hassin (2016) on this topic.

Under the unobservable setting, it is found that equilibrium and system throughput are not monotone in the service reward R, which departs from the conclusions of priority queueing models in the existing literature without balking. This is driven by customer's self-interested nature and the intervention of their joining/balking and pay-for-priority decisions. Specifically, when R is sufficiently small (or large, respectively), all the joining customers choose to opt out (or in, respectively) for priority. But when R is in the middle range, $R \uparrow$ induces higher probability of priority purchase which discourages non-priority customers by increasing their expected waiting cost. Accordingly, the total system throughput decreases in R due to less joining customers. Furthermore, because the server revenue is a function of both the system throughput (on which it collects basic service fees) and the number of priority purchases (on which it collects priority premiums), increasing service rewards for the customers ($R \uparrow$) may in fact lead to reduced firm revenue. Under the observable setting, we show again that the equilibrium threshold and the server's optimal price do not need to be monotone in R. In addition, this theoretical work identifies the necessary and sufficient condition for an equilibrium to hold. In the presence of multiple equilibria, one can nail down the unique Pareto-dominant equilibrium analytically. (2) The second contribution, which has not been done in literature before, is to compare an observable priority queue with an unobservable priority queue. By comparing the server's optimal revenues in the two different settings, it is found that the service provider is better off with the observable setting when the system load is either low or high, but with the unobservable setting when the system load is medium.

This work has some limitations. For example, the analyses are based on M/M/1 queues in steady state. In Appendix B of Wang et al. (2019), it is shown that the analysis for the unobservable case can be generalized to the M/G/1 setting, but unfortunately the observable case is not tractable. A dynamic pricing or decision model would be also an excellent future research direction. In the literature, Afèche and Sarhangian (2015) consider rational abandonment of the lower priority customers in an observable queue. The priority discipline is preemptive and customers' priority levels are exogenous. Yang et al. (2017) study an unobservable referral priority queue where a customer can get into the priority queue by referring new customers. Finally, it may be an interesting direction to consider heterogeneous customers in our model setting.

References

Adiri, I., and U. Yechiali. 1974. Optimal priority purchasing and pricing decisions in non-monopoly and monopoly queues. *Operations Research* 22: 1051–1066.

Afèche, Philipp and Vahid Sarhangian. 2015. Rational abandonment from priority queues: Equilibrium strategy and pricing implications. Columbia Business School Research Paper.

Afèche, Philipp, and Haim Mendelson. 2004. Pricing and priority auctions in queueing systems with a generalized delay cost structure. *Management Science* 50 (7): 869–882.

Afimeimounga, Heti, Wiremu Solomon, and Ilze Ziedins. 2005. The Downs-Thomson paradox: Existence, uniqueness and stability of user equilibria. *Queueing Systems* 49: 321–334.

Altman, Eitan, and Nahum Shimkin. 1998. Individual equilibrium and learning in processor sharing systems. *Operations Research* 46: 776–784.

Altman, Eitan, Urtzi Ayesta, and Balakrishna J. Prabhu. 2011. Load balance in processor sharing systems. *Telecommunication Systems* 47: 35–48.

Balachandran, K. R. 1972. Purchasing priorities in queues. *Management Science* 18: 319–326.

Burnetas, A., and Antonis Economou. 2007. Equilibrium customer strategies in a single server Markovian queue with setup times. *Queueing Systems* 56: 213–228.

Chambers, Chester, and Panagiotis Kouvelis. 2006. Modeling and managing the percentage of satisfied customers in hidden and revealed waiting line systems. *Production and Operations Management* 15 (1): 103–116.

Chen, Hong, and Murray Frank. 2004. Monopoly pricing when customers queue. *IIE Transactions* 36: 569–581.

Cui, Shiliang, Zhongbin Wang, and Luyi Yang. 2023. *Innovative Priority Mechanisms in Service Operations Theory and Applications*. Springer Nature.

Cui, Shiliang, Zhongbin Wang, and Luyi Yang. 2020. The economics of line-sitting. *Management Science* 66 (1): 227–242.

Dolan, R. J. 1978. Incentive mechanisms for priority queueing problems. *Bell Journal of Economics* 9: 421–436.

Edelson, Noel M., and David K. Hildebrand. 1975. Congestion tolls for Poisson queueing processes. *Econometrica* 43: 81–92.

Gavirneni, S., and V. G. Kulkarni. 2016. Self-selecting priority queues with Burr distributed waiting costs. *Production and Operations Management* 25 (6): 979–992.

Gerber, H. U., and G. Pafum. 1998. Utility functions: From risk theory to finance. *North American Actuarial Journal* 2 (3): 74–91.

Guo, Pengfei, and Refael Hassin. 2011. Strategic behavior and social optimization in Markovian vacation queues. *Operations Research* 59: 986–997.

Guo, Pengfei, and Zhe George Zhang. 2013. Strategic queueing behavior and its impact on system performance in service systems with the congestion-based staffing policy. *Manufacturing & Service Operations Management* 15: 118–131.

Guo, Pengfei, Robin Lindsey, and Zhe George Zhang. 2014. On the Downs-Thomson paradox in a self-financing two-tier queuing system. *Manufacturing & Service Operations Management* 16 (2): 315–322.

Hanna, Alperstein. 1988. Optimal pricing policy for the service facility offering a set of priority prices. *Management Science* 34 (5): 666–671.

Harchol-Balter, Mor. 2013. *Performance Modeling and Design of Computer Systems: Queueing Theory in Action*. New York, USA: Cambridge University Press.

Hassin, Refael. 2016. *Rational Queueing*. Chapman and Hall/CRC Press.

Hassin, Refael. 1986. Consumer information in markets with random product quality: The case of queues and balking. *Econometrica* 54: 1185–1195.

Hassin, Refael. 1995. Decentralized regulation of a queue. *Management Science* 41: 163–173.

Hassin, Refael, and Moshe Haviv. 1997. Equilibrium threshold strategies: The case of queues with priorities. *Operations Research* 45: 966–973.

Hassin, Refael, and Moshe Haviv. 2003. *To Queue or Not to Queue: Equilibrium Behavior in Queueing Systems*. Boston: Kluwer Academic Publishers.

Hassin, Refael, and Ricky Roet-Green. 2017. The impact of inspection cost on equilibrium, revenue, and social welfare in a single-server queue. *Operations Research* 65 (3): 804–820.

Haviv, Moshe. 2014. Regulating an M/G/1 queue when customers know their demand. *Performance Evaluation* 77: 57–71.

Haviv, Moshe, and J. van der Wal. 1997. Equilibrium strategies for processor sharing and queues with relative priorities. *Probability in the Engineering and Informational Sciences* 11: 403–412.

Hu, Ming (ed.). 2019. *Sharing Economics: Making Supply Meet Demand*, Springer Nature.

Hu, Ming, Yang Li, and Jianfu Wang. 2018. Efficient ignorance: Information heterogeneity in a queue. *Management Science* 64 (6): 2650–2671.

Jouini, O., Y. Dallery, and O. Z. Akşin. 2009. Queueing models for multiclass call centers with real-time anticipated delays. *International Journal of Production Economics* 120: 389–399.

Lillo, R. E. 2001. Optimal control of an M/G/1 queue with impatient priority customers. *Naval Research Logistics* 48: 201–209.

Mendelson, H., and S. Whang. 1990. Optimal incentive-compatible priority pricing for the M/M/1 queue. *Operations Research* 38: 870–883.

Naor, Pinhas. 1969. The regulation of queue size by levying tolls. *Econometrica* 37: 15–24.

Shone, R., V. Knight, and J. Williams. 2013. Comparisons between observable and unobservable $M/M/1$ queues with respect to optimal customer behavior. *European Journal of Operational Research* 227: 133–141.

Stidham, S., Jr. 2009. *Optimal Design of Queueing Systems*. Taylor and Francis, Boca Raton: CRC Press.

Tilt, B., and K. R. Balachandran. 1979. Stable and superstable customer policies with balking and priority options. *European Journal of Operational Research* 3: 485–498.

Wang, Jinting, Shiliang Cui, and Zhongbin Wang. 2019. Equilibrium strategies in M/M/1 priority queues with balking. *Production and Operations Management* 28 (1): 43–62.

Wang, Zhongbin, and Jinting Wang. 2019. Pre-commitment or post-payment: Which worsens a line-sitting firm's revenue? *Operations Research Letters* 47 (5): 447–451.

Yang, Luyi, Laurens G. Debo, and Varun Gupta. 2017. Trading time in a congested environment. *Management Science* 63 (7): 2377–2395.

Yang, Luyi, Zhongbin Wang, and Shiliang Cui. 2021. A model of queue-scalping. *Management Science* 67 (11): 6803–6821.

Chapter 7
Repairable Queueing Systems

In real life, it is often encountered that the unreliable server fails to provide service for customers. At this time, repairmen need to repair the service facility. After the service facility is repaired, it can continue to perform its mission and to serve customers. This kind of queueing system which is repairable after breaking down called repairable queueing system.

As early as in the 1960s, Avi-Itzhak and Naor (1962), Thiruvenydan (1963) and Mitrany and Avi-Itzhak (1968) begin to study the queueing system that may fail and be repairable, and got the relevant quantitative indicators of queueing theory. Cao and Cheng (1982) study the reliability problem of M/G/1 repairable queueing system for the first time, and obtained some important reliability indexes such as the first failure time distribution of the server, the failure probability, the failure frequency and the failure time of the server at time t.

In the framework of game-theoretic analysis of repairable queueing systems, Economou and Kanta (2008) first consider the equilibrium balking behavior of customers in the M/M/1 repairable queueing system. In their model, when the server fails, it is immediately repaired and fully restored. In both fully observable and almost unobservable cases, customers follow the corresponding pure threshold equilibrium strategy. Li et al. (2014) present the equilibrium queueing strategy of customers in the M/M/1 repairable queueing system under two other observable situations. Jagannathan et al. (2012) analyze dynamic spectrum access systems in cognitive radio. They consider the primary user uses the spectrum randomly and the secondary user must wait in line as "the server is not reliable". They obtain the secondary user's choice of equilibrium strategy in the non-cooperative game, and prove the existence and uniqueness of the symmetric Nash equilibrium solution. Li and Han (2011) indicate that in the observable case, the individual optimal threshold joining strategy of sub-users was inconsistent with the social optimal strategy. Do et al. (2012) study the queueing model with server unreliability in cognitive radio systems, and obtained secondary user's social optimal strategy in the observable case. Yang et al. (2013) consider the equilibrium joining strategy of customers in a discrete repairable

J. Wang, *Fundamentals of Queueing-Game Models*,
https://doi.org/10.1007/978-981-95-0261-5_7

queueing system. Wang and Zhang (2011) analyze the equilibrium threshold joining strategy of customers in the repairable queuing system with delayed repair. Li et al. (2013) discuss the equilibrium strategy analysis in an M/M/1 queueing system with partial failures. Zhang and Wang (2023) explore the strategic shield against external shocks in a Markovian queue with vulnerable server.

Boudali and Economou (2012) study the Markov queueing system with disaster arrival. When a disaster occurs, all customers in the system are forced to leave the queue, and the server will break down due to the arrival of the disaster, and any customers who arrive will not be allowed to enter the system until the server is repaired. They finally get the Nash equilibrium of customers and the social optimal strategy. Boudali and Economou (2013) consider the more general case in which arriving customers were allowed to queue up after a disaster had caused the server to break down and until the server was repaired. Economou and Mamou (2013) discuss customer equilibrium strategies in a queueing system with unreliable server with random customer empty-out function, where both customer arrival rate and empty-out rate are controlled by external mechanisms. Dimitrakopoulos and Burnetas (2011) investigate an M/M/1 queueing system with dynamic service control, in which the service rate is dynamically switched between high-speed and low-speed according to the congestion degree of the system. They show that the customer's equilibrium strategy is no longer unique due to the change of service rate, and obtain an upper bound on the number of possible equilibrium solutions. Zhang et al. (2014) consider a repairable queueing system with a single server and a constant retrial rate. Wang et al. (2015) analyze the equilibrium behavior of customers in a constant rate retrial queue with single deleted negative customers. Gao et al. (2016) explore a repairable retrial queue under Bernoulli schedule and general retrial mechanism.

7.1 M/M/1 Repairable Queueing Systems

7.1.1 Model Description

Consider a single server queueing system with an infinite capacity for waiting space. Customers arrive as a Poisson process with parameter λ and service time is exponential distributed with parameter μ. The server's life follows an exponential distribution, and the failure rate is ζ. The repair time is exponential distributed with parameter θ. We represent the state at time t by pair $(N(t), I(t))$, where $N(t)$ denotes the number of customers in the system and $I(t)$(0: under repair, 1: work normally) denotes the state of server. It is clear that the process $\{(N(t), I(t)), t \geq 0\}$ is a continuous time Markov chain.

Assume that customers can decide whether to join the queue or not according to the system information they have upon arrival. Every customer receives a reward R for completing service. This may reflect his satisfaction and/or the added value of being served. On the other hand there exists a waiting cost of C units per time unit that the customer remains in the system (in queue or in service). Customers are risk neutral and maximize their expected net benefit. Upon arrival at the system, they need to estimate their excepted cost of stay and then make a decision whether to join the queue or not. Once a customer has made a choice, there is no renege, that is, they can neither quit in the middle of the line nor arrive after they have balked.

7.1.2 Equilibrium Threshold Strategies for the Fully Observable Case

In the fully observable case, customer observe both the state of sever $I(t)$ and the number of customer in the system $N(t)$.

Theorem 7.1.1 *In the fully observable $M/M/1$ repairable queueing system, there exist unique threshold*

$$(n_e(0), n_e(1)) = \left(\left\lfloor \frac{R\mu\theta}{C(\theta+\zeta)} - \frac{\mu}{\theta+\zeta}\right\rfloor - 1, \left\lfloor \frac{R\mu\theta}{C(\theta+\zeta)}\right\rfloor - 1\right), \tag{7.1.1}$$

such that the customer arrive at time t and observe $(N(t), I(t))$, enter if $N(t) \leq n_e(I(t))$ and balk otherwise.

Proof For a new customer, if he choose to join the queue, his expected net benefit is

$$S(n,i) = R - CT(n,i), \tag{7.1.2}$$

where $T(n,i)$ denote the expected sojourn time of customer who observe the state (n,i) and choose to join the queue. Thus, the balance equations are as followed:

$$T(n,0) = \frac{1}{\theta} + T(n,1), \quad n = 0, 1, 2, \dots, \tag{7.1.3}$$

$$T(0,1) = \frac{1}{\mu+\zeta} + \frac{\zeta}{\mu+\zeta}T(0,0), \tag{7.1.4}$$

$$T(n,1) = \frac{1}{\mu+\zeta} + \frac{\mu}{\mu+\zeta}T(n-1,1) + \frac{\zeta}{\mu+\zeta}T(n,0), \quad n = 1, 2, 3, \dots. \tag{7.1.5}$$

Solve the above equations, and obtain:

$$T(n,i) = (n+1)\left(1+\frac{\zeta}{\theta}\right)\frac{1}{\mu} + (1-i)\frac{1}{\theta}, \quad i = 0, 1. \tag{7.1.6}$$

If $S(n,i) > 0$, i.e., if the reward by receiving service is larger than the total cost, the customer is willing to join the queue. If $S(n,i) = 0$, i.e., if the reward by receiving service equals the total cost, it's different to join or balk. Solving inequality $S(n, i) \geq 0$ by taking into account (7.1.2) and (7.1.6), we can conclude that if and only if $n \leq n_e(I(t))$ holds, the customer choose to join the queue, where $(n_e(0), n_e(1)$ is shown as Eq. (7.1.1). Note that the behavior of customers after choosing this strategy has nothing to do with the behavior choice of other customers, so it is also a weakly dominant strategy. □

7.1.3 Equilibrium Strategies for the Almost Observable Case

In the almost observable case, where arriving customers only observe the number of customers in the system $N(t)$ while the server's state $I(t)$ is unobservable. In order to find the equilibrium pure threshold strategy of customers, the steady state distribution of the system needs to be solved first.

Lemma 7.1.1 *In the $M/M/1$ almost observable repairable queueing system, assume that all customers use the same threshold for entrance n_e. The stationary distribution of the system is*

$$p(n, 0) = A(\rho_1^{n+1} - \rho_2^{n+1}), \quad n = 0, 1, 2, \ldots, n_e, \tag{7.1.7}$$

$$p(n, 1) = A(\nu_1 \rho_1^{n+1} - \nu_2 \rho_2^{n+1}), \quad n = 0, 1, 2, \ldots, n_e, \tag{7.1.8}$$

$$p(n_e + 1, 0) = \frac{\lambda A}{\theta}\left(\left(1 + \frac{\zeta}{\mu}(1 + \nu_1)\right)\rho_1^{n_e+1} - \left(1 + \frac{\zeta}{\mu}(1 + \nu_2)\right)\rho_2^{n_e+1}\right), \tag{7.1.9}$$

where A can be obtained by the normalization condition, and

$$\rho_{1,2} = \frac{\lambda}{2\mu(\lambda + \theta)}\left(\mu + \zeta + \lambda + \theta \pm \sqrt{(\mu + \zeta + \lambda + \theta)^2 - 4\mu(\lambda + \theta)}\right), \tag{7.1.10}$$

$$\nu_i = \frac{(\lambda + \theta)\rho_i - \lambda}{\zeta \rho_i}, \quad i = 1, 2. \tag{7.1.11}$$

Proof List the balance equations

$$(\lambda + \theta)p(0, 0) = \zeta p(0, 1), \tag{7.1.12}$$

$$(\lambda + \theta)p(n, 0) = \lambda P(n - 1, 0) + \zeta p(n, 1), \quad n = 1, 2, \ldots, n_e, \tag{7.1.13}$$

$$\theta p(n_e + 1, 0) = \lambda p(n_e, 0) + \zeta p(n_e + 1, 1), \tag{7.1.14}$$

$$\mu p(n + 1, 1) = \lambda p(n, 0) + \lambda p(n, 1), \quad n = 0, 1, 2, \ldots, n_e. \tag{7.1.15}$$

Solving the above equations we can obtain the conclusions in the Theorem. The details are omitted here. □

Lemma 7.1.2 *Assuming that all customers use the same threshold strategy n_e, it is readily seen that the benefit of a new customer who observes n customers in the system and chooses to join the queue is*

$$S(n) = R - C\frac{n+1}{\mu}\left(1+\frac{\zeta}{\theta}\right) - \frac{C}{\theta}\frac{\sigma^{n+1}-1}{(1+\nu_1)\sigma^{n+1}-(1+\nu_2)},$$
$$n = 0, 1, 2, \ldots, n_e, \tag{7.1.16}$$
$$S(n_e+1) = R - C\frac{n_e+2}{\mu}\left(1+\frac{\zeta}{\theta}\right) - \frac{C}{\theta}\frac{(\mu+\zeta(1+\nu_1))\sigma^{n_e+1}-(\mu+\zeta(1+\nu_2))}{(\mu+(\zeta+\theta)(1+\nu_1))\sigma^{n_e+1}-(\mu+(\zeta+\theta)(1+\nu_2))}, \tag{7.1.17}$$

where $\sigma = \frac{\rho_1}{\rho_2}$.

Proof If a new customer who observe n customers in the system and choose to join the queue, so his benefit is

$$S(n) = R - CT(n), \tag{7.1.18}$$

where $T(n) = \mathbb{E}[S\,|N^- = n]$, represents the excepted sojourn time under the condition that the customer observe n customers in the system, and

$$T(n) = T(n,1) + \frac{1}{\theta}\Pr(I^- = 0 \mid N^- = n) = T(n,1) + \frac{1}{\theta}\frac{p(n,0)}{p(n,0)+p(n,1)}. \tag{7.1.19}$$

By computing, we can obtain

$$T(n) = \frac{n+1}{\mu}\left(1+\frac{\zeta}{\theta}\right) + \frac{1}{\theta}\frac{\sigma^{n+1}-1}{(1+\nu_1)\sigma^{n+1}-(1+\nu_2)}, \; n = 0, 1, 2, \ldots, n_e, \tag{7.1.20}$$

$$T(n_e+1) = \frac{n_e+2}{\mu}\left(1+\frac{\zeta}{\theta}\right) + \frac{1}{\theta}\frac{(\mu+\zeta(1+\nu_1))\sigma^{n_e+1}-(\mu+\zeta(1+\nu_2))}{(\mu+(\zeta+\theta)(1+\nu_1))\sigma^{n_e+1}-(\mu+(\zeta+\theta)(1+\nu_2))}. \tag{7.1.21}$$

By substituting (7.1.20) and (7.1.21) in (7.1.18), we obtain (7.1.16) and (7.1.17). □

To ensure that the reward for service exceeds the expected cost for a customer who finds the system empty, the following conditions have to be satisfied

$$R > \frac{C}{\mu}\left(1+\frac{\zeta}{\theta}\right) + \frac{C}{\theta}\frac{\zeta}{\lambda+\theta+\zeta}. \tag{7.1.22}$$

Next, we discuss the pure equilibrium threshold strategy of customers.

Theorem 7.1.2 *Define two sequences* $(f_1(n) : n = 0, 1, \ldots)$ *and* $(f_2(n) : n = 0, 1, \ldots)$*:*

$$f_1(n) = R - C\frac{n+1}{\mu}\left(1+\frac{\zeta}{\theta}\right) - \frac{C}{\theta}\frac{\sigma^{n+1}-1}{(1+\nu_1)\sigma^{n+1}-(1+\nu_2)}, \quad n = 0, 1, \ldots, \tag{7.1.23}$$

$$\begin{aligned} f_2(n) = {} & R - C\frac{n+1}{\mu}\left(1+\frac{\zeta}{\theta}\right) \\ & - \frac{C}{\theta}\frac{(\mu+\zeta(1+\nu_1))\sigma^n - (\mu+\zeta(1+\nu_2))}{(\mu+(\zeta+\theta)(1+\nu_1))\sigma^n - (\mu+(\zeta+\theta)(1+\nu_2))}, \quad n = 0, 1, \ldots. \end{aligned} \tag{7.1.24}$$

There exists a finite non-negative integer $n_L \leq n_U$ *that makes*

$$f_1(0), f_1(1), \ldots, f_1(n_U) > 0, \quad f_1(n_U+1) \leq 0, \tag{7.1.25}$$

and

$$f_2(n_U+1), f_2(n_U), f_2(n_U-1), \ldots, f_2(n_L+1) \leq 0, \quad f_2(n_L) > 0, \tag{7.1.26}$$

or

$$f_2(n_U+1), f_2(n_U), f_2(n_U-1), \ldots, f_2(0) \leq 0. \tag{7.1.27}$$

So in the $M/M/1$ *almost observable repairable queueing system, all customers use the same threshold for entrance* n_e*, for* $n_e \in \{n_L, n_L+1, \ldots, n_U\}$ *are equilibrium strategies.*

Proof According to (7.1.22), $f_1(0) > 0$ and $\lim_{n\to\infty} f_1(n) = -\infty$ hold. So if n_U makes $f_1(n_U)$ be the first non-positive term in the sequence $(f_1(n))$, thus the condition (7.1.25) holds.

On the other hand, because of $f_1(n) > f_2(n)$, so $f_2(n_{\mathrm{U}}+1) < f_1(n_{\mathrm{U}}+1) \leq 0$ holds. Then, look forward from n_U to 0, find the first positive integer n_L that makes the item in sequence $(f_2(n))$ a positive number, as shown in (7.1.26). If from n_U to 0, all terms in the sequence $(f_2(n))$ are non-positive, then (7.1.34) holds.

We tag a new arrival customer, and assume that other customers use the n_e threshold joining strategy, where $n_e \in \{n_L, n_L+1, \ldots, n_U\}$. If the tagged customer observes $n \leq n_e$ customers in the system and decides to join, so according to (7.1.16), (7.1.23) and (7.1.25), we know his expected net benefit is $f_1(n) > 0$. So in this case he will choose to join. If the tagged customer observes $n = n_e + 1$ customers in the system and decides to join the queue, so according to (7.1.17), (7.1.24) and (7.1.26) or (7.1.34), we know his expected net benefit is $f_2(n_e+1) \leq 0$. So in this case he will choose to balk. Therefore, for $n_e \in \{n_L, n_L+1, \ldots, n_U\}$ are equilibrium strategies. □

When $n_L < n_U$, there exist multiple equilibrium threshold strategies. What's more, this is observed generally in "Follow-The-Crowd (FTC)" situations in almost observable queueing system, that is one s optimal response to a strategy x adopted by all others is increasing in x. Consider the function $S_{n_e}(n)$ as shown in (7.1.18), and tag a customer who observe n customers in the system and others use the same threshold for entrance n_e. When other customers use the threshold for entrance $n_e + 1$, for all $n = 0, 1, \ldots, n_e - 1$, the benefit function of tagged customer is $S_{n_e+1}(n) = S_{n_e}(n)$, but $S_{n_e+1}(n_e) = f_1(n_e) > f_2(n_e) = S_{n_e}(n_e)$. This means that when other customers use the threshold for entrance $n_e + 1$, the optimal threshold of the tagged customer is larger than the threshold of the tagged customer when others use the threshold for entrance n_e. In other words, when other customers adopt a larger threshold joining strategy, tagged customers are more likely to choose to enter the queue.

7.1.4 Equilibrium Strategies for the Almost Unobservable Case

In the almost unobservable case, assume that all customers use a mixed strategy $(q(0), q(1))$, where $q(i)$ is the probability of joining when the server is in state i. Thus, the arrival rate equals $\lambda_i = \lambda q(i)$ for states where the server is in state i.

Lemma 7.1.3 *In the $M/M/1$ almost unobservable repairable queueing system, assume that $\mu\theta > \lambda(\zeta + \theta)$, and the expected net benefit of customer is*

$$S(0, q(0), q(1)) = R - C\Big[\Big(\frac{\zeta\theta\lambda_0 + \zeta^2\lambda_1 + \mu\theta\lambda_0 - \theta\lambda_0\lambda_1}{\theta(\mu\theta - \zeta\lambda_0 - \theta\lambda_1)} + 1\Big) \times \Big(1 + \frac{\zeta}{\theta}\Big)\frac{1}{\mu} + \frac{1}{\theta}\Big], \tag{7.1.28}$$

$$S(1, q(0), q(1)) = R - C\Big(\frac{\theta\lambda_0^2 + \theta^2\lambda_0 + \theta\zeta\lambda_1}{\mu\theta\zeta - \zeta^2\lambda_0 - \theta\zeta\lambda_1} + 1\Big)\Big(1 + \frac{\zeta}{\theta}\Big)\frac{1}{\mu}. \tag{7.1.29}$$

Proof We give the balance equations as follows,

$$(\lambda_0 + \theta)p(0, 0) = \zeta p(0, 1), \tag{7.1.30}$$
$$(\lambda_0 + \theta)p(n, 0) = \lambda_0 p(n-1, 0) + \zeta p(n, 1), \quad n = 1, 2, 3, \ldots, \tag{7.1.31}$$
$$(\zeta + \lambda_1)p(0, 1) = \mu p(1, 1) + \theta p(0, 0), \tag{7.1.32}$$
$$(\mu + \zeta + \lambda_1)p(n, 1) = \mu p(n+1, 1) + \theta p(n, 0) + \lambda_1 p(n-1, 1), \quad n = 1, 2, 3, \ldots, \tag{7.1.33}$$
$$\mu p(n+1, 1) = \lambda_0 p(n, 0) + \lambda_1 p(n, 1), \quad n = 0, 1, 2, \ldots. \tag{7.1.34}$$

Define the probability generating function $P_i(z) = \sum_{n=0}^{\infty} z^n p(n, i)$, $|z| \le 1, i = 0, 1$, from the above equations we can obtain

$$P_0(1) = \frac{\theta}{\theta + \zeta}, \tag{7.1.35}$$

$$P_1(1) = \frac{\zeta}{\theta + \zeta}, \tag{7.1.36}$$

$$P_0'(1) = \frac{\theta\zeta\lambda_0 + \zeta^2\lambda_1 + \mu\theta\lambda_0 - \theta\lambda_0\lambda_1}{(\theta + \zeta)(\mu\theta - \zeta\lambda_0 - \theta\lambda_1)}, \tag{7.1.37}$$

$$P_1'(1) = \frac{\theta(\lambda_0^2 + \theta\lambda_0 + \zeta\lambda_1)}{(\theta + \zeta)(\mu\theta - \zeta\lambda_0 - \theta\lambda_1)}. \tag{7.1.38}$$

Let $\mathbb{E}(N^-|I^- = i)$ be the expected number of system in state i, and

$$\mathbb{E}(N^-|I^- = 0) = \frac{\sum_{n=0}^{\infty} np(n,0)}{\sum_{k=0}^{\infty} p(k,0)} = \frac{P_0'(1)}{P_0(1)} = \frac{\theta\zeta\lambda_0 + \zeta^2\lambda_1 + \mu\theta\lambda_0 - \theta\lambda_0\lambda_1}{\theta(\mu\theta - \zeta\lambda_0 - \theta\lambda_1)}, \tag{7.1.39}$$

$$\mathbb{E}(N^-|I^- = 1) = \frac{\sum_{n=0}^{\infty} np(n,1)}{\sum_{k=0}^{\infty} p(k,1)} = \frac{P_1'(1)}{P_1(1)} = \frac{(\lambda_0^2 + \theta\lambda_0 + \zeta\lambda_1)\theta}{(\mu\theta - \zeta\lambda_0 - \theta\lambda_1)\zeta}. \tag{7.1.40}$$

According to (7.1.6), the customer's expected net benefit of joining is

$$S(0, q(0), q(1)) = R - C\Big[\Big(\mathbb{E}(N^-|I^- = 0) + 1\Big)\Big(1 + \frac{\zeta}{\theta}\Big)\frac{1}{\mu} + \frac{1}{\theta}\Big], \tag{7.1.41}$$

$$S(1, q(0), q(1)) = R - C\Big(\mathbb{E}(N^-|I^- = 1) + 1\Big)\Big(1 + \frac{\zeta}{\theta}\Big)\frac{1}{\mu}. \tag{7.1.42}$$

Substituting (7.1.39) and (7.1.40) in (7.1.41) and (7.1.42), we can obtain (7.1.28) and (7.1.29). □

Next, we discuss the Nash equilibrium strategy of customers.

Theorem 7.1.3 *In the $M/M/1$ almost unobservable repairable queueing system, assume that $\mu\theta > \lambda(\zeta + \theta)$, and there exists mixed strategy $(q_e(0), q_e(1))$ "observe $I(t)$ and enter with probability $q_e(I(t))$" where the vector $(q_e(0), q_e(1))$ is given as follows*

Case 1: If $\frac{R}{C} < \frac{\theta+\zeta}{\theta\mu}$, *so* $(q_e(0), q_e(1)) = (0, 0)$;

Case 2: If $\frac{\theta+\zeta}{\theta(\mu-\lambda)} < \frac{R}{C} < \frac{\theta+\zeta}{\theta(\mu-\lambda)} + \frac{1}{\theta}$, *so* $(q_e(0), q_e(1)) = (0, 1)$;

Case 3: If $\frac{R}{C} > \frac{(\mu\lambda-\lambda^2+\mu\theta)(\theta+\zeta)}{\mu\theta(\mu\theta-\zeta\lambda-\theta\lambda)} + \frac{1}{\theta}$, *so* $(q_e(0), q_e(1)) = (1, 1)$;

Case 4: If $\frac{\theta+\zeta}{\theta\mu} \le \frac{R}{C} \le \frac{\theta+\zeta}{\theta(\mu-\lambda)}$, *so* $(q_e(0), q_e(1)) = (0, \frac{R\mu\theta - C\theta - C\zeta}{R\theta\lambda})$;

Case 5: If $\frac{\theta+\zeta}{\theta(\mu-\lambda)} + \frac{1}{\theta} \le \frac{R}{C} \le \frac{(\mu\lambda-\lambda^2+\mu\theta)(\theta+\zeta)}{\mu\theta(\mu\theta-\zeta\lambda-\theta\lambda)} + \frac{1}{\theta}$, *so* $(q_e(0), q_e(1)) = (q_e^*(0), 1)$, *where* $q_e^*(0) = \frac{(\mu\theta-\theta\lambda)(R\mu\theta-C\mu)-(\theta+\zeta)C\mu\theta}{(\mu\lambda-\lambda^2)(\theta+\zeta)C+(R\mu\theta-C\mu)\zeta\lambda}$.

Proof Assume that all customers who find the system state i enter with probability $q(i)$. Under the condition of $S(i, q(0), q(1)) > 0$, the tagged customer will choose to enter. Similarly, when $S(i, q(0), q(1)) = 0$, he is indifferent between entering and balking. When $S(i, q(0), q(1)) < 0$, he decides to balk. For fixed $q(0) \in [0, 1]$ and $q(1) \in [0, 1]$,

$$S(1, q(0), q(1)) = R - C\frac{(\mu\theta^2 + \zeta\lambda_0^2)(\theta + \zeta)}{(\mu\theta - \zeta\lambda_0 - \theta\lambda_1)\mu\theta^2},$$

and

$$S(0, q(0), q(1)) = R - C\Big[\frac{(\mu\lambda_0 + \mu\theta - \lambda_0\lambda_1)(\theta + \zeta)}{(\mu\theta - \zeta\lambda_0 - \theta\lambda_1)\mu\theta} + \frac{1}{\theta}\Big],$$

decrease monotonically with respect to $q(1)$ and $q(0)$, respectively. The following five cases are discussed:

Case 1: $\frac{R}{C} < \frac{\theta+\zeta}{\theta\mu} \Leftrightarrow S(1, 0, 0) < 0$.

In this case, for any $q(0)$ and $q(1)$, $S(1, q(0), q(1)) \le S(1, 0, 0) < 0$. So regardless of the strategy used by other customers, the expected benefit of tagged customers who find the server busy and join is negative. And $q_e(1) = 0$ is equilibrium.

On the other hand, $S(0, 0, 0) = R - C\Big(\frac{\zeta+\theta}{\mu\theta} + \frac{1}{\theta}\Big) < 0$. Because $S(0, q(0), 0)$ decrease monotonically with $q(0) \in [0, 1]$, regardless of the strategy used by other customers, the expected benefit of tagged customers who find the server under repair and join is negative. And $q_e(0) = 0$ is equilibrium.

Case 2: $\frac{\theta+\zeta}{\theta(\mu-\lambda)} < \frac{R}{C} < \frac{\theta+\zeta}{\theta(\mu-\lambda)} + \frac{1}{\theta} \Leftrightarrow S(1, 0, 1) > 0$ and $S(0, 0, 1) < 0$.

In this case, for any $q(1) \in [0, 1]$, $S(1, 0, q(1)) \ge S(1, 0, 1) > 0$. If other customers use $q(0) = 0q(1) \le 1$, so the tagged customer who find server busy has a positive expected net benefit if he decides to enter. Hence, he will join with probability 1.

On the other hand, because $q(0) \in [0, 1]$ decrease monotonically, $S(0, q(0), 1) \le S(0, 0, 1) < 0$. If other customer use $q(0) \ge 0, q(1) = 1$, and the expected benefit of tagged customers who find the server break down and join is negative, so he will balk.

To summarize, in this case, the equilibrium joining strategy of customer is $(q_e(0), q_e(1)) = (0, 1)$.

Case 3: $\frac{R}{C} > \frac{(\mu\lambda-\lambda^2+\mu\theta)(\theta+\zeta)}{\mu\theta(\mu\theta-\zeta\lambda-\theta\lambda)} + \frac{1}{\theta} \Leftrightarrow S(1, 1, 1) > 0$ and $S(0, 1, 1) > 0$.

In this case, for any $q(0) \in [0, 1]$ and $q(1) \in [0, 1]$, we have $S(i, q(0), q(1)) \ge S(i, 1, 1) > 0$. So regardless of the strategy used by other customer and server state, he gains a positive benefit when enters the queue. Hence, $(q_e(0), q_e(1)) = (1, 1)$.

Case 4: $\frac{\theta+\zeta}{\theta\mu} \le \frac{R}{C} \le \frac{\theta+\zeta}{\theta(\mu-\lambda)} \Leftrightarrow S(1, 0, 1) \le 0 \le S(1, 0, 0)$.

In this case, equation $S(1, 0, q(1)) = 0$ has unique solution $q(1) = \frac{R\mu\theta-C\theta-C\zeta}{R\theta\lambda}$ when $q(1) \in [0, 1]$. If other customers use $(q(0), q(1)) = (0, \frac{R\mu\theta-C\theta-C\zeta}{R\theta\lambda})$, so according to $S(0,0,q(1)) \leq S(1,0,q(1))$, the expected benefit of tagged customer when he find server under repair and enter is $S(0,0,\frac{R\mu\theta-C\theta-C\zeta}{R\theta\lambda}) < 0$, so he optimal strategy is balking. When tagged customer find server busy and enter with any probability q, his expected benefit is zero. Hence, $(q_e(0), q_e(1)) = (0, \frac{R\mu\theta-C\theta-C\zeta}{R\theta\lambda})$ is an equilibrium strategy.

Case 5: $\frac{\theta+\zeta}{\theta(\mu-\lambda)} + \frac{1}{\theta} \leq \frac{R}{C} \leq \frac{(\mu\lambda-\lambda^2+\mu\theta)(\theta+\zeta)}{\mu\theta(\mu\theta-\zeta\lambda-\theta\lambda)} + \frac{1}{\theta} \Leftrightarrow S(0,1,1) \leq 0 \leq S(0,0,1)$.

In this case, we can find equation $S(0,q(0),1) = 0$ has unique solution $q_e^*(0) = \frac{(\mu\theta-\theta\lambda)(R\mu\theta-C\mu)-(\theta+\zeta)C\mu\theta}{(\mu\lambda-\lambda^2)(\theta+\zeta)C+(R\mu\theta-C\mu)\zeta\lambda}$ when $q(0) \in [0,1]$. If other customer use joining strategy $(q(0), q(1)) = (q_e^*(0), 1)$, and according to $S(0,q(0),1) \leq S(1,q(0),1)$, the tagged customer's expected benefit $S(1,q_e^*(0),1)$ is non-negative when he find server busy and enter, so his optimal strategy is joining. When tagged customer find server under repair and enter with any probability q, his expected benefit is zero. Hence, $(q_e(0), q_e(1)) = (q_e^*(0), 1)$
is an equilibrium strategy. □

If the condition $\mu\theta > \lambda(\zeta+\theta)$ in Theorem 7.1.3 is not satisfied, so for the case of $\lambda \geq \mu$, a similar analytical approach can be used and obtain the Nash equilibrium strategy $(q_e(0), q_e(1))$ given as follows:

Case 1: If $\frac{R}{C} < \frac{\theta+\zeta}{\theta\mu}$, so $(q_e(0), q_e(1)) = (0,0)$;

Case 2: If $\frac{R}{C} \geq \frac{\theta+\zeta}{\theta\mu}$, so $(q_e(0), q_e(1)) = (0, \frac{R\mu\theta-C\theta-C\zeta}{R\theta\lambda})$.

For $\lambda < \mu \leq \frac{\lambda(\theta+\zeta)}{\theta}$, we can use the similar analytical approach and obtain the Nash equilibrium strategy $(q_e(0), q_e(1))$ given as follows:

Case 1: If $\frac{R}{C} < \frac{\theta+\zeta}{\theta\mu}$, so $(q_e(0), q_e(1)) = (0,0)$;

Case 2: If $\frac{\theta+\zeta}{\theta\mu} \leq \frac{R}{C} \leq \frac{\theta+\zeta}{\theta(\mu-\lambda)}$, so $(q_e(0), q_e(1)) = (0, \frac{R\mu\theta-C\theta-C\zeta}{R\theta\lambda})$;

Case 3: If $\frac{\theta+\zeta}{\theta(\mu-\lambda)} < \frac{R}{C} < \frac{\theta+\zeta}{\theta(\mu-\lambda)} + \frac{1}{\theta}$, so $(q_e(0), q_e(1)) = (0,1)$;

Case 4: If $\frac{R}{C} \geq \frac{\theta+\zeta}{\theta(\mu-\lambda)} + \frac{1}{\theta}$, so $(q_e(0), q_e(1)) = (q_e^*(0), 1)$, where $q_e^*(0) = \frac{(\mu\theta-\theta\lambda)(R\mu\theta-C\mu)-(\theta+\zeta)C\mu\theta}{(\mu\lambda-\lambda^2)(\theta+\zeta)C+(R\mu\theta-C\mu)\zeta\lambda}$.

7.1.5 Equilibrium Strategy for the Fully Unobservable Case

In the fullyume that the customers who f at all customers join the queue with the probability q, and the arrive rate is λq. In this case, the stability of the system is $\mu\theta > \lambda q(\zeta+\theta)$. We consider $\mu\theta > \lambda(\zeta+\theta)$ first. In Lemma 7.1.3, let $q(0) = q(1) = q$

and can obtain the steady-state distribution of the system. Thus, the expected number of customers is

$$\begin{aligned}\mathbb{E}(N) &= \mathbb{E}(N^-|I^- = 0)P_0(1) + \mathbb{E}(N^-|I^- = 1)P_1(1)\\ &= \frac{\lambda q((\theta+\zeta)^2+\mu\theta)}{(\theta+\zeta)(\mu\theta-\lambda q\zeta-\lambda q\theta)}.\end{aligned} \tag{7.1.43}$$

According to the Little's formula, the average sojourn time of customers is

$$\mathbb{E}(W) = \frac{\mathbb{E}(N)}{\lambda q} = \frac{(\theta+\zeta^2+\mu\theta)}{(\theta+\zeta)(\mu\theta-\lambda q\zeta-\lambda q\theta)}. \tag{7.1.44}$$

Theorem 7.1.4 *In the fully unobservable M/M/1 repairable queueing system, assume that* $\mu\theta > \lambda(\zeta+\theta)$*, so there exists a unique mixed equilibrium strategy "enter with probability* q_e*", where* q_e *is given by*

$$q_e = \begin{cases} 0, & \text{if } \frac{R}{C} < \frac{(\theta+\zeta)^2+\mu\theta}{(\theta+\zeta)\mu\theta},\\ q_e^*, & \text{if } \frac{(\theta+\zeta)^2+\mu\theta}{(\theta+\zeta)\mu\theta} \le \frac{R}{C} \le \frac{(\theta+\zeta)^2+\mu\theta}{(\theta+\zeta)(\mu\theta-\lambda\zeta-\lambda\theta)},\\ 1, & \text{if } \frac{R}{C} > \frac{(\theta+\zeta)^2+\mu\theta}{(\theta+\zeta)(\mu\theta-\lambda\zeta-\lambda\theta)}, \end{cases} \tag{7.1.45}$$

where $q_e^* = \frac{R\mu\theta(\theta+\zeta)-C(\theta+\zeta)^2-C\mu\theta}{\lambda R(\theta+\zeta)^2}$.

Proof We consider a tagged customer at his arrival instant. If he decides to enter his expected net benefit is

$$S(q) = R - C\frac{(\theta+\zeta)^2+\mu\theta}{(\theta+\zeta)(\mu\theta-\lambda q\zeta-\lambda q\theta)}. \tag{7.1.46}$$

Obviously, $S(q)$ decreases monotonically when $q \in [0, 1]$, which maximum and minimum values are respectively

$$S(0) = R - C\frac{(\theta+\zeta)^2+\mu\theta}{(\theta+\zeta)\mu\theta}, \tag{7.1.47}$$

$$S(1) = R - C\frac{(\theta+\zeta)^2+\mu\theta}{(\theta+\zeta)(\mu\theta-\lambda\zeta-\lambda\theta)}. \tag{7.1.48}$$

- If $\frac{(\theta+\zeta)^2+\mu\theta}{(\theta+\zeta)\mu\theta} \le \frac{R}{C} \le \frac{(\theta+\zeta)^2+\mu\theta}{(\theta+\zeta)(\mu\theta-\lambda\zeta-\lambda\theta)}$, there exists a unique solution $q_e^* = \frac{R\mu\theta(\theta+\zeta)-C(\theta+\zeta)^2-C\mu\theta}{\lambda R(\theta+\zeta)^2}$ for equation $S(q) = 0$ when $q \in [0, 1]$.
- If $\frac{R}{C} > \frac{(\theta+\zeta)^2+\mu\theta}{(\theta+\zeta)(\mu\theta-\lambda\zeta-\lambda\theta)}$, so for any $q \in [0, 1]$, the expected net benefit of tagged customer $S(q) > 0$, he will choose to enter.

- If $\frac{R}{C} < \frac{(\theta+\zeta)^2+\mu\theta}{(\theta+\zeta)\mu\theta}$, so for any $q \in [0, 1]$, the expected net benefit of tagged customer is negative, he decides to balk.

□

Since $S(q)$ is a monotonically decreasing function when the probability of other customers joining the queue $q < q_e$, the expected net benefit of tagged customer is positive, he will choose to enter. Similarly, when the probability of other customers joining the queue $q > q_e$, tagged customer decides to balk. And when $q = q_e$, it is indifferent for tagged customer to join or balk. This indicates that the optimal strategy of an individual is the reduction function of the strategy of other customers, that is, when the probability of other customers joining the queue is greater, the probability of tagged customers joining the queue is smaller, so this is the ATC situation.

If $\mu\theta \leq \lambda(\zeta + \theta)$, according to the method of Theorem 7.1.4, we can obtain a unique equilibrium probability for entrance

$$q_e = \begin{cases} 0, & \text{if } \frac{R}{C} < \frac{(\theta+\zeta)^2+\mu\zeta}{(\theta+\zeta)\mu\theta}, \\ \frac{R\mu\theta(\theta+\zeta)-C(\theta+\zeta)^2-C\mu\zeta}{\lambda R(\theta+\zeta)^2}, & \text{if } \frac{R}{C} \geq \frac{(\theta+\zeta)^2+\mu\zeta}{(\theta+\zeta)\mu\theta}. \end{cases}$$

7.2 M/M/1 Queueing Systems with Disaster

7.2.1 Model Description

Consider an M/M/1 queueing system. Customers arrive as a Poisson process with parameter λ. The service time is exponential distributed with parameter μ, the system is subject to catastrophes/failures according to a Poisson process at rate ξ. When a catastrophe occurs all customers are forced to abandon the system prematurely, without being served. The system is rendered inoperative and a repair process is set on. The length of a repair time is exponentially distributed at rate η. During a repair time, arrivals are not accepted. Assume that interarrival times, service times, intercatastrophe times and repair times are mutually independent.

We represent the state of the system at time t by a pair $(Q(t), I(t))$, where $Q(t))$ records the number of customers at the system and $I(t)$ denotes the server state, with 1 describing a system in operation and 0 describing a system under repair. Note that whenever $I(t)$ is zero, $Q(t)$ should be necessarily zero too. Thus, the stochastic process $\{(Q(t), I(t)) : t \geq 0\}$ is a continuous time Markov chain with state space $S = \{(n, 1), n \geq 0\} \cup \{(0, 0)\}$. Assuming that each customer receives either a reward of R_s units for completing service or a compensation of R_f units in case that he is forced to abandon the system due to a failure. Moreover, a customer is charged a cost of C units per time unit that he remains in the system. According to the different information disclosed by the system, two situations are studied: the observable case (the length of queue can be seen upon arrival) and the unobservable case (the length of queue cannot be seen upon arrival).

7.2.2 *Joining Strategy Analysis in Observable Case*

In the observable case, suppose that the customers who find the server active can observe the number of customers in the system.

7.2.2.1 Nash Equilibrium Strategy

Lemma 7.2.1 *In the $M/M/1$ observable queue with catastrophes, causing complete removals of customers. The expected net benefit of a customer that observes n customers in the system upon arrival and decides to enter is given by*

$$S_{obs}(n) = R_s\left(\frac{\mu}{\mu+\xi}\right)^{n+1} + \left(R_f - \frac{C}{\xi}\right)\left[1-\left(\frac{\mu}{\mu+\xi}\right)^{n+1}\right], \quad n \geq 0. \tag{7.2.1}$$

Proof Consider a tagged customer that finds the system at state $(n, 1)$ upon arrival and decides to enter. This customer may leave the system either due to its service completion or due to a catastrophe that will force him to abandon prematurely the system. For his service completion, he has to wait for a sum of $n+1$ independent exponentially distributed times with parameter μ. For the next catastrophe, he has to wait for an exponentially distributed time with parameter ξ. Therefore, the sojourn time of such a customer in the system is given as $Z = \min(Y_n, X)$, where Y_n follows a Gamma distribution with parameters $n+1$, μ and X is an exponentially distributed random variable with rate ξ, independent of Y_n. Moreover, the tagged customer will be served with probability $\Pr[Y_n < X]$, while he will be forced to abandon the system due to a catastrophe with the complementary probability $\Pr[Y_n \geq X]$. Therefore his net benefit will be

$$S_{obs}(n) = R_s \Pr[Y_n < X] + R_f \Pr[Y_n \geq X] - C\mathbb{E}[Z]. \tag{7.2.2}$$

Note that

$$\Pr[Y_n < X] = \int_0^\infty e^{-\xi y}\frac{\mu^{n+1}}{n!}y^n e^{-\mu y}\mathrm{d}y = \left(\frac{\mu}{\mu+\xi}\right)^{n+1}, \tag{7.2.3}$$

and

$$\mathbb{E}[Z] = \int_0^\infty e^{-\xi z}\int_z^\infty \frac{\mu^{n+1}}{n!}u^n e^{-\mu u}\mathrm{d}u\, \mathrm{d}z = \frac{1}{\xi}\left[1-\left(\frac{\mu}{\mu+\xi}\right)^{n+1}\right]. \tag{7.2.4}$$

Plugging (7.2.3) and (7.2.4) in (7.2.2) yields (7.2.1) □

Since arrivals are not permitted during the repair period, if a customer observes the system at state $(0, 0)$, he is not allowed to enter and thus there is no decision. So, we only consider the case where a customer observes the system at a specified state $(n, 1)$.

Theorem 7.2.1 *In the observable model of the $M/M/1$ queue with catastrophes, a unique individual optimal pure strategy exists. There are three cases:*

Case 1: *If $R_f < \frac{C}{\xi} - \frac{\mu R_s}{\xi}$, the unique individual optimal strategy is always to balk.*

Case 2: *If $\frac{C}{\xi} - \frac{\mu R_s}{\xi} \leq R_f < \frac{C}{\xi}$, there exists a unique equilibrium threshold strategy.*

$$n_e = \left\lfloor \frac{\ln K}{\ln S} - 1 \right\rfloor, \tag{7.2.5}$$

where

$$K = \frac{\frac{C}{\xi} - R_f}{R_s - R_f + \frac{C}{\xi}}, \quad S = \frac{\mu}{\mu + \xi}. \tag{7.2.6}$$

Case 3: *If $R_f \geq \frac{C}{\xi}$, the unique individual optimal strategy is always to enter.*

Proof Consider a tagged customer that observes the system upon arrival. If he finds the system at state $(n, 1)$ and decides to enter, then his expected net benefit is given by (7.2.1). The customer will prefer to enter if $S_{obs}(n) \geq 0$, which is written easily as

$$\left(R_s - R_f + \frac{C}{\xi}\right) \cdot \left(\frac{\mu}{\mu + \xi}\right)^{n+1} \geq \frac{C}{\xi} - R_f. \tag{7.2.7}$$

Since $R_s - R_f + \frac{C}{\xi} \geq \frac{C}{\xi} - R_f$, we have the following three cases.
Case A: $\frac{C}{\xi} - R_f > 0 \Leftrightarrow R_f < \frac{C}{\xi}$.

By solving (7.2.7) with respect to n_e given by (7.2.5). When $R_f < \frac{C}{\xi} - \frac{\mu R_s}{\xi}$, n_e is negative and the optimal strategy is balking and and we conclude with Case 1. When $\frac{C}{\xi} - \frac{\mu R_s}{\xi} \leq R_f < \frac{C}{\xi}$, the threshold n_e is non-negative and we conclude with Case 2.
Case B: $R_s - R_f + \frac{C}{\xi} > 0 \geq \frac{C}{\xi} - R_f \Leftrightarrow \frac{C}{\xi} \leq R_f < \frac{C}{\xi} + R_s$.

In this case the inequality (7.2.7) is always true. Therefore the tagged customer is always willing to enter.
Case C: $0 \geq R_s - R_f + \frac{C}{\xi} > \frac{C}{\xi} - R_f \Leftrightarrow R_f \geq \frac{C}{\xi} + R_s$.

In this case the threshold n_e is negative. Therefore the tagged customer should to enter. □

7.2.2.2 Social Optimal Strategies

Given that the customers follow a threshold strategy with threshold n, we consider the problem of social optimization next, which is to find the threshold n_{soc} that maximizes the social benefit $S_{obs}^{soc}(n)$ per unit time.

Theorem 7.2.2 *In the $M/M/1$ observable queue with catastrophes, if all customers follow a threshold strategy with threshold n, the expected net social benefit per time unit is given by*

$$\begin{aligned} S_{obs}^{soc}(n) = & \frac{\lambda(R_s - R_f)}{(\mu+\xi)^{n+1}} \left\{ \frac{\mu d_1(n)[(\mu+\xi)^{n+1} - (\mu x_1)^{n+1}]}{\mu+\xi-\mu x_1} \right. \\ & + \left. \frac{\mu d_2(n)[(\mu+\xi)^{n+1} - (\mu x_2)^{n+1}]}{\mu+\xi-\mu x_2} \right\} \\ & + \lambda R_f \left(\frac{\eta}{\xi+\eta} - d_1(n){x_1}^{n+1} - d_2(n){x_2}^{n+1} \right) \\ & - \frac{C\mu^2}{\xi^2} d_1(n) x_1 (1-x_2)^2 [1-(n+2){x_1}^{n+1} + (n+1){x_1}^{n+2}] \\ & - \frac{C\mu^2}{\xi^2} d_2(n) x_2 (1-x_1)^2 [1-(n+2){x_2}^{n+1} + (n+1){x_2}^{n+2}], \quad n \ge 0, \end{aligned} \tag{7.2.8}$$

where

$$x_{1,2} = \frac{(\lambda+\mu+\xi) \pm \sqrt{(\lambda+\mu+\xi)^2 - 4\lambda q \mu}}{2\mu}, \tag{7.2.9}$$

$$\begin{aligned} d_1(n) = & -\eta\xi[(\mu+\xi)x_2 - \lambda]{x_2}^n \Big\{ \xi+\eta)\{(\lambda+\xi-\mu x_2)[(\mu+\xi)x_1-\lambda]{x_1}^n \\ & - (\lambda+\xi-\mu x_1)[(\mu+\xi)x_2-\lambda]{x_2}^n\} \Big\}^{-1}, \end{aligned} \tag{7.2.10}$$

$$\begin{aligned} d_2(n) = & -\eta\xi[(\mu+\xi)x_1 - \lambda]{x_1}^n \Big\{ \xi+\eta)\{(\lambda+\xi-\mu x_2)[(\mu+\xi)x_1-\lambda]{x_1}^n \\ & - (\lambda+\xi-\mu x_1)[(\mu+\xi)x_2-\lambda]{x_2}^n\} \Big\}^{-1}. \end{aligned} \tag{7.2.11}$$

Proof Let $p_{obs}(k, i)$ denote the stationary probabilities, we can obtain balance equations as follows.

$$\eta p_{obs}(0,0) = \xi \sum_{k=0}^{n+1} p_{obs}(k,1), \tag{7.2.12}$$

$$(\lambda+\xi)p_{obs}(0,1) = \mu p_{obs}(1,1) + \eta p_{obs}(0,0), \tag{7.2.13}$$

$$(\lambda+\mu+\xi)p_{obs}(k,1) = \lambda p_{obs}(k-1,1) + \mu p_{obs}(k+1,1), \quad 1 \le k \le n, \tag{7.2.14}$$

$$(\mu+\xi)p_{obs}(n+1,1) = \lambda p_{obs}(n,1). \tag{7.2.15}$$

Solving (7.2.12)–(7.2.15) yield

$$p_{obs}(0,0) = \frac{\xi}{\xi + \eta} \tag{7.2.16}$$

$$p_{obs}(k,1) = d_1(n){x_1}^k + d_2(n){x_2}^k, \quad 0 \le k \le n+1. \tag{7.2.17}$$

The expected net social benefit per time unit is

$$S_{obs}^{soc}(n) = \lambda P_{obs}^{ser} R_s + \lambda P_{obs}^{cat} R_f - C\mathbb{E}_{obs}[Q], \tag{7.2.18}$$

where P_{obs}^{ser} and P_{obs}^{cat} are the fractions of customers that leave the system due to service and catastrophes respectively and $\mathbb{E}_{obs}[Q]$ is the mean number of customers in system. Using (7.2.16), (7.2.17) and (7.2.3), we can obtain

$$P_{obs}^{ser} = \sum_{k=0}^{n} p_{obs}(k,1)\Big(\frac{\mu}{\mu+\xi}\Big)^{k+1} = \sum_{k=0}^{n}(d_1(n)x_1^k + d_2(n)x_2^k)\Big(\frac{\mu}{\mu+\xi}\Big)^{k+1}, \tag{7.2.19}$$

$$\begin{aligned} P_{obs}^{cat} &= \sum_{k=0}^{n} p_{obs}(k,1)\Big[1 - \Big(\frac{\mu}{\mu+\xi}\Big)^{k+1}\Big] \\ &= \sum_{k=0}^{n}(d_1(n)x_1^k + d_2(n)x_2^k)\Big[1 - \Big(\frac{\mu}{\mu+\xi}\Big)^{k+1}\Big], \end{aligned} \tag{7.2.20}$$

$$\mathbb{E}_{obs}[Q] = \sum_{k=0}^{n+1} k p_{obs}(k,1) = \sum_{k=0}^{n+1} k(d_1(n)x_1^k + d_2(n)x_2^k). \tag{7.2.21}$$

Computing (7.2.19)–(7.2.21) and substituting in (7.2.18) and yields (7.2.8). □

If the biggest n satisfies the $S_{obs}^{soc}(n)$ shown in (7.2.8), then it is the optimal threshold n_{soc}. According to numerical experiments, $n_{soc} \le n_e$.

7.2.3 Joining Strategy Analysis in Unobservable Case

In unobservable case, assume that the customers who find the server active can't observe the number of customers in the system, so we consider mixed strategy q of customer. In other words, when the customers find the server active, they enter with the probability q.

Lemma 7.2.2 *In the $M/M/1$ observable queue with catastrophes, if all customers follow a mixed strategy with joining probability q, then the steady state probability of the system $p_{un}(k,i)$ is*

$$p_{un}(0,0) = \frac{\xi}{\xi+\eta}, \tag{7.2.22}$$

$$p_{un}(k,1) = \frac{\eta(1-x_2(q))x_2(q)^k}{\xi+\eta}, \quad k \geq 0, \tag{7.2.23}$$

where

$$x_2(q) = \frac{(\lambda q + \mu + \xi) - \sqrt{(\lambda q + \mu + \xi)^2 - 4\lambda q \mu)}}{2\mu}. \tag{7.2.24}$$

And when other customers follow a mixed strategy with joining probability q, the expected net benefit of a customer that enters is

$$S_{un}(q) = \Big(R_s - R_f + \frac{C}{\xi}\Big)\frac{\mu(1-x_2(q))}{\mu+\xi-\mu x_2(q)} + R_f - \frac{C}{\xi}. \tag{7.2.25}$$

Proof The balance equations for the stationary distribution are given as follows

$$\eta p_{un}(0,0) = \xi \sum_{k=0}^{\infty} p_{un}(k,1), \tag{7.2.26}$$

$$(\lambda q + \xi) p_{un}(0,1) = \mu p_{un}(1,1) + \eta p_{un}(0,0), \tag{7.2.27}$$

$$(\lambda q + \mu + \xi) p_{un}(k,1) = \lambda q p_{un}(k-1,1) + \mu p_{un}(k+1,1), \quad k \geq 1 \tag{7.2.28}$$

By solving above equations, (7.2.22) and (7.2.23) can be obtained

Due to PASTA property, the probability that there are k customers in the system upon a customer's arrival, given that he finds the server operative is

$$p_{un}^{arr(\cdot,1)}(k,1) = \frac{p_{un}(k,1)}{\sum_{i=0}^{\infty} p_{un}(i,1)} = (1-x_2(q))x_2(q)^k, \quad k \geq 0. \tag{7.2.29}$$

Such a customer receives on the average $S_{obs}(k)$ units, given by (7.2.1). Therefore, the expected net benefit of a customer that decides to enter given that he has found an active system is

$$\begin{aligned} S_{un}(q) &= \sum_{k=0}^{\infty} p_{un}^{arr(\cdot,1)}(k,1) S_{obs}(k) \\ &= \sum_{k=0}^{\infty} (1-x_2(q))x_2(q)^k \Big\{R_s\Big(\frac{\mu}{\mu+\xi}\Big)^{n+1} + \Big(R_f - \frac{C}{\xi}\Big)\Big[1-\Big(\frac{\mu}{\mu+\xi}\Big)^{n+1}\Big]\Big\} \\ &= \Big(R_s - R_f + \frac{C}{\xi}\Big)\frac{\mu(1-x_2(q))}{\mu+\xi-\mu x_2(q)} + R_f - \frac{C}{\xi}. \end{aligned} \tag{7.2.30}$$

□

7.2.3.1 Nash Equilibrium Strategy

Theorem 7.2.3 *In the $M/M/1$ unobservable queue with catastrophes, a unique Nash equilibrium mixed strategy exists, with joining probability q_e given by*

$$q_e = \begin{cases} 0, & if R_f \leq \frac{C}{\xi} - \frac{\mu R_s}{\xi}; \\ \frac{(C-\xi R_f+\xi R_s)(\mu R_s+\xi R_f-C)}{\lambda(C-\xi R_f)R_s}, & if \frac{C}{\xi} - \frac{\mu R_s}{\xi} < R_f < \frac{C}{\xi} - \frac{\mu R_s(1-x_2)}{\xi}, \\ 1, & if R_f \geq \frac{C}{\xi} - \frac{\mu R_s(1-x_2)}{\xi}, \end{cases} \quad (7.2.31)$$

where $x_2 = x_2(1)$.

Proof Suppose that customers who find the server operative enter with probability q and consider a tagged arriving customer. Then, the tagged customer prefers to enter if $S_{un}(q) > 0$, he is indifferent between entering and balking if $S_{un}(q) = 0$ and he prefers to balk if $S_{un}(q) < 0$. We solve the equation $S_{un}(q) = 0$ for $x_2(q)$. It has a unique solution given by

$$x_{2e} = \frac{\mu R_s + \xi R_f - C}{\mu R_s}. \quad (7.2.32)$$

Let $x_2(q) = x_{2e}$ and substitute in (7.2.24). This yields q_e

$$q_e = \frac{x_{2e}[x_{2e}(1 - x_{2e}) + \xi]}{\lambda(1 - x_{2e})} = \frac{(C - \xi R_f + \xi R_s)(\mu R_s + \xi R_f - C)}{\lambda(C - \xi R_f)R_s}. \quad (7.2.33)$$

According to

$$\frac{d}{dq}x_2(q) = \frac{\lambda}{2\mu}\left(1 - \frac{\lambda q + \xi - \mu}{\sqrt{(\lambda q + \xi - \mu)^2 + 4\xi\mu}}\right) > 0, \quad q \in [0, 1], \quad (7.2.34)$$

the function $x_2(q)$ is strictly increasing for $q \in [0, 1]$. Therefore, if and only if $x_2(q_e))$ lies in the interval $(0, x_2)$, or equivalently $R_f \in \left(\frac{C}{\xi} - \frac{\mu R_s}{\xi}, \frac{C}{\xi} - \frac{\mu R_s(1-x_2)}{\xi}\right)$, given by (7.2.33), q_e lies in the interval $(0, 1)$, thus we obtain the second case of (7.2.31).

In addition, when $R_f \leq \frac{C}{\xi} - \frac{\mu R_s}{\xi}$ or $R_f \geq \frac{C}{\xi} - \frac{\mu R_s(1-x_2)}{\xi}$, the function $S_{un}(q)$ keeps the same sign. Taking into account equation (7.2.25) we have that

$$S_{un}(1) = R_s\frac{\mu(1 - x_2)}{\mu + \xi - \mu x_2} + \left(R_f - \frac{C}{\xi}\right)\frac{\xi}{\mu + \xi - \mu x_2}. \quad (7.2.35)$$

when $R_f \leq \frac{C}{\xi} - \frac{\mu R_s}{\xi}$,

$$S_{un}(1) \leq \frac{\mu R_s x_2}{\mu + \xi - \mu x_2} < 0. \quad (7.2.36)$$

thus $S_{un}(q)$ keeps a negative sign for any $q \in [0, 1]$. Thus, the option of "balk" is the unique Nash equilibrium strategy and we obtain the first case of (7.2.31).

Similarly, when $R_f \geq \frac{C}{\xi} - \frac{\mu R_s(1-x_2)}{\xi}$, (7.2.25) yields

$$S_{un}(0) = R_s \frac{\mu}{\mu+\xi} + \left(R_f - \frac{C}{\xi}\right)\frac{\xi}{\mu+\xi} \geq \frac{\mu R_s x_2}{\mu+\xi} > 0. \tag{7.2.37}$$

and $S_{un}(q)$ keeps a positive sign for any $q \in [0, 1]$. Thus, the option of "enter" is the unique Nash equilibrium strategy and we obtain the third case of (7.2.31). □

7.2.3.2 Social Optimal Strategies

Lemma 7.2.3 *In the $M/M/1$ unobservable queue with catastrophes, if all customers follow a mixed strategy with joining probability q, then the expected net social benefit per time unit is*

$$S_{un}^{soc}(q) = \frac{\eta x_2(q)[\mu R_s(1 - x_2(q)) + \xi R_f - C]}{(\xi + \eta)(1 - x_2(q))}, \tag{7.2.38}$$

with $x_2(q)$ given by (7.2.24).

Proof The expected net social benefit per time unit is given by

$$S_{un}^{soc}(n) = \lambda P_{un}^{ser} R_s + \lambda P_{un}^{cat} R_f - C\mathbb{E}_{un}[Q], \tag{7.2.39}$$

where P_{un}^{ser} and P_{un}^{cat} are the fractions of customers that join but leave the system due to service and catastrophes respectively and $\mathbb{E}_{un}[Q]$ is the mean number of customers in system. Using (7.2.22), (7.2.23) and (7.2.3), we can obtain

$$P_{un}^{ser} = \sum_{k=0}^{\infty} p_{un}(k,1) q\Big(\frac{\mu}{\mu+\xi}\Big)^{k+1} = \sum_{k=0}^{\infty} \frac{\eta(1 - x_2(q))x_2(q)^k}{\xi+\eta} q\Big(\frac{\mu}{\mu+\xi}\Big)^{k+1}, \tag{7.2.40}$$

$$P_{un}^{cat} = \sum_{k=0}^{\infty} p_{un}(k,1)\Big[1 - \Big(\frac{\mu}{\mu+\xi}\Big)^{k+1}\Big] = \sum_{k=0}^{\infty} \frac{\eta(1 - x_2(q))x_2(q)^k}{\xi+\eta} q\Big[1 - \Big(\frac{\mu}{\mu+\xi}\Big)^{k+1}\Big], \tag{7.2.41}$$

$$\mathbb{E}_{un}[Q] = \sum_{k=0}^{\infty} k p_{un}(k,1) = \sum_{k=0}^{\infty} k \frac{\eta(1 - x_2(q))x_2(q)^k}{\xi+\eta}. \tag{7.2.42}$$

By computing (7.2.40)–(7.2.42) and substituting in (7.2.39) yields (7.2.38). □

We obtain the optimal joining probability q_{soc} and also compare it with the corresponding equilibrium probability q_e. The results are summarized in the following theorem.

Theorem 7.2.4 *In the $M/M/1$ unobservable queue with catastrophes, a unique social optimal strategy exists, with joining probability q_{soc} given by*

$$q_{soc} = \begin{cases} 0, & \text{if } R_f \le \frac{C}{\xi} - \frac{\mu R_s}{\xi}, \\ \frac{\sqrt{D}(\mu R_s - \sqrt{D})(\xi R_s + \sqrt{D})}{\lambda D R_s}, & \text{if } \frac{C}{\xi} - \frac{\mu R_s}{\xi} < R_f < \frac{C}{\xi} - \frac{\mu R_s (1-x_2)^2}{\xi}, \\ 1, & \text{if } R_f \ge \frac{C}{\xi} - \frac{\mu R_s (1-x_2)^2}{\xi}, \end{cases} \quad (7.2.43)$$

where $D = \mu R_s (C - \xi R_f)$ and $x_2 = x_2(1)$. Moreover, the social optimal joining probability is always smaller than the individual one, i.e.,

$$q_{soc} \le q_e. \quad (7.2.44)$$

Proof We observe that the function $S_{un}^{soc}(q)$ given by (7.2.38) can be written as $f(x_2(q))$, where

$$f(x) = \frac{\eta x [\mu R_s (1-x) + \xi R_f - C]}{(\xi + \eta)(1-x)}. \quad (7.2.45)$$

Note that the function $x_2(q)$ is strictly increasing for $q \in [0, 1]$ and it takes values in $[0, x_2]$. For any $q \in [0, 1]$, we solve the equation

$$S_{un}^{soc\prime}(q) = f'(x_2(q)) x_2'(q) = 0. \quad (7.2.46)$$

$x_2'(q) \ne 0$ for $q \in [0, 1]$. Therefore $f'(x_2(q)) = 0$. So we have to solve $f'(x) = 0$, which is also written after some straightforward algebra in the form

$$\mu R_s x^2 - 2\mu R_s x + (\mu R_s + \xi R_f - C) = 0. \quad (7.2.47)$$

There are two cases to discuss.

Case 1: $R_f \ge \frac{C}{\xi}$. In this case, the quadratic polynomial in (7.2.47) is non-positive, thus $f(x)$ is monotonically increasing, and consequently $S_{un}^{soc}(q)$ is also increasing. The social optimal joining probability is $q_{soc} = 1$.

Case 2: $R_f < \frac{C}{\xi}$. In this case, the equation $f'(x) = 0$ has two distinct roots

$$x_2^- = 1 - \frac{\sqrt{D}}{\mu R_s}, \quad x_2^+ = 1 + \frac{\sqrt{D}}{\mu R_s}. \quad (7.2.48)$$

Since $x_2 < 1 < x_2^+$, there are only 3 subcases.

Case 2a: $x_2^- \le 0$. In this case, $x_2^- \le 0 < x < x_2^+$, then $f(x)$ is decreasing in $[0, x_2]$ and consequently $S_{un}^{soc}(q)$ is decreasing in $[0, 1]$. The social optimal joining probability is $q_{soc} = 0$.

Case 2b: $0 < x_2^- < x_2$. In this case, $f'(x)$ is positive in $(0, x_2^-)$ and negative in (x_2^-, x_2). Therefore, the maximum of $S_{un}^{soc}(q)$ is attained for q such that $x_2(q) =$

x_2^-. The social optimal joining probability is found by substituting $x_2(q) = x_2^-$ in (7.2.24)

$$q_{soc} = \frac{x_2^-[\mu(1-x_2^-)+\xi]}{\lambda(1-x_2^-)} = \frac{\sqrt{D}(\mu R_s - \sqrt{D})(\xi R_s + \sqrt{D})}{\lambda D R_s}. \quad (7.2.49)$$

Case 2c: $x_2^- \geq x_2$. In this case, $f(x)$ is increasing in $[0, x_2]$ and consequently $S_{un}^{soc}(q)$ is increasing in $[0, 1]$. The social optimal joining probability is $q_{soc} = 1$.

To summarize, we obtain (7.2.43).

Next, we compare q_{soc} and q_e. Since (7.2.31) and (7.2.43), when $R_f \leq \frac{C}{\xi} - \frac{\mu R_s}{\xi}$, $q_e = q_{soc} = 0$, whereas $q_e = q_{soc} = 1$ for $R_f \geq \frac{C}{\xi} - \frac{\mu R_s(1-x_2)^2}{\xi}$. Moreover, $q_e = 1$ and $q_{soc} \in (0, 1)$ when $\frac{C}{\xi} - \frac{\mu R_s(1-x_2)}{\xi} \leq R_f < \frac{C}{\xi} - \frac{\mu R_s(1-x_2)^2}{\xi}$. When $\frac{C}{\xi} - \frac{\mu R_s}{\xi} < R_f < \frac{C}{\xi} - \frac{\mu R_s(1-x_2)}{\xi}$, $\mu^2 \xi R_s^3 + D\sqrt{D} \geq 0$ holds. According to (7.2.33) and (7.2.49), we can obtain $q_{soc} \leq q_e$.

□

References

Avi-Itzhak, Benjamin, and Pinhas Naor. 1962. Some queueing problems with the service station subject to breakdown. *Operations Research* 10: 303–320.

Boudali, O., and Antonis Economou. 2012. Optimal and equilibrium balking strategies in the single server Markovian queue with catastrophes. *European Journal of Operational Research* 218: 708–715.

Boudali, O., and Antonis Economou. 2013. The effect of catastrophes on the strategic customer behavior in queueing systems. *Naval Research Logistics* 60: 571–587.

Cao, Jinhua, and Kan Cheng. 1982. Analysis of M/G/1 queueing system with repairable service station (in Chinese). *Acta Mathematicae Applicatae Sinica* 5: 113–127.

Dimitrakopoulos, Y., and A. Burnetas. 2011. Customer equilibrium and optimal strategies in an M/M/1 queue with dynamic service control. *European Journal of Operational Research* 252: 477–486.

Do, C., N. Tran, M. V. Nguyen, C. S. Hong, and S. Lee. 2012. Social optimization strategy in unobserved queueing systems in cognitive radio networks. *Communications Letters, IEEE* 16: 1944–1947.

Economou, Antonis, and Spyridoula Kanta. 2008. Equilibrium balking strategies in the observable single-server queue with breakdowns and repairs. *Operations Research Letters* 36: 696–699.

Economou, Antonis, and Athanasia Manou. 2013. Equilibrium balking strategies for a clearing queueing system in alternating environment. *Annals of Operations Research* 208: 489–514.

Gao, Shan, Jinting Wang, and Tien Van Do. 2016. A repairable retrial queue under Bernoulli schedule and general retrial policy. *Annals of Operations Research* 247 (1): 169–192.

Jagannathan, K., I. Menache, E. Modiano, and G. Zussman. 2012. Non-cooperative spectrum access-the dedicated versus free spectrum choice. *IEEE Journal on Selected Areas in Communications* 30: 2251–2261.

Li, Le., F. Jinting Wang, and Zhang. 2013. Equilibrium customer strategies in Markovian queues with partial breakdowns. *Computers & Industrial Engineering* 66: 751–757.

Li, H., and Zhu Han. 2011. Socially optimal queuing control in CR networks subject to service interruptions: to queue or not to queue? *IEEE Transactions on Wireless Communications* 10: 1656–1666.

Li, Xiangyu, Jinting Wang, and F. Zhang. 2014. New results on equilibrium balking strategies in the single-server queue with breakdowns and repairs. *Applied Mathematics and Computation* 241: 380–388.

Mitrany, I. L., and B. Avi-Itzhak. 1968. A many server queue with service interruptions. *Operations Research* 16: 628–638.

Thiruvenydan, K. 1963. Queueing with breakdown. *Operations Research* 11: 62–71.

Wang, Fang, F. Jinting Wang, and Zhang. 2015. Strategic behavior in the single-server constant retrial queue with individual removal. *Quality Technology and Quantitative Management* 2015 (12): 325–342.

Wang, Jinting, and F. Zhang. 2011. Equilibrium analysis of the observable queues with balking and delayed repairs. *Applied Mathematics and Computation* 218: 2716–2729.

Yang, Tengteng, Jinting Wang, and Feng Zhang. 2013. Equilibrium balking strategies in the Geo/Geo/1 queues with server breakdowns and repairs. *Quality Technology and Quantitative Management* 11: 231–243.

Zhang, Z., Jinting Wang, F. Zhang. 2014. Equilibrium customer strategies in the single-server constant retrial queue with breakdowns and repairs. *Mathematical Problems in Engineering* 2014: 14 p., Article ID 379572

Zhang, Lingjiao, and Jinting Wang. 2023. Strategic shield against external shocks in a Markovian queue with vulnerable server. *Journal of Industrial and Management Optimization* 19 (5): 3483–3508.

Chapter 8
Vacation Queueing Systems

The queueing system that takes advantage of the idle period to adjust or maintain the service facilities, or the attendants go on vacation during the idle period, or engage in auxiliary work, is called the vacation queueing system. Since the middle of the 20th century, because of the important role in computer systems, communication systems, management engineering and other fields, this kind of queueing system has been widely concerned. Takagi (1991) and Tian and Zhang (2006) have made in-depth and extensive studies on the vacation queueing system, and have achieved rich results in both theoretical and practical applications.

In the past decades, there has emerged research work on the behavior strategy of customers in the vacation queueing system. Burnetas and Economou (2007) first study the M/M/1 vacation queueing system. They analyzed a single-server Markov vacation queueing system with set-up time, and obtained the balanced joining strategies of customers under different information situations. Sun et al. (2010) study the equilibrium threshold strategies in observable queueing systems with setup/closedown times. Zhang et al. (2013) investigate Nash equilibrium strategies of customers with four different information situations in Markov queueing system with working vacation. Economou et al. (2011) consider the M/G/1 queueing system with generally distributed vacation time, and obtained Nash equilibrium and socially optimal queueing strategy in completely unobservable and almost unobservable cases. Zhang et al. (2013) discuss the M/G/1 queueing system with general distributed setup time, and give equilibrium queueing probability of customers in fully observable and almost observable cases. Zhang and Wang (2017) discuss the equilibrium pricing issue in an M/G/1 retrial queue with reserved idle time and setup time. Zhang and Wang (2021) discuss the strategic joining and information disclosing in Markovian queues with an unreliable server and working vacations.

Guo and Hassin (2011) consider the vacation queueing model under the N-policy, and they obtained the customer equilibrium strategy and social optimal strategy in both fully observable and completely unobservable situations. Guo and Li (2013) discuss two other situations, that is, almost observable and almost unobservable.

J. Wang, *Fundamentals of Queueing-Game Models*,
https://doi.org/10.1007/978-981-95-0261-5_8

Guo and Hassin (2012) extend this model to the N-policy vacation queues with different types of customers. Wang et al. (2017) explore the strategic behavior and social optimization in a double-ended queue with gated policy. Wang et al. (2017) discuss the strategic behavior and social optimization in the constant retrial queue with N-policy.

Liu et al. (2012) and Ma et al. (2012) respectively study the equilibrium queueing strategy of customers in a discrete queueing system with single and multiple vacations. Wang et al. (2014) consider the Geo/Geo/1 queueing model with single vacation. Gao and Wang (2016) explore the equilibrium balking strategies in the observable Geo/Geo/1 queue with delayed multiple vacations.

Zhang et al. (2013) introduce the game-theoretic analysis to the queueing systems with working vacations. Li et al. (2016) consider the equilibrium joining strategies in M/M/1 queues with working vacation and vacation interruptions. Liu and Wang (2017) present the strategic joining rules in a single server Markovian queue with Bernoulli-type vacation. Zhang and Wang (2023) study customer strategic behavior and operational optimization of an M/M/1 queue with N-policy and dynamic service-rate control in general, and of a make-to-order production system with a strategic manufacturer as a practical example in particular. Wang et al. (2024) consider a stochastic clearing system that the server is subject to a Poisson-generated catastrophe and a follow-up repair process. Whenever a fatal shock (catastrophe) occurs, all customers are cleared from the system and the server fails. A repair is rendered immediately to fix the server with an exponential repair time. The customer's equilibrium strategy and social benefit of the system under four different information scenarios are obtained. Wang et al. (2021) obtain customers' equilibrium joining strategies in an M/M/K queue with asynchronous and synchronous multiple vacations. It is found that the equilibrium social welfare under an asynchronous vacation policy is higher than that under a synchronous vacation policy when the traffic density is low, the opposite relation exists when the traffic density is high. Furthermore, when the traffic density is high enough, the difference between the two vacation policies in terms of social welfare can be negligible. If all customers follow the equilibrium strategies, compared to the situation with no information provided for customers, the social planner would prefer to reveal the queue length.

8.1 M/M/1 Vacation Queueing Systems with Setup Periods

8.1.1 Model Description

Suppose a single service server queueing system with infinite waiting space. Customers arrive as a Poisson process with parameter λ and service time is exponential distributed with parameter μ. The server is deactivated as soon as the queue becomes empty. When a new customer arrives at an empty system, a setup process starts for the server to be reactivated. The time required for setup is also exponentially distributed with rate θ. During the setup customers continue to arrive. We assume that interarrival times, service times and setup times are mutually independent. We represent $N(t)$ denotes the number of customers in the system and $I(t)$ denotes the state of the server at time t. Thus, it is clear that the process $\{N(t), I(t) : t \geq 0\}$ is a continuous time Markov chain with state space $S = \{(n, i) \,|n \geq i, i = 0, 1\}$.

Assume that every customer receives a reward of R units for completing service. On the other hand there exists a waiting cost of C units per time unit that the customer remains in the system. Customers are risk neutral and maximize their expected net benefit. When arriving at the system, customers estimate their average waiting cost through their own system information, and then make a decision whether to join the queue or not. Finally, the decisions are irrevocable: retrials of balking customers and reneging of entering customers are not allowed. To ensure that the reward for service exceeds the expected cost for a customer who finds the system empty, we assume

$$R > \frac{C}{\mu} + \frac{C}{\theta}. \tag{8.1.1}$$

8.1.2 *Equilibrium Threshold Strategies in the Fully Observable Case*

Theorem 8.1.1 *In the fully observable $M/M/1$ queue with setup times there exist thresholds*

$$(ne\,(0)\,, ne\,(1)) = \left(\left\lfloor \frac{R\mu}{C} - \frac{\mu}{\theta} \right\rfloor - 1, \left\lfloor \frac{R\mu}{C} \right\rfloor - 1\right), \tag{8.1.2}$$

such that the strategy "observe $(N(t), (I(t))$, enter if $N(t) \leq ne(I(t))$ and balk otherwise" is a unique equilibrium in the class of threshold strategies. Moreover, it is also a weakly dominant strategy.

Proof A customer who joins the system when he observes state (n, i) has mean sojourn time equal to $\frac{n+1}{\mu} + \frac{1-i}{\theta}$, thus his expected net benefit is

$$R - \frac{C(n+1)}{\mu} - \frac{C(1-i)}{\theta}. \tag{8.1.3}$$

Thus, customer enters if and only if $n + 1 \leq \frac{R\mu}{C} - \frac{\mu(1-i)}{\theta}$. We thus obtain the conclusion. □

8.1.3 Equilibrium Threshold Strategies in the Almost Unobservable Case

In the almost observable case, arriving customers only observe the number of customers in the system $N(t)$ while the server's state $I(t)$ is unobservable. In order to obtain the customer's pure threshold joining strategy, the steady-state distribution of the system state should be calculated first.

Lemma 8.1.1 *In the almost observable $M/M/1$ queue with setup times, assume $\sigma \neq 1 \neq \rho$. When customers follow the threshold policy n_e, the stationary probabilities $(p_{ao}(n,i):(n,i) \in \{(0,0)\} \cup \{1,2,\ldots,n_e+1\} \times \{0,1\})$ is*

$$p_{ao}(1,1) = \Big[\frac{1}{\rho(1-\sigma)(1-\rho)} - \frac{\sigma^{n_e+1}}{(1-\sigma)(1-\rho)} + \frac{\rho^{n_e+2}}{(1-\rho)(\sigma-\rho)}\Big(1-(\frac{\sigma}{\rho})^{n_e+1}\Big)\Big]^{-1}, \tag{8.1.4}$$

$$p_{ao}(n,0) = \frac{1}{\rho}\sigma^n p_{ao}(1,1), n = 0,1,\ldots,n_e, \tag{8.1.5}$$

$$p_{ao}(n_e+1,0) = \frac{\sigma^{n_e(0)+1}}{\rho(1-\sigma)} p_{ao}(1,1), \tag{8.1.6}$$

$$p_{ao}(n,1) = \frac{1}{\sigma-\rho}(\sigma^n - \rho^n)p_{ao}(1,1), \quad n = 1,2,\ldots,n_e+1, \tag{8.1.7}$$

Proof We first write the balance equations as follows

$$\lambda p(0,0) = \mu p(1,1), \tag{8.1.8}$$

$$(\lambda+\theta)p(n,0) = \lambda p(n-1,0), n = 1,2,\ldots,n_e, \tag{8.1.9}$$

$$\theta p(n_e+1,0) = \lambda p(n_e,0), \tag{8.1.10}$$

$$(\lambda+\mu)p(1,1) = \theta p(1,0) + \mu p(2,1), \tag{8.1.11}$$

$$(\lambda+\mu)p(n,1) = \lambda p(n-1,1) + \theta p(n,0) + \mu p(n+1,1), \quad n = 2,3,\ldots,n_e, \tag{8.1.12}$$

$$\mu p(n_e+1,1) = \lambda p(n_e,1) + \theta p(n_e+1,0). \tag{8.1.13}$$

Solve the above equations and denote $\rho = \frac{\lambda}{\mu}$, $\sigma = \frac{\lambda}{\lambda+\theta}$, the stationary probabilities can be obtained. □

The expected sojourn time of a new customer who observe n customers in the system and choose to join the queue is $\frac{n+1}{\mu} + \frac{\pi_{\bar{I}|N}(0|n)}{\theta}$, where $\pi_{\bar{I}|N}(0 \mid n)$ represents the probability of customer who find n customers in system and $I(t) = 0$. His expected net benefit is

$$R - \frac{C(n+1)}{\mu} - \frac{C\pi_{\bar{I}|N}(0 \mid n)}{\theta}, \tag{8.1.14}$$

where

$$\pi_{\bar{I}|N}(0 \mid n) = \frac{\lambda p_{ao}(n,0)}{\lambda p_{ao}(n,0) + \lambda p_{ao}(n,1)\mathbb{I}\{n \geq 1\}}, \quad n = 0, 1, \ldots, n_e + 1, \tag{8.1.15}$$

and $\mathbb{I}\{\cdot\}$ is the indicator function.

According to (8.1.4)–(8.1.7),

$$\pi_{\bar{I}|N}(0 \mid n) = \left[1 + \frac{\lambda + \theta}{\mu - \lambda - \theta}\left(1 - \left(\frac{\rho}{\sigma}\right)^n\right)\right]^{-1}, \quad n = 0, 1, \ldots, n_e, \tag{8.1.16}$$

$$\pi_{\bar{I}|N}(0 \mid n_e + 1) = \left[1 + \frac{\theta}{\mu - \lambda - \theta}\left(1 - \left(\frac{\rho}{\sigma}\right)^{n_e+1}\right)\right]^{-1}. \tag{8.1.17}$$

Construct the following function

$$f(x,n) = R - \frac{C(n+1)}{\mu} - \frac{C}{\theta}\left[1 + \frac{\lambda x + \theta}{\mu - \lambda - \theta}\left(1 - \left(\frac{\rho}{\sigma}\right)^n\right)\right]^{-1}, \quad x \in [0,1], \quad n = 0, 1, \ldots. \tag{8.1.18}$$

Let

$$f_U(n) = f(1,n), \quad f_L(n) = f(0,n), \quad n = 0, 1, 2, \ldots. \tag{8.1.19}$$

It is easy to see $f_U(0) = f_U(0) = R - \frac{C}{\mu} - \frac{C}{\theta} > 0$. In addition, $\lim_{n\to\infty} f_U(n) = \lim_{n\to\infty} f_L(n) = -\infty$. Because of $f_L(n) \leq f_U(n)$, there exists $n_L \leq n_U$ such that

$$f_U(0), f_U(1), \ldots, f_U(n_U) > 0, \quad f_U(n_U + 1) \leq 0, \tag{8.1.20}$$

and

$$f_L(n_U + 1), f_L(n_U), \ldots, f_L(n_L + 1) \leq 0, \quad f_L(n_L) > 0. \tag{8.1.21}$$

Theorem 8.1.2 *In the almost observable $M/M/1$ queue with setup times all pure threshold strategies $n_e \in \{n_L, n_L + 1, \ldots, n_U\}$ are equilibrium strategies.*

Proof Consider a tagged customer at his arrival instant and assume that all other customers follow the same threshold strategy $n_e \in \{n_L, n_L + 1, \ldots, n_U\}$.

If the tagged customer finds $n \leq n_e$ customers and decides to enter, his expected net benefit is equal to

$$R - \frac{C(n+1)}{\mu} - \frac{C}{\theta}\left[1 + \frac{\lambda + \theta}{\mu - \lambda - \theta}\left(1 - \left(\frac{\rho}{\sigma}\right)^n\right)\right]^{-1} = f_U(n) > 0.$$

because of (8.1.14), (8.1.16), (8.1.18)–(8.1.20). So the customer will enter.

If the tagged customer finds $n = n_e + 1$ customers and decides to enter, his expected net benefit is

$$R - \frac{C(n_e+2)}{\mu} - \frac{C}{\theta}\Big[1 + \frac{\theta}{\mu-\lambda-\theta}\Big(1-\Big(\frac{\rho}{\sigma}\Big)^{n_e+1}\Big)\Big]^{-1} = f_L(n_e+1) \le 0.$$

because of (8.1.14), (8.1.17)–(8.1.19) and (8.1.21) So the customer will balk. □

When $n_L < n_U$, there exists multiple equilibrium threshold strategies. This is observed generally in "follow-the-crowd (FTC)" situations. We seek equilibrium strategies in the more general class of mixed threshold strategies (n_e, q_e) next. The following formulas can be derived for the probabilities $\pi^-_{I|N}(0 \mid n)$

$$\pi^-_{I|N}(0 \mid n) = \Big[1 + \frac{\lambda+\theta}{\mu-\lambda-\theta}\Big(1-\Big(\frac{\rho}{\sigma}\Big)^{n}\Big)\Big]^{-1}, \quad n = 0, 1, \ldots, n_e - 1, \tag{8.1.22}$$

$$\pi^-_{I|N}(0 \mid n_e) = \Big[1 + \frac{\lambda q_e+\theta}{\mu-\lambda-\theta}\Big(1-\Big(\frac{\rho}{\sigma}\Big)^{n_e}\Big)\Big]^{-1}, \tag{8.1.23}$$

$$\pi^-_{I|N}(0 \mid n_e+1) = \Big[1 + \frac{\theta}{\mu}\frac{\lambda q_e+\theta}{\mu-\lambda-\theta}\Big(1-\Big(\frac{\rho}{\sigma}\Big)^{n_e}\Big) + \frac{\theta}{\mu}\Big]^{-1}. \tag{8.1.24}$$

For the case $n_L < n_U$, the following theorem shows that there is one mixed threshold equilibrium policy between every two consecutive pure threshold equilibrium policies.

Theorem 8.1.3 *In the almost observable $M/M/1$ queue with setup times all mixed threshold strategies (n_e, q_e) are equilibrium strategies for $n_e \in \{n_L, n_L+1, \ldots, n_U\}$ and*

$$q_e = \frac{1}{\lambda}\Big\{\frac{\mu-\lambda-\theta}{1-(\frac{\rho}{\sigma})^{n_e}}\Big[\frac{C}{\theta(R-\frac{C(n_e+1)}{\mu})} - 1\Big] - \theta\Big\}. \tag{8.1.25}$$

Proof Fix an $n_e \in \{n_L, n_L+1, \ldots, n_U\}$ and define the corresponding q_e by (8.1.25), where q_e is the unique solution of $f(x, n_e) = 0$. We now consider a tagged customer at his arrival instant and assume that all other customers follow the same mixed threshold strategy (n_e, q_e).

If the tagged customer finds $n \le n_e - 1$ customers and decides to enter, his expected net benefit is

$$R - \frac{C(n+1)}{\mu} - \frac{C}{\theta}\Big[1 + \frac{\lambda+\theta}{\mu-\lambda-\theta}\Big(1-\Big(\frac{\rho}{\sigma}\Big)^{n}\Big)\Big]^{-1} = f_U(n) > 0.$$

because of (8.1.14), (8.1.22) and (8.1.20) Therefore, the tagged customer prefers to enter.

If the tagged customer finds $n = n_e$ customers and decides to enter, his expected net benefit is equal to

$$R - \frac{C(n_e+1)}{\mu} - \frac{C}{\theta}\left[1 + \frac{\lambda q_e + \theta}{\mu - \lambda - \theta}\left(1 - \left(\frac{\rho}{\sigma}\right)^{n_e}\right)\right]^{-1} = f(q_e, n_e) = 0.$$

because of (8.1.14), (8.1.23) and (8.1.25) In this case, the initial expression is valid and any joining probability is optimal, so $q_e$is optimal.

If the tagged customer finds $n = n_e + 1$ customers and decides to enter, his expected net benefit is equal to

$$\begin{aligned} &R - \frac{C(n_e+2)}{\mu} - \frac{C}{\theta}\left[1 + \frac{\theta}{\mu}\frac{\lambda q_e}{\mu - \lambda - \theta}\left(1 - \left(\frac{\rho}{\sigma}\right)^{n_e}\right) + \frac{\theta}{\mu}\right]^{-1} \\ &\leq R - \frac{C(n_e+2)}{\mu} - \frac{C}{\theta}\left[1 + \frac{\theta}{\mu - \lambda - \theta}\left(1 - \left(\frac{\rho}{\sigma}\right)^{n_e+1}\right)\right]^{-1} = f_L(n_e+1) \leq 0. \end{aligned}$$

because of(8.1.14), (8.1.24) and (8.1.21). Thus, the customer prefers to balk. □

The equilibrium pure threshold strategies are ESS while the equilibrium mixed threshold strategies are not ESS.

8.1.4 Equilibrium Mixed Strategies for the Almost Unobservable Case

In the almost unobservable case, customers observe the state $I(t)$ of the server at their arrival instant and are not informed about the number of customers $N(t)$. Assume that all customers follow the same mixed strategy$(q(0), q(1))$, where $q(i)$ is the probability of joining when the server is in state i. So the arrival rate equals $\lambda(i) = \lambda q(i)$ for states where the server is in state i. If and only if $\lambda(1) < \mu$, the system is stable. Let $(p_{au}(n, i){:}(n, i) \in \{(0, 0)\} \cup \{1, 2, \ldots\} \times \{0, 1\})$ be the stationary distribution of the corresponding system.

Lemma 8.1.2 *In the almost unobservable $M/M/1$ queue with setup times, assume $\sigma(0) \neq \rho(1)$. If all customers follow the same mixed strategy $(q(0), q(1))$, the corresponding stationary probabilities are*

$$p_{au}(n, 0) = \frac{(1 - \sigma(0))(1 - \rho(1))}{1 - \rho(1) + \rho(0)}\sigma(0)^n, \quad n = 0, 1, \ldots, \tag{8.1.26}$$

$$p_{au}(n, 1) = \frac{(1 - \sigma(0))(1 - \rho(1))\rho(0)}{1 - \rho(1) + \rho(0)(\sigma(0) - \rho(1))}(\sigma(0)^n - \rho(1)^n), \quad n = 1, 2, \ldots, \tag{8.1.27}$$

where $\rho(0) = \frac{\lambda(0)}{\mu}$ $\rho(1) = \frac{\lambda(1)}{\mu}$ $\sigma(0) = \frac{\lambda(0)}{\lambda(0)+\theta}$

Proof The lemma can be obtained by solving the following equilibrium equation:

$$\begin{aligned}
\lambda(0)p(0,0) &= \mu p(1,1),\\
(\lambda(0)+\theta)p(n,0) &= \lambda(0)p(n-1,0), n=1,2,\ldots,\\
(\lambda(1)+\mu)p(1,1) &= \theta p(1,0)+\mu p(2,1),\\
(\lambda(1)+\mu)p(n,1) &= \lambda(1)p(n-1,1)+\theta p(n,0)+\mu p(n+1,1), \quad n=2,3,\ldots.
\end{aligned}$$

The details of derivation are omitted here. □

Consider a customer observe the state i of the server upon arrival, his mean sojourn time is $\frac{\mathbb{E}[N^-|i]+1}{\mu}+\frac{1-i}{\theta}$, where $\mathbb{E}[N^- \mid i]$ is the expected number of customers in system found by an arrival, given that the server is found at state i. The expected benefit of such a customer who decides to enter is

Theorem 8.1.4 *In the almost unobservable $M/M/1$ queue with setup times and $\lambda<\mu$, there exists a unique mixed equilibrium strategy $q_e(0), q_e(1)$ is given as follows:*

Case I: $\frac{1}{\theta}<\frac{1}{\mu}$.

$$(q_e(0),q_e(1))=\begin{cases}
\left(\frac{1}{\lambda}\left(\frac{\mu\theta R}{C}-\mu-\theta\right),0\right), & R\in\left(\frac{C}{\mu}+\frac{C}{\theta},\frac{C(\lambda+\theta)}{\mu\theta}+\frac{C}{\theta}\right);\\
(1,0), & R\in\left[\frac{C(\lambda+\theta)}{\mu\theta}+\frac{C}{\theta},\frac{C(\lambda+\theta)}{\mu\theta}+\frac{C}{\mu}\right);\\
\left(1,\frac{1}{\lambda}\left(\mu-\frac{C}{R-\frac{C(\lambda+\theta)}{\mu\theta}}\right)\right), & R\in\left[\frac{C(\lambda+\theta)}{\mu\theta}+\frac{C}{\mu},\frac{C(\lambda+\theta)}{\mu\theta}+\frac{C}{\mu-\lambda}\right);\\
(1,1), & R\in\left[\frac{C(\lambda+\theta)}{\mu\theta}+\frac{C}{\mu-\lambda},\infty\right).
\end{cases}$$

Case II: $\frac{1}{\mu}\le\frac{1}{\theta}\le\frac{1}{\mu-\lambda}$.

$$(q_e(0),q_e(1))=\begin{cases}
\left(\frac{1}{\lambda}\left(\frac{\mu\theta R}{C}-\mu-\theta\right),\frac{\mu-\theta}{\lambda}\right), & R\in\left(\frac{C}{\mu}+\frac{C}{\theta},\frac{C(\lambda+\theta)}{\mu\theta}+\frac{C}{\theta}\right);\\
\left(1,\frac{1}{\lambda}\left(\mu-\frac{C}{R-\frac{C(\lambda+\theta)}{\mu\theta}}\right)\right), & R\in\left[\frac{C(\lambda+\theta)}{\mu\theta}+\frac{C}{\theta},\frac{C(\lambda+\theta)}{\mu\theta}+\frac{C}{\mu-\lambda}\right);\\
(1,1), & R\in\left[\frac{C(\lambda+\theta)}{\mu\theta}+\frac{C}{\mu-\lambda},\infty\right).
\end{cases}$$

Case III: $\frac{1}{\mu-\lambda}<\frac{1}{\theta}$.

$$(q_e(0),q_e(1))=\begin{cases}
\left(\frac{1}{\lambda}\left(\frac{\mu\theta R}{C}-\mu-\theta\right),1\right), & R\in\left(\frac{C}{\mu}+\frac{C}{\theta},\frac{C(\lambda+\theta)}{\mu\theta}+\frac{C}{\theta}\right);\\
(1,1), & R\in\left[\frac{C(\lambda+\theta)}{\mu\theta}+\frac{C}{\theta},\infty\right).
\end{cases}$$

Proof Consider a tagged customer who finds the server at state 0 upon arrival. Because of

$$\mathbb{E}[N^- \mid 0] = \sum_{n=0}^{\infty} \frac{\lambda(0)p(n,0)n}{\sum_{k=0}^{\infty}\lambda(0)p(n,0)} = \frac{\sigma(0)}{1-\sigma(0)}, \tag{8.1.28}$$

$$\mathbb{E}[N^- \mid 1] = \sum_{n=1}^{\infty} \frac{\lambda(1)p(n,1)n}{\sum_{k=1}^{\infty}\lambda(1)p(n,1)} = \frac{\rho(1)}{1-\rho(1)} + \frac{1}{1-\sigma(0)}, \tag{8.1.29}$$

If he decides to enter, his expected net benefit is

$$R - \frac{C(\mathbb{E}[N^- \mid 0]+1)}{\mu} - \frac{C}{\theta} = R - \frac{C(\lambda(0)+\theta)}{\mu\theta} - \frac{C}{\theta}. \tag{8.1.30}$$

Consider two cases:

Case 1: $\frac{C}{\mu} + \frac{C}{\theta} < R \le \frac{C(\lambda+\theta)}{\mu\theta} + \frac{C}{\theta}$. In this case if all customers who find the system empty enter with probability $q_e(0) = 1$, then the tagged customer suffers a negative expected benefit if he decides to enter. Hence, $q_e(0) = 1$ does not lead to an equilibrium. Similarly, $q_e(0) = 0$ also cannot be part of an equilibrium mixed strategy. Therefore, there exists a unique $q_e(0) = \frac{1}{\lambda}\left(\frac{\mu\theta R}{C} - \mu - \theta\right)$, satisfying

$$R - \frac{C(\lambda q_e(0)+\theta)}{\mu\theta} - \frac{C}{\theta} = 0,$$

for which customers are indifferent between entering and balking.

Case 2: $\frac{C(\lambda+\theta)}{\mu\theta} + \frac{C}{\theta} < R$. In this case, for every strategy of the other customers, the tagged customer has a positive expected net benefit if he decides to enter. Hence, $q_e(0) = 1$

We tag a customer who finds the server at state 1 upon arrival. If he decides to enter his expected net benefit is equal to

$$R - \frac{C(\mathbb{E}[N^- \mid 1]+1)}{\mu} = \begin{cases} \frac{C}{\theta} - \frac{C}{\mu-\lambda(1)}, & \text{case 1;} \\ R - \frac{C}{\mu-\lambda(1)} - \frac{C(\lambda+\theta)}{\mu\theta}, & \text{case 2.} \end{cases} \tag{8.1.31}$$

Similarly, we examine Cases 1 and 2 separately and find $q_e(1)$ in equilibrium, and then get the theorem. This completes the proof. □

When $\lambda \geq \mu$, the mixed equilibrium strategy $(q_e(0), q_e(1))$ is

Case I: $\frac{1}{\theta} < \frac{1}{\mu}$.

$$(q_e(0), q_e(1)) = \begin{cases} \left(\frac{1}{\lambda}\left(\frac{\mu\theta R}{C} - \mu - \theta\right), 0\right), & R \in \left(\frac{C}{\mu} + \frac{C}{\theta}, \frac{C(\lambda+\theta)}{\mu\theta} + \frac{C}{\theta}\right); \\ (1, 0), & R \in \left[\frac{C(\lambda+\theta)}{\mu\theta} + \frac{C}{\theta}, \frac{C(\lambda+\theta)}{\mu\theta} + \frac{C}{\mu}\right); \\ \left(1, \frac{1}{\lambda}\left(\mu - \frac{C}{R - \frac{C(\lambda+\theta)}{\mu\theta}}\right)\right), & R \in \left[\frac{C(\lambda+\theta)}{\mu\theta} + \frac{C}{\mu}, \infty\right). \end{cases}$$

Case II: $\frac{1}{\mu} \leq \frac{1}{\theta}$.

$$(q_e(0), q_e(1)) = \begin{cases} \left(\frac{1}{\lambda}\left(\frac{\mu\theta R}{C} - \mu - \theta\right), \frac{\mu-\theta}{\lambda}\right), & R \in \left(\frac{C}{\mu} + \frac{C}{\theta}, \frac{C(\lambda+\theta)}{\mu\theta} + \frac{C}{\theta}\right); \\ \left(1, \frac{1}{\lambda}\left(\mu - \frac{C}{R - \frac{C(\lambda+\theta)}{\mu\theta}}\right)\right), & R \in \left[\frac{C(\lambda+\theta)}{\mu\theta} + \frac{C}{\theta}, \infty\right). \end{cases}$$

8.1.5 Equilibrium Threshold Strategies in the Fully Unobservable Case

In fully unobservable case, customers enter with the probability q. The stationary distribution of the system state is given by Lemma 8.1.2 by taking $q(0) = q(1) = q$.

Theorem 8.1.5 *In the fully unobservable $M/M/1$ queue with setup times and $\lambda < \mu$, there exists a unique mixed equilibrium strategy q_e is given by*

$$q_e = \begin{cases} \frac{1}{\lambda}\left(\mu - \frac{C}{R - \frac{C}{\theta}}\right), & R \in \left(\frac{C}{\mu} + \frac{C}{\theta}, \frac{C}{\mu-\lambda} + \frac{C}{\theta}\right); \\ 1, & R \in \left[\frac{C}{\mu-\lambda} + \frac{C}{\theta}, \infty\right). \end{cases} \tag{8.1.32}$$

Proof We consider a tagged customer at his arrival instant. If he decides to enter his expected net benefit is

$$R - \frac{C}{\mu}(\mathbb{E}[N^-] + 1) - \frac{C}{\theta}\Pr[I^- = 0] = R - \frac{C}{\mu - \lambda q} - \frac{C}{\theta}. \tag{8.1.33}$$

When $R \in (\frac{C}{\mu} + \frac{C}{\theta}, \frac{C}{\mu-\lambda q} + \frac{C}{\theta})$, we find that $R - \frac{C}{\mu-\lambda q} - \frac{C}{\theta} = 0$ has a unique root in (0, 1) which gives the first branch of (8.1.32). When $R \in (\frac{C}{\mu-\lambda q} + \frac{C}{\theta}, \infty)$, the customers' expected net benefit is positive for any q, and the unique equilibrium point is $q_e = 1$, which gives the second branch of (8.1.32). □

For the case $\lambda \geq \mu$, there exists a unique mixed equilibrium strategy q_e is

$$q_e = \frac{1}{\lambda}\Big(\mu - \frac{C}{R - \frac{C}{\theta}}\Big), \quad R \in \Big(\frac{C}{\mu} + \frac{C}{\theta}, \infty\Big). \tag{8.1.34}$$

8.2 M/M/1 Queueing Systems with Working Vacation

8.2.1 Model Description

We consider a single server queueing system with an infinite waiting room in which the arrival process is Poisson with rate λ and the service order is first-come first-served (FCFS). The server begins a working vacation of the average duration $\frac{1}{\theta}$ as soon as the system becomes empty of customers. If there are customers arriving during the vacation period, the server continues to work at a lower rate. If the system is empty when a working vacation ends, another begins. Otherwise, it switches to the normal working level. During the working vacation period customers are served at a mean rate of μ_0. When the server is not on a working vacation the service rate is μ_1. It is assumed that the service times and vacation times are both exponentially distributed, and $\mu_1 > \mu_0$.

We represent the state at time t by the pair $(N(t), I(t))$, where $N(t)$ and $N(t)$ denote the number of customers and the state of the server (0: on working vacation; 1: normal working state) respectively. It is clear that the process $\{(N(t), I(t)), t \geq 0\}$ is a two-dimensional continuous time Markov chain.

We assume that the customers are allowed to decide whether to join or balk upon their arrival based on the information they have. After service, every customer receives a reward of R units. Moreover, there exists a waiting cost C units per time unit when the customers remain in the system including the time of waiting in the queue and being served. Customers are risk neutral and wish to maximize their expected net benefit. At their arrival instants, they have to assess their expected waiting costs against their reward associated with receiving service, given the information regarding the system, and decide accordingly whether or not to join. Their decisions are irrevocable, meaning that neither reneging of entering customers nor retrials of balking customers are allowed.

8.2.2 Equilibrium Strategies for the Fully Observable Case

In the fully observable case where customers are informed both the state of the server $I(t)$ and the number of the present customers $N(t)$ at arrival time t, a pure threshold strategy is specified by a pair $(n_e(0), n_e(1))$ and the balking strategy has the form "While arriving at time t, observe $(N(t), I(t))$; enter if $N(t) \leq n_e(I(t))$ and balk otherwise".

Theorem 8.2.1 *In the fully observable $M/M/1$ queue with working vacations there exist thresholds $(n_e(0), n_e(1))$ which are given by*

$$n_e(0) = \lfloor x_e \rfloor, \tag{8.2.1}$$

$$n_e(1) = \left\lfloor \frac{R\mu_1}{C} \right\rfloor - 1, \tag{8.2.2}$$

where x_e is the unique root of equation

$$\frac{x+1}{\mu_1} + \frac{\mu_1 - \mu_0}{\mu_1 \theta}\left(1 - \left(\frac{\mu_0}{\mu_0 + \theta}\right)^{x+1}\right) = \frac{R}{C}.$$

Moreover, it is a weakly dominant strategy.

Proof Based on the reward-cost structure which is imposed on the system, we conclude that for an arriving customer, his expected net benefit is

$$S(n, i) = R - CT(n, i), \tag{8.2.3}$$

where $T(n, i)$ denotes his expected mean sojourn time given that he finds the system at state (n, i) upon his arrival. Then, we have the system

$$T(0, 0) = \frac{1}{\mu_0 + \theta} + \frac{\theta}{\mu_0 + \theta}\frac{1}{\mu_1}, \tag{8.2.4}$$

$$T(n, 0) = \frac{1}{\mu_0 + \theta} + \frac{\theta}{\mu_0 + \theta}T(n, 1) + \frac{\mu_0}{\mu_0 + \theta}T(n-1, 0), \quad n = 1, 2, \ldots \tag{8.2.5}$$

$$T(n, 1) = \frac{n+1}{\mu_1}, \quad n = 1, 2, \ldots. \tag{8.2.6}$$

By iterating (8.2.5) and taking into account (8.2.4) and (8.2.6), we obtain

$$T(n, 0) = \frac{n+1}{\mu_1} + \frac{\mu_1 - \mu_0}{\mu_1 \theta}\left(1 - \left(\frac{\mu_0}{\mu_0 + \theta}\right)^{n+1}\right), \quad n = 0, 1, \ldots. \tag{8.2.7}$$

We can easily check that $T(n, 0)$ is strictly increasing for n. A customer strictly prefers to enter if the reward for service exceeds the expected cost for waiting, i.e., $S(n, i) > 0$ and is indifferent between entering and balking if the reward is equal to the cost, i.e. $S(n, i) = 0$

We assume that

$$R > C\left(\frac{1}{\mu_0 + \theta} + \frac{\theta}{\mu_0 + \theta}\frac{1}{\mu_1}\right),$$

which ensures that the reward for service exceeds the expected cost for a customer who finds the system empty. By solving $S(n, i) \geq 0$ for n, using (8.2.3), (8.2.6) and (8.2.7), we obtain that $(n_e(0), n_e(1))$ given by (8.2.1) and (8.2.2). This strategy is preferable, independently of what the other customers do, i.e., it is a weakly dominant strategy. □

8.2.3 *Equilibrium Strategies for the Almost Observable Case*

In the almost observable case where the arriving customers observe the number of the present customers $N(t)$ in the system upon their arrival, but not the state of the server $I(t)$.

Lemma 8.2.1 *In the almost observable* $M/M/1$ *queue with working vacations where the customers enter the system according to the same threshold strategy* n_e, *the stationary distribution* $(p_{ao}(n, i){:}(n, i) \in \{(0, 0)\} \cup \{1, 2, \ldots, n_e + 1\} \times \{0, 1\})$ *is given as follows:*

$$p_{ao}(n, 0) = c_1 z_1^n + c_2 z_2^n, \quad n = 0, 1, \ldots, n_e + 1, \tag{8.2.8}$$

$$p_{ao}(n, 1) = v_1 z_1^n + v_2 z_2^n - (v_1 + v_2)\rho^n, \quad n = 1, 2, \ldots, n_e + 1, \tag{8.2.9}$$

where

$$\rho = \frac{\lambda}{\mu_1},$$

$$z_{1,2} = \frac{(\lambda + \mu_0 + \theta) \pm \sqrt{(\lambda + \mu_0 + \theta)^2 - 4\lambda\mu_0}}{2\mu_0},$$

$$c_i = -\frac{(\mu_0 z_j - \lambda) z_j}{\lambda(z_i - z_j)} \left(\frac{1}{z_i}\right)^{n_e} p(n_e + 1, 0), \quad i = 1, 2, i \neq j,$$

$$v_i = -\frac{\mu_0 z_i - \lambda}{\mu_1 z_i - \lambda} c_i, \quad i = 1, 2,$$

and $p_{ao}(n_e + 1, 0)$ *is computed using the normalization equation.*

Proof The corresponding stationary distribution is obtained as the unique positive normalized solution of the following system of balance equations:

$$(\lambda + \mu_1)p(1, 1) = \theta p(1, 0) + \mu_1 p(2, 1), \tag{8.2.10}$$

$$(\lambda + \mu_1)p(n, 1) = \lambda p(n-1, 1) + \theta p(n, 0) + \mu_1 p(n+1, 1), \quad n = 2, 3, \dots, n_e, \tag{8.2.11}$$

$$\mu_1 p(n_e + 1, 1) = \lambda p(n_e, 1) + \theta p(n_e + 1, 0), \tag{8.2.12}$$

$$\lambda p(0, 0) = \mu_1 p(1, 1) + \mu_0 p(1, 0), \tag{8.2.13}$$

$$(\lambda + \mu_0 + \theta)p(n, 0) = \lambda p(n-1, 0) + \mu_0 p(n+1, 0), \quad n = 1, 2, \dots, n_e, \tag{8.2.14}$$

$$(\mu_0 + \theta)p(n_e + 1, 0) = \lambda p(n_e, 0). \tag{8.2.15}$$

Solving the above equation and according to the normalization condition $\sum_{n=0}^{n_e+1} p(n, 0) + \sum_{n=1}^{n_e+1} p(n, 1) = 1$, we obtain the conclusion. □

We now proceed to find the expected net benefit of a customer who observes n customers ahead of him and decides to enter. We have the following lemma.

Lemma 8.2.2 *In the almost observable $M/M/1$ queue with working vacations where all customers enter the system according to a threshold strategy n_e, the expected net benefit of a customer that observes n customers and decides to enter is given by*

$$S_{n_e}(n) = R - C\Bigg(\frac{n+1}{\mu_1} + \left(1 - \left(\frac{\mu_0}{\mu_0+\theta}\right)^{n+1}\right)\frac{\mu_1 - \mu_0}{\mu_1\theta} \times \frac{z_1^n + \gamma z_2^n}{\frac{\mu_1-\mu_0}{\mu_1 z_1-\lambda} z_1^{n+1} + \frac{\mu_1-\mu_0}{\mu_1 z_2-\lambda}\gamma z_2^{n+1} + \left(\frac{\mu_0 z_1-\lambda}{\mu_1 z_1-\lambda} + \frac{\mu_0 z_2-\lambda}{\mu_1 z_2-\lambda}\gamma\right)\rho^n}\Bigg), \quad n = 0, 1, \dots, n_e + 1, \tag{8.2.16}$$

where

$$\gamma = \frac{c_2}{c_1} = -\frac{1-z_2}{1-z_1}\left(\frac{z_1}{z_2}\right)^{n_e+2}.$$

Moreover, $S_{n_e}(n)$ decreases with respect to n for $0 \le n \le n_e + 1$ and any fixed n_e.

Proof The expected net benefit, if a customer who observes n customers ahead of him upon arrival and decides to enter, is

$$S_{n_e}(n) = R - CT(n), \tag{8.2.17}$$

where $T(n) = \mathbb{E}[S \mid N^- = n]$ denotes his expected mean sojourn time given that he finds n customers in the system just before his arrival. Let $\Pr(I^- = i \mid N^- = n)$ be the probability that the state of the server is i when he observes n customers in the system upon his arrival. Conditioning on the state of the server that he finds upon arrival we obtain

$$T(n) = T(n,1)\Pr(I^- = 1\,|N^- = n) + T(n,0)\Pr(I^- = 0\,|N^- = n)\,. \quad (8.2.18)$$

Due to PASTA property, we have

$$\Pr(I^- = 1\,|N^- = n) = \frac{p_{ao}(n,1)}{p_{ao}(n,0) + p_{ao}(n,1)}, \quad n = 0, 1, \ldots, n_e + 1,$$
$$\Pr(I^- = 0\,|N^- = n) = \frac{p_{ao}(n,0)}{p_{ao}(n,0) + p_{ao}(n,1)}, \quad n = 0, 1, \ldots, n_e + 1,$$

where $p_{ao}(0,1) \equiv 0$.

Using the stationary probabilities obtained in 8.2.1 we can get the conditional probabilities $\Pr(I^- = 1\,|N^- = n)$ and $\Pr(I^- = 0\,|N^- = n)$. Thus we conclude by taking into account (8.2.6), (8.2.7) and (8.2.18) that

$$T(n) = \frac{n+1}{\mu_1} + \left(1 - \left(\frac{\mu_0}{\mu_0+\theta}\right)^{n+1}\right)\frac{\mu_1-\mu_0}{\mu_1\theta} \times \frac{z_1^n + \gamma z_2^n}{\frac{\mu_1-\mu_0}{\mu_1 z_1-\lambda} z_1^{n+1} + \frac{\mu_1-\mu_0}{\mu_1 z_2-\lambda}\gamma z_2^{n+1} + \left(\frac{\mu_0 z_1-\lambda}{\mu_1 z_1-\lambda} + \frac{\mu_0 z_2-\lambda}{\mu_1 z_2-\lambda}\gamma\right)\rho^n}, \quad n = 0, 1, \ldots, n_e + 1. \quad (8.2.19)$$

Then (8.2.16) can be obtained by substituting (8.2.19) in (8.2.17)

We tag a customer who finds $j(1 \le j \le n_e + 1)$ customers in the system upon his arrival and decides to enter. After the first customer being served, the residual sojourn time of the tagged customer is equal to the time from the epoch that he arrives observing $j-1$ customers ahead of him and decides to enter the system to his service completion epoch. Therefore, the expected mean sojourn time when a customer observes j customers $T(j)$, is greater than that, $T(j-1)$, when he observes $j-1$ customers. Because of (8.2.17), we have $S_{n_e}(j) < S_{n_e}(j-1)$ and conclude that $S_{n_e}(n)$ decreases with respect to n. □

It should be noted that since we assume $R > C\left(\frac{1}{\mu_0+\theta} + \frac{\theta}{\mu_0+\theta}\frac{1}{\mu_1}\right)$, or equivalently, $S_{n_e}(0) > 0$, so a customer who finds the system empty, the customer prefers to enter.

Substituting $n = n_e$ and $n = n_e + 1$ respectively into (8.2.16) yields

$$S_{n_e}(n_e) = R - C\left(\frac{n_e+1}{\mu_1} + \left(1 - \left(\frac{\mu_0}{\mu_0+\theta}\right)^{n_e+1}\right)\frac{\mu_1-\mu_0}{\mu_1\theta} \times \frac{\mu_0+\theta}{\frac{\lambda(\mu_1-\mu_0)(\mu_0(\mu_1-\lambda)+\mu_1\theta)}{\mu_0(\mu_1 z_1-\lambda)(\mu_1 z_2-\lambda)} + \frac{\lambda\theta}{z_1-z_2}\left(\frac{z_2}{\mu_1 z_1-\lambda}\left(\frac{\rho}{z_1}\right)^{n_e} - \frac{z_1}{\mu_1 z_2-\lambda}\left(\frac{\rho}{z_2}\right)^{n_e}\right)}\right),$$

$$S_{n_e}(n_e+1) = R - C\left(\frac{n_e+2}{\mu_1} + \left(1 - \left(\frac{\mu_0}{\mu_0+\theta}\right)^{n_e+2}\right)\frac{\mu_1-\mu_0}{\mu_1\theta} \times \frac{\mu_0}{\frac{\lambda(\mu_1-\mu_0)(\mu_1-\lambda)}{(\mu_1 z_1-\lambda)(\mu_1 z_2-\lambda)} + \frac{\lambda\theta}{z_1-z_2}\left(\frac{1}{\mu_1 z_1-\lambda}\left(\frac{\rho}{z_1}\right)^{n_e+1} - \frac{1}{\mu_1 z_2-\lambda}\left(\frac{\rho}{z_2}\right)^{n_e+1}\right)}\right).$$

Define the sequences $(f_1(n) : n = 0, 1, \ldots)$ and $(f_2(n) : n = 1, 2, \ldots)$ by

$$\begin{aligned} f_1(n) = R - C\Bigg(&\frac{n+1}{\mu_1} + \Big(1 - \Big(\frac{\mu_0}{\mu_0+\theta}\Big)^{n+1}\Big)\frac{\mu_1-\mu_0}{\mu_1\theta} \\ &\times \frac{\mu_0+\theta}{\frac{\lambda(\mu_1-\mu_0)(\mu_0(\mu_1-\lambda)+\mu_1\theta)}{\mu_0(\mu_1 z_1-\lambda)(\mu_1 z_2-\lambda)} + \frac{\lambda\theta}{z_1-z_2}\Big(\frac{z_2}{\mu_1 z_1-\lambda}\Big(\frac{\rho}{z_1}\Big)^n - \frac{z_1}{\mu_1 z_2-\lambda}\Big(\frac{\rho}{z_2}\Big)^n\Big)}\Bigg), \quad n = 0, 1, \ldots, \end{aligned} \tag{8.2.20}$$

$$\begin{aligned} f_2(n) = R - C\Bigg(&\frac{n+1}{\mu_1} + \Big(1 - \Big(\frac{\mu_0}{\mu_0+\theta}\Big)^{n+1}\Big)\frac{\mu_1-\mu_0}{\mu_1\theta} \\ &\times \frac{\mu_0}{\frac{\lambda(\mu_1-\mu_0)(\mu_1-\lambda)}{(\mu_1 z_1-\lambda)(\mu_1 z_2-\lambda)} + \frac{\lambda\theta}{z_1-z_2}\Big(\frac{1}{\mu_1 z_1-\lambda}\Big(\frac{\rho}{z_1}\Big)^n - \frac{1}{\mu_1 z_2-\lambda}\Big(\frac{\rho}{z_2}\Big)^n\Big)}\Bigg), \quad n = 1, 2, \ldots. \end{aligned} \tag{8.2.21}$$

By definition, $f_1(n) = S_n(n)$ for $n \geq 0$ and $f_2(n) = S_{n-1}(n)$ for $n \geq 1$. In other words, $f_1(n)$ is the expected net benefit of a customer that observes n customers and decides to enter if given that all other customers follow n threshold strategy, and $f_2(n)$ is the expected net benefit of a customer that observes n customers and decides to enter if given that all other customers follow $n-1$ threshold strategy. Moreover, we have $f_1(n) > f_2(n)$ for $n = 1, 2, \ldots$.

Next we describe the equilibrium balking pure threshold strategies in the almost observable case with the help of the sequences $f_1(n)$ and $f_2(n)$ and have the following result.

Theorem 8.2.2 *For the sequences $f_1(n)$ and $f_2(n)$, there exist finite non-negative integers $n_L \leq n_U$ such that*

$$f_1(0), f_1(1), \ldots, f_1(n_U) > 0, \quad f_1(n_U+1) \leq 0, \tag{8.2.22}$$

and

$$f_2(n_U+1), f_2(n_U), f_2(n_U-1), \ldots, f_2(n_L+1) \leq 0, \quad f_2(n_L) > 0, \tag{8.2.23}$$

or

$$f_2(n_U+1), f_2(n_U), f_2(n_U-1), \ldots, f_2(0) \leq 0. \tag{8.2.24}$$

In the almost observable $M/M/1$ queue with working vacations the pure threshold strategies of the form "While arriving at time t enter if $N(t) \leq n_e$ and balk otherwise", for $n_e \in \{n_L, n_L+1, \ldots, n_U\}$ are equilibrium strategies.

Proof We have that $f_1(0) > 0$ because assume that $S_0(0) > 0$ and $\lim_{n\to\infty} f_1(n) = -\infty$, so if $n_U + 1$ is the subscript of the first non-positive term of the sequence $\{f_1(n)\}$, as shown in Eq. (8.2.22). On the other hand, we have $f_1(n) > f_2(n)$ for $n = 0, 1, \ldots$. We conclude that $f_2(n_U+1) < f_1(n_U+1) \leq 0$. Then, we begin to go backwards, starting from the subscript $n_U + 1$ to 1, and we let n_L be the subscript of the first positive term of the sequence $\{f_2(n)\}$ and we have (8.2.23). If all the terms of $\{f_2(n)\}$ going backwards from $n_U + 1$ to 1 are non-positive we have (8.2.24).

We can now establish the existence of equilibrium threshold policies in the almost observable case. In this model we consider a tagged customer at his arrival instant and assume that all other customers follow the same threshold strategy n_e for some fixed $n_e \in \{n_L, n_L + 1, \dots, n_U\}$.

If the tagged customer finds $n \le n_e$ customers in front of him and decides to enter, his expected net benefit is equal to $S_{n_e}(n) \ge S_{n_e}(n_e) = f_1(n_e) > 0$ because of Lemma 8.2.2, (8.2.20) and (8.2.22). So in this case the customer prefers to enter.

If the tagged customer finds $n = n_e + 1$ customers in front of him and decides to enter, his expected net benefit is equal to $S_{n_e}(n_e + 1) = f_2(n_e) \le 0$ because of Lemma (8.2.21) and (8.2.23) or (8.2.24). So in this case the customer prefers to balk.

In summary, every $n_e \in \{n_L, n_L + 1, \dots, n_U\}$ is an equilibrium threshold and we have a "Follow-The-Crowd" (FTC) situation. □

8.2.4 Equilibrium Strategies for the Almost Unobservable Case

In the almost unobservable case, assume that all customers follow a mixed strategy $(q(0), q(1))$, where $q(i)$ is the probability of joining when the server is in state i. So when the server is in state i, the arrival rate is $\lambda_i = \lambda q(i)$. The system is stable if and only if $\lambda_1 < \mu_1$. Let $(p_{au}(n, i):(n, i) \in \{(0, 0)\} \cup \{1, 2, \dots\} \times \{0, 1\})$ be the stationary distribution of the corresponding system. We have the following conclusion.

Lemma 8.2.3 *In the almost unobservable queue with working vacations in which all customers adopt a mixed balking strategy $(q(0), q(1))$, the stationary distribution is given as follows*

$$p_{au}(n,0) = \frac{(1-x_2)^2(\mu_1-\lambda_1)}{(1-x_2)(\mu_1-\lambda_1)+\theta x_2} x_2^n, \quad n = 0, 1, \dots, \tag{8.2.25}$$

$$p_{au}(n,1) = \frac{(1-x_2)^2(\mu_1-\lambda_1)}{((1-x_2)(\mu_1-\lambda_1)+\theta x_2)(\mu_1 x_2-\lambda_1)} \times \left(\frac{\theta}{1-x_2} x_2^{n+1} + (\mu_0 x_2 - \lambda_0)\left(\frac{\lambda_1}{\mu_1}\right)^n\right), \quad n = 1, 2, \dots, \tag{8.2.26}$$

where

$$x_2 = \frac{(\lambda_0+\mu_0+\theta) - \sqrt{(\lambda_0+\mu_0+\theta)^2 - 4\lambda_0\mu_0}}{2\mu_0}. \tag{8.2.27}$$

Proof The balance equations for the stationary distribution are given as follows:

$$(\lambda_1+\mu_1)p(1,1)=\theta p(1,0)+\mu_1 p(2,1), \tag{8.2.28}$$

$$(\lambda_1+\mu_1)p(n,1)=\lambda_1 p(n-1,1)+\theta p(n,0)+\mu_1 p(n+1,1),\quad n=2,3,\ldots, \tag{8.2.29}$$

$$\lambda_0 p(0,0)=\mu_1 p(1,1)+\mu_0 p(1,0), \tag{8.2.30}$$

$$(\lambda_0+\mu_0+\theta)p(n,0)=\lambda_0 p(n-1,0)+\mu_0 p(n+1,0),\quad n=1,2,\ldots. \tag{8.2.31}$$

Rewriting (8.2.31) as

$$\mu_0 p(n+1,0)-(\lambda_0+\mu_0+\theta)p(n,0)+\lambda_0 p(n-1,0)=0. \tag{8.2.32}$$

which can be considered as a homogeneous linear difference equation of order 2 with constant coefficients and characteristic equation

$$\mu_0 x^2-(\lambda_0+\mu_0+\theta)x+\lambda_0=0, \tag{8.2.33}$$

which has two roots, given by

$$x_{1,2}=\frac{(\lambda_0+\mu_0+\theta)\pm\sqrt{(\lambda_0+\mu_0+\theta)^2-4\lambda_0\mu_0}}{2\mu_0}. \tag{8.2.34}$$

Therefore from the standard theory of homogeneous linear difference equations (see Sect. 2.3 of Elaydi (2005)), we can see that $p(n,0)=d_1x_1^n+d_2x_2^n$ for any $n\geq 0$, where d_1 and d_2 are constants to be determined. Note that $x_1>1$, hence d_1 should be necessarily 0, for $p(n,0)$ are probabilities, i.e.,

$$p(n,0)=d_2x_2^n,\quad n=0,1,\ldots. \tag{8.2.35}$$

By substituting (8.2.35) into (8.2.29), yields

$$p(n+1,1)-p(n,1)+\frac{\theta d_2}{\mu_1x_2-\lambda_1}x_2^{n+1}=\frac{\lambda_1}{\mu_1}\Big(p(n,1)-p(n-1,1)+\frac{\theta d_2}{\mu_1x_2-\lambda_1}x_2^n\Big),\quad n=2,3,\ldots, \tag{8.2.36}$$

By iterating, yields

$$p(n+1,1)-p(n,1)=\Big(\frac{\lambda_1}{\mu_1}\Big)^{n-1}\Big(p(2,1)-p(1,1)+\frac{\theta d_2}{\mu_1 x_2-\lambda_1}x_2^2\Big)-\frac{\theta d_2}{\mu_1 x_2-\lambda_1}x_2^{n+1},\quad n=1,2,\ldots, \tag{8.2.37}$$

where $p(1,1)$ and $p(2,1)$ can be obtained with the help of (8.2.28), (8.2.30) and (8.2.35). After some algebra (8.2.37) can be rewritten as

$$p(n+1,1)-p(n,1)=d_2\Big(\frac{\lambda_1}{\mu_1}\Big)^{n}\Big(\frac{\lambda_0-\mu_0 x_2}{\mu_1}+\frac{\theta x_2}{\mu_1 x_2-\lambda_1}\Big)-\frac{\theta d_2}{\mu_1 x_2-\lambda_1}x_2^{n+1},\quad n=1,2,\ldots. \tag{8.2.38}$$

and rewrite above equation, we have

$$\begin{aligned}p(n+1,1)&+\frac{\mu_1 d_2}{\mu_1-\lambda_1}\Big(\frac{\lambda_0-\mu_0 x_2}{\mu_1}+\frac{\theta x_2}{\mu_1 x_2-\lambda_1}\Big)\Big(\frac{\lambda_1}{\mu_1}\Big)^{n+1}\\&-\frac{\theta d_2}{(1-x_2)(\mu_1 x_2-\lambda_1)}x_2^{n+2}=p(n,1)+\frac{\mu_1 d_2}{\mu_1-\lambda_1}\Big(\frac{\lambda_0-\mu_0 x_2}{\mu_1}+\frac{\theta x_2}{\mu_1 x_2-\lambda_1}\Big)\Big(\frac{\lambda_1}{\mu_1}\Big)^{n}\\&-\frac{\theta d_2}{(1-x_2)(\mu_1 x_2-\lambda_1)}x_2^{n+1},\quad n=1,2,\ldots.\end{aligned} \tag{8.2.39}$$

Then we can get by iterating (8.2.39) and taking into account (8.2.33) and $p(1,1)$ that

$$p(n,1)=\frac{\theta d_2}{(1-x_2)(\mu_1 x_2-\lambda_1)}x_2^{n+1}+\frac{(\mu_0 x_2-\lambda_0)d_2}{\mu_1 x_2-\lambda_1}\Big(\frac{\lambda_1}{\mu_1}\Big)^{n},\quad n=1,2,\ldots. \tag{8.2.40}$$

Then, the unique unknown constant d_2 is computed using the normalization equation

$$d_2=\frac{(1-x_2)^2(\mu_1-\lambda_1)}{(1-x_2)(\mu_1-\lambda_1)+\theta x_2}. \tag{8.2.41}$$

Substituting (8.2.41) into (8.2.35) and (8.2.40), we derive (8.2.25) and (8.2.26). □

We now tag a customer who finds the server at state i upon arrival. The expected mean sojourn time of such a customer that decides to enter is

$$W(i,q(0),q(1))=\frac{\sum_{n=i}^{\infty}T(n,i)p_{au}(n,i)}{\sum_{k=i}^{\infty}p_{au}(k,i)},\quad i=0,1. \tag{8.2.42}$$

Substituting (8.2.6), (8.2.7), (8.2.25) and (8.2.26) into (8.2.42), we obtain

$$W(0, q(0), q(1)) = \frac{1}{\mu_1(1-x_2)} + \frac{\mu_1 - \mu_0}{\mu_1(\theta + \mu_0(1-x_2))}, \tag{8.2.43}$$

$$W(1, q(0), q(1)) = \frac{1}{\mu_1} + \frac{\mu_1 - \lambda_1 x_2}{\mu_1(1-x_2)(\mu_1 - \lambda_1)}. \tag{8.2.44}$$

Based on the reward-cost structure, the expected net benefit of an arriving customer is

$$S_{au}(0, q(0), q(1)) = R - C\Big(\frac{1}{\mu_1(1-x_2)} + \frac{\mu_1 - \mu_0}{\mu_1(\theta + \mu_0(1-x_2))}\Big), \tag{8.2.45}$$

$$S_{au}(1, q(0), q(1)) = R - C\Big(\frac{1}{\mu_1} + \frac{\mu_1 - \lambda_1 x_2}{\mu_1(1-x_2)(\mu_1 - \lambda_1)}\Big). \tag{8.2.46}$$

We now can proceed to determine mixed equilibrium strategies of a customer in the almost unobservable case and have the following theorem.

Theorem 8.2.3 *In the almost unobservable model of the $M/M/1$ queue with working vacations and $\lambda < \mu_1$, a unique Nash equilibrium mixed strategy $(q_e(0), q_e(1))$ "observe i and enter with probability $q_e(i)$" exists, where the vector $(q_e(0), q_e(1))$ is given as follows:*

Case 1a $C\Big(\frac{1}{\mu_0+\theta} + \frac{\theta}{\mu_0+\theta}\frac{1}{\mu_1}\Big) < R \le C\Big(\frac{1}{\mu_1(1-x_2(1))} + \frac{\mu_1-\mu_0}{\mu_1(\theta+\mu_0(1-x_2(1)))}\Big)$ *and* $R < C\Big(\frac{1}{\mu_1} + \frac{1}{\mu_1(1-x_{2e})}\Big)$.
$(q_e(0), q_e(1)) = \Big(\frac{x_{2e}(\mu_0(1-x_{2e})+\theta)}{\lambda(1-x_{2e})}, 0\Big)$.

Case 1b $C\Big(\frac{1}{\mu_0+\theta} + \frac{\theta}{\mu_0+\theta}\frac{1}{\mu_1}\Big) < R \le C\Big(\frac{1}{\mu_1(1-x_2(1))} + \frac{\mu_1-\mu_0}{\mu_1(\theta+\mu_0(1-x_2(1)))}\Big)$ *and* $C\Big(\frac{1}{\mu_1} + \frac{1}{\mu_1(1-x_{2e})}\Big) \le R \le C\Big(\frac{1}{\mu_1-\lambda} + \frac{1}{\mu_1(1-x_{2e})}\Big)$.
$(q_e(0), q_e(1)) = \Big(\frac{x_{2e}(\mu_0(1-x_{2e})+\theta)}{\lambda(1-x_{2e})}, \frac{\mu_1 - \frac{C\mu_1(1-x_{2e})}{R\mu_1(1-x_{2e})-C}}{\lambda}\Big)$.

Case 1c $C\Big(\frac{1}{\mu_0+\theta} + \frac{\theta}{\mu_0+\theta}\frac{1}{\mu_1}\Big) < R \le C\Big(\frac{1}{\mu_1(1-x_2(1))} + \frac{\mu_1-\mu_0}{\mu_1(\theta+\mu_0(1-x_2(1)))}\Big)$ *and* $C\Big(\frac{1}{\mu_1-\lambda} + \frac{1}{\mu_1(1-x_{2e})}\Big) < R$.
$(q_e(0), q_e(1)) = \Big(\frac{x_{2e}(\mu_0(1-x_{2e})+\theta)}{\lambda(1-x_{2e})}, 1\Big)$.

Case 2a $C\Big(\frac{1}{\mu_1(1-x_2(1))} + \frac{\mu_1-\mu_0}{\mu_1(\theta+\mu_0(1-x_2(1)))}\Big) < R$ *and* $R < C\Big(\frac{1}{\mu_1(1-x_2(1))} + \frac{1}{\mu_1}\Big)$.
$(q_e(0), q_e(1)) = (1, 0)$.

Case 2b $C\Big(\frac{1}{\mu_1(1-x_2(1))} + \frac{\mu_1-\mu_0}{\mu_1(\theta+\mu_0(1-x_2(1)))}\Big) < R$ *and* $C\Big(\frac{1}{\mu_1(1-x_2(1))} + \frac{1}{\mu_1}\Big) \le R \le C\Big(\frac{1}{\mu_1(1-x_2(1))} + \frac{1}{\mu_1-\lambda}\Big)$.
$(q_e(0), q_e(1)) = \Big(1, \frac{\mu_1 - \frac{C\mu_1(1-x_2(1))}{R\mu_1(1-x_2(1))-C}}{\lambda}\Big)$.

Case 2c $C\left(\frac{1}{\mu_1(1-x_2(1))}+\frac{\mu_1-\mu_0}{\mu_1(\theta+\mu_0(1-x_2(1)))}\right)<R$ *and* $C\left(\frac{1}{\mu_1(1-x_2(1))}+\frac{1}{\mu_1-\lambda}\right)<R$. $(q_e(0),q_e(1))=(1,1)$, *where*

$$x_{2e}=1+\frac{\mu_1(R\theta-C)-\sqrt{\mu_1^2(R\theta-C)^2+4RC\theta\mu_1\mu_0}}{2R\mu_1\mu_0}, \tag{8.2.47}$$

$$x_2(1)=\frac{(\lambda+\mu_0+\theta)-\sqrt{(\lambda+\mu_0+\theta)^2-4\lambda\mu_0}}{2\mu_0}. \tag{8.2.48}$$

Proof Suppose that customers who find the server at state i upon arrival enter with probability $q(i)$ and consider a tagged arriving customer. Then, the tagged customer prefers to enter if $S_{au}(i,q(0),q(1))>0$, he is indifferent between entering and balking if $S_{au}(i,q(0),q(1))=0$ and he prefers to balk if $S_{au}(i,q(0),q(1))<0$. We note that x_2 is a function of $q(0)$ and since

$$\frac{d}{dq(0)}x_2(q(0))=\frac{\lambda}{2\mu_0}\left(1-\frac{\lambda q(0)+\theta-\mu_0}{\sqrt{(\lambda q(0)+\theta-\mu_0)^2+4\mu_0\theta}}\right)>0,\quad q(0)\in[0,1],$$

x_2 is strictly increasing for $q(0)\in[0,1]$.

On the other hand, $S_{au}(0,q(0),q(1))$ is strictly decreasing for $x_2\in[0,x_2(1)]$, where$0<x_2(1)<1$. Thus, we conclude that $S_{au}(0,q(0),q(1))$ is strictly decreasing for $q(0)\in[0,1]$. We solve the equation $S_{au}(0,q(0),q(1))=0$ for $x_2(q(0))$. Let x_{2e} be the unique solution of equation for $x_{2e}<1$ and x_{2e} is given by

$$x_{2e}=1+\frac{\mu_1(R\theta-C)-\sqrt{\mu_1^2(R\theta-C)^2+4RC\theta\mu_1\mu_0}}{2R\mu_1\mu_0}.$$

The corresponding unique $q_e^*(0)$ is found by considering (8.2.33) for $x=x_{2e}$. Solving this linear equation with respect to $q(0)$, we obtain

$$q_e^*(0)=\frac{x_{2e}(\mu_0(1-x_{2e})+\theta)}{\lambda(1-x_{2e})}. \tag{8.2.49}$$

Therefore, we have two cases:

Case 1 $C\left(\frac{1}{\mu_0+\theta}+\frac{\theta}{\mu_0+\theta}\frac{1}{\mu_1}\right)<R\le C\left(\frac{1}{\mu_1(1-x_2(1))}+\frac{\mu_1-\mu_0}{\mu_1(\theta+\mu_0(1-x_2(1)))}\right)$.

In this case if all customers who find the server at state 0 enter with probability 1, then the tagged customer suffers a non-positive expected benefit if he decides to enter. Hence, $q_e(0)=1$ does not lead to an equilibrium. Similarly, $q_e(0)=0$ also cannot be part of an equilibrium mixed strategy. Therefore, there exists a unique $q_e^*(0)$ given by (8.2.49), for which customers are indifferent between entering and balking.

Case 2 $C\left(\frac{1}{\mu_1(1-x_2(1))}+\frac{\mu_1-\mu_0}{\mu_1(\theta+\mu_0(1-x_2(1)))}\right)<R$.

In this case, for every strategy of the other customers, the tagged customer has a positive expected net benefit if he decides to enter. Thus, his best response is 1, i.e., entering is a dominant strategy and in particular the unique Nash equilibrium strategy.

We next tag a customer who finds the server at state 1 upon arrival. From (8.2.46) we have that his expected net benefit is equal to

$$R - C\Big(\frac{1}{\mu_1 - \lambda_1} + \frac{1}{\mu_1(1 - x_2(q_e(0)))}\Big) = \begin{cases} R - C\Big(\frac{1}{\mu_1-\lambda_1} + \frac{1}{\mu_1(1-x_{2e})}\Big), & \text{case 1;} \\ R - C\Big(\frac{1}{\mu_1-\lambda_1} + \frac{1}{\mu_1(1-x_2(1))}\Big), & \text{case 2.} \end{cases} \tag{8.2.50}$$

Therefore, to find $q_e(1)$ in equilibrium, we must examine Cases 1 and 2 separately and consider several subcases in each by a similar argument in determining $q_e(0)$. Finally, we obtain Cases (1a)–(2c) in the statement of this theorem. □

In Theorem 8.2.3 we assume that $\lambda < \mu_1$. For the opposite case, an analogous result holds by following a similar analysis, that, when $\lambda \geq \mu_1$, a unique Nash equilibrium mixed strategy $(q_e(0), q_e(1))$ exists.

Case 1a $C\Big(\frac{1}{\mu_0+\theta} + \frac{\theta}{\mu_0+\theta}\frac{1}{\mu_1}\Big) < R \leq C\Big(\frac{1}{\mu_1(1-x_2(1))} + \frac{\mu_1-\mu_0}{\mu_1(\theta+\mu_0(1-x_2(1)))}\Big)$ and $R < C\Big(\frac{1}{\mu_1} + \frac{1}{\mu_1(1-x_{2e})}\Big)$.

$(q_e(0), q_e(1)) = \Big(\frac{x_{2e}(\mu_0(1-x_{2e})+\theta)}{\lambda(1-x_{2e})}, 0\Big)$.

Case 1b $C\Big(\frac{1}{\mu_0+\theta} + \frac{\theta}{\mu_0+\theta}\frac{1}{\mu_1}\Big) \leq R \leq C\Big(\frac{1}{\mu_1(1-x_2(1))} + \frac{\mu_1-\mu_0}{\mu_1(\theta+\mu_0(1-x_2(1)))}\Big)$ and $C\Big(\frac{1}{\mu_1} + \frac{1}{\mu_1(1-x_{2e})}\Big) \leq R$.

$(q_e(0), q_e(1)) = \Big(\frac{x_{2e}(\mu_0(1-x_{2e})+\theta)}{\lambda(1-x_{2e})}, \frac{\mu_1 - \frac{C\mu_1(1-x_{2e})}{R\mu_1(1-x_{2e})-C}}{\lambda}\Big)$.

Case 2a $C\Big(\frac{1}{\mu_1(1-x_2(1))} + \frac{\mu_1-\mu_0}{\mu_1(\theta+\mu_0(1-x_2(1)))}\Big) < R$ and $R < C\Big(\frac{1}{\mu_1(1-x_2(1))} + \frac{1}{\mu_1}\Big)$.

$(q_e(0), q_e(1)) = (1, 0)$.

Case 2b $C\Big(\frac{1}{\mu_1(1-x_2(1))} + \frac{\mu_1-\mu_0}{\mu_1(\theta+\mu_0(1-x_2(1)))}\Big) < R$ and $C\Big(\frac{1}{\mu_1(1-x_2(1))} + \frac{1}{\mu_1}\Big) \leq R$.

$(q_e(0), q_e(1)) = \Big(1, \frac{\mu_1 - \frac{C\mu_1(1-x_2(1))}{R\mu_1(1-x_2(1))-C}}{\lambda}\Big)$.

8.2.5 Equilibrium Strategies for the Fully Unobservable Case

In the fully unobservable case, where the customers do not observe the state of the system at all. Consider symmetric equilibrium strategies in this case. With a mixed strategy, an arriving customer joins with a certain probability, q. To identical the equilibrium strategies of the customers, we should first investigate the stationary distribution of the system when all customers follow a given strategy q, which can be obtained by Lemma 8.2.3 by taking $q(0) = q(1) = q$. So we have

$$p_{fu}(n,0) = \frac{(1-x_2(q))^2(\mu_1-\lambda q)}{(1-x_2(q))(\mu_1-\lambda q)+\theta x_2(q)} x_2(q)^n, \quad n = 0, 1, \ldots, \tag{8.2.51}$$

$$p_{fu}(n,1) = \frac{(1-x_2(q))^2(\mu_1-\lambda q)}{((1-x_2(q))(\mu_1-\lambda q)+\theta x_2(q))(\mu_1 x_2(q)-\lambda q)} \times\Big(\frac{\theta}{1-x_2(q)} x_2(q)^{n+1} + (\mu_0 x_2(q)-\lambda q)\Big(\frac{\lambda q}{\mu_1}\Big)^n\Big), \quad n = 1, 2, \ldots, \tag{8.2.52}$$

where

$$x_2(q) = \frac{(\lambda q+\mu_0+\theta) - \sqrt{(\lambda q+\mu_0+\theta)^2 - 4\lambda q\mu_0}}{2\mu_0}. \tag{8.2.53}$$

Then the mean number of the customers in the system

$$\begin{aligned}\mathbb{E}[N] &= \sum_{i=1}^{\infty} n(p_{fu}(n,0)+p_{fu}(n,1)) \\ &= \frac{x_2(q)}{1-x_2(q)} + \frac{\theta\mu_1 x_2(q)}{((1-x_2(q))(\mu_1-\lambda q)+\theta x_2(q))(\mu_1-\lambda q)}.\end{aligned} \tag{8.2.54}$$

Noting the following relationship $(\lambda q+\theta)x_2(q) = \lambda - \mu_0 x_2(q)(1-x_2(q))$, (8.2.54) can be rewritten as

$$\mathbb{E}[N] = \frac{x_2(q)}{1-x_2(q)}\Big(1+\frac{\theta\mu_1}{(\mu_1-\mu_0 x_2(q))(\mu_1-\lambda q)}\Big). \tag{8.2.55}$$

Hence, the expected mean sojourn time of a customer who decides to enter upon his arrival can be obtained by using Little's law

$$\mathbb{E}[W] = \frac{x_2(q)}{\lambda q(1-x_2(q))}\Big(1+\frac{\theta\mu_1}{(\mu_1-\mu_0 x_2(q))(\mu_1-\lambda q)}\Big). \tag{8.2.56}$$

Lemma 8.2.4 *In the fully unobservable model of the $M/M/1$ queue with working vacations and $\lambda < \mu_1$, $\mathbb{E}[W]$, the expected mean sojourn time of a customer who decides to enter, is strictly increasing for $q \in [0, 1]$.*

Proof Let

$$g_1(q) = \frac{x_2(q)}{\lambda q(1-x_2(q))}, \tag{8.2.57}$$

$$g_2(q) = 1+\frac{\theta\mu_1}{(\mu_1-\mu_0 x_2(q))(\mu_1-\lambda q)}, \tag{8.2.58}$$

and (8.2.56) can be rewritten as

$$\mathbb{E}[W] = g_1(q)g_2(q). \tag{8.2.59}$$

We respectively differentiate $g_1(q)$ and $g_2(q)$ with respect to q

$$\frac{dg_1(q)}{dq} = \frac{\lambda^2 q(\sqrt{(\lambda_0+\theta-\mu_0)^2+4\mu_0\theta}-(\lambda_0+\theta-\mu_0))}{2\mu_0(\lambda q(1-x_2(q)))^2}\times$$
$$\frac{(\lambda_0+\mu_0+\theta)-\sqrt{(\lambda_0+\mu_0+\theta)^2-4\lambda_0\mu_0}}{\sqrt{(\lambda_0+\mu_0+\theta)^2-4\lambda_0\mu_0}(\sqrt{(\lambda_0+\mu_0+\theta)^2-4\lambda_0\mu_0}+(\lambda_0+\mu_0+\theta))}, \tag{8.2.60}$$

$$\frac{dg_2(q)}{dq} = \frac{\theta\mu_1}{(\mu_1-\mu_0 x_2(q))^2(\mu_1-\lambda q)^2}(\lambda(\mu_1-\mu_0 x_2(q))+\mu_0\frac{dx_2(q)}{dq}(\mu_1-\lambda q)). \tag{8.2.61}$$

Since $\frac{dx_2(q)}{dq} = \frac{\lambda}{2\mu_0}\left(1-\frac{\lambda q+\theta-\mu_0}{\sqrt{(\lambda q+\theta-\mu_0)^2+4\mu_0\theta}}\right) > 0$ and $0 \le x_2(q) < 1$, by considering the condition $\lambda < \mu_1$, we can obtain that $\frac{dg_1(q)}{dq} > 0$ and $\frac{dg_2(q)}{dq} > 0$, i.e., both $g_1(q)$ and $g_2(q)$ are strictly increasing for $q \in [0, 1]$. Thus, for both $g_1(q)$ and $g_2(q)$ are positive, we conclude that $\mathbb{E}[W]$ is strictly increasing for $q \in [0, 1]$. □

The equilibrium behavior of the customers is described in the following.

Theorem 8.2.4 *In the fully unobservable model of the M/M/1 queue with working vacations and $\lambda < \mu_1$, a unique Nash equilibrium mixed strategy "enter with probability q_e" exists, where q_e is given by*

$$q_e = \begin{cases} q_e^*, \ \text{if} \ \frac{C(\mu_1+\theta)}{\mu_1(\mu_0+\theta)} < R < \frac{Cx_2(1)}{\lambda(1-x_2(1))}\left(1+\frac{\theta\mu_1}{(\mu_1-\mu_0x_2(1))(\mu_1-\lambda)}\right), \\ 1, \ \ \text{if} \ R \ge \frac{Cx_2(1)}{\lambda(1-x_2(1))}\left(1+\frac{\theta\mu_1}{(\mu_1-\mu_0x_2(1))(\mu_1-\lambda)}\right), \end{cases} \tag{8.2.62}$$

where q_e^* is the unique solution of equation for $q \in (0, 1)$

$$R - \frac{Cx_2(q)}{\lambda q(1-x_2(q))}\left(1+\frac{\theta\mu_1}{(\mu_1-\mu_0x_2(q))(\mu_1-\lambda q)}\right) = 0.$$

Proof We tag a customer at his arrival instant. If he decides to enter his expected net benefit is

$$S_{fu}(q) = R - \frac{Cx_2(q)}{\lambda q(1-x_2(q))}\left(1+\frac{\theta\mu_1}{(\mu_1-\mu_0x_2(q))(\mu_1-\lambda q)}\right), \tag{8.2.63}$$

and we have

$$S_{fu}(0) = R - \frac{C(\mu_1+\theta)}{\mu_1(\mu_0+\theta)}, \tag{8.2.64}$$

$$S_{fu}(1) = R - \frac{Cx_2(1)}{\lambda(1-x_2(1))}\left(1+\frac{\theta\mu_1}{(\mu_1-\mu_0x_2(1))(\mu_1-\lambda)}\right). \tag{8.2.65}$$

It follows from Lemma 8.2.4 that, when $R \in \left(\frac{C(\mu_1+\theta)}{\mu_1(\mu_0+\theta)}, \frac{Cx_2(1)}{\lambda(1-x_2(1))}\left(1+\frac{\theta\mu_1}{(\mu_1-\mu_0x_2(1))(\mu_1-\lambda)}\right)\right)$, there exists a unique solution of the equation $S_{fu}(q)=0$ which lies in the interval (0,1), denoted by q_e^*. Thus we obtain the first branch of (8.2.62). When $R \in \left[\frac{Cx_2(1)}{\lambda(1-x_2(1))}\left(1+\frac{\theta\mu_1}{(\mu_1-\mu_0x_2(1))(\mu_1-\lambda)}\right), \infty\right)$, $S_{fu}(q)$ is positive for every q. In other words, the tagged customer s best response is 1 in this case. Thus, entering is the unique Nash equilibrium strategy (in fact a dominant strategy) and we obtain the second branch of (8.2.62). □

For the case $\lambda \geq \mu_1$, when $R \geq \frac{Cx_2(1)}{\lambda(1-x_2(1))}\left(1+\frac{\theta\mu_1}{(\mu_1-\mu_0x_2(1))(\mu_1-\lambda)}\right)$, the analog of Theorem 8.2.4 is that a unique Nash equilibrium mixed strategy "enter with probability q_e" exists, where

$$q_e = q_e^*. \tag{8.2.66}$$

It is easy to verify that the ATC situation exists in fully unobservable scenarios.

8.3 M/M/1 Queueing System with N-Policy

8.3.1 Model Description

Consider an M/M/1 queueing system with N-policy. Potential customers arrive in accordance with a Poisson process with rate Λ and the service times are independent and exponentially distributed with μ. The server shuts down when the system becomes empty of customers and resumes service after N arrivals. Unless otherwise stated, assume that $N > 1$; otherwise, it becomes a regular M/M/1 queue. B and I represent that the server is "busy" and "idle", respectively.

Based on the information, the customer decides whether to join or not. R is the reward from receiving the service and there exists a waiting cost of C units per time unit that the customer remains in the system. Customers are risk neutral and wish to maximize their expected net benefit. Their decisions are irrevocable, meaning that neither reneging of entering customers nor retrials of balking customers are allowed. Let $\nu = \frac{R\mu}{C}$ be the upper bound on the number of service epochs and $\rho = \frac{\Lambda}{\mu}$ be the system utilization factor when every potential customer joins.

8.3.2 Equilibrium Strategies for the Fully Observable Case

In the fully observable case we assume that customers have information both on the state of server and the queue length. We use superscripts $-$ and $+$ to distinguish between states with idle and busy server, respectively, when both states are possible. Thus, the set of the states is

$$\{0, 1^-, \ldots, (N-1)^-, 1^+, 2^+, \ldots, (N-1)^+, N, N+1, \ldots\},$$

where m^- means that the system occupancy is m and the server is idle, and m^+ means that the system occupancy is m and the server is busy.

In general, the strategy to never join is always an equilibrium when $N > 1$, since if this policy is adopted by others, the expected wait for a customer who joins at state 0 is infinite, and thus balking at this state is the best response. We concentrate on the existence of other equilibrium strategies in which the server is busy for at least a positive fraction of the time. We refer to such as a solution with an active server.

8.3.2.1 Equilibrium Strategy

When $\Lambda > \mu$, the longest expected waiting time for a customer who arrives when the server is idle is when there are $N-1$ customers waiting in front of him/her; hence, the waiting time is $\frac{N}{\mu}$. When $\Lambda < \mu$, the longest waiting happens when a customer joins an empty system; the waiting time is $\frac{N-1}{\lambda} + \frac{1}{\mu}$. Customers join an idle queue if their reward is at least as great as the worst waiting cost. We summarize the sufficient conditions for a server to be active in the following theorem.

Theorem 8.3.1 *There exists an equilibrium solution with an active server if and only if either (1) $\rho \geq 1$ and $\nu \geq N$, or (2) $\rho < 1$ and $\nu \geq \frac{N-1}{\rho} + 1$.*

We now assume that the conditions for an active server as given in Theorem 8.3.1 are satisfied. Observe that an arrival at state m^+ is associated with a lower expected waiting time than arriving at m^-. It follows that in equilibrium with an active server, all join when the number of customers observed upon arrival is at most $N-1$. The equilibrium strategy is therefore characterized by a threshold value $n_e \geq N$. A customer who observes state m^+ is motivated to join if $R \geq \frac{C(m+1)}{\mu}$ or $m \leq \frac{R\mu}{C} - 1$. Therefore, $n_e = \lfloor \frac{RC}{\mu} \rfloor$. The conditions of Theorem 8.3.1 indeed guarantee that this value is at least N. We summarize the equilibrium threshold solution in the following proposition.

Theorem 8.3.2 *Suppose that the conditions of Theorem 8.3.1 are satisfied. Then the unique equilibrium threshold $n_e = \lfloor \nu \rfloor \geq N$.*

8.3.2.2 Social Optimal Strategy

By the assumption, a customer joins if his utility is 0. Since it is assumed that the customers are identical, the state-dependent arrival rates are either 0 or Λ. In the idle states, obviously, it is never optimal to set any arrival rate to be 0; otherwise, the system will stay idle forever. Therefore, the social planner is left to decide which states will have Λ when the server is busy. Consider two adjacent states n^+ and $(n+1)^+$. Obviously, when $n \geq N$, it will never be possible that the arrival rate on

n^+ is 0 while the one on $(n+1)^+$ is Λ as the state $(n+1)^+$ cannot be reached. However, when $n < N$, state $(n+1)^+$ is reachable from state $(n+2)^+$ even though the arrival rate at n^+ is 0. But the case with the arrival rate on n+ to be 0 and on $(n+1)^+$ to be Λ is not optimal, as we can simply consider a similar system with the arrival rate being Λ on both n^+ and $(n+1)^+$. The latter system can generate higher average utility for customers.

Denote the optimal threshold by n^*. If social welfare is positive, then there are two cases: 1. $n^* \geq N$; 2. $n^* \in \{1 \ldots N-1\}$. Case 2 means that if the server is idle, customers always join, and when the server is busy, customers are not allowed to join if the system occupancy is n^* or larger. The maximum social welfare obtained in cases 1 and 2 is denoted by SW_1 and SW_2, respectively. The optimal solution value is then $SW = \max\{SW_1, SW_2\}$. In the following analysis, we present the results with $\rho \neq 1$. One can easily derive the results for the degenerate case with $\rho = 1$.

Case 1. $n^* \geq N$.

The corresponding balance equations for n is

$$p_0\Lambda = p_{1^+}\mu, \tag{8.3.1}$$

$$p_{m^-} = p_{(m+1)^-}, \quad 0 \leq m \leq N-2, \tag{8.3.2}$$

$$p_0\Lambda + p_{m^+}\Lambda = p_{(m+1)^+}\mu, \quad 0 < m < N, \quad (p_{N^+} \equiv p_N) \tag{8.3.3}$$

$$p_m\Lambda = p_{m+1}\mu, \quad N < m < n. \tag{8.3.4}$$

By solving the above equations, the social welfare per time unit is

$$SW_1(n) = R\Lambda(1 - p_n) - CL,$$

where

$$p_n = \frac{\rho^{n-N+1}(1-\rho^N)}{1-\rho}p_0,$$

$$p_0 = \frac{(1-\rho)^2}{N - N\rho - \rho^{n-N+2} + \rho^{n+2}},$$

the expected number of customers in the system is

$$L = \frac{p_0}{1-\rho} \cdot \frac{(N-1)N}{2} + p_0\rho\frac{(1-\rho)N + \rho^{n+1}(1-\rho^{-N})(n(1-\rho)+1)}{(1-\rho)^3}. \tag{8.3.5}$$

Case 2. $n^* \in \{1, \dots, N-1\}$.

The corresponding balance equations for n is

$$p_0\Lambda = p_{1^+}\mu, \tag{8.3.6}$$

$$p_{m^-} = p_{(m+1)^-}, \quad 0 \le m \le N-2, \tag{8.3.7}$$

$$p_0\Lambda + p_{m^+}\Lambda = p_{(m+1)^+}\mu, \quad 0 < m < n, \tag{8.3.8}$$

$$p_0\Lambda = p_{m^+}\mu, \quad n < m \le N. \tag{8.3.9}$$

By solving the above equations, the social welfare per time unit is

$$\begin{aligned} SW_2(n) &= (R - CW)\Lambda(1 - p_{n^+} - p_{(n+1)^+} - \cdots - p_N) \\ &= R\Lambda\Big(1 - \frac{\rho(1-\rho^n)}{1-\rho}p_0 - \rho(N-n)p_0\Big) - CL, \end{aligned}$$

where

$$p_0 = \frac{(1-\rho)^2}{N - N\rho - (N-n+1)\rho^2 + (N-n)\rho^3 + \rho^{n+2}},$$

the expected number of customers in the system is

$$\begin{aligned} L = {} & \frac{p_0(N-1)N}{2} + \frac{p_0\rho}{1-\rho}\Big(\frac{n(n+1)}{2} + \frac{-\rho + (n+1)\rho^{n+1} - n\rho^{n+2}}{(1-\rho)^2}\Big) \\ & + \rho p_0\frac{(n+N+1)(N-n)}{2}. \end{aligned} \tag{8.3.10}$$

When $n \ge N$, define $SW(n)$ as $SW_1(n)$ and $SW_2(n)$ otherwise.

Theorem 8.3.3 *The social welfare function $SW(n)$ is unimodal, so the optimal threshold n^* is unique.*

Proof Consider two systems that customers take n and $n+1$ as the queueing threshold respectively, so there is a relationship

$$SW(n+1) = \frac{p_0(n+1)}{p_0(n)}SW(n) + p_n(n+1)\Lambda(R - CW_n),$$

where W_n is the average sojourn time at state i.

We can demonstrate that in $n \ge N$, $n < N-1$ and $n = N-1$, $SW(n+1) - SW(n) \le 0$ if $SW(n) - SW(n-1) \le 0$, so $SW(n)$ is unimodal. The social optimal threshold n^* is unique from the unimodal property of the function. □

8.3.2.3 Optimal Value of the Threshold N

Assuming that the value of N is adjustable and the operating cost per unit of time when the server is busy is c_b, the average operating cost per unit of time is $\theta(N) = c_b P_B$, where P_B represents the probability of server in busy. The social problem is to compute an activation value N^* which maximizes social welfare

$$(R - CW)\Lambda(1 - p_{n_e}) - \theta(N)$$

where W is expected sojourn time in the system and p_{n_e} is the probability for the state to be n_e.

Similar to case 1 in 2, we can obtain

$$\begin{aligned} p_{n_e} &= \frac{(1-\rho)(\rho^{n_e-N+1} - \rho^{n_e+1})}{N(1-\rho) - \rho^{n_e-N+2} + \rho^{n_e+2}}, \\ P_B &= 1 - \frac{N(1-\rho)^2}{N - N\rho - \rho^{n_e-N+2} + \rho^{n_e+2}} = \Lambda\frac{1-p_e}{\mu}. \end{aligned} \tag{8.3.11}$$

the social welfare can be rewritten as

$$\Lambda(1 - p_{n_e})\left(R - CW - \frac{c_b}{\mu}\right) \tag{8.3.12}$$

assume that $R \geq CW + \frac{c_b}{\mu}$, otherwise, the server will not work at all.

Lemma 8.3.1 *In an $M/M/1$ queue with N-policy, both p_n and W are increasing in N.*

Proof When $\rho \neq 1$,

$$p_n = \frac{1}{\frac{N}{\rho^{n+1}(\rho^{-N}-1)} - \frac{\rho}{1-\rho}}. \tag{8.3.13}$$

Take the derivative of the above equation, we can obtain p_n is increasing in N. When $\rho = 1$, $p_n = \frac{1}{n-N/2+1.5}$ is also increasing with N.

Due to Little's law, the expected sojourn time is

$$W = \frac{L}{\Lambda(1 - p_n)}, \tag{8.3.14}$$

where L and p_n is shown by (8.3.5) and (8.3.13), respectively. By taking the derivative of the above equation, we can obtain W is increasing with N. □

Lemma 8.3.1 implies that the welfare function (8.3.12) is decreasing in N. Thus, the decision maker should make N the smallest, that is, $N^* = 1$.

Theorem 8.3.4 *When the queue is observable, $N^* = 1$.*

8.3.3 Equilibrium Strategies for the Almost Observable Case

In the almost observable case, customers can only observe the queue length. Clearly, when an incoming customer sees an empty system, the customer knows that the server is inactive. Therefore, "all balking" is always an equilibrium. Let $P_{k,s}$ be the stable probability of system where queue length is k and server's state is s ($s = B, I$). In the following analysis, we consider the positive equilibrium, under which the server can be active.

8.3.3.1 Equilibrium Strategy

We consider a threshold strategy with threshold n. With this strategy, the customers join the system if and only if there are, at most, $n - 1$ customers in the system. As the server s status is unobservable, the threshold should be at least N, i.e., $n \geq N$, as otherwise the server will never be activated once it is idle. To ensure that all customers join in seeing states $\{0, 1, \ldots, N - 1\}$, we need to find the largest expected waiting time on those states and make sure that the net customer utility is still nonnegative.

Lemma 8.3.2 *(1) The conditional expected waiting time for state k, W_k, satisfies*

$$W_k = \begin{cases} \frac{k+1}{\mu} + \frac{(1-\rho)[N-(k+1)]}{(1-\rho^{k+1})\Lambda}, & \text{if } 0 \leq k \leq N-1, \\ \frac{k+1}{\mu}, & \text{if } k \geq N. \end{cases}$$

(2) The conditional expected waiting time W_k is convex in k if $0 \leq k \leq N - 1$.

Proof (1) Under the threshold strategy with threshold n, The steady state probabilities of this system are

$$\begin{aligned} P_{k,B} &= \frac{\rho - \rho^{k+1}}{1-\rho} P_0, \quad 1 \leq k \leq N, \\ P_{k,I} &= P_0, \quad 1 \leq k \leq N-1, \\ P_{N+k,B} &= \rho^j P_{N,B}, \quad 1 \leq k \leq n-N. \end{aligned}$$

If an arriving customer observes state $k(k \leq N-1)$, the expected waiting time of him is

$$\begin{aligned} W_k &= \frac{P_{k,B}}{p_{k,B} + P_{k,I}} \frac{k+1}{\mu} + \frac{P_{k,I}}{P_{k,B} + P_{k,I}} \left[\frac{N-(k+1)}{\Lambda} + \frac{k+1}{\mu} \right] \\ &= \frac{k+1}{\mu} + \frac{1-\rho}{1-\rho^{k+1}} \frac{N-(1+k)}{\Lambda}. \end{aligned}$$

If an arriving customer observes state $k(k \geq N)$, the server must be busy and the expected waiting time of the customer is $W_k = \frac{k+1}{\mu}$.

(2) For any real number x, we define $W(x) = \frac{x}{\mu} + \frac{1-\rho}{1-\rho^x}\frac{N-x}{\Lambda}$. Take the second derivative of this

$$\frac{d^2W(x)}{dx^2} = \frac{-2(1-\rho)\rho^x \ln \rho}{\Lambda(1-\rho^x)^2} + \frac{(1-\rho)(N-x)\rho^x(1+\rho^x)(\ln(\rho))^2}{\Lambda(1-\rho^x)^3}.$$

If $x \in [1, N]$, whether $\rho > 1$ or $\rho < 1$ can obtain $\frac{d^2W(x)}{dx^2} > 0$. □

When $0 \le k \le N-1$, from Lemma 8.3.2, the longest waiting time is either W_0 or W_{N-1}. It is straightforward to show that

$$W_0 = \frac{1}{\mu} + \frac{N-1}{\Lambda}, \quad W_{N-1} = \frac{N}{\mu}.$$

If $\Lambda > \mu$, then $W_{N-1} > W_0$; if $\Lambda < \mu$, then $W_{N-1} \le W_0$. When state 0 is observed, the customer knows that the server must be idle; when state $N-1$ is observed, the customer knows that the server will be busy once he joins the system. Hence, the expected waiting times in the two cases are identical to those in the corresponding cases in the fully observable model.

Theorem 8.3.5 *For the almost observable vacation queues with N-policy, there exists an equilibrium solution with an active server if and only if (1) $\rho \ge 1$ and $\nu \ge N$ or (2) $\rho < 1$ and $\nu \ge \frac{N-1}{\rho} + 1$, the unique equilibrium threshold is $\lfloor \nu \rfloor \ge N$.*

Proof From Lemma 8.3.2, the equilibrium threshold $n \ge N$ exists only if $R \ge CW_0$ and $R \ge CW_{N-1}$. If $\Lambda \ge N$, then $\rho \ge 1$ $W_{N-1} \ge W_0$. In this case, it suffices to have $R \ge CW_{N-1} = C\frac{N}{\mu}$, which implies $\nu \ge N$. If $\Lambda \le \mu$, we have $\rho \le 1$ and $W_{N-1} \le W_0$. In this case, it suffices to have $R \ge CW_0 = C\left(\frac{1}{\mu} + \frac{N-1}{\Lambda}\right)$, which implies $\nu \ge 1 + \frac{N-1}{\rho}$. If the equilibrium threshold $n \ge N$ exists, then $n = \lfloor \nu \rfloor$. □

Therefore, as long as customers know that the system adopts N-policy, providing customers with partial information on queue length actually generates the same effect as providing customers with full information.

8.3.3.2 Social Optimal Strategy

We denote $SW(n)$ as the social welfare when the threshold is n. In the fully observable case, the optimal n^* could be larger or smaller than N. If $n^* \ge N$, the optimal thresholds for the two information scenarios are the same. Consequently, managers can reduce one dimension of information (server s status), which might greatly reduce administrative costs. If $n^* < N$, the manager allows customers to join an idle-server queue even though the queue is larger than n^*, and once the server is activated, no arrivals are allowed to enter the system until the queue length drops below n^*. To make the control policy fair and plausible, it is reasonable to consider the threshold to satisfy $n \geq N$ to ensure that the system can be activated. Under the fully observable case $SW(n)$ is unimodal in n when $n \geq 1$. Thus, when the optimal threshold with fully observable information is less than N, $SW(n)$ decreases in $[N, \infty]$, and the optimal threshold for almost observable information must be N.

8.3.4 Equilibrium Strategies for the Almost Unobservable Case

In the almost unobservable case, customers only observe the state of system. In general, a customers' strategy upon seeing state s can be represented by a joining probability α_s, where $s = B, I$. The effective arrival rates are $\lambda_s = \alpha_s \Lambda$.

8.3.4.1 Equilibrium Strategy

Instead of working on joining probabilities α_B and α_I, we consider the arrival rates (λ_B, λ_I) adopted by customers.

Lemma 8.3.3 *Given arrival rates* (λ_B, λ_I)*, the expected waiting times are*

$$W_B = \frac{1}{\mu - \lambda_B} + \frac{N+1}{2\mu}, \tag{8.3.15}$$

$$W_I = \frac{N-1}{2\lambda_I} + \frac{N+1}{2\mu} \tag{8.3.16}$$

Proof Let $\rho_s = \frac{\lambda_s}{\mu}$ and P_0 be the probability that system is empty. Clearly, for any $k = 0, 1, \ldots, P_{N+k,I} = 0$. The balance equations are

$$\begin{aligned}
\lambda_I P_0 &= \mu P_{1,B}, \\
\lambda_I P_0 + \lambda_B P_{k,B} &= \mu P_{k+1,B}, \quad k = 1, 2, \ldots, N-1, \\
\lambda_I P_{k,I} &= \lambda_I P_{k+1,I}, \\
\lambda_B P_{N+k,B} &= \mu P_{N+k+1,B}, \quad k = 0, 1, \ldots.
\end{aligned}$$

Solving the above equality set, we obtain

$$\begin{aligned}
P_{k,I} &= \frac{1-\rho_B}{N(1-\rho_B+\rho_I)}, \quad k = 1, 2, \ldots, N-1, \\
P_{k,B} &= \frac{\rho_I(1-\rho_B^k)}{N(1-\rho_B+\rho_I)}, \quad k = 1, 2, \ldots, N-1, \\
P_{N+k,B} &= \frac{\rho_I \rho_B^k (1-\rho_B^N)}{N(1-\rho_B+\rho_I)}, \quad k = 1, 2, \ldots.
\end{aligned}$$

Let $\Pr(-|s)$ be the probability that an arriving customer observes state s. According to the property of PASTA, we have

$$\Pr(-|I) = P_0 + \sum_{k=1}^{N-1} P_{k,I} = NP_0 = \frac{1-\rho_B}{1-\rho_B+\rho_I}, \tag{8.3.17}$$

$$\Pr(-|B) = 1 - \Pr(-|I) = \frac{\rho_I}{1-\rho_B+\rho_I}. \tag{8.3.18}$$

Let $P(k|s)$ be the probability that the queue length is k conditional on observing the server's status s. If the arriving customer observes a busy server, the conditional probabilities are

$$\Pr(k|B) = \frac{P_{k,B}}{\Pr(-|B)} = \frac{1-\rho_B^k}{N}, \quad 1 \le k \le N,$$

$$\Pr(N+k|B) = \frac{P_{N+k,B}}{\Pr(-|B)} = \frac{\rho_B^k(1-\rho_B^N)}{N}, \quad k \ge 0.$$

Note that, if the queue length is k and the server is busy, the expected waiting time of the arriving customer, denoted as $W(k|B)$, is the expected service time of the $k+1$ customer in the system, i.e., $\frac{k+1}{\mu}$. Thus, the expected waiting time upon seeing a busy server is

$$W_B = \sum_{k=1}^{\infty} W(k|B)\Pr(k|B) = \frac{1}{\mu-\lambda_B} + \frac{N+1}{2\mu}.$$

If the arriving customer observes an idle server, the respective conditional probabilities and expected waiting times are

$$\Pr(k|I) = \frac{P_{k,I}}{\Pr(-|I)} = \frac{1}{N}, \quad 0 \le k \le N-1,$$

$$W(k|I) = \frac{N-(k+1)}{\lambda_I} + \frac{k+1}{\mu}, \quad 0 \le k \le N-1.$$

Furthermore, if the arriving customer observes an idle server, the expected sojourn times is

$$W_I = \sum_{k=0}^{N-1} W(k|I)\Pr(k|I) = \frac{N+1}{2\mu} + \frac{N-1}{2\lambda_I}.$$

□

From (8.3.15) and (8.3.16), we can see that the expected waiting time $W_B(W_I)$ is independent of the arrival rate $\lambda_I(\lambda_B)$. Thus, we can determine the equilibrium arrival rates upon seeing busy and idle servers separately.

We first consider the equilibrium arrival rate upon seeing a busy server, denoted as λ_B^e. From (8.3.15), the expected waiting time W_B increases with λ_B. This demonstrates "avoid-the-crowd" customer behavior, and consequently, at most one equilibrium arrival rate exists. If

$$R \leq C(\frac{1}{\mu} + \frac{N+1}{2\mu}),$$

or equivalently,

$$\nu \leq \frac{N+3}{2},$$

there exists no positive equilibrium. Otherwise, there exists an arrival rate that satisfies

$$R = C\Big(\frac{1}{\mu - \lambda_B} + \frac{N+1}{2\mu}\Big).$$

The above equation has a unique solution

$$\lambda_B = \mu - \frac{2\mu C}{2\mu R - (N+1)C} = \frac{2\nu - (N+3)}{2\nu - (N+1)}\mu.$$

which is positive. The equilibrium arrival rate upon seeing a busy server is summarized in the following theorem.

Theorem 8.3.6 *(1) If $\nu \leq \frac{N+3}{2}$, there exists no positive equilibrium arrival rate. (2) If $\nu > \frac{N+3}{2}$, there exists a unique positive arrival rate $\lambda_B^e = min \left\{\frac{2\nu-(N+3)}{2\nu-(N+1)}\mu, \Lambda\right\}$*

Denote λ_I^e as the equilibrium arrival rate upon seeing an idle server. From (8.3.16), the expected waiting time W_I decreases in λ_I This demonstrates "follow-the-crowd" customer behavior, and consequently, multiple equilibria could exist. Clearly, "all balking upon seeing an idle server" is always a pure equilibrium strategy. In the following equations, we consider the positive equilibrium arrival rates. If

$$R < C\Big(\frac{N-1}{2\Lambda} + \frac{N+1}{2\mu}\Big),$$

or equivalently, if

$$\nu < \frac{N-1}{2\rho} + \frac{N+1}{2},$$

there exists no positive equilibrium, otherwise, "all join" is an equilibrium. If $\nu > \frac{N-1}{2\rho} + \frac{N+1}{2}$, there also exists an equilibrium arrival rate $0 < \lambda_I < \Lambda$, which satisfies

$$R = C\Big(\frac{N-1}{2\lambda_I} + \frac{N+1}{2\mu}\Big),$$

The above equation has a unique solution:

$$\lambda_I = \frac{(N-1)\mu C}{2\mu R - (N+1)C} = \frac{N-1}{2\nu - (N+1)}\mu$$

The equilibrium arrival rate upon seeing an idle customer is summarized in the following theorem

Theorem 8.3.7 *(1) If $\nu < \frac{N-1}{2\rho} + \frac{N+1}{2}$, there exists no positive equilibrium arrival rate.*
(2) If $\nu = \frac{N-1}{2\rho} + \frac{N+1}{2}$, there exists a unique positive equilibrium arrival rate $\lambda_I^e = \Lambda$.
(3) If $\nu > \frac{N-1}{2\rho} + \frac{N+1}{2}$, there exist two positive equilibrium arrival rates: $\lambda_I^e = \frac{N-1}{2\nu-(N+1)}\mu < \Lambda$ and $\lambda_I^e = \Lambda$.

Note that W_I decreases in λ_I, the smaller equilibrium arrival rate in case (3) is not stable. The waiting time is reduced if more customers join, which, in turn, will attract even more customers to join.

8.3.4.2 Social Optimal Strategy

Given any (λ_B, λ_I), the social welfare per time unit is

$$\begin{aligned} SW(\lambda_B, \lambda_I) &= \lambda_B[R - CW_B(\lambda_B, \lambda_I)]\Pr(-|B) + \lambda_I[R - CW_I(\lambda_B, \lambda_I)]\Pr(-|I) \\ &= \frac{C\lambda_I(\nu - \frac{N+1}{2})}{\mu - \lambda_B + \lambda_I} - \frac{C[\lambda_I\lambda_B + \frac{N-1}{2}(\mu - \lambda_B)^2]}{(\mu - \lambda_B + \lambda_I)(\mu - \lambda_B)}. \end{aligned} \tag{8.3.19}$$

Lemma 8.3.4 *(1) Given any λ_B such that $\nu > \frac{1}{1-\lambda_B/\mu}$, $SW(\lambda_B, \lambda_I)$ is strictly increasing in λ_I.*
(2) Given any $\lambda_I > 0$, $SW(\lambda_B, \lambda_I)$ is quasi-concave in λ_B.

Proof By calculating the partial-first-order $SW(\lambda_B, \lambda_I)$ for λ_I, we have

$$\frac{\partial SW(\lambda_B, \lambda_I)}{\partial \lambda_I} = \frac{C(\mu - \lambda_B)}{(\mu - \lambda_B + \lambda_I)^2}\left[\nu - \frac{1}{1 - \lambda_B/\mu}\right].$$

Note that $\mu > \lambda_B$ is a necessary condition for system stability. Thus, if $\nu > \frac{1}{1-\lambda_B/\mu}$, then $\frac{\partial SW(\lambda_B,\lambda_I)}{\partial \lambda_I} > 0$

To show $SW(\lambda_B, \lambda_I)$ is quasi-concave in λ_B, it suffices to show that

$$\left.\frac{\partial^2 SW(\lambda_B, \lambda_I)}{\partial \lambda_B^2}\right|_{\frac{\partial SW(\lambda_B,\lambda_I)}{\partial \lambda_B}=0} < 0.$$

First, we consider the partial-first-order of $SW(\lambda_B, \lambda_I)$ about λ_B

$$\frac{\partial SW(\lambda_B, \lambda_I)}{\partial \lambda_B} = \frac{\mu\lambda_I[(\mu-\lambda_B)^2 R - 2(\mu-\lambda_B)C - \lambda_I C]}{(\mu-\lambda_B)^2(\mu-\lambda_B+\lambda_I)^2}. \tag{8.3.20}$$

Therefore, if $\lambda_I = 0$ or $(\mu-\lambda_B)^2 R - 2(\mu-\lambda_B)C - \lambda_I C = 0$ (or both), $\frac{\partial SW(\lambda_B,\lambda_I)}{\partial \lambda_B} = 0$. If $\lambda_I = 0$, the system shall never be active once on vacation. We only consider those $\lambda_I > 0$ and take

$$(\mu-\lambda_B)^2 R - 2(\mu-\lambda_B)C - \lambda_I C = 0, \tag{8.3.21}$$

as the condition for the partial-first-order condition $\frac{\partial SW(\lambda_B,\lambda_I)}{\partial \lambda_B} = 0$. Next, we consider the partial-second-order of $SW(\lambda_B, \lambda_I)$ about λ_B.

$$\frac{\partial^2 SW(\lambda_B, \lambda_I)}{\partial \lambda_B^2} = \frac{2\mu\lambda_I}{(\mu-\lambda_B+\lambda_I)^3(\mu-\lambda_B)^2}\left[R(\mu-\lambda_B)^2 - 3C(\mu-\lambda_B+\lambda_I) - \frac{\lambda_I^2 C}{\mu-\lambda_B}\right].$$

This, together with (8.3.21), implies that

$$\frac{\partial^2 SW(\lambda_B, \lambda_I)}{\partial \lambda_B^2}\Big|_{\frac{\partial SW(\lambda_B,\lambda_I)}{\partial \lambda_B}=0} = -\frac{2\mu\lambda_I C}{(\mu-\lambda_B+\lambda_I)^3(\mu-\lambda_B)^2}\left(\mu-\lambda_B+2\lambda_I+\frac{\lambda_I^2}{\mu-\lambda_B}\right) < 0.$$

Therefore, for any given λ_I, $SW(\lambda_B, \lambda_I)$ is quasi-concave in λ_B. □

Theorem 8.3.8 *(1) The optimal arrival rates are* $\lambda_B^* = min\left\{\left(\mu - \frac{C+\sqrt{C^2+R\Lambda C}}{R}\right)^+, \Lambda\right\}$, *where* $x^+ = max\{x, 0\}$, *and* $\lambda_I^* = \Lambda$.
(2) There exists a threshold value $\tilde{N}$ *such that* $\lambda_B^* \le \lambda_B^e$ *if* $N \le \tilde{N}$, *and* $\lambda_B^* \ge \lambda_B^e$ *if* $N \ge \tilde{N}$.
(3) $\lambda_I^* \ge \lambda_I^e$.

Proof (1) From part (2) of 8.3.4, for any given $\lambda_I > 0$, the optimal arrival rate $\lambda_B^*(\lambda_I)$ is the unique solution of $\partial SW(\lambda_B, \lambda_I)/\partial\lambda_B = 0$, that is, $\lambda_B^*(\lambda_I)$ is the solution of (8.3.21):

$$\lambda_B^*(\lambda_I) = \mu - \frac{C+\sqrt{C^2+R\lambda_I C}}{R}.$$

It is easy to check that $\lambda_B^*(\lambda_I)$ satisfies

$$\nu = \frac{1+\sqrt{1+\lambda_I/C}}{1-\lambda_B^*(\lambda_I)/\mu} > \frac{1}{1-\lambda_B^*(\lambda_I)/\mu}.$$

From part (1) of Lemma 8.3.4, $SW(\lambda_B^*(\lambda_I), \lambda_I)$ increase in λ_I. Therefore, $\lambda_I^* = \Lambda$ and $\lambda_B^* = \min\left\{\left(\mu - \frac{C+\sqrt{C^2+R\Lambda C}}{R}\right)^+, \Lambda\right\}$, where $x^+ = \max\{x, 0\}$.

(2) Let $\tilde{\lambda}_B = \mu - \frac{C+\sqrt{C^2+R\Lambda C}}{R}$. Then the first derivative of ν is

$$\frac{\partial \tilde{\lambda}_B}{\partial \nu} = \mu\left(\frac{1}{\nu^2} + \frac{2\nu^{-3} + \rho\nu^{-2}}{\sqrt{\frac{1}{\nu^2} + \frac{\rho}{\nu}}}\right) > 0.$$

Thus, $\tilde{\lambda}_B$ increases in ν. Next, we prove the case $\rho < 1$. The case $\rho \geq 1$ can be proved similarly. If $\rho < 1$, by solving $\tilde{\lambda}_B = 0$ and $\tilde{\lambda}_B = \Lambda$, we obtain $\nu = \rho + 2$ and $\nu = \frac{2-\rho}{(1-\rho)^2}$, respectively. Thus, the social optimal arrival rate λ_B^* is

$$\lambda_B^* = \begin{cases} 0, & \text{if } \nu \leq \rho + 2, \\ \mu\left[1 - \frac{1}{\nu} - \sqrt{\frac{1}{\nu^2} + \frac{\rho}{\nu}}\right], & \text{if } \rho + 2 < \nu < \frac{2-\rho}{(1-\rho)^2}, \\ \Lambda, & \text{if } \nu \geq \frac{2-\rho}{(1-\rho)^2}. \end{cases} \tag{8.3.22}$$

Note that the value of λ_B^e can also be specified as

$$\lambda_B^e = \begin{cases} \Lambda, & \text{if } N \leq 2\nu - 1 - \frac{2}{1-\rho}, \\ \mu\left[1 + \frac{2}{N-(2\nu-1)}\right], & \text{if } 2\nu - 1 - \frac{2}{1-\rho} < N < 2\nu - 3, \\ 0, & \text{if } N \geq 2\nu - 3. \end{cases} \tag{8.3.23}$$

By comparing (8.3.22) and (8.3.23), can obtain the part (2) of this theorem.

(3) Clearly, it is readily seen that $\lambda_I^* \geq \lambda_I^e$. This completes the proof. □

8.3.5 *Equilibrium Strategies for the Fully Unobservable Case*

In the fully unobservable case, consider symmetric equilibrium strategies. For the pure strategies, either all of the customers join the queue or all balk. With a mixed strategy, an arriving customer joins with a certain probability, α, and the effective arrival rate is $\lambda = \Lambda\alpha$. Obviously, "all balk" is always an equilibrium strategy, i.e., $\lambda = 0$. Below, we restrict our attention to positive equilibrium arrival rates.

8.3.5.1 Equilibrium Strategy

The stability condition of the system is $\lambda < \mu$. In the fully unobservable case, let $\lambda_B = \lambda_I = \lambda$, so the expected sojourn time of customer is

$$W(\lambda) = \Pr(-|B)W_B + \Pr(-|I)W_I = \frac{1}{\mu - \lambda} + \frac{N-1}{2\lambda}. \tag{8.3.24}$$

The function $W(\lambda)$ is strictly convex in λ. When $\tilde{\lambda} = \frac{\mu\sqrt{\frac{N-1}{2}}}{1+\sqrt{\frac{N-1}{2}}}$, $W(\lambda)$ obtain the minimum.

$$W(\tilde{\lambda}) = \frac{1}{\mu}\left(1+\sqrt{\frac{N-1}{2}}\right)^2. \tag{8.3.25}$$

According to (8.3.24), if $\Lambda < \mu$ and $R \geq C\left(\frac{1}{\mu-\Lambda} + \frac{N-1}{2\Lambda}\right)$, or equivalently, $\nu \geq \frac{1}{\rho} + \frac{N-1}{2\rho}$, then Λ is an equilibrium arrival rate. Otherwise, the solution of

$$R = C\left(\frac{1}{\mu-\lambda} + \frac{N-1}{2\lambda}\right), \tag{8.3.26}$$

is $0 < \lambda < \Lambda$.

Equation (8.3.26) may have no, one, or two solution. If the two solutions λ_1 and λ_2 satisfy $0 \leq \lambda_1 \leq \tilde{\lambda} \leq \lambda_2$ where

$$\lambda_{1,2} = \frac{R\mu - \frac{(3-N)C}{2} \mp \sqrt{R^2\mu^2 + \frac{(3-N)^2C^2}{4} - (N+1)\mu C R}}{2R}. \tag{8.3.27}$$

Using (8.3.25), the condition $R > (=, <) CW(\tilde{\lambda})$ is equivalent to $\nu > (=, <)\left(1+\sqrt{\frac{N-1}{2}}\right)^2$. This gives the following characterization of the equilibrium solutions:

Theorem 8.3.9 *(1) If $\nu < \left(1+\sqrt{\frac{N-1}{2}}\right)^2$, there exists no positive equilibrium arrival rate;*

(2) If $\nu = \left(1+\sqrt{\frac{N-1}{2}}\right)^2$, there exist one positive equilibrium arrival rate $\lambda_e = \tilde{\lambda}$ iff $\tilde{\lambda} \leq \Lambda$;

(3) If $\nu > \left(1+\sqrt{\frac{N-1}{2}}\right)^2$, λ_1, λ_2 and Λ could all be an equilibrium arrival rate. Specifically, there exist two positive equilibrium arrival rates $\lambda_e = (\lambda_1, \lambda_2)$ if $\lambda_2 \leq \Lambda$, two positive equilibrium arrival rates $\lambda_e = (\lambda_1, \Lambda)$ if $\lambda_1 < \Lambda < \lambda_2$, (which reduces to one if $\lambda_1 = \Lambda$), and no positive equilibrium arrival rate if $\Lambda < \lambda_1$.

For the two equilibrium arrival rates in case (3) of Theorem 8.3.9, the solution with λ_2 or Λ are stable. However, the equilibrium with λ_1 is unstable. Obviously, "all customers choose to balk" is a stable equilibrium strategy.

8.3.5.2 Socially Optimal Strategy

We first assume a fixed N and consider that the decision maker will set an arrival rate λ so that social welfare, $SW(\lambda)$, is maximized, where

$$SW(\lambda) = \lambda\Big[R - C\Big(\frac{1}{\mu - \lambda} + \frac{N-1}{2\lambda}\Big)\Big] \tag{8.3.28}$$

This social welfare function is concave and the optimality condition is

$$SW'(\lambda) = R - \frac{\mu C}{(\mu - \lambda)^2} = 0. \tag{8.3.29}$$

Solving this equation yields the unique optimal solution

$$\bar{\lambda} = \mu - \sqrt{\frac{\mu C}{R}}. \tag{8.3.30}$$

It is interesting that $\bar{\lambda}$ does not depend on N. We now give bounds on $\bar{\lambda}$.

Theorem 8.3.10 *If* $R > CW(\tilde{\lambda})$, *then* $\tilde{\lambda} < \bar{\lambda} < \lambda_2$.

Proof According to

$$SW'(\tilde{\lambda}) = R - CW(\tilde{\lambda}) - \tilde{\lambda}CW'(\tilde{\lambda}) = R - CW(\tilde{\lambda}) > 0,$$

and

$$SW'(\lambda_2) = R - CW(\lambda_2) - \lambda_2 CW'(\lambda_2) = -\lambda_2 CW'(\lambda_2) < 0,$$

then $\tilde{\lambda} < \bar{\lambda} < \lambda_2$. □

Social welfare at $\bar{\lambda}$ is

$$\begin{aligned} SW(\bar{\lambda}) &= \mu R - \sqrt{\mu C R} + C\Big(1 - \sqrt{\frac{\mu R}{C}}\Big) - C\frac{N-1}{2} \\ &= C\Big(\frac{\mu R}{C} - 2\sqrt{\frac{\mu R}{C}} - \frac{N-3}{2}\Big) \end{aligned} \tag{8.3.31}$$

We can see that $SW(\bar{\lambda})$ is monotone decreasing in N. Therefore, even though the optimal solution $\bar{\lambda}$ does not depend in N, the social welfare N does.

Theorem 8.3.11 *(1) If* $\nu < \Big(1 + \sqrt{\frac{N-1}{2}}\Big)^2$, *there exists a unique optimal arrival rate* $\lambda^* = 0$;

(2) If $\nu = \Big(1 + \sqrt{\frac{N-1}{2}}\Big)^2$, *there exists two optimal arrival rates* $\lambda^* = \{0, \tilde{\lambda}\}$;

(3) If $\nu > \Big(1 + \sqrt{\frac{N-1}{2}}\Big)^2$, *there exists a unique optimal arrival rate* $\lambda^* = \min\{\bar{\lambda}, \Lambda\}$.

Proof (1) If $\nu < \Big(1 + \sqrt{\frac{N-1}{2}}\Big)^2$, then the net benefit of customer is negative, so the optimal arrival rate is $\lambda^* = 0$.

(2) If $\nu = \left(1+\sqrt{\frac{N-1}{2}}\right)^2$, then the net benefit of customer is non-positive, so the optimal arrival rate is 0 or $\tilde{\lambda}$.

(3) If $\nu > \left(1+\sqrt{\frac{N-1}{2}}\right)^2$, the conclusion is easy to understand.

□

8.3.5.3 Optimal Threshold *N*

Like the assumptions in the fully observable, we consider a busy-period cost, c_b. Thus, the average operating cost per unit time can be expressed as $\theta(N) = c_b P_B$, where P_B is the busy probability for the server. Next we maximize social benefits $SW(N) - \theta(N)$ by adjusting N.

Lemma 8.3.9 has three cases. In the first case, $\lambda_e = 0$. The second case assumes that ν and N exactly satisfy a specific relation, which is unlikely to hold with general data. In the third case, $\nu > \left(1+\sqrt{\frac{N-1}{2}}\right)^2$ or equivalently, $N \le \bar{N}$, where $\bar{N} = \lfloor 2(\sqrt{\nu}-1)^2 \rfloor + 1$. In this case, there are three equilibrium solutions. Among them, 0 is obviously not interesting and λ_1 is unstable. A small perturbation will continue until a stable equilibrium $\min(\lambda_2, \Lambda)$. Therefore, we assume the third equilibrium when we compute an optimal N. The optimization problem becomes

$$\max_N \lambda_e \Big[R - C\Big(\frac{1}{\mu-\lambda_e} + \frac{N-1}{2\lambda_e}\Big)\Big] - \theta(N), \tag{8.3.32}$$

where $\lambda_e = \min(\lambda_2, \Lambda)$, λ_2 is shown as (8.3.27). The busy probability for the $M/M/1$ with an N-policy and arrival rate λ_e is $P_B = \frac{\lambda_e}{\mu}$, which is independent of N. Thus, $\theta(N) = c_b \frac{\lambda_e}{\mu}$.

Theorem 8.3.12 *If* $\rho \le 1 - \frac{1}{\sqrt{\nu}}$, *then* $N^* = 1$; *if* $\rho \ge 1 - \frac{1}{\sqrt{\nu}}$, *then* $N^* = \bar{N}$.

Proof Note that $\rho \le 1 - \frac{1}{\sqrt{\nu}}$ is equivalent to $\Lambda < \tilde{\lambda}(\bar{N})$. Because of $N \le \bar{N}$ and $\lambda_2(N)$ is decreasing in N, then $\Lambda < \tilde{\lambda}(\bar{N}) \le \lambda_2(\bar{N}) \le \lambda_2(N)$ and $\lambda_e(N) = \min\{\lambda_2(N), \Lambda\} = \Lambda$. Therefore, the busy probability ρ is independent of N, and $\theta(N)$ is a constant and independent of N. Using (8.3.32), we know the optimal N is $N^* = 1$.

When $N = 1$, $\frac{\lambda_2}{\mu} = 1 - \frac{1}{\nu}$. So $\rho \ge 1 - \frac{1}{\sqrt{\nu}}$ shows that $\lambda_e = \lambda_2 \le \Lambda$. In this case, all customers' benefit is 0. So maximizing (8.3.32) is minimizing $\theta(N)$, then $N^* = \bar{N}$. □

8.4 M/G/1 Queues with Multiple Vacations

8.4.1 Model Description

We consider a single server queueing system with an infinite waiting room in which the arrival process is Poisson with rate λ. The server has two states: 0 (working) or 1(vacation). The server can only provide service to customers when they are working and not when on vacation. The service times are independent, identically distributed random variables with a common probability distribution function $B(x)$, and the first and second moments are respectively $\mathbb{E}(B) < \infty$ and $\mathbb{E}(B^2) < \infty$. Denote by R_B the residual service time in steady state. Once the number of customers in the system becomes zero, i.e., the system is empty, the server enters a mode of multiple vacations. The vacation time is independent of each other and follows a general distribution. Its distribution function is $V(x)$, and the first-order moment and the second-order moment are $\mathbb{E}(V) < \infty$, $\mathbb{E}(V^2) < \infty$. Denote by R_V the residual vacation time in the steady state.

After the service is completed, the service reward received by each customer is K. At the same time, the cost per unit time of stay (including service time and waiting time in queue) is C. Customers are risk neutral and aim to maximize their benefits. When they arrive at the system, they need to estimate their expected waiting cost, after which make a decision whether to join the queue. Once the customer makes a choice, he can no longer go back, that is, he can neither exit in the middle of the queue nor re arrive after stopping. Before considering the almost invisible and completely invisible situation, assume

$$K > C(\mathbb{E}[R_V] + \mathbb{E}[B]). \tag{8.4.1}$$

When a arriving customer finds that the server is on vacation, if he joins the queue, he will have to wait for the remaining vacation time plus his own service time. So if the condition of (8.4.1) can not be satisfied, then if the server is on vacation, no customers will enter the queue, and the system will remain empty.

8.4.2 Equilibrium Analysis of Joining Strategies in the Fully Unobservable Case

In the fully unobservable case, assume that all arriving customers join the system with probability q, we know that the effective arrival rate is λq By the results in Fuhrmann and Cooper (1985), the stability condition of the system is $\lambda q \mathbb{E}[B] < 1$ and the expected sojourn time is given by

$$\mathbb{E}[S] = \mathbb{E}[R_V] + \mathbb{E}[B] + \frac{\lambda q \mathbb{E}[B]}{1 - \lambda q \mathbb{E}[B]} \mathbb{E}[R_B]. \tag{8.4.2}$$

8.4.2.1 Nash Equilibrium Strategies

Let's tag a new arriving customer, if his expected incoming is

$$U_e(q) = K - C\Big(\mathbb{E}[R_V] + \mathbb{E}[B] + \frac{\lambda q \mathbb{E}[B]}{1 - \lambda q \mathbb{E}[B]}\mathbb{E}[R_B]\Big). \tag{8.4.3}$$

It is easy to verify that $U_e(q)$ is monotone increasing with q, and there exists a unique root

$$q_e^* = \frac{1}{\lambda \mathbb{E}[B]}\Big(1 - \frac{\mathbb{E}[R_B]}{\frac{K}{C} - \mathbb{E}[R_V] - \mathbb{E}[B] + \mathbb{E}[R_B]}\Big). \tag{8.4.4}$$

Theorem 8.4.1 *In the fully unobservable case the unique Nash equilibrium joining probability q_e is given by*

$$q_e = min\{q_e^*, 1\}, \tag{8.4.5}$$

where q_e^* is given by (8.4.4)

Proof By the monotone property of $U_e(q)$ we know that for all $q < q_e^*$ there exists $U_e(q) > 0$; For all $q > q_e^*$ we have $U_e(q) < 0$ If $q_e^* < 1$, then the optimal solution of tagged customer is q_e^*, and therefore it is the unique Nash equilibrium strategy If $q_e^* \geq 1$, then for all $q \in [0, 1]$ joining with probability 1 is a weekly dominated strategy and also the unique Nash equilibrium strategy □

We can verify that it is ATC and the Nash equilibrium strategy q_e is ESS

8.4.2.2 Socially-Optimal Strategies

By virtue of Little's law we can obtain the social welfare per time unit as follows.

$$U_s(q) = \lambda q K - C\mathbb{E}[L] = \lambda q\Big(K - C\Big(\mathbb{E}[R_V] + \mathbb{E}[B] + \frac{\lambda q \mathbb{E}[B]}{1 - \lambda q \mathbb{E}[B]}\mathbb{E}[R_B]\Big)\Big). \tag{8.4.6}$$

Theorem 8.4.2 *In the case of fully observable case the unique socially optimal joining strategy q_s is given by*

$$q_s = min\{q_s^*, 1\}, \tag{8.4.7}$$

where q_s^* is

$$q_s^* = \frac{1}{\lambda \mathbb{E}[B]}\Big(1 - \sqrt{\frac{\mathbb{E}[R_B]}{\frac{K}{C} - \mathbb{E}[R_V] - \mathbb{E}[B] + \mathbb{E}[R_B]}}\Big). \tag{8.4.8}$$

Proof It is not difficult to verify that if q_s^* provided by (8.4.8) is the solution of the equation $U_s'(q) = 0$ Since $\lambda q \mathbb{E}[B] < 1$, for any $q \in [0, 1/(\lambda \mathbb{E}[B]))$ we have $U_s''(q) < 0$ As a result $U_s(q)$ is strictly concave in this interval and there exists a unique maximum q_s^* for $U_s(q)$ in the interval If $q_s^* < 1$, then $U_s(q)$ attains the maximum at the point $q = q_s^*$; Otherwise, the maximum will be attained at $q = 1$ □

By comparing the results in Theorems 8.4.1 and 8.4.2, we can get $q_s \leq q_e$

8.4.3 Equilibrium Analysis of Joining Strategies in the Almost Unobservable Case

Assume customers can only observe the status of the server upon arrival, then there exists four kinds of pure strategies: Never join the queue (0,0), only balk when the server is on vacation (0,1), only balk when the server is working (1,0), never balk the queue (1,1) The mixed strategy of customers can be denoted by (q_0, q_1), where q_i is the joining probability of customers when the server is in the state i

When all customers adopt a mixed strategy (q_0, q_1), the stability condition of the system is

$$\lambda q_1 \mathbb{E}[B] < 1. \tag{8.4.9}$$

Denoted by L the number of customers in the system, I the status of the server (1: working; 0: on vacation), S the sojourn time of joining customers, respectively. Let p_i be the steady-state probability of the server's status being i Denoted by L_i the number of customers in the system given that the server's status is i, and S_i the sojourn time of customers who arrive at the system finding the server status is i and decide to join the queue. Use $\bar{\lambda}$ to represent the effective arrival rate, which is expressed as

$$\bar{\lambda} = \lambda(q_0 p_0 + q_1 p_1). \tag{8.4.10}$$

By Little's law we have

$$p_1 = \bar{\lambda}\mathbb{E}[B], \tag{8.4.11}$$
$$\mathbb{E}[L] = \bar{\lambda}\mathbb{E}[S], \tag{8.4.12}$$

where $\mathbb{E}[L] = p_0\mathbb{E}[L_0] + p_1\mathbb{E}[L_1]$ Solving (8.4.10) and (8.4.11), combining $p_0 + p_1 = 1$, we can obtain that

$$p_0 = \frac{1 - \lambda q_1 \mathbb{E}[B]}{1 - \lambda(q_1 - q_0)\mathbb{E}[B]}, \tag{8.4.13}$$

$$p_1 = \frac{\lambda q_0 \mathbb{E}[B]}{1 - \lambda(q_1 - q_0)\mathbb{E}[B]}, \tag{8.4.14}$$

$$\bar{\lambda} = \frac{\lambda q_0}{1 - \lambda(q_1 - q_0)\mathbb{E}[B]}. \tag{8.4.15}$$

From (8.4.10)-(8.4.15) we get

$$\mathbb{E}[S] = (1 - \lambda q_1 \mathbb{E}[B])\mathbb{E}[S_0] + \lambda q_1 \mathbb{E}[B]\mathbb{E}[S_1]. \tag{8.4.16}$$

In virtue of the property of PASTA, we have

$$\mathbb{E}[S_0] = \mathbb{E}[R_V] + (\mathbb{E}[L_0] + 1)\mathbb{E}[B], \tag{8.4.17}$$

$$\mathbb{E}[S_1] = \mathbb{E}[R_B] + \mathbb{E}[L_1]\mathbb{E}[B]. \tag{8.4.18}$$

When the server is on vacation at time t, the queue length increases with the rate λq_0 during the corresponding vacation period In virtue of the property of PASTA, we obtain that

$$\mathbb{E}[L_0] = \lambda q_0 \mathbb{E}[R_V]. \tag{8.4.19}$$

By solving (8.4.12) and (8.4.16)-(8.4.19) we can reach the following lemma, where (8.4.19) and (8.4.20) can be derived from (8.4.32)

Lemma 8.4.1 *In almost unobservable situations, assuming the system is stable and all customers adopt a* (q_0, q_1) *mixed strategy, the conditional average sojourn time of customers after joining the queue is*

$$\mathbb{E}[S_0] = \mathbb{E}[R_V] + (\lambda q_0 \mathbb{E}[R_V] + 1)\mathbb{E}[B], \tag{8.4.20}$$

$$\mathbb{E}[S_1] = (\lambda q_0 \mathbb{E}[R_V] + 1)\mathbb{E}[B] + \frac{\mathbb{E}[R_B]}{1 - \lambda q_1 \mathbb{E}[B]}, \tag{8.4.21}$$

and furthermore, the expected queue length is

$$\mathbb{E}[L] = \lambda q_0 \Big(\mathbb{E}[R_V] + \Big(\frac{\lambda q_1 \mathbb{E}[R_B]}{1 - \lambda q_1 \mathbb{E}[B]} + 1\Big)\frac{\mathbb{E}[B]}{1 - \lambda(q_1 - q_0)\mathbb{E}[B]}\Big). \tag{8.4.22}$$

8.4.3.1 Nash Equilibrium Strategy

Tag an arriving customer and his average payoff is irrelevant to q_1 given that the tagged customer finds the server is on vacation and decides to join the system. The average payoff is given by

$$U_e(0; q_0) = K - C\mathbb{E}[S_0] = K - C(\mathbb{E}[R_V] + (\lambda q_0 \mathbb{E}[R_V] + 1)\mathbb{E}[B]). \quad (8.4.23)$$

While if the tagged customer finds the server is on working and decides to join the system, then for any $q_1 \in [0, \frac{1}{\lambda \mathbb{E}[B]})$ the average payoff is given by

$$U_e(1; q_0, q_1) = K - C\mathbb{E}[S_1] = K - C\Big((\lambda q_0 \mathbb{E}[R_V] + 1)\mathbb{E}[B] + \frac{\mathbb{E}[R_B]}{1 - \lambda q_1 \mathbb{E}[B]}\Big), \quad (8.4.24)$$

otherwise $U_e(1; q_0, q_1) = -\infty$

Theorem 8.4.3 *In almost unobservable situations, assuming that the condition (8.4.1) is satisfied, there exists a unique Nash equilibrium mixed strategy* $(q_e(0), q_e(1))$

$$q_e(0) = min\{q_e^*(0), 1\}, \quad (8.4.25)$$

$$q_e(1) = max\{min\{q_e^*(1), 1\}, 0\}, \quad (8.4.26)$$

where

$$q_e^*(0) = \frac{1}{\lambda \mathbb{E}[R_V]\mathbb{E}[B]}\Big(\frac{K}{C} - \mathbb{E}[R_V] - \mathbb{E}[B]\Big), \quad (8.4.27)$$

$$q_e^*(1) = \frac{1}{\lambda \mathbb{E}[B]}\Big(1 - \frac{\mathbb{E}[R_B]}{max\{\mathbb{E}[R_V], \frac{K}{C} - (\lambda \mathbb{E}[R_V] + 1)\mathbb{E}[B]\}}\Big). \quad (8.4.28)$$

Proof It is readily seen that $U_e(0; q_0)$ is strictly monotone decreasing with q_0, and the unique root of equation $U_e(0; q_0) = 0$ is provided by (8.4.27) The condition of (8.4.1) guarantees $q_e^*(0) > 0$ By the virtue of monotonicity of $U_e(0; q_0)$, if $q_e^*(0) < 1$ then we conclude that $q_e^*(0)$ is the self optimal strategy. That is, if the customers find the server is on vacation upon arrival, $q_e^*(0)$ is the unique equilibrium joining probability If $q_e^*(0) \geq 1$, for any $q_0 \in [0, 1]$, $U_e(0; q_0) \geq 0$, and therefore the unique equilibrium joining probability is 1. To summarize, when the server is found on vacation upon arrival, customers will join the queue with equilibrium joining strategy $q_e(0)$ given by (8.4.25)

After determining $q_e(0)$, we determine $q_e(1)$ in the following. With the stability condition (8.4.9), the equilibrium solution $q_e(1)$ should satisfy $0 \leq q_e(1) < \frac{1}{\lambda \mathbb{E}[B]}$ Plugging $q_e(0)$ in (8.4.25) to (8.4.24) and denoting the resulting function as $\bar{U}_e(1; q_1)$ When the server is busy, let's tag a new arriving customer regarding an equilibrium $q_e(0)$ If $\bar{U}_e(1; q_1) > 0$, the customer prefers to join the queue; If $\bar{U}_e(1; q_1) < 0$, the customer will balk from the system; If $\bar{U}_e(1; q_1) = 0$, the customer chooses joining the queue or chooses balking.

When $\mathbb{E}[R_V] + \mathbb{E}[B] < \frac{K}{C} < \mathbb{E}[R_V] + (\lambda \mathbb{E}[R_V] + 1)\mathbb{E}[B]$, $q_e(0) = q_e^*(0)$, and we can obtain the following equation by plugging it into (8.4.24)

$$\bar{U}_e(1;q_1) = C\Big(\mathbb{E}[R_V] - \frac{\mathbb{E}[R_B]}{1-\lambda q_1 \mathbb{E}[B]}\Big).$$

If $q_0 = q_e^*(0)$, $U_e(0;q_0) = 0$ Combining (8.4.23) we know that $\bar{U}_e(1;q_1) = 0$ has a unique solution

$$\hat{q}_e^*(1) = \frac{1}{\lambda \mathbb{E}[B]}\Big(1 - \frac{\mathbb{E}[R_B]}{\mathbb{E}[R_V]}\Big). \tag{8.4.29}$$

If $\mathbb{E}[R_V] + (\lambda \mathbb{E}[R_V] + 1)\mathbb{E}[B] \le \frac{K}{C}$, $q_e(0) = 1$, we plug it into (8.4.24) and obtain

$$\bar{U}_e(1;q_1) = C\Big((\lambda \mathbb{E}[R_v] + 1)\mathbb{E}[R_B] + \frac{\mathbb{E}[R_B]}{1-\lambda q_1 \mathbb{E}[B]}\Big).$$

The equation $\bar{U}_e(1;q_1) = 0$ has a unique solution

$$\hat{q}_e^*(1) = \frac{1}{\lambda \mathbb{E}[B]}\Big(1 - \frac{\mathbb{E}[R_B]}{\frac{K}{C} - (\lambda \mathbb{E}[R_V] + 1)\mathbb{E}[B]}\Big). \tag{8.4.30}$$

By (8.4.29) and (8.4.30) we know that $\bar{U}_e(1;q_1)$ is strictly monotone decreasing with q_1, and the unique solution is $q_e^*(1)$ provided by (8.4.28)

Tag a newly arriving customer given that the server is busy at the arriving epoch, we focus on the following three cases:

(1) If $q_e^*(1) \le 0$, then for any $q_1 \in [0,1]$, $\bar{U}_e(1;q_1) \le 0$ Accordingly, the tagged customer chooses balking.

(2) If $q_e^*(1) \in (0,1)$, then $q_e^*(1)$ is an optimal strategy against himself, and therefore this is the unique equilibrium joining probability

(3) If $q_e^*(1) \ge 1$, then for any $q_1 \in [0,1]$, $\bar{U}_e(1;q_1) \ge 0$ Therefore, the tagged customer chooses joining the queue.

Incorporating case (1)–case (3) we can obtain (8.4.26) Note that it is an ATC, and there exists a unique Nash equilibrium mixed strategy $(q_e(0), q_e(1))$ □

8.4.3.2 Socially Optimal Strategy

For any mixed strategy $(q_0, q_1) \in \Re = (0,1] \times [0,1]$, the social welfare per time unit is $U_s(q_0,q_1) = \bar{\lambda} K - C\mathbb{E}[L]$ The following discussion is on the basis of assumption $\lambda \mathbb{E}[B] < 1$, while for the case of $\lambda \mathbb{E}[B] \ge 1$ the corresponding results can be derived in a similar way. By (8.4.15) and (8.4.22) we have

$$U_s(q_0,q_1) = \frac{\lambda q_0 C}{1-\lambda(q_1-q_0)\mathbb{E}[B]}\Big(\frac{K}{C} - \Big(\frac{\lambda q_1 \mathbb{E}[R_B]}{1-\lambda q_1 \mathbb{E}[B]} + 1\Big)\mathbb{E}[B]\Big) - \lambda q_0 C \mathbb{E}[R_V]. \tag{8.4.31}$$

We divide $\mathfrak{R}$ into four regions: $\mathfrak{R}_1 = (0, 1) \times (0, 1), \mathfrak{R}_2 = \{1\} \times [0, 1], \mathfrak{R}_3 = (0, 1] \times \{0\}$, $\mathfrak{R}_4 = (0, 1] \times \{1\}$ In the sequel we consider the social welfare and seek to obtain the extremum point in each region. As such, the socially optimal strategy $(q_s(0), q_s(1))$ is the optimal value among these extremum points

For region $\mathfrak{R}_1$, the first-order derivative optimal condition for (8.4.31) is given by

$$\mathbb{E}[R_V] = \frac{1 - \lambda q_1 \mathbb{E}[B]}{(1 - \lambda(q_1 - q_0)\mathbb{E}[B])^2} \Big(\frac{K}{C} - \Big(\frac{\lambda q_1 \mathbb{E}[R_B]}{1 - \lambda q_1 \mathbb{E}[B]} + 1\Big)\mathbb{E}[B]\Big), \tag{8.4.32}$$

$$\frac{\mathbb{E}[R_B]}{(1 - \lambda q_1 \mathbb{E}[B])^2} = \frac{1}{1 - \lambda(q_1 - q_0)\mathbb{E}[B]} \Big(\frac{K}{C} - \Big(\frac{\lambda q_1 \mathbb{E}[R_B]}{1 - \lambda q_1 \mathbb{E}[B]} + 1\Big)\mathbb{E}[B]\Big), \tag{8.4.33}$$

from which one can obtain that

$$(1 - \lambda(q_1 - q_0)\mathbb{E}[B])\mathbb{E}[R_V] = \frac{\mathbb{E}[R_B]}{1 - \lambda q_1 \mathbb{E}[B]}. \tag{8.4.34}$$

Substituting (8.4.34) into (8.4.32) and (8.4.33), we have

$$\frac{1}{\lambda \mathbb{E}[B]} \Big(\frac{\mathbb{E}[R_B]}{(1 - \lambda q_1 \mathbb{E}[B])\mathbb{E}[R_V]} - (1 - \lambda q_1 \mathbb{E}[B])\Big) = q_0, \tag{8.4.35}$$

$$\frac{\mathbb{E}[R_V](1 - \lambda q_1 \mathbb{E}[B])^3}{(\mathbb{E}[R_B])^2} \Big(\frac{K}{C} - \Big(\frac{\lambda q_1 \mathbb{E}[R_B]}{1 - \lambda q_1 \mathbb{E}[B]} + 1\Big)\mathbb{E}[B]\Big) = 1, \tag{8.4.36}$$

This is a necessary condition for strategy $(q_0, q_1) \in \mathfrak{R}_1$ to be socially optimal.

Denoted by $y = 1 - \lambda q_1 \mathbb{E}[B]$, we know from (8.4.35) that $0 < q_0 < 1$ can be rewritten as

$$y < \frac{\mathbb{E}[R_B]}{y\mathbb{E}[R_V]} < y + \lambda \mathbb{E}[B]. \tag{8.4.37}$$

(8.4.36) can be rewritten as

$$P_1(y) = \Big(1 + \frac{1}{\mathbb{E}[R_B]}\Big(\frac{K}{C} - \mathbb{E}[B]\Big)\Big)y^3 - y^2 - \frac{\mathbb{E}[R_B]}{\mathbb{E}[R_V]}. \tag{8.4.38}$$

(8.4.37) can also be rewritten as

$$1 - \lambda \mathbb{E}[B] < y < \min\left\{1, \sqrt{\frac{\mathbb{E}[R_B]}{\mathbb{E}[R_V]}}\right\} \text{and } y^2 + \lambda \mathbb{E}[B]y - \frac{\mathbb{E}[R_B]}{\mathbb{E}[R_V]} > 0. \tag{8.4.39}$$

To find the root of $P_1(y) = 0$ with $y \in (1 - \lambda \mathbb{E}[B], 1)$, we need to consider the range of y. In the following we discuss several cases of this problem.

Case 1: If $(1-\lambda\mathbb{E}[B])^2 < \frac{\mathbb{E}[R_B]}{\mathbb{E}[R_V]} < 1-\lambda\mathbb{E}[B]$, $y \in (1-\lambda\mathbb{E}[B], \sqrt{\mathbb{E}[R_B]/\mathbb{E}[R_V]}) \subset (0,1)$ There exists a unique solution for (8.4.36) with $q_1 \in (0,1)$ that satisfies

$$\frac{1}{\lambda\mathbb{E}[B]}\left(1-\sqrt{\frac{\mathbb{E}[R_B]}{\mathbb{E}[R_V]}}\right) < q_1 < 1.$$

Case 2: If $1-\lambda\mathbb{E}[B] \le \frac{\mathbb{E}[R_B]}{\mathbb{E}[R_V]} < 1$, $y \in (y_1, \sqrt{\mathbb{E}[R_B]/\mathbb{E}[R_V]}) \subset (0,1)$, where

$$y_1 = \frac{-\lambda\mathbb{E}[B]}{2} + \sqrt{\left(\frac{\lambda\mathbb{E}[B]}{2}\right)^2 + \frac{\mathbb{E}[R_B]}{\mathbb{E}[R_V]}}, \tag{8.4.40}$$

and it is the positive root of $y^2 + \lambda\mathbb{E}[B]y - \mathbb{E}[R_B]/\mathbb{E}[R_V] = 0$ There exists a unique solution for (8.4.36) with $q_1 \in (0,1)$ that satisfies

$$\frac{1}{\lambda\mathbb{E}[B]}\left(1-\sqrt{\frac{\mathbb{E}[R_B]}{\mathbb{E}[R_V]}}\right) < q_1 < \frac{1}{\lambda\mathbb{E}[B]}(1-y_1).$$

Case 3: If $1 \le \frac{\mathbb{E}[R_B]}{\mathbb{E}[R_V]} < 1+\lambda\mathbb{E}[B]$, $y \in (y_1, 1) \subset (0,1)$, where y_1 is provided by (8.4.40) There exists a unique solution for (8.4.36) with $q_1 \in (0,1)$ that satisfies

$$0 < q_1 < \frac{1}{\lambda\mathbb{E}[B]}(1-y_1).$$

In addition, if $\mathbb{E}[R_B]/\mathbb{E}[R_V] \le (1-\lambda\mathbb{E}[B])^2$ or $(1-\lambda\mathbb{E}[B] \le \mathbb{E}[R_B]/\mathbb{E}[R_V]$, there does not exist strategies satisfying (8.4.35) and (8.4.36) with $(q_0, q_1) \in \Re_1$ If and only if $1 + \frac{1}{\mathbb{E}[R_B]}\left(\frac{K}{C} - \mathbb{E}[B]\right)$ is positive, the equation $P_1(y) = 0$ described in (8.4.38) has a unique positive root. Otherwise, there is no positive root for this equation Note that under the condition (8.4.1),

$$\mathbb{E}[B] - \mathbb{E}[R_B] < \frac{K}{C}. \tag{8.4.41}$$

Theorem 8.4.4 *If the condition of (8.4.1) is satisfied we have the following conclusions*

(1) In Case 1, if and only if

$$\mathbb{E}[B] - \mathbb{E}[R_B] + 2\sqrt{\mathbb{E}[R_B]\mathbb{E}[R_V]} < \frac{K}{C} < \mathbb{E}[B] - \mathbb{E}[R_B] + \frac{\mathbb{E}[R_B]}{(1-\lambda\mathbb{E}[B])^3} \times\left(\frac{\mathbb{E}[R_B]}{\mathbb{E}[R_V]} + (1-\lambda\mathbb{E}[B])^2\right), \tag{8.4.42}$$

and $U_s(q_0, q_1)$ *has a unique extremum point* $(q_s^*(0), q_s^*(1))$ *in* $\mathfrak{R}_1$

(2) In Case 2, if and only if

$$\mathbb{E}[B] - \mathbb{E}[R_B] + 2\sqrt{\mathbb{E}[R_B]\mathbb{E}[R_V]} < \frac{K}{C} < \mathbb{E}[B] - \mathbb{E}[R_B] + \frac{\mathbb{E}[R_B]}{y_1^3}\Big(\frac{\mathbb{E}[R_B]}{\mathbb{E}[R_V]} + y_1^2\Big), \tag{8.4.43}$$

$U_s(q_0, q_1)$ *has a unique extremum point* $(q_s^*(0), q_s^*(1))$ *in* $\mathfrak{R}_1$.

(3) In Case 3, if and only if

$$\mathbb{E}[B] + \frac{(\mathbb{E}[R_B])^2}{\mathbb{E}[R_V]} < \frac{K}{C} < \mathbb{E}[B] - \mathbb{E}[R_B] + \frac{\mathbb{E}[R_B]}{y_1^3}\Big(\frac{\mathbb{E}[R_B]}{\mathbb{E}[R_V]} + y_1^2\Big), \tag{8.4.44}$$

$U_s(q_0, q_1)$ *has a unique extremum point* $(q_s^*(0), q_s^*(1))$ *in* $\mathfrak{R}_1$

Proof In the case 1 since $P_1(0) < 0$, if and only if $P_1(\sqrt{\mathbb{E}[R_B]/\mathbb{E}[R_V]}) > 0$ and $P_1(1 - \lambda\mathbb{E}[B]) < 0$, $P_1(y) = 0$ has a unique solution with $y \in (1 - \lambda\mathbb{E}[B], \sqrt{\mathbb{E}[R_B]/\mathbb{E}[R_V]})$ Therefore, it follows that

$$P_1(1 - \lambda\mathbb{E}[B]) < 0 \Longleftrightarrow \frac{K}{C} < \mathbb{E}[B] - \mathbb{E}[R_B] + \frac{\mathbb{E}[R_B]}{(1 - \lambda\mathbb{E}[B])^3}\Big(\frac{\mathbb{E}[R_B]}{\mathbb{E}[R_V]} + (1 - \lambda\mathbb{E}[B])^2\Big),$$
$$P_1\Big(\sqrt{\frac{\mathbb{E}[R_B]}{\mathbb{E}[R_V]}}\Big) > 0 \Longleftrightarrow \mathbb{E}[B] - \mathbb{E}[R_B] + 2\sqrt{\mathbb{E}[R_B]\mathbb{E}[R_V]} < \frac{K}{C}. \tag{8.4.45}$$

In a similar way, in Case 2,

$$P_1(y_1) < 0 \Longleftrightarrow \frac{K}{C} < \mathbb{E}[B] - \mathbb{E}[R_B] + \frac{\mathbb{E}[R_B]}{y_1^3}\Big(\frac{\mathbb{E}[R_B]}{\mathbb{E}[R_V]} + y_1^2\Big). \tag{8.4.46}$$

From (8.4.45) and (8.4.46) we obtain (8.4.43)

In Case 3,

$$P_1(1) > 0 \Longleftrightarrow \mathbb{E}[B] + \frac{(\mathbb{E}[R_B])^2}{\mathbb{E}[R_V]} < \frac{K}{C}.$$

And therefore $P_1(y) = 0$ has a unique extremum point with $y \in (y_1, 1)$ □

To sum up, there exists a unique socially optimal strategy $(q_s^*(0), q_s^*(1))$ in $\mathfrak{R}_1$, where $q_s^*(0)$ is given by (8.4.35), $q_s^*(1)$ is the solution of $y^* = 1 - \lambda q_s^*(1)\mathbb{E}[B]$, and y^* is the unique solution of $P_1(y) = 0$ in the Case 1 with $y \in (1 - \lambda\mathbb{E}[B], \sqrt{\mathbb{E}[R_B]/\mathbb{E}[R_V]})$, or the unique solution of $y \in (y_1, \sqrt{\mathbb{E}[R_B]/\mathbb{E}[R_V]})$ in Case 2, or the unique solution of $y \in (y_1, 1)$ where y_1 is given by (8.4.40)

In $\Re_2$, arriving customers join with probability 1 if the server is found on vacation upon arrival. The first order derivative optimal condition of $U_s(1, q_1)$ is $P_2(y) = 0$, where $y = 1 - \lambda q_1 \mathbb{E}[B]$

$$P_2(y) = \Big(1 + \frac{1}{\mathbb{E}[R_B]}\Big(\frac{K}{C} - \mathbb{E}[B]\Big)\Big)y^2 - 2y - \lambda\mathbb{E}[B].$$

Under the condition $\lambda\mathbb{E}[B] < 1$, the range of values for y is $[1 - \lambda\mathbb{E}[B], 1] \subset [0, 1]$ We will discuss two scenarios in the following.

Case 1: When $K/C \le \mathbb{E}[B] - \mathbb{E}[R_B]$, for all $y \in [1 - \lambda\mathbb{E}[B], 1]$, $P_2(y) < 0$ So $U_s(1, q_1)$ is monotone decreasing with $q_1 \in [0, 1]$ and $\max_{q_1\in[0,1]} \{U_s(1, q_1)\} = U_s(1, 0) < 0$

Case 2: When $\mathbb{E}[B] - \mathbb{E}[R_B] < K/C$, the equation $P_2(y) = 0$ has a unique positive root

$$\bar{y}_1 = \frac{1 + \sqrt{1 + \lambda\mathbb{E}[B]\Big(1 + \frac{1}{\mathbb{E}[R_B]}\Big(\frac{K}{C} - \mathbb{E}[B]\Big)\Big)}}{1 + \frac{1}{\mathbb{E}[R_B]}\Big(\frac{K}{C} - \mathbb{E}[B]\Big)}. \tag{8.4.47}$$

If and only if $P_2(1 - \lambda\mathbb{E}[B]) \le 0$ $P_2(1) \ge 0$, $\bar{y}_1 \in [1 - \lambda\mathbb{E}[B], 1]$ Since $P_2(0) < 0$, such a condition equals

$$\mathbb{E}[B] - \mathbb{E}[R_B] + (2 + \lambda\mathbb{E}[B])\mathbb{E}[R_B] \le \frac{K}{C} \le \ \mathbb{E}[B] - \mathbb{E}[R_B] + \frac{2 - \lambda\mathbb{E}[B]}{(1 - \lambda\mathbb{E}[B])^2}\mathbb{E}[R_B]. \tag{8.4.48}$$

Therefore for all K/C that meet the condition (8.4.48), if $y \in [1 - \lambda\mathbb{E}[B], \bar{y}_1)$ then we have $P_2(y) < 0$, if $y \in (\bar{y}_1, 1]$ then $P_2(y) > 0$ Under (8.4.48), we have

$$\max_{q_1\in[0,1]} \{U_s(1, q_1)\} = U_s\Big(1, \frac{1 - \bar{y}_1}{\lambda\mathbb{E}[B]}\Big).$$

Similarly, when K/C satisfies

$$\mathbb{E}[B] - \mathbb{E}[R_B] < \frac{K}{C} < \mathbb{E}[B] - \mathbb{E}[R_B] + (2 + \lambda\mathbb{E}[B])\mathbb{E}[R_B],$$

we have $\max_{q_1\in[0,1]} \{U_s(1, q_1)\} = U_s(1, 0) < 0$ If

$$\mathbb{E}[B] - \mathbb{E}[R_B] + (2 + \lambda\mathbb{E}[B])\mathbb{E}[R_B] < \frac{K}{C},$$

it follows that $\max_{q_1\in[0,1]} \{U_s(1, q_1)\} = U_s(1, 1)$

The conclusions for these two situations can be summarized as the following theorem.

Theorem 8.4.5 *If the condition (8.4.1) is satisfied, assume that customers join the system with probability 1 if the server is found on vacation upon arrival, then their socially optimal probability of joining the queue when the server is found busy upon arrival is given by*

$$q_s(1) = \begin{cases} 0, & \text{if } \frac{K}{C} < Z_1, \\ \frac{1-\bar{y}_1}{\lambda E[B]}, & \text{if } Z_1 \leq \frac{K}{C} \leq Z_2, \\ 1, & \text{if } Z_2 < \frac{K}{C}, \end{cases}$$

where $\bar{y}_1$ is given by (8.4.47), and

$$Z_1 = \mathbb{E}[B] - \mathbb{E}[R_B] + (2 + \lambda\mathbb{E}[B])\mathbb{E}[R_B],$$
$$Z_2 = \mathbb{E}[B] - \mathbb{E}[R_B] + \frac{2 - \lambda\mathbb{E}[B]}{(1 - \lambda\mathbb{E}[B])^2}\mathbb{E}[R_B].$$

In region $\Re_3$, none of the arriving customers join the queue if the server is found busy upon arrival. For $q_0 \in (0, 1]$, the optimal condition for the first-order derivative of social benefits is

$$\frac{K}{C} = \mathbb{E}[B] + (1 + \lambda q_0\mathbb{E}[B])^2\mathbb{E}[R_V].$$

We will discuss three scenarios in the following.

Case 1: When $K/C \leq \mathbb{E}[B] + \mathbb{E}[R_V]$, $U_s(q_0, 0)$ is strictly monotone decreasing with $q_0 \in (0, 1]$. As a result, $\sup_{q_0 \in (0,1]} \{U_s(q_0, 0)\} = U_s(0, 0) = 0$

Case 2: When $\mathbb{E}[B] + \mathbb{E}[R_V] < K/C \leq \mathbb{E}[B] + (1 + \lambda\mathbb{E}[B])^2\mathbb{E}[R_V]$, it is readily seen that $U_s(q_0, 0)$ is monotone increasing with $q_0 \in (0, (y_0 - 1)/(\lambda E[B]))$, and is monotone decreasing with $q_0 \in ((y_0 - 1)/(\lambda\mathbb{E}[B]), 1]$, where

$$y_0 = \sqrt{\frac{1}{\mathbb{E}[R_V]}\left(\frac{K}{C} - \mathbb{E}[B]\right)}. \tag{8.4.49}$$

Therefore we have $\max_{q_0 \in (0,1]} \{U_s(q_0, 0)\} = U_s((y_0 - 1)/(\lambda E[B]), 0)$

Case 3: When $\mathbb{E}[B] + (1 + \lambda\mathbb{E}[B])^2\mathbb{E}[R_V] < K/C$, $U_s(q_0, 0)$ is strictly monotone increasing with $q_0 \in (0, 1]$ Consequently $\max_{q_0 \in (0,1]} \{U_s(q_0, 0)\} = U_s(1, 0)$

The above conclusions for these three situations can be summarized as the following theorem.

Theorem 8.4.6 *If the condition (8.4.1) is satisfied, assume that customers balk from the system upon arrival if the server is found busy, then their socially optimal probability of joining the queue when the server is found on vacation upon arrival is given by*

$$q_s(0) = \begin{cases} \frac{y_0-1}{\lambda\mathbb{E}[B]}, & \text{if } \mathbb{E}[B]+\mathbb{E}[R_V] < \frac{K}{C} \le \mathbb{E}[B]+(1+\lambda\mathbb{E}[B])^2\mathbb{E}[R_V], \\ 1, & \text{if } \mathbb{E}[B]+(1+\lambda\mathbb{E}[B])^2\mathbb{E}[R_V] < \frac{K}{C}, \end{cases}$$

where y_0 is given by (8.4.49).

Regarding $\Re_4$, arriving customers join the system with probability 1 upon arrival if the server is found busy. For $q_0 \in (0, 1]$, the optimal condition for the first-order derivative of social benefits is

$$\frac{1-\lambda\mathbb{E}[B]}{\mathbb{E}[R_V]}\Big(\frac{K}{C} - \Big(\frac{\lambda\mathbb{E}[R_B]}{1-\lambda\mathbb{E}[B]}+1\Big)\mathbb{E}[B]\Big) = u^2, \tag{8.4.50}$$

where $u = 1-\lambda(1-q_0)\mathbb{E}[B] \in (1-\lambda\mathbb{E}[B], 1] \subset (0, 1]$ We will discuss three scenarios below.

Case 1: When

$$\frac{K}{C} \le \mathbb{E}[B]+(1-\lambda\mathbb{E}[B])\mathbb{E}[R_V]+\frac{\lambda\mathbb{E}[B]}{1-\lambda\mathbb{E}[B]}\mathbb{E}[R_B],$$

it is evident that $U_s(q_0, 1)$ is strictly monotone decreasing with $q_0 \in (0, 1]$ and consequently, $\sup_{q_0\in(0,1]} \{U_s(q_0, 1)\} = U_s(0, 1)$

Case 2: When

$$\mathbb{E}[B]+(1-\lambda\mathbb{E}[B])\mathbb{E}[R_V]+\frac{\lambda\mathbb{E}[B]}{1-\lambda\mathbb{E}[B]}\mathbb{E}[R_B] < \frac{K}{C} \le \mathbb{E}[B]+\frac{1}{1-\lambda\mathbb{E}[B]}(\mathbb{E}[R_V]+\lambda\mathbb{E}[B]\mathbb{E}[R_B]),$$

$U_s(q_0, 1)$ is monotone increasing with $q_0 \in (0, 1-(1-\bar{y}_0)/(\lambda\mathbb{E}[B]))$, and is monotone decreasing with $q_0 \in (1-(1-\bar{y}_0)/(\lambda\mathbb{E}[B]), 1]$, where

$$\bar{y}_0 = \sqrt{\frac{1-\lambda\mathbb{E}[B]}{\mathbb{E}[R_V]}\Big(\frac{K}{C} - \Big(\frac{\lambda\mathbb{E}[R_B]}{1-\lambda\mathbb{E}[B]}+1\Big)\mathbb{E}[B]\Big)}. \tag{8.4.51}$$

Therefore, we have $\max_{q_0\in(0,1]} \{U_s(q_0, 1)\} = U_s(1-(1-\bar{y}_0)/(\lambda\mathbb{E}[B]), 1)$

Case 3: When $\mathbb{E}[B]+\frac{1}{1-\lambda\mathbb{E}[B]}(\mathbb{E}[R_V]+\lambda\mathbb{E}[B]\mathbb{E}[R_B]) < K/C$, $U_s(q_0, 1)$ is strictly monotone increasing with $q_0 \in (0, 1]$ Therefore, $\max_{q_0\in(0,1]} \{U_s(q_0, 1)\} = U_s(1, 1)$

We summarize the above conclusions regarding three cases as the following theorem.

Theorem 8.4.7 *Under the condition (8.4.1), assume customers join with probability 1 provided that the server is found busy upon arrival, then the socially-optimal joining probability of customers if they found that the server is on vacation upon arrival is given by*

$$q_s(0) = \begin{cases} 1 - \frac{1-\bar{y}_0}{\lambda \mathbb{E}[B]}, & \text{if } X_1 < \frac{K}{C} \le X_2, \\ 1, & \text{if } X_2 < \frac{K}{C}, \end{cases}$$

where $\bar{y}_0$ is given by (8.4.51), and

$$X_1 = \mathbb{E}[B] + (1 - \lambda \mathbb{E}[B])\mathbb{E}[R_V] + \frac{\lambda \mathbb{E}[B]}{1 - \lambda \mathbb{E}[B]}\mathbb{E}[R_B],$$

$$X_2 = \mathbb{E}[B] + \frac{1}{1 - \lambda \mathbb{E}[B]}(\mathbb{E}[R_V] + \lambda \mathbb{E}[B]\mathbb{E}[R_B]).$$

References

Burnetas, A., and Antonis Economou. 2007. Equilibrium customer strategies in a single server Markovian queue with setup times. *Queueing Systems* 56: 213–228.

Economou, Antonis, Antonio Gómez-Corral, and Spyridoula Kanta. 2011. Optimal balking strategies in single-server queues with general service and vacation times. *Performance Evaluation* 68: 967–982.

Elaydi, S. 2005. *An Introduction to Difference Equations*. New York: Springer Science & Business Media.

Fuhrmann, S. W., and Robert B. Cooper. 1985. Stochastic decompositions in the M/G/1 queue with generalized vacations. *Operations Research* 33: 1117–1129.

Gao, Shan, and Jinting Wang. 2016. Equilibrium balking strategies in the observable Geo/Geo/1 queue with delayed multiple vacations. *RAIRO-Operations Research* 50 (1): 119–129.

Guo, Pengfei, and Refael Hassin. 2011. Strategic behavior and social optimization in Markovian vacation queues. *Operations Research* 59: 986–997.

Guo, Pengfei, and Refael Hassin. 2012. Strategic behavior and social optimization in Markovian vacation queues: The case of heterogeneous customers. *European Journal of Operational Research* 222: 278–286.

Guo, Pengfei, and Qingying Li. 2013. Strategic behavior and social optimization in partially-observable Markovian vacation queue. *Operations Research Letters* 41: 277–284.

Li, Kaili, Jinting Wang, Yanjia Ren, and Jingwei Chang. 2016. Equilibrium joining strategies in M/M/1 queues with working vacation and vacation interruptions. *RAIRO-Operations Research* 50 (3): 451–471.

Liu, Jie, and Jinting Wang. 2017. Strategic joining rules in a single server Markovian queue with Bernoulli vacation. *Operational Research: An International Journal* 17 (2): 413–434.

Liu, Weiqi, Yan Ma, and Jihong Li. 2012. Equilibrium threshold strategies in observable queueing systems under single vacation policy. *Applied Mathematical Modelling* 36: 6186–6202.

Ma, Yan, Weiqi Liu, and Jihong Li. 2012. Equilibrium balking behavior in the Geo/Geo/1 queueing system with multiple vacations. *Applied Mathematical Modelling* 37: 3861–3878.

Sun, Wei, Pengfei Guo, and Naishuo Tian. 2010. Equilibrium threshold strategies in observable queueing systems with setup/closedown times. *Central European Journal of Operational Research* 18: 241–268.

Takagi, H. 1991. *Queueing Analysis: A Foundation of Performance Evaluation, Volume 1: Vacation and priority systems*, Part I, North-Holland, Amsterdam.

Tian, N., and Zhe George Zhang. 2006. *Vacation Queueing Models: Theory and Applications*. New York: Springer.

Wang, Fang, Jinting Wang, and F. Zhang. 2014. Equilibrium customer strategies in the Geo/Geo/1 queues with single working vacation. *Discrete Dynamics in Nature and Society* 2014, Article ID 309489, 9 p.

Wang, Jinting, Xuelu Zhang, and Ping Huang. 2017. Strategic behavior and social optimization in the constant retrial queue with N-policy. *European Journal of Operational Research* 256 (3): 841–849.

Wang, Fang, Jinting Wang, and Zhe George Zhang. 2017. Strategic behavior and social optimization in a double-ended queue with gated policy. *Computers & Industrial Engineering* 114: 264–273.

Wang, Jinting, Yu. Zhang, and Zhe George Zhang. 2021. Strategic joining in an M/M/K queue with asynchronous and synchronous multiple vacations. *Journal of the Operational Research Society* 72 (1): 161–179.

Wang, Yilin, Jinting Wang, and Zhe George Zhang. 2024. The effect of information on the strategic behavior in a Markovian queue with catastrophes and working vacations. *Quality Technology & Quantitative Management* 21 (5): 688–721.

Zhang, Lingjiao and Jinting Wang. 2023. Strategic behavior and optimization of an M/M/1 queue with N-policy and hysteretic control. *Methodology and Computing in Applied Probability*, 2023, Volume 25, article number 79.

Zhang, Yu., and Jinting Wang. 2017. Equilibrium pricing in an M/G/1 retrial queue with reserved idle time and setup time. *Applied Mathematical Modelling* 49: 514–530.

Zhang, Yu., and Jinting Wang. 2021. Strategic joining and information disclosing in Markovian queues with an unreliable server and working vacations. *Quality Technology and Quantitative Management* 18 (3): 298–325.

Zhang, F., Jinting Wang, and Bin Liu. 2013. Equilibrium balking strategies in Markovian queues with working vacations. *Applied Mathematical Modelling* 37: 8264–8282.

Zhang, F., Jinting Wang, and Bin Liu. 2013. Equilibrium joining probabilities in observable queues with general service and setup times. *Journal of Industrial and Management Optimization* 9: 901–917.

Chapter 9
Retrial Queueing Systems

In a retrial queue, customers who observe a busy system upon arrival, must leave temporarily and return later for service. Between trials, the customer is said to be in orbit. Retrial queueing model has been widely used in telephone switching systems, telecommunication networks and computer systems. Early retrial queueing models were mainly used for telephone call problems.

In retrial queues, every customer in orbit can ask for service repeatedly, so the service discipline is no longer FCFS. They are influenced not only by competition among themselves, but also by the queuing strategies of customers who arrive later. Therefore, it is much more complicated to study the customer equilibrium behavior and social optimal strategy in the retrial queueing system. Aguir Karaesmen, Akşin and Chauvet (2004), Artalejo (1995), Artalejo and Gómez-Corral (1997, 1998, 2008), Falin (2008), Falin and Templeton (1997), etc. They have done a lot of research work on the performance analysis of traditional retrial queueing system.

One of the earliest studies on customer behavior in retrial queues is Kulkarni (1983a, b) in which competition and equilibrium retrial strategy of customers in a single server retrial queueing system with two types of customers are studied. Subsequently, Elcan (1994) calculates the social optimal and Nash equilibrium retrial rates in an M/M/1 retrial queue. Hassin and Haviv (1996) extend the service time to follow a general probability distribution. Zhang et al. (2012) consider a single server queueing system with two types of customers, in which the server is unreliable and work with vacation. They discuss and compare the optimal and equilibrium retrial rates when customers are cooperative and non-cooperative.

However, in these articles, retrial customers are considered while balking behavior is prohibited. Economou and Kanta (2011) first study the balking equilibrium strategy of customers and the problem of social and benefit maximization in retrial queue. The retrial rate is constant in their model, that is, customers waiting in the orbit follow FCFS. Wang and Zhang (2013) study the customer balking strategy in the traditional retrial queue for the first time, and obtain the Nash equilibrium and social optimal queuing strategy under two information conditions respectively. Cui, Su and

J. Wang, *Fundamentals of Queueing-Game Models*,
https://doi.org/10.1007/978-981-95-0261-5_9

Veeraraghavan (2013) assume that customers could choose to join the queue, balk or retry upon arrival in observable queue. According to the value range of retrial cost, they discussed three cases and obtained the mixed Nash equilibrium strategy of customers in each case. Zhang and Wang (2016) deliberate equilibrium pricing strategies in retrial queueing systems with complementary service. Wang and Wang (2019) consider information heterogeneity in a retrial queue and its influence on the system throughput and social welfare maximization. Li and Wang (2021) explore the equilibrium balking strategies in the single-server retrial queue with constant retrial rate and catastrophes. Zhang and Wang (2023) consider the research issue that how to manage retrial queueing systems with boundedly rational customers. Zhang and Wang (2022) also study quality-speed trade-offs in customer-intensive services with boundedly rational customers and retrials. Wang and Wang (2021) focus on exploring the strategic priority-purchasing and joining rules in a retrial queue.

9.1 M/M/1 Retrial Queues

9.1.1 Model Description

Assume that there is a single server without waiting space in the system, customers arrive according to a Poisson process with rate λ. When a new customer arrives at the system and finds the server idle, he immediately occupies the server and receives service. Service time is exponential distributed with parameter μ. And if he observes a busy server, he enters a virtual retial orbit and tries again until receiving service. Assume that the time lengths between retrials are independent and exponentially distributed with parameter θ. Assume that the successive arrival time, retrial interval and service time of customers are independent of each other.

$I(t)$ denotes the state of the server at time t. $I(t) = 0, 1$ represents server is idle or busy at time t. $N(t)$ denotes the number of customers(not including customers in service) in the orbit. It is clear that the process $\{(I(t), N(t)), t \geq 0\}$ is is a continuous time Markov chain with state space $\{0, 1\} \times \{0, 1, 2, \ldots\}$.

For stability, $\lambda < \mu$ holds. Let $p(i, j)$ be the steady-state probability of server state i and j customers in the orbit. And write $p_0(z) = \sum_{j=0}^{\infty} z^j p(0, j)$ and $p_1(z) = \sum_{j=0}^{\infty} z^j p(1, j)$ as the corresponding generating functions.

Theorem 9.1.1 *If stability condition $\lambda < \mu$ is satisfied, the steady-state probability is*

$$p(0, j) = \frac{\lambda^j}{j!\mu^j\theta^j} \prod_{i=0}^{j-1}(\lambda + i\theta) \cdot \left(1 - \frac{\lambda}{\mu}\right)^{\frac{\lambda}{\theta}+1}, \quad j = 0, 1, 2, \ldots, \tag{9.1.1}$$

$$p(1, j) = \frac{\lambda^{j+1}}{j!\mu^{j+1}\theta^j} \prod_{i=1}^{j}(\lambda + i\theta) \cdot \left(1 - \frac{\lambda}{\mu}\right)^{\frac{\lambda}{\theta}+1}, \quad j = 0, 1, 2, \ldots, \tag{9.1.2}$$

For $j < k$, *define* $\prod_{i=k}^{j} f(i) = 1$. *The corresponding generating function is*

$$p_0(z) = \frac{\mu - \lambda}{\mu}\left(\frac{1 - \frac{\lambda}{\mu}}{1 - \frac{\lambda}{\mu}z}\right)^{\frac{\lambda}{\theta}}, \tag{9.1.3}$$

$$p_1(z) = \frac{\lambda}{\mu}\left(\frac{1 - \frac{\lambda}{\mu}}{1 - \frac{\lambda}{\mu}z}\right)^{\frac{\lambda}{\theta}+1}. \tag{9.1.4}$$

Proof Under statistical equilibrium, the balance equations are listed

$$(\lambda + j\theta)p(0, j) = \mu p(1, j), \quad j = 0, 1, 2, \dots, \tag{9.1.5}$$

$$(\lambda + \mu)p(1, j) = \lambda p(0, j) + (j + 1)\theta p(0, j + 1) + \lambda p(1, j - 1), \quad j = 0, 1, 2, \dots, \tag{9.1.6}$$

where $p(1, -1) = 0$

Multiplying (9.1.5) and (9.1.6) by z^j, then sum all j:

$$\lambda p_0(z) + \theta z \frac{\mathrm{d}}{\mathrm{d}z} p_0(z) = \mu p_1(z), \tag{9.1.7}$$

$$(\lambda(1 - z) + \mu)p_1(z) = \lambda p_0(z) + \theta \frac{\mathrm{d}}{\mathrm{d}z} p_0(z). \tag{9.1.8}$$

Cancel out $p_1(z)$ and get the differential equation:

$$\frac{\mathrm{d}}{\mathrm{d}z} p_0(z) = \frac{\lambda^2}{\theta(\mu - \lambda z)} p_0(z),$$

The solution is

$$p_0(z) = \frac{Const}{\left(1 - \frac{\lambda}{\mu}z\right)^{\frac{\lambda}{\theta}}}. \tag{9.1.9}$$

Substituting (9.1.9) into (9.1.7), we obtain

$$p_1(z) = \frac{\frac{\lambda}{\mu}Const}{\left(1 - \frac{\lambda}{\mu}z\right)^{\frac{\lambda}{\theta}+1}}. \tag{9.1.10}$$

From the normalization conditions $\sum_{j=0}^{\infty}(p(0, j) + p(1, j)) = p_0(1) + p_1(1) = 1$, we obtain

$$Const = \left(1 - \frac{\lambda q}{\mu}\right)^{\frac{\lambda}{\theta}+1}. \tag{9.1.11}$$

According to (9.1.9)–(9.1.11), we can get the generating function immediately (9.1.3) and (9.1.4).

Use the binomial formula

$$(1 + x)^m = \sum_{n=0}^{\infty} \frac{x^n}{n!} \prod_{i=0}^{n-1}(m - i),$$

We can expand (9.1.3) and (9.1.4) to get the steady-state probabilities (9.1.1) and (9.1.2). □

9.1.2 Retrial Strategy Analysis

Suppose the customer arrives and must enter the system, either to receive service immediately, or enter the orbit to retry. Without loss of generality, each retrial costs c, and the cost while orbiting is w per unit of time. Assume that the time lengths between retrials are independent distributed with parameter θ. In particular, any value of θ corresponds to a strategy. So for the social manager and the individual customer, what strategy can be chosen to minimize the total cost of the customer?

For the social manager, he wants to minimize the cost of all customers per unit of time.

Theorem 9.1.2 *In the case that all customers cooperate, the social optimal retrial rate is*

$$\theta^* = \sqrt{\frac{w\mu}{c}}. \tag{9.1.12}$$

Proof The expected number of customers waiting in the retrial space is

$$\mathbb{E}(N) = \frac{\mathrm{d}}{\mathrm{d}z} p_0(z)\Big|_{z=1} + \frac{\mathrm{d}}{\mathrm{d}z} p_1(z)\Big|_{z=1} = \frac{\lambda^2(\theta + \mu)}{\theta\mu(\mu - \lambda)}. \tag{9.1.13}$$

The expected waiting time of customers in the retrial space can be obtained by the Little's formula

$$W_1 = \frac{\lambda(\theta + \mu)}{\theta\mu(\mu - \lambda)}. \tag{9.1.14}$$

Then the expected number of retries per customer in the retrial space is θW_1.

The expected total cost for each individual is S, which includes the total wait cost wW_1 and total retrial cost $c\theta W_1$ during the retry period. Thus, The expected total cost is:

$$S = (w + c\theta)W_1 = \frac{\lambda}{\mu - \lambda}\left(\frac{c\theta}{\mu} + \frac{w}{\theta}\right) + \frac{\lambda}{\mu - \lambda}\left(\frac{w}{\mu} + c\right). \qquad (9.1.15)$$

When the retrial rate is θ^*, the customer's total cost reaches the minimum. □

Note that the social optimal retrial rate is independent of the arrival rate. When the arrival rate λ increases, the utilization rate of the server increases. Therefore, for any value of θ, the excepted number of retries and the waiting time will increase accordingly. Surprisingly, these changes maintain a balance between these two opposing influences, so that θ^* does not change.

In the case of non-cooperation, customers only consider their own interests, so competition is bound to occur. To get the Nash equilibrium retrial rate, we mark one customer in the retrial space with a retrial rate of γ and all other customers with a retrial rate of θ. Mark ϕ_i as the expected wait time of customers when the server is busy and i other customers are in the retrial space.

Lemma 9.1.1 *For any integer $i \geq 0$,*

$$\phi_i = ai + b, \qquad (9.1.16)$$

where

$$a = \frac{\theta}{(\mu - \lambda)\theta + \gamma\mu}, \qquad (9.1.17)$$

$$b = \frac{1}{\mu} + \frac{\lambda + \mu}{\gamma\mu} + \frac{\lambda\theta(\lambda + \gamma)}{\gamma\mu[(\mu - \lambda)\theta + \gamma\mu]}. \qquad (9.1.18)$$

Proof First, the relationship between ϕ_i and the independent variable i is analyzed. Compare the two situations: one is that the customer is waiting in the retrial space with other i individuals; the other is that the customer is waiting in the retrial space with other $i + 1$ individuals. In contrast, an additional customer in the latter case increases the expected waiting time of the tagged customer. Let's analyze that the increased waiting time is independent of i. In the retrial space, the tagged customer retries at rate γ and the redundant customer retries at rate θ. Therefore, the probability of success of the first retry of the redundant customer is $\frac{\theta}{\theta+\gamma}$(independent with i). The excepted increased waiting time is equal to the excepted service time $\frac{1}{\mu}$ of the additional customers, plus the service time of those customers who arrived during that customer service period and ultimately preceded the tagged customer service, and so on. All of these values are independent of i. Therefore, the value of ϕ_i can be expressed in the linear form (9.1.16).

Note that a is the average increase in waiting time due to additional customers. The probability that this delay time is 0 is $\frac{\gamma}{\gamma+\theta}$, the probability of being positive is $\frac{\theta}{\gamma+\theta}$. In the latter case, the average delay time is equal to the average service time $\frac{1}{\mu}$ plus the average number of customers arriving during the service period $\frac{\lambda}{\mu}$, times the delay a for each new customer arriving to the marked customer. So to sum it up, it is

$$a = \frac{\theta}{\gamma+\theta}\Big(\frac{1}{\mu} + \frac{\lambda}{\mu}a\Big),$$

The unique solution is shown by (9.1.17).

b is the average waiting time when the tagged customer is alone in the orbit and the server is busy. Let W be the average waiting time that tagged customer alone in the orbit and the server is free. Thus,

$$b = \frac{1}{\lambda+\mu} + \frac{\lambda}{\lambda+\mu}(a+b) + \frac{\mu}{\lambda+\mu}W, \tag{9.1.19}$$

$$W = \frac{1}{\lambda+\gamma} + \frac{\lambda}{\lambda+\gamma}b. \tag{9.1.20}$$

Solve (9.1.19) and (9.1.20), obtain b is shown by (9.1.18). □

By virtue of the total expectation formula, the average total revenue of tagged customers is derived as

$$\begin{aligned} W_2 &= \sum_{i=0}^{\infty}(ai+b)p(1,i) = a\frac{\mathrm{d}}{\mathrm{d}z}p_1(z)\Big|_{z=1} + bp_1(1) \\ &= \frac{\lambda}{\mu-\lambda}\Big[\frac{1}{\mu} + \frac{1}{\gamma} + \frac{\lambda}{\mu}\frac{\theta-\gamma}{(\mu-\lambda)\theta+\mu\gamma}\Big]. \end{aligned} \tag{9.1.21}$$

Thus, the excepted waiting time of tagged customer is

$$f(\gamma,\theta) = (w+c\gamma)W_2 = \frac{\lambda(w+c\gamma)}{\mu-\lambda}\Big[\frac{1}{\mu} + \frac{1}{\gamma} + \frac{\lambda}{\mu}\frac{\theta-\gamma}{(\mu-\lambda)\theta+\mu\gamma}\Big]. \tag{9.1.22}$$

Theorem 9.1.3 *In the case of non-cooperation, the customer's Nash equilibrium retrial rate is*

$$\theta_e = \frac{\lambda w + \sqrt{\lambda^2 w^2 + 8\mu w c(\mu-\lambda)(2\mu-\lambda)}}{4c(\mu-\lambda)}. \tag{9.1.23}$$

Proof Take the partial of (9.1.22) with respect to γ and set $\gamma = \theta$, we obtain

$$\frac{\partial}{\partial\gamma}f(\gamma,\theta)\Big|_{\gamma=\theta} = \frac{\lambda}{\mu-\lambda}\Big[c\Big(\frac{1}{\theta}+\frac{1}{\mu}\Big) + (w+c\theta)\Big(-\frac{1}{\theta^2} + \frac{\lambda}{\theta\mu(\lambda-2\mu)}\Big)\Big]. \tag{9.1.24}$$

The necessary condition for the retry rate θ to be a Nash equilibrium solution is $\frac{\partial}{\partial \gamma} f(\gamma, \theta)\Big|_{\gamma=\theta} = 0$, that is

$$2c(\mu - \lambda)\theta^2 - \lambda w\theta - w\mu(2\mu - \lambda) = 0.$$

The unique solution to the above equation θ_e is shown by Eq. (9.1.23). □

The individually-optimal and socially optimal retrial rates are compared below.

Theorem 9.1.4 *In the $M/M/1$ retrial queue, the equilibrium retrial rate θ_e is larger than the social optimal retrial rate θ_e.*

Proof According to $2\mu - \lambda > 2(\mu - \lambda)$, it is readily obtained that

$$\theta_e > \frac{\sqrt{16\mu wc(\mu - \lambda)^2}}{4c(\mu - \lambda)} = \theta^*. \tag{9.1.25}$$

□

In the non-cooperative case, when a customer can choose his own retrial rate, he will ignore the behavior of other customers. Therefore, the social optimal and Nash equilibrium retrial rates will be different. Customer of the orbit will retry to prevent other customers from receiving the service before him. A new customer arrival, or a retrial customer, is a negative externalizing effect for that customer. And when considering the social optimal problem, it tries to eliminate the negative influence between customers due to competition, so there is a result $\theta_e > \theta^*$.

If the difference between the social optimal and Nash equilibrium retrial rates is to be eliminated, two measures can be considered. On the one hand, a certain amount of compensation for the actual waiting cost of the customer. One the other hand, a certain additional charge for each retry. So the customers will reduce their retrial frequency and avoid paying high retrial fees.

Theorem 9.1.5 *If a certain compensation $P < w$ is given for each unit of waiting time so that the equilibrium retrial rate is consistent with the social optimal retrial rate θ^*, the compensation can be determined by the following formula*

$$\frac{\lambda(w - P) + \sqrt{\lambda^2(w - P)^2 + 8\mu(w - P)c(\mu - \lambda)(2\mu - \lambda)}}{4c(\mu - \lambda)} = \sqrt{\frac{w\mu}{c}}. \tag{9.1.26}$$

The same effect can be achieved by charging an additional retrial fee $T = c\frac{P}{w-P}$. Then the retial cost becomes $c + T$.

Proof According to (9.1.25) we know $\theta_e > \theta^*$. Note that θ_e is continuously monotonically increasing with respect to w, then for a sufficiently small w, the equilibrium retrial rate is smaller than θ^*. And as $w - P$ goes to 0, the value of θ_e goes to 0. So some value $P \in (0, w)$ must exist which make the equation (9.1.26) hold.

Since the social optimal and Nash equilibrium retrial rates depend only on the ratio of c to w, the additional retrial charge T can be obtained by solving the equation $\frac{c+T}{w} = \frac{c}{w-P}$. □

9.1.3 Equilibrium Strategy in Unobservable Retrial Queue

We assume that the customers are allowed to decide whether to join or balk upon their arrival based on the information they have. After service, every customer receives a reward of R units. This may reflect his satisfaction or the added value of being served. Moreover, each customer incurs a waiting cost C units per time unit when he remains in the system (this cost is accumulated both in the retrial pool and in the service area). We observe that the expected waiting cost incurred in the service area is $\frac{C}{\mu}$, independently of customers' strategy, and one can subtract it from R. Without loss of generality, we assume that waiting costs are incurred only for the time in the retrial pool. Customers are risk neutral and wish to maximize their expected benefit. At their arrival instants, they have to assess their expected waiting costs against their reward associated with receiving service, given the information regarding the system, and decide accordingly whether to join or not. Specifically, the customers strictly prefer to enter if the reward for service exceeds the expected cost for waiting, and are indifferent between entering and balking if the reward is equal to the cost. Moreover, customers decisions are irrevocable, meaning that neither reneging of entering customers nor retrials of balking customers are allowed.

We begin with the unobservable case in which the customers are informed about the state of the server but not the number of customers in the pool. Since all customers are assumed indistinguishable, we can consider the situation as a symmetric game among them. As we have already noted, a customer that finds an idle server upon arrival prefers to enter, because he has a positive expected net benefit R. Thus, entering is a dominant strategy of the customer which need not take into account the strategies of any other customers. We next concentrate our study on the behavior of customers that find the server busy upon arrival. In the present model, there are two pure strategies, to join or to balk. And a mixed strategy is specified by the joining probability of an arriving customer that finds the server busy. Our goal is to identify the Nash equilibrium mixed balking strategy.

9.1.3.1 Nash Equilibrium Strategy

Suppose that all customers follow a mixed strategy "enter with probability q while observing a busy server". The system is stable if $\lambda < \mu$. Letting $p_{un}(i, j)$ be the steady-state probability of state (i, j) and defining the corresponding partial generating functions as $p_0(z) = \sum_{j=0}^{\infty} z^j p_{un}(0, j)$ and $p_1(z) = \sum_{j=0}^{\infty} z^j p_{un}(1, j)$, we have the following Lemma.

Lemma 9.1.2 *The partial generating functions* $p_0(z)$ *and* $p_1(z)$ *which are with respect to the number of customers in the retrial pool, are given by*

$$p_0(z) = \frac{\mu - \lambda q}{\mu + \lambda(1-q)} \left(\frac{1 - \frac{\lambda q}{\mu}}{1 - \frac{\lambda}{\mu} z} \right)^{\frac{\lambda}{\theta}}, \tag{9.1.27}$$

$$p_1(z) = \frac{\lambda}{\mu + \lambda(1-q)} \left(\frac{1 - \frac{\lambda q}{\mu}}{1 - \frac{\lambda q}{\mu} z} \right)^{\frac{\lambda}{\theta}+1}. \tag{9.1.28}$$

Proof The proof is similar to that of Theorem 9.1.1, so it is omitted here. □

We now tag a customer in the pool. Define $T(1, k)$ as the expected time until service begins for the tagged customer, given that k other customers are in the retrial pool and the server is busy.

Lemma 9.1.3 *For* $k \geq 0$*, the expected waiting time for a tagged customer in the retrial pool is given by*

$$T(1, k) = mk + b, \tag{9.1.29}$$

where

$$m = \frac{1}{2\mu - \lambda q}, \tag{9.1.30}$$

$$k = \frac{2(\lambda + \theta + \mu)}{\theta(2\mu - \lambda q)}. \tag{9.1.31}$$

Proof The proof is similar to that of Lemma 9.1.1, so it is omitted here. □

We now consider a customer that arrives to find the server busy. Due to PASTA property, the probability that the customer observes j other customers in the pool is the stationary probability that the server is busy and j customers are in the pool $p_{au}(i, j)$. If this customer decides to enter, his expected waiting time in the pool is equal to $T(1, k)$.

Theorem 9.1.6 *The expected time that an arriving customer waits until his service begins if he decides to enter given that the server is busy, is given by*

$$T(q) = \frac{1}{\theta} + \frac{\lambda + \theta}{\theta(\mu - \lambda q)}. \tag{9.1.32}$$

Proof When the system is in steady state, the expected waiting time for an arriving customer given that he finds the server busy and he decides to enter is

$$T(q) = \frac{\sum_{j=0}^{\infty} p_{un}(1, j) T(1, j)}{\sum_{j=0}^{\infty} p_{un}(1, j)} = \frac{\frac{\mathrm{d}}{\mathrm{d}z} p_1(z)\Big|_{z=1}}{p_1(1)} m + b = \frac{1}{\theta} + \frac{\lambda + \theta}{\theta(\mu - \lambda q)}.$$

□

Based on the reward–cost structure we have mentioned before, the expected net benefit of an arriving customer if he finds the server busy and decides to enter is

$$S(q) = R - C\Big(\frac{1}{\theta} + \frac{\lambda + \theta}{\theta(\mu - \lambda q)}\Big). \tag{9.1.33}$$

Theorem 9.1.7 *In the unobservable retrial queue with $\lambda < \mu$, a unique Nash equilibrium mixed strategy enter with probability q_e while observing a busy server exists, where q_e is given by*

$$q_e = \begin{cases} 0, & if 0 < \frac{R}{C} \le \frac{\lambda+\theta+\mu}{\theta\mu}, \\ \frac{\mu - \frac{(\lambda+\theta)C}{R\theta - C}}{\lambda}, & if \frac{\lambda+\theta+\mu}{\theta\mu} < \frac{R}{C} < \frac{\theta+\mu}{\theta(\mu-\lambda)}, \\ 1, & if \frac{R}{C} \ge \frac{\theta+\mu}{\theta(\mu-\lambda)}. \end{cases} \tag{9.1.34}$$

Proof We observe that $S(q)$ is strictly decreasing for q and has a unique maximum

$$S(0) = R - C\frac{\lambda + \theta + \mu}{\theta\mu},$$

and a unique minimum

$$S(1) = R - C\frac{\theta + \mu}{\theta(\mu - \lambda)}.$$

Therefore, when $\frac{R}{C} \in (0, \frac{\lambda+\theta+\mu}{\theta\mu}]$, $S(q)$ is non-positive for every q, the best response is balking and the unique equilibrium point is $q_e = 0$. When $\frac{R}{C} \in (\frac{\lambda+\theta+\mu}{\theta\mu}, \frac{\theta+\mu}{\theta(\mu-\lambda)})$, there exists a unique solution of the equation $S(q) = 0$ which lies in the interval $(0, 1)$ and gives the second branch of (9.1.34). When $\frac{R}{C} \in [\frac{\theta+\mu}{\theta(\mu-\lambda)}, \infty)$, $S(q)$ is non-negative for every q. In other words, the best response is $q_e = 1$ in this case. Therefore, we have an ATC situation in unobservable case. □

9.1.3.2 Social Optimal Strategy

Theorem 9.1.8 *In the unobservable retrial queue with $\lambda < \mu$, a unique mixed strategy "enter with probability q^* while observing a busy server" that maximizes the social net benefit per time unit exists and is given by*

$$q^* = \begin{cases} 0, & if 0 < \frac{R}{C} \le \frac{(\lambda+\mu)(\lambda+\theta+\mu)}{\theta\mu^2}, \\ \frac{\mu - x_2}{\lambda}, & if \frac{(\lambda+\mu)(\lambda+\theta+\mu)}{\theta\mu^2} < \frac{R}{C} < \frac{\mu^2(\theta+\mu)+\lambda\theta(\mu-\lambda)}{\theta\mu(\mu-\lambda)^2}, \\ 1, & if \frac{R}{C} \ge \frac{\mu^2(\theta+\mu)+\lambda\theta(\mu-\lambda)}{\theta\mu(\mu-\lambda)^2}, \end{cases} \tag{9.1.35}$$

where

$$x_2 = \frac{\mu C(\lambda+\theta) + \sqrt{(\mu C(\lambda+\theta))^2 + \lambda\mu C(\lambda+\theta)(C(\theta-\mu)+\mu\theta R)}}{(C(\theta-\mu)+\mu\theta R)}.$$

Proof The social benefit per time unit when all customers follow a mixed policy "enter with probability q while observing a busy server" is

$$S_{soc}^{un}(q) = \lambda^*(q)R - C\mathbb{E}[N(q)], \tag{9.1.36}$$

where $\lambda^*(q)$ is the mean arrival rate for the customers that decide to enter the system and $\mathbb{E}[N(q)]$ is the mean number of customers in the retrial pool. We have

$$\lambda^*(q) = \lambda p_0(1) + \lambda q p_1(1) = \frac{\lambda\mu}{\mu+\lambda(1-q)}, \tag{9.1.37}$$

$$\mathbb{E}[N(q)] = \frac{\mathrm{d}}{\mathrm{d}z}p_0(z)\Big|_{z=1} + \frac{\mathrm{d}}{\mathrm{d}z}p_1(z)\Big|_{z=1} = \frac{\lambda^2 q(\theta+\mu+\lambda(1-q))}{\theta(\mu-\lambda q)(\mu+\lambda(1-q))}. \tag{9.1.38}$$

Substituting (9.1.37) and (9.1.38) into (9.1.36), we obtain

$$S_{soc}^{un}(q) = \frac{\lambda\mu}{\mu+\lambda(1-q)}R - \frac{\lambda^2 q(\theta+\mu+\lambda(1-q))}{\theta(\mu-\lambda q)(\mu+\lambda(1-q))}C. \tag{9.1.39}$$

Let $x = \mu - \lambda q$, then (9.1.39) can be rewritten as a function of x

$$S_{soc}(x) = \frac{\lambda}{\theta}\Big(C + \frac{((\theta-\mu)C+\mu\theta R)x - \mu C(\lambda+\theta)}{x^2+\lambda x}\Big). \tag{9.1.40}$$

So, the problem of maximizing $S_{soc}^{un}(q)$ for $q \in [0, 1]$ is equivalent to the problem of maximizing $S_{soc}(x)$ for $x \in [\mu-\lambda, \mu]$. Differentiating $S_{soc}(x)$ we have that

$$\frac{\mathrm{d}}{\mathrm{d}x}S_{soc}(x) = \frac{\lambda}{\theta}\Big(\frac{-((\theta-\mu)C+\mu\theta R)x^2 + 2\mu C(\lambda+\theta)x + \lambda\mu C(\lambda+\theta)}{(x^2+\lambda x)^2}\Big). \tag{9.1.41}$$

We consider two cases.

Case 1: $\frac{R}{C} \le \frac{\mu-\theta}{\theta\mu} \Leftrightarrow (\theta-\mu)C + \mu\theta R \le 0$.

In this case, we observe $\frac{\mathrm{d}}{\mathrm{d}x}S_{soc}(x) > 0$ for any $x \in [\mu-\lambda, \mu]$, which implies that $S_{soc}(x)$ is strictly increasing in x and the unique maximum is attained at $x = \mu$. Thus, the best response is $q^* = 0$.

Case 2: $\frac{R}{C} > \frac{\mu-\theta}{\theta\mu} \Leftrightarrow (\theta-\mu)C + \mu\theta R > 0$.

In this case, $S_{soc}(x)$ may not be monotone in $x \in [\mu-\lambda, \mu]$. Solving $\frac{\mathrm{d}}{\mathrm{d}x}S_{soc}(x) = 0$, we obtain two roots

$$x_1 = \frac{\mu C(\lambda+\theta) - \sqrt{(\mu C(\lambda+\theta))^2 + \lambda\mu C(\lambda+\theta)(C(\theta-\mu)+\mu\theta R)}}{(C(\theta-\mu)+\mu\theta R)} < 0, \quad (9.1.42)$$

$$x_2 = \frac{\mu C(\lambda+\theta) + \sqrt{(\mu C(\lambda+\theta))^2 + \lambda\mu C(\lambda+\theta)(C(\theta-\mu)+\mu\theta R)}}{(C(\theta-\mu)+\mu\theta R)} > 0. \quad (9.1.43)$$

Therefore, $S_{soc}(x)$ is decreasing in $(-\infty, x_1)$ and $(x_2, +\infty)$ respectively, and increasing in $[x_1, x_2]$.

Case 2a: $\mu < x_2 \Leftrightarrow \frac{R}{C} < \frac{(\lambda+\mu)(\lambda+\mu+\theta)}{\theta\mu^2}$. In this case, $S_{soc}(x)$ is increasing for $x \in [\mu-\lambda, \mu]$, the unique maximum is attained at $x = \mu$, i.e., $q^* = 0$.

Case 2b: $\mu - \lambda \le x_2 \le \mu \Leftrightarrow \frac{(\lambda+\mu)(\lambda+\mu+\theta)}{\theta\mu^2} \le \frac{R}{C} \le \frac{\mu^2(\theta+\mu)+\lambda\theta(\mu-\lambda)}{(\mu-\lambda)^2\theta\mu}$ In this case, $S_{soc}(x)$ is increasing for $x \in [\mu-\lambda, x_2]$ and decreasing for $x \in [x_2, +\infty]$. So the optimal solution is $x = x_2$, i.e., $q^* = \frac{\mu - x_2}{\lambda}$.

Case 2c: $\mu - \lambda > x_2 \Leftrightarrow \frac{R}{C} > \frac{\mu^2(\theta+\mu)+\lambda\theta(\mu-\lambda)}{(\mu-\lambda)^2\theta\mu}$ In this case, $S_{soc}(x)$ is decreasing for $x \in [\mu-\lambda, \mu]$, so it takes its maximum at $x = \mu - \lambda$, i.e., $q^* = 1$. □

To conclude, we compare the equilibrium solution q_e with the socially optimal solution q^* which maximizes social net benefit per time unit.

Theorem 9.1.9 $q^* \le q_e$.

Proof According to (9.1.34) and (9.1.35), we consider two cases:

(i) $\mu^2 < \lambda(\lambda+\theta+\mu)$. In this case, $\frac{(\lambda+\mu)(\lambda+\theta+\mu)}{\theta\mu^2} < \frac{\theta+\mu}{\theta(\mu-\lambda)}$. We discuss the following five subcases.

Case 1a: If $0 < \frac{R}{C} \le \frac{\lambda+\theta+\mu}{\theta\mu}$, then $q^* = q_e = 0$.

Case 1b: If $\frac{\lambda+\theta+\mu}{\theta\mu} < \frac{R}{C} \le \frac{(\lambda+\mu)(\lambda+\theta+\mu)}{\theta\mu^2}$, then $q^* = 0$ and $q_e = \frac{\mu - \frac{(\lambda+\theta)C}{R\theta - C}}{\lambda} \in (0, 1)$. So the inequality $q^* < q_e$ holds.

Case 1c: If $\frac{(\lambda+\mu)(\lambda+\theta+\mu)}{\theta\mu^2} < \frac{R}{C} \le \frac{\theta+\mu}{\theta(\mu-\lambda)}$, then $q^* = \frac{\mu - x_2}{\lambda}$ and $q_e = \frac{\mu - \frac{(\lambda+\theta)C}{R\theta - C}}{\lambda}$. After some algebra, we obtain $q^* < q_e$.

Case 1d: If $\frac{\theta+\mu}{\theta(\mu-\lambda)} < \frac{R}{C} \le \frac{\mu^2(\theta+\mu)+\lambda\theta(\mu-\lambda)}{\theta\mu(\mu-\lambda)^2}$, then $q^* = \frac{\mu - x_2}{\lambda} \in (0, 1]$ and $q_e = 1$, and obviously $q^* \le q_e$.

Case 1e: If $\frac{R}{C} > \frac{\mu^2(\theta+\mu)+\lambda\theta(\mu-\lambda)}{\theta\mu(\mu-\lambda)^2}$, then $q^* = q_e = 1$.

(ii) $\mu^2 \ge \lambda(\lambda+\theta+\mu)$. In this case, $\frac{(\lambda+\mu)(\lambda+\theta+\mu)}{\theta\mu^2} \ge \frac{\theta+\mu}{\theta(\mu-\lambda)}$. We also discuss five subcases.

Case 2a: If $0 < \frac{R}{C} \le \frac{\lambda+\theta+\mu}{\theta\mu}$, then $q^* = q_e = 0$.

Case 2b: If $\frac{\lambda+\theta+\mu}{\theta\mu} < \frac{R}{C} \le \frac{\theta+\mu}{\theta(\mu-\lambda)}$, then $q^* = 0$ and $q_e = \frac{\mu - \frac{(\lambda+\theta)C}{R\theta - C}}{\lambda} \in (0, 1]$. So the inequality $q^* < q_e$ holds.

Case 2c: If $\frac{\theta+\mu}{\theta(\mu-\lambda)} < \frac{R}{C} \le \frac{(\lambda+\mu)(\lambda+\theta+\mu)}{\theta\mu^2}$, then $q^* = 0$ and $q_e = 1$.

Case 2d: If $\frac{(\lambda+\mu)(\lambda+\theta+\mu)}{\theta\mu^2} < \frac{R}{C} \le \frac{\mu^2(\theta+\mu)+\lambda\theta(\mu-\lambda)}{\theta\mu(\mu-\lambda)^2}$, then $q^* = \frac{\mu-x_2}{\lambda} \in (0, 1]$ and $q_e = 1$, and obviously $q^* \le q_e$.

Case 2e: If $\frac{R}{C} > \frac{\mu^2(\theta+\mu)+\lambda\theta(\mu-\lambda)}{\theta\mu(\mu-\lambda)^2}$, then $q^* = q_e = 1$.

So, the inequality of $q^* \le q_e$ holds in any case. □

All of the results derived above in this section is under the condition $\lambda < \mu$. For the opposite case $\lambda \ge \mu$, the equilibrium joining probability and the socially optimal joining probability can be obtained by following a similar analysis

$$q_e = \begin{cases} 0, & \text{if } 0 < \frac{R}{C} \le \frac{\lambda+\theta+\mu}{\theta\mu}, \\ \frac{\mu-\frac{(\lambda+\theta)C}{R\theta-C}}{\lambda}, & \text{if } \frac{R}{C} > \frac{\lambda+\theta+\mu}{\theta\mu}. \end{cases} \tag{9.1.44}$$

$$q^* = \begin{cases} 0, & \text{if } 0 < \frac{R}{C} \le \frac{(\lambda+\mu)(\lambda+\theta+\mu)}{\theta\mu^2}, \\ \frac{\mu-x_2}{\lambda}, & \text{if } \frac{R}{C} > \frac{(\lambda+\mu)(\lambda+\theta+\mu)}{\theta\mu^2}. \end{cases} \tag{9.1.45}$$

where

$$x_2 = \frac{\mu C(\lambda+\theta) + \sqrt{(\mu C(\lambda+\theta))^2 + \lambda\mu C(\lambda+\theta)(C(\theta-\mu)+\mu\theta R)}}{(C(\theta-\mu)+\mu\theta R)},$$

and these two probabilities are also ordered as $q^* \le q_e$.

9.1.4 *Equilibrium Strategy in Observable Retrial Queue*

We assume that in the observable case where the arriving customers observe not only the state of the server but also the number of customers already waiting in the pool. Similar to the unobservable case, this information is useful only for the customers who find the server busy at their arrival instants since if the server is found idle, the customers receive service immediately. For those who arrive finding a busy server, this additional information may help them assess their expected net benefit more accurately.

9.1.4.1 Nash Equilibrium Strategy

In contrast to the FCFS queueing systems in which the customers are not affected by the decisions of further customers, however, in the retrial queue the customers' mean waiting time in the pool is influenced by the behavior of the other customers that find the server busy. We shall find that there exists an equilibrium strategy of threshold type $[n, p]$ in the observable queue, where n is an integer and $0 \le p \le 1$. More concretely, when they find the server busy upon arrival, they always enter if the queue length in the pool is at most $n-1$, always balk if the queue length is at least $n+1$, and enter with probability p when the queue length is n. A threshold policy

$[n, p]$ can be denoted by $[x]$ where $x = n + p$. If $[x]$ is an integer, the strategy is pure. Otherwise, it is mixed. Note that the threshold strategy $[n, 1]$ and $[n+1, 0]$ are identical.

Define $\phi_{[n,p]}(i, k)$ as the expected time until service begins for the tagged customer in the pool, given that there are $i\,(i = 0, 1)$ customers receiving service and k other customers waiting in the pool, and all other customers follow $[n, p]$ threshold balking strategy. If he finds the system at state $(1, k)$ and decides to enter, his expected waiting time until service begins is $\phi_{[n,p]}(i, k)$. In the following theorem we calculate the expected waiting time $\phi_{[n,p]}(i, k)$ of the tagged customer.

Theorem 9.1.10 *In the observable retrial queue, if all other customers follow $[n, p]$ threshold strategy when they find the server busy at their arrival instants, the expected waiting time of the tagged customer on condition that he finds the system at state $(1, k)$ and decides to enter is given as follows*

1. For $n \geq 1$,

$$\phi_{[n,p]}(1,k) = a(k)\phi_{[n,p]}(1,0) + b(k), \quad 0 \leq k \leq n-1, \tag{9.1.46}$$

$$\phi_{[n,p]}(1,n) = \frac{\lambda+\mu+(n+1)\theta}{(n+1)\theta\mu} + \frac{n}{n+1}(a(n-1)\phi_{[n,p]}(1,0) + b(n-1)), \tag{9.1.47}$$

where

$$\begin{aligned}\phi_{[n,p]}(1,0) = &\left\{\left(\frac{\lambda p(\lambda+n\theta)}{n+1} + n\theta\mu\right)a(n-1) - (n-1)\theta\mu a(n-2)\right\}^{-1} \\ &\times\left\{\lambda+\mu+n\theta + \frac{\lambda p(\lambda+n\theta)(\lambda+\mu+(n+1)\theta)}{(n+1)\theta\mu}\right. \\ &\left. -\left(\frac{\lambda p(\lambda+n\theta)}{n+1} + n\theta\mu\right)b(n-1) + (n-1)\theta\mu b(n-2)\right\},\end{aligned} \tag{9.1.48}$$

$a(k)$ and $b(k)$ are recursively computed by

$$a(0) = 1, \quad a(1) = 1 + \frac{\mu\theta}{\lambda(\lambda+\theta)}, \tag{9.1.49}$$

$$a(k+1) = f_1(k)a(k) + f_2(k)a(k-1), \quad 1 \leq k \leq n-2, \tag{9.1.50}$$

$$b(0) = 0, \quad b(1) = -\frac{\lambda+\theta+\mu}{\lambda(\lambda+\theta)}, \tag{9.1.51}$$

$$b(k+1) = f_1(k)b(k) + f_2(k)b(k-1) + g(k), \quad 1 \leq k \leq n-2, \tag{9.1.52}$$

where

$$f_1(k) = 1 + \frac{(k+1)\mu\theta}{\lambda(\lambda+(k+1)\theta)}, \quad k \geq 1, \tag{9.1.53}$$

$$f_2(k) = -\frac{k\mu\theta}{\lambda(\lambda+(k+1)\theta)}, \quad k \geq 1, \tag{9.1.54}$$

$$g(k) = -\frac{\lambda+(k+1)\theta+\mu}{\lambda(\lambda+(k+1)\theta)}, \quad k \geq 1. \tag{9.1.55}$$

2. *For* $n = 0$,

$$\phi_{[0,p]}(1,0) = \frac{\lambda+\mu+\theta}{\theta\mu}. \tag{9.1.56}$$

Proof For the case $n \geq 1$, by a first step argument, we have the following system of linear equations

$$\phi_{[n,p]}(0,0) = \frac{1}{\lambda+\theta} + \frac{\lambda}{\lambda+\theta}\phi_{[n,p]}(1,0), \tag{9.1.57}$$

$$\begin{aligned}\phi_{[n,p]}(0,k) = &\frac{1}{\lambda+(k+1)\theta} + \frac{\lambda}{\lambda+(k+1)\theta}\phi_{[n,p]}(1,k)\\ &+\frac{k\theta}{\lambda+(k+1)\theta}\phi_{[n,p]}(1,k-1), \quad k = 1,2,\ldots,n, \end{aligned}\tag{9.1.58}$$

$$\begin{aligned}\phi_{[n,p]}(1,k) = &\frac{1}{\lambda+\mu} + \frac{\lambda}{\lambda+\mu}\phi_{[n,p]}(1,k+1) + \frac{\mu}{\lambda+\mu}\phi_{[n,p]}(0,k),\\ &k = 0,1,\ldots,n-2,\end{aligned} \tag{9.1.59}$$

$$\phi_{[n,p]}(1,n-1) = \frac{1}{\lambda p+\mu} + \frac{\lambda p}{\lambda p+\mu}\phi_{[n,p]}(1,n) + \frac{\mu}{\lambda p+\mu}\phi_{[n,p]}(0,n-1), \tag{9.1.60}$$

$$\phi_{[n,p]}(1,n) = \frac{1}{\mu} + \phi_{[n,p]}(0,n). \tag{9.1.61}$$

Substituting (9.1.58) into (9.1.59), we obtain

$$\phi_{[n,p]}(1,k+1) = f_1(k)\phi_{[n,p]}(1,k) + f_2(k)\phi_{[n,p]}(1,k-1) + g(k), \quad 1 \leq k \leq n-2, \tag{9.1.62}$$

where $f_1(k)$, $f_2(k)$ and $g(k)$ are given by (9.1.68)–(9.1.70). By (9.1.62), $\phi_{[n,p]}(1,k)$ can be represented by

$$\phi_{[n,p]}(1,k) = a(k)\phi_{[n,p]}(1,0) + b(k), \quad 0 \leq k \leq n-1, \tag{9.1.63}$$

where the coefficients $a(k)$ and $b(k)$ are obtained recursively by substituting (9.1.75) into (9.1.62) which yields (9.1.50) and (9.1.66). By substituting(9.1.72) into (9.1.59) for $k = 0$ and then substituting $\phi_{[n,p]}(1,1)$ from (9.1.75), we obtain (9.1.49) and (9.1.51). Substituting $\phi_{[n,p]}(0,n)$ from (9.1.58) into (9.1.74) and then substituting $\phi_{[n,p]}(1,n-1)$ from (9.1.62) yields $\phi_{[n,p]}(1,n)$ given by (9.1.47). Finally, substituting $\phi_{[n,p]}(0, n-1)$ from (9.1.58) into (9.1.73) we obtain after some algebraic manipulation $\phi_{[n,p]}(1, 0)$ given by (9.1.48).

For $n = 0$, we can easily get $\phi_{[0,p]}(1, 0)$ that is given by (9.1.71). □

To find the equilibrium threshold strategy of the customers in the observable case, we first introduce two lemmas.

Lemma 9.1.4 *For every threshold policy $[n, p]$ and $k \in [0, n]$, $\phi_{[n,p]}(1, k)$ is strictly increasing in k.*

Proof We tag a customer in the pool and consider two cases under the condition that the server is busy, the first when he is with k customers in the pool, and the second when he is with $k+1$ other customers. The extra customer in the second case receives service prior to the tagged customer with probability $\frac{1}{2}$ and hence increases the delay of the tagged one. Therefore, the expected waiting time in the second case $\phi_{[n,p]}(1, k+1)$ is greater than that in the first case $\phi_{[n,p]}(1, k)$. □

Lemma 9.1.5 *For every $n \geq 0$ and $0 \leq k \leq n$, $\phi_{[n,p]}(1, k)$ is strictly increasing and continuous in $p \in [0, 1]$.*

Proof Tag a customer in the pool with k customers when the server is busy. Consider two cases, the first is that all other customers follow $[n, p_1]$ strategy and the second is with $[n, p_2]$ strategy with $0 \leq p_1 < p_2 \leq 1$. When the queue lengths reach n simultaneously before the tagged customer receives service, a new primary customer that arrives and finds a busy server will enter with higher probability in the second case and hence increases more delay of the tagged customer. Therefore, $\phi_{[n,p]}(1, k)$ is strictly increasing in $p \in [0, 1]$.

It follows from (9.1.50), (9.1.68) and (9.1.69) that

$$a(k+1) = a(k) + \frac{\mu\theta}{\lambda(\lambda + (k+1)\theta)}a(k) + \frac{k\mu\theta}{\lambda(\lambda + (k+1)\theta)}(a(k) - a(k-1)),$$
$$1 \leq k \leq n-2.$$

Together with the fact $a(1) > a(0) > 0$, we get that $a(k)$ is strictly increasing in $k \in [0, n-1]$. By this result, the denominator of $\phi_{[n,p]}(1, 0)$ is positive for any $p \in [0, 1]$. It is clear that $\phi_{[n,p]}(1, 0)$ is continuous in $p \in [0, 1]$. We obtain that $\phi_{[n,p]}(1, k)$ for $k \in [0, n]$ is continuous in $p \in [0, 1]$. □

Due to the fact that $\phi_{[n,1]}(1, k) = \phi_{[n+1,0]}(1, k)$, Lemma 9.1.5 tells us that $\phi_{[x]}(1, k)$ is strictly increasing and continuous in x. Given that all other customers follow $[n, p]$ threshold strategy. Define

$$S_{[n,p]}(k) = R - C\phi_{[n,p]}(1, k), \quad n \geq 0, \quad 0 \leq k \leq n, \tag{9.1.64}$$

as the expected net benefit for a customer who observes a busy server and k customers in the pool upon arrival and decides to enter, We determine the unique equilibrium threshold $[n_e, p_e]$ in the following theorem.

Theorem 9.1.11 *The equilibrium threshold $[n_e, p_e]$ exists and is unique, which may be determined by the following procedure.*

Let $n_e \triangleq max\{n \geq 0 : S_{[n,0]}(n) > 0\}$.

Case 1: If $S_{[n_e,1]}(n_e) \geq 0$, then the equilibrium threshold is [$[n_e, 1]$.

Case 2: If $S_{[n_e,1]}(n_e) < 0$, then the equilibrium threshold is $[n_e, p_e]$, where p_e $(0 < p_e < 1)$ is the unique solution of

$$S_{[n_e,p]}(n_e) = 0,$$

and is given by

$$p_e = \frac{\theta\mu((n_e-1)b(n_e-2) - n_e b(n_e-1) - (n_e a(n_e-1) - (n_e-1)a(n_e-2))\psi)}{\frac{\lambda(\lambda+n_e\theta)}{n_e+1}\left(a(n_e-1)\psi + b(n_e-1) - \frac{\lambda+\mu+(n_e+1)\theta}{\theta\mu}\right)} + \frac{\lambda+\mu+n_e\theta}{\frac{\lambda(\lambda+n_e\theta)}{n_e+1}\left(a(n_e-1)\psi + b(n_e-1) - \frac{\lambda+\mu+(n_e+1)\theta}{\theta\mu}\right)}, \quad (9.1.65)$$

where

$$\psi = \frac{\left(\frac{R}{C} - \frac{\lambda+\mu+(n_e+1)\theta}{(n_e+1)\theta\mu}\right)\frac{n_e+1}{n_e} - b(n_e-1)}{a(n_e-1)}.$$

Proof From Lemmas 9.1.4 and 9.1.5 and (9.1.76), we obtain $S_{[n,p]}(k)$ is strictly decreasing in k and p respectively, and is continuous in $p \in [0, 1]$. It follows that $S_{[n,0]}(n) > S_{[n,1]}(n) > S_{[n,1]}(n+1) = S_{[n+1,0]}(n+1)$ for any n. By definition, we have $S_{[n_e,0]}(n_e) > 0$ and $S_{[n_e+1,0]}(n_e+1) \leq 0$. We tag a customer when he arrives finding the server busy and consider the following two cases.

Case 1: $S_{[n_e,1]}(n_e) \geq 0$. In this case, all other customers follow threshold strategy $[n_e, 1]$. If he observes k $(0 \leq k \leq n_e)$ customers in the pool and enters, his expected net benefit is $S_{[n_e,1]}(k) \geq S_{[n_e,1]}(n_e) \geq 0$. So he prefers to enter. If he observes $n_e + 1$ customers in the pool, his expected benefit is $S_{[n_e,1]}(n_e+1) = S_{[n_e+1,0]}(n_e+1) \leq 0$. So he prefers to balk when he observes $n_e + 1$ customers in the pool. Therefore, $[n_e, 1]$ is the optimal strategy of the tagged customer, i.e., $[n_e, 1]$ is the equilibrium threshold strategy.

Case 2: $S_{[n_e,1]}(n_e) < 0$. In this case, by the defined conditions $S_{[n_e,0]}(n_e) > 0$ and $S_{[n_e,1]}(n_e) < 0$, there exits a unique $0 < p_e < 1$ such that $S_{[n_e,p_e]}(n_e) = 0$ since $S_{[n_e,p]}(n_e)$ is strictly decreasing and continuous in p. Suppose all other customers use $[n_e, p_e]$ as their threshold strategy. If the tagged customer observes $0 \leq k \leq n_e - 1$ customers in the pool and enters, his expected net benefit is $S_{[n_e,p_e]}(k) > S_{[n_e,p_e]}(n_e) = 0$. So he prefers to enter. If he observes $n_e + 1$ customers in the pool, his expected benefit is $S_{[n_e,p_e]}(n_e+1) < S_{[n_e,p_e]}(n_e) = 0$. So he prefers to balk. If n_e customers are found in the pool, the expected benefit of the tagged one is 0 no matter what action he chooses. Therefore, $[n_e, p_e]$ is the equilibrium threshold strategy.

It remains to prove the uniqueness of the equilibrium threshold.

Denote $S_{[x]}(k) = S_{[n,p]}(k)$, where $x = n + p$. From Lemmas 9.1.4 and 9.1.5 and the fact $S_{[n,1]}(k) = S_{[n+1,0]}(k)$, we have that $S_{[x]}(k)$ is strictly decreasing in x and k, respectively. Define $L(x)$ as the set of "optimal thresholds" against the threshold strategy $[x]$, namely,

$$L(x) = \{x^{'} : [x^{'}] \text{is optimal against} [x]\}.$$

Let $l(x)$ denote the maximal integer $k \geq 0$ for which $S_{[x]}(k) \geq 0$. $L(x)$ can be expressed as follows:

$$L(x) = \begin{cases} \{l(x) + p : 0 \leq p \leq 1\}, & \text{if } S_{[x]}(l(x)) = 0, \\ \{l(x) + 1\}, & \text{if } S_{[x]}(l(x)) > 0. \end{cases}$$

Tag a customer and consider two cases when other customers follow the strategy $[x_1]$ and $[x_2]$ $(x_1 < x_2)$. We first determine the order of $l(x_1)$ and $l(x_2)$. If $l(x_1) < l(x_2)$, we have that $S_{[x_1]}(l(x_2)) > S_{[x_2]}(l(x_2)) \geq 0$ because of the monotonicity property of $S_{[x]}(k)$, which implies that $l(x_1)$ is not the maximal integer for which $S_{[x_1]}(k) \geq 0$ and is contrary to the definition of $l(x_1)$. Hence, $l(x_1) \geq l(x_2)$.

If $l(x_1) > l(x_2)$, for any $0 \leq p_1, p_2 \leq 1$, we get $l(x_1) + 1 \geq l(x_1) + p_1 \geq l(x_2) + 1 \geq l(x_2) + p_2$. Thus, the optimal threshold selected by the tagged customer when other customers follow $[x_2]$ threshold strategy is not smaller than that when $[x_1]$ is adopted by other customers.

If $l(x_1) = l(x_2)$, $S_{[x_1]}(l(x_1)) > 0$. Otherwise, if $S_{[x_1]}(l(x_1)) = 0$, then $S_{[x_2]}(l(x_2)) < S_{[x_1]}(l(x_2)) = S_{[x_1]}(l(x_1)) = 0$, which is contrary to the definition of $l(x_2)$. In this case, the optimal threshold against $[x_1]$ is $l(x_1) + 1$. Since $S_{[x_2]}(l(x_2)) \geq 0$, the optimal threshold selected by the tagged customer when all others use $[x_2]$ as their threshold strategy is $l(x_2) + 1$ or $l(x_2) + p_2$. We observe that $l(x_1) + 1 = l(x_2) + 1 \geq l(x_2) + p_2$.

Above all, for any case and $x_1 < x_2$, the optimal threshold of the tagged customer against $[x_1]$ is not smaller than that against $[x_2]$. That is, one's optimal response to a threshold x adopted by all other customers is decreasing in x. This shows that we have an ATC situation. in the ATC case at most one equilibrium exists, $[n_e, p_e]$ is the unique symmetric equilibrium threshold strategy. □

9.1.4.2 Socially Optimal Strategy

Let n^* be the value of the threshold that maximizes the social net benefit per time unit. When the customers arrive finding a busy server, they are admitted to enter if and only if there are less than n^* customers in the pool. To find the socially optimal threshold n^*, we first need to determine the stationary behavior of the system.

Suppose that the threshold $[n, 0]$ is used by all customers when they observe a busy server upon arrival. That is, they join the system as long as the number of customers in the pool is less than n and balk otherwise. It is clear that the stochastic process $\{(I(t), N(t)), t \geq 0\}$ is a two-dimensional continuous time Markov chain with state space $\{0, 1\} \times \{0, 1, 2, \ldots, n\}$. The stationary distribution of the system is given in the following theorem.

Theorem 9.1.12 *For the observable retrial queue in the steady state, in which all customers follow* $[n, 0]$ *threshold strategy when they find the server busy upon arrival, the state probabilities* $p(i, j)$*, where* $(i, j) \in \{0, 1\} \times \{0, 1, 2, \ldots, n\}$*, are given by*

$$p^{obs}_{[n,0]}(1, 0) = \left\{ \sum_{j=1}^{n} \left(\frac{\lambda}{\theta\mu}\right)^j \left(1 + \frac{\mu}{\lambda + j\theta}\right) \frac{\prod_{i=1}^{j}(\lambda + i\theta)}{j!} + \frac{\lambda + \mu}{\lambda} \right\}^{-1}, \tag{9.1.66}$$

$$p^{obs}_{[n,0]}(0, 0) = \frac{\mu}{\lambda} p(1, 0), \tag{9.1.67}$$

$$p^{obs}_{[n,0]}(1, j) = \left(\frac{\lambda}{\theta\mu}\right)^j \frac{\prod_{i=1}^{j}(\lambda + i\theta)}{j!} p(1, 0), \quad j = 1, 2, \ldots, n, \tag{9.1.68}$$

$$p^{obs}_{[n,0]}(0, j) = \left(\frac{\lambda}{\theta\mu}\right)^j \frac{\mu \prod_{i=1}^{j-1}(\lambda + i\theta)}{j!} p(1, 0), \quad j = 1, 2, \ldots, n. \tag{9.1.69}$$

Proof The balance equations for the stationary distribution are given as follows:

$$(\lambda + j\theta)p(0, j) = \mu p(1, j), \quad j = 0, 1, \ldots, n \tag{9.1.70}$$

$$(\lambda + \mu)p(1, j) = \lambda p(0, j) + (j + 1)\theta p(0, j + 1) + \lambda p(1, j - 1),$$
$$j = 0, 1, \ldots, n - 1, \tag{9.1.71}$$

$$\mu p(1, n) = \lambda p(0, n) + \lambda p(1, n - 1), \tag{9.1.72}$$

where $p(1, -1) = 0$.

By solving (9.1.70)–(9.1.72), we can obtain

$$p(1, j) = \left(\frac{\lambda}{\theta\mu}\right)^j \frac{\prod_{i=1}^{j}(\lambda + i\theta)}{j!} p(1, 0), \quad j = 1, 2, \ldots, n. \tag{9.1.73}$$

$$p(0, 0) = \frac{\mu}{\lambda} p(1, 0), \tag{9.1.74}$$

$$p(0, j) = \left(\frac{\lambda}{\theta\mu}\right)^j \frac{\mu \prod_{i=1}^{j-1}(\lambda + i\theta)}{j!} p(1, 0), \quad j = 1, 2, \ldots, n. \tag{9.1.75}$$

The remaining probability $p(1, 0)$ can be found with the help of the normalization condition $\sum_{j=0}^{n}(p(0, j) + p(1, j)) = 1$ and is given by (9.1.66). □

Because the probability of balking is $p^{obs}_{[n,0]}(1,n)$, the social benefit per time unit is

$$S^{obs}_{soc}(n) = \lambda R(1 - p^{obs}_{[n,0]}(1,n)) - C\sum_{j=1}^{n} j(p^{obs}_{[n,0]}(0,j) + p^{obs}_{[n,0]}(1,j)), \quad n \geq 0. \quad (9.1.76)$$

n^* is obtained as the integer that maximizes $S^{obs}_{soc}(n)$.

Because of the complexity of the expression of $S^{obs}_{soc}(n)$, an analytic characterization of the optimal solution n^* is not tractable. According to numerical analysis, we observe that the socially optimal threshold n^* is lower than the Nash equilibrium threshold $x_e = n_e + p_e$. This is due to that if the customers are noncooperative, they may join the system even if the retrial pool is congested, without considering other customers' welfare. Otherwise, if they are cooperative, they make decisions to maximize their overall welfare by taking into account the negative externalities imposed on each other.

9.2 The M/M/1 Constant Retrial Queue

9.2.1 *Model Description*

We consider an $M/M/1$ retrial queue with constant retrial rate in which customers arrive according to a Poisson process at rate λ. The service times are assumed to be exponentially distributed with parameter μ. Customers that find the server idle upon arrival will occupy the server immediately and start being served. The customers that find the server busy upon arrival, they can waiting in retrial orbit or balk. The time required to find a customer from the retrial orbit is assumed exponentially distributed with rate α. We assume that the interarrival, service and seeking times are mutually independent. We represent the state of the system at time t by a random vector $(I(t)X(t))$, where $I(t)$ denotes the state of the server (0: idle, 1: busy) and $X(t)$ is the number of customers in the orbit. It is obvious that the stochastic process $(I(t)X(t))$ is a continuous-time Markov chain with state space $S = \{0, 1\} \times \{0, 1, 2, \ldots\}$.

We assume that every customer receives a reward of R units for completing service. Moreover, there exists a waiting cost of C units per time unit. We assume that $R > \frac{C}{\mu}$, which ensures that the customers who find the server idle do enter in the system. Customers decide whether to join to retrial orbit or balk upon their arrival when server is busy.

9.2.2 Equilibrium Strategy in Unobservable Retrial Queue

In the unobservable case, customers can only observe the state of server while not observe the number of customers in the retrial orbit. Customers that find the server idle upon arrival will occupy the server immediately and start being served. A (mixed) strategy of a customer is completely specified by the probability r of joining whenever he sees the server busy.

Lemma 9.2.1 *In the unobservable M/M/1 constant retrial queue, the system is stable if and only if*

$$\rho = \frac{\lambda r(\lambda + \alpha)}{\mu\alpha} < 1, \tag{9.2.1}$$

In this case, the stationary probabilities of idle and busy server are given, respectively, by

$$p_{0.} = \frac{\mu - \lambda r}{\mu + \lambda(1 - r)}, \tag{9.2.2}$$

$$p_{1.} = \frac{\lambda}{\mu + \lambda(1 - r)}. \tag{9.2.3}$$

Moreover, given that he sees upon arrival the server busy, the expected sojourn time in the retrial orbit of an arriving customer and the expected sojourn time of a customer in the system, are given respectively, by

$$\mathbb{E}(S|I = 1) = \frac{\lambda + \alpha + \mu}{\mu\alpha - \lambda r(\lambda + \alpha)} + \frac{1}{\mu}, \tag{9.2.4}$$

$$\mathbb{E}(S) = \frac{\mu(\alpha + \lambda r)}{(\mu\alpha - \lambda r(\lambda + \alpha))(\mu + \lambda(1 - r))}. \tag{9.2.5}$$

Proof Let $(p_{ij} : (ij) \in S)$ be the stationary distribution of system. The balance equations for the stationary distribution are

$$\lambda p_{0,0} = \mu p_{1,0}, \tag{9.2.6}$$

$$(\lambda + \alpha)p_{0,j} = \mu p_{1,j}, \quad j = 1, 2, \ldots, \tag{9.2.7}$$

$$\lambda r p_{1,j} = \alpha p_{0,j+1}, \quad j = 0, 1, \ldots. \tag{9.2.8}$$

By solving these equations, we can obtain

$$p_{0,j} = \frac{\mu}{\mu + \lambda(1 - r)} \cdot \frac{\lambda}{\lambda + \alpha(1 - \delta_{j0})} \cdot (1 - \rho)\rho^j, \quad j = 0, 1, 2, \ldots, \tag{9.2.9}$$

$$p_{1,j} = \frac{\mu}{\mu + \lambda(1 - r)} \cdot (1 - \rho)\rho^j, \quad j = 0, 1, 2, \ldots, \tag{9.2.10}$$

where ρ is given by (9.2.1) and $\delta_{ij} = 1$ if and only if $j = i$ and $\delta_{ij} = 0$ otherwise. Summing (9.2.9) and (9.2.10) for all j yield (9.2.2) and (9.2.3).

Because of the PASTA property, the probability $p_{1.}^{\text{arr}}$ that a customer finds the server busy upon arrival coincides with the stationary probability of a busy server $p_{1.}$, so the total arrival rate in the retrial orbit is

$$\lambda_{ret} = \lambda r p_{1.}^{\text{arr}} = \frac{\lambda^2}{\mu + \lambda(1-r)}. \tag{9.2.11}$$

On the other hand, the stationary mean number of customers in the retrial orbit is

$$\mathbb{E}[X] = \sum_{j=1}^{\infty} j(p_{0,j} + p_{1,j}) = \frac{\lambda^2 r(\lambda + \alpha + \mu)}{(\mu + \lambda(1-r))(\mu\alpha - \lambda r(\lambda + \alpha))}. \tag{9.2.12}$$

By applying Little's law in the retrial orbit, we obtain the mean waiting time at the retrial orbit of a customer that enters to the retrial orbit as $\frac{\mathbb{E}[X]}{\lambda_{ret}}$. Summing the mean service time $\frac{1}{\mu}$ yields $\mathbb{E}(S|i=1)$ shown as (9.2.4). The mean sojourn time of a customer in the system can be obtained by solving

$$\mathbb{E}(S) = p_{0.}\frac{1}{\mu} + p_{1.}((1-r)\cdot 0 + r \cdot \mathbb{E}(S|i=1)), \tag{9.2.13}$$

□

In what follows, we will assume that

$$\rho = \frac{\lambda(\lambda + \alpha)}{\mu\alpha} < 1. \tag{9.2.14}$$

The results for the case where (9.2.14) does not hold can be easily concluded from this case as we comment later on.

9.2.2.1 Nash Equilibrium Strategy

Theorem 9.2.1 *In the unobservable $M/M/1$ constant retrial queue, there exists a unique mixed equilibrium joining strategy "enter the retrial orbit with probability r_e whenever finding the server busy". The probability r_e is given by*

$$r_e = \begin{cases} 0, & \frac{R}{C} \le t_{Le}, \\ \frac{\alpha\mu}{\lambda(\lambda+\alpha)} - \frac{\lambda+\alpha+\mu}{\lambda(\lambda+\alpha)}\left(\frac{R}{C} - \frac{1}{\mu}\right)^{-1}, & t_{Le} < \frac{R}{C} < t_{Ue}, \\ 1, & \frac{R}{C} \ge t_{Ue}, \end{cases} \tag{9.2.15}$$

where

$$t_{Le} = \frac{\lambda + \alpha + \mu}{\alpha\mu} + \frac{1}{\mu}, \tag{9.2.16}$$

$$t_{Ue} = \frac{\lambda + \alpha + \mu}{(\mu\alpha - \lambda(\lambda + \alpha)} + \frac{1}{\mu}. \tag{9.2.17}$$

Proof Suppose that all the customers follow a given joining strategy r, so the expected net benefit of a tagged customer that finds the server busy and decides to join the retrial orbit is

$$S_e(r) = R - C\mathbb{E}(S|I = 1) = R - C\Big(\frac{\lambda + \alpha + \mu}{\mu\alpha - \lambda r(\lambda + \alpha)} + \frac{1}{\mu}\Big). \tag{9.2.18}$$

A proof similar to Theorem 9.1.7 leads to a conclusion, we have an "Avoid-The-Crowd" (ATC) situation. □

9.2.2.2 Socially Optimal Strategy

Theorem 9.2.2 *In the unobservable $M/M/1$ constant retrial queue, there exists a unique mixed joining strategy "join the retrial orbit with probability r_{soc} whenever finding the server busy" that maximizes the social net benefit per time unit. The probability r_{soc} is given by*

$$r_{soc} = \begin{cases} 0, & \frac{R}{C} \leq t_{Lsoc}, \\ \frac{\mu+\lambda-A\mu\alpha}{\lambda(1-A(\lambda+\alpha))}, & t_{Lsoc} < \frac{R}{C} < t_{Usoc}, \\ 1, & \frac{R}{C} \geq t_{Usoc}, \end{cases} \tag{9.2.19}$$

where

$$A = \sqrt{\frac{\lambda R + C}{C\alpha(\lambda + \alpha)}}, \tag{9.2.20}$$

$$t_{Lsoc} = \frac{1}{\mu} + \frac{(\lambda + \alpha + \mu)(\lambda + \mu)}{\alpha\mu^2}, \tag{9.2.21}$$

$$t_{Usoc} = \frac{\mu\alpha(\lambda + \alpha + \mu) + (\lambda + \alpha)(\mu\alpha - \lambda(\lambda + \alpha))}{(\mu\alpha - \lambda(\lambda + \alpha))^2}. \tag{9.2.22}$$

Proof For a given joining strategy r for the customers who find the server busy, the social net benefit per time unit is

$$S_{soc}(r) = \lambda^*(r)R - C\lambda\mathbb{E}[S(r)], \tag{9.2.23}$$

where $\lambda^*(r)$ is the mean arrival rate for the customers that decide to enter the system, and $\mathbb{E}[S(r)]$ is the mean sojourn time of a customer of the system. From

$$\lambda^*(r) = \lambda p_{0.} + \lambda r p_{1.} = \frac{\lambda\mu}{\mu + \lambda(1-r)}, \tag{9.2.24}$$

we obtain

$$S_{soc}(r) = \frac{(\lambda R + C)\mu}{\mu + \lambda(1-r)} - \frac{C\mu\alpha}{\mu\alpha - \lambda r(\lambda + \alpha)}. \tag{9.2.25}$$

Differentiating that and we have

$$\frac{d}{dr}S_{soc}(r) = 0 \Leftrightarrow \frac{\mu + \lambda(1-r)}{\mu\alpha - \lambda r(\lambda + \alpha)} = \sqrt{\frac{\lambda R + C}{C\alpha(\lambda + \alpha)}}. \tag{9.2.26}$$

and solving for r we obtain that the unique maximum of $S_{soc}(r)$ is attained at $r = \frac{\mu+\lambda-A\mu\alpha}{\lambda(1-A(\lambda+\alpha))}$. If $\frac{\mu+\lambda-A\mu\alpha}{\lambda(1-A(\lambda+\alpha))} \in (0, 1)$, i.e., $t_{Lsoc} < \frac{R}{C} < t_{Usoc}$, then the optimal probability is $r_{soc} = \frac{\mu+\lambda-A\mu\alpha}{\lambda(1-A(\lambda+\alpha))}$; if $\frac{\mu+\lambda-A\mu\alpha}{\lambda(1-A(\lambda+\alpha))} \leq 0$, i.e., $\frac{R}{C} \leq t_{Lsoc}$, then the optimal probability is $r_{soc} = 0$; if $\frac{\mu+\lambda-A\mu\alpha}{\lambda(1-A(\lambda+\alpha))} \geq 1$, i.e., $\frac{R}{C} \geq t_{Lsoc}$, then the optimal probability is $r_{soc} = 1$. □

9.2.2.3 Strategy of Profit Maximization

Suppose that the administrator of the system imposes an entrance fee p. By imposing this fee the reward for the customers changes from R to $R - p$, and therefore, a new strategy $r_{prof}(p)$ is their new equilibrium strategy. We consider the strategy r_{prof} that maximizes the administrator s profit per time unit.

Theorem 9.2.3 *In the unobservable $M/M/1$ constant retrial queue, there exists a unique mixed joining strategy maximizes the administrator's net reward per time unit. The probabilityr_{prof} is given by*

$$r_{prof} = \begin{cases} 0, & \frac{R}{C} \leq t_{Lprof}, \\ \frac{\mu+\lambda-B\mu\alpha}{\lambda(1-B(\lambda+\alpha))}, & t_{Lprof} < \frac{R}{C} < t_{Uprof}, \\ 1, & \frac{R}{C} \geq t_{Uprof}, \end{cases} \tag{9.2.27}$$

where

$$B = \frac{1}{\lambda + \alpha}\sqrt{\frac{\lambda R}{C} - \frac{\lambda}{\mu} + 1}, \tag{9.2.28}$$

$$t_{Lprof} = \frac{1}{\mu} - \frac{1}{\lambda} + \frac{(\lambda+\alpha)^2(\lambda+\mu)^2}{\lambda\mu^2\alpha^2}, \tag{9.2.29}$$

$$t_{Uprof} = \frac{1}{\mu} - \frac{1}{\lambda} + \frac{\mu^2(\lambda+\alpha)^2}{\lambda(\mu\alpha - \lambda(\lambda+\alpha))^2}. \tag{9.2.30}$$

Proof For a given joining strategy r, define $S_{prof}(r)$ to be the administrator's net profit per time unit. According to Theorem 9.2.1, we have to solve the equation

$$R - p = C\Big(\frac{1}{\mu} + \frac{\lambda + \mu + \alpha}{\mu\alpha - \lambda r(\lambda + \alpha)}\Big), \tag{9.2.31}$$

from which we obtain

$$p(r) = R - \frac{C}{\mu} - \frac{(\lambda + \mu + \alpha)C}{\mu\alpha - \lambda r(\lambda + \alpha)}, \quad r \in [0, 1]. \tag{9.2.32}$$

Therefore, we have that

$$S_{prof}(r) = \lambda^*(r)p(r) = \Big(R - \frac{C}{\mu} + \frac{C}{\lambda}\Big)\frac{\lambda\mu}{\mu + \lambda(1-r)} - C\frac{\mu(\lambda + \alpha)}{\mu\alpha - \lambda r(\lambda + \alpha)}. \tag{9.2.33}$$

Differentiating the above equation, we have

$$\frac{d}{dr}S_{prof})(r) = 0 \Leftrightarrow \frac{\mu + \lambda(1-r)}{\mu\alpha - \lambda r(\lambda + \alpha)} = \frac{1}{\lambda + \alpha}\sqrt{\frac{\lambda R}{C} - \frac{\lambda}{\mu} + 1}. \tag{9.2.34}$$

r_{prof} can be obtained along the same lines with the proof of 9.2.2. □

We now proceed to the comparison of the optimal joining probabilities for the three different aspects of the problem.

Theorem 9.2.4 *The optimal joining probabilities r_e, r_{soc} and r_{prof} are ordered as*

$$r_{prof} \leq r_{soc} \leq r_e. \tag{9.2.35}$$

Proof First, we will prove that $r_{soc} \leq r_e$. According to $t_{Le} < t_{Lsoc} < t_{Ue} < t_{Usoc}$, consider the following cases:

1. If $\frac{R}{C} \leq t_{Le}$, then $r_e = r_{soc} = 0$.
2. If $t_{Le} < \frac{R}{C} \leq t_{Lsoc}$, then $r_e \in (0, 1)$, $r_{soc} = 0$. So, $r_{soc} < r_e$.
3. If $t_{Lsoc} < \frac{R}{C} < t_{Ue}$, then $r_e \in (0, 1)$ $r_{soc} \in (0, 1)$. After some algebra, we obtain that $r_{soc} < r_e$.
4. If $t_{Ue} \leq \frac{R}{C} < t_{Usoc}$, then $r_e = 1$ $r_{soc} \in (0, 1)$. So, $r_{soc} < r_e$.
5. If $t_{Usoc} \leq \frac{R}{C}$, then $r_e = r_{soc} = 1$.

Next, we prove $r_{prof} \leq r_{soc}$. It can be easily seen that $t_{Lsoc} < t_{Lprof}$ and $t_{Usoc} < t_{Uprof}$, but the relationship between t_{Lprof} and t_{Usoc} is uncertain. For $t_{Lprof} < t_{Usoc}$ and $t_{Lprof} \geq t_{Usoc}$, the rest of the proof proceeds along the same lines. We obtain that $r_{soc} \geq r_{prof}$ holds in both cases. □

9.2.3 Equilibrium Strategy in Observable Retrial Queue

In the observable case, the customers get informed not only about the state of the server but also about the exact number of customers in the orbit in front of them. Let $T(ij)$ be the mean sojourn time of a tagged customer that he is at the jth position in the orbit, given that the server is at state i, when all the customers follow the general strategy. We have that

$$T(1,0) = \frac{1}{\mu}, \tag{9.2.36}$$

$$T(1,j) = \frac{1}{\mu} + T(0,j), \quad j = 1, 2, \ldots, \tag{9.2.37}$$

$$T(0,j) = \frac{1}{\lambda+\alpha} + \frac{\lambda}{\lambda+\alpha} T(1,j) + \frac{\alpha}{\lambda+\alpha} T(1,j-1), \quad j = 1, 2, \ldots, \tag{9.2.38}$$

Note that $T(1,0)$ denotes the mean residual service time of a customer. By solving the equations, we obtain

$$T(0,j) = j\frac{\lambda+\alpha+\mu}{\mu\alpha} + \frac{1}{\mu}, \quad j = 1, 2, \ldots, \tag{9.2.39}$$

$$T(1,j) = j\frac{\lambda+\alpha+\mu}{\mu\alpha}, \quad j = 0, 1, \ldots. \tag{9.2.40}$$

Assume that

$$\frac{R}{C} > \frac{\lambda+\alpha+\mu}{\mu\alpha} + \frac{1}{\mu}. \tag{9.2.41}$$

otherwise, a customer that finds the server busy does not join the orbit even if the orbit is empty.

9.2.3.1 Nash Equilibrium Strategy

Theorem 9.2.5 *In the observable $M/M/1$ constant retrial queue, there exists the threshold equilibrium joining strategy "enter the retrial orbit if there are at most n_e customers in the orbit, whenever finding the server busy". The threshold n_e is given from*

$$n_e = \left\lfloor \frac{\mu\alpha}{\lambda+\alpha+\mu}\left(\frac{R}{C} - \frac{1}{\mu}\right) - 1 \right\rfloor. \tag{9.2.42}$$

Proof Consider a tagged customer that finds upon arrival the system at state $(1, j)$. If he decides to join, he goes directly to the orbit where he is assigned at the $j+1$th position. Then, his expected profit is

$$S_e(j) = R - CT(1, j+1) = R - C\Big((j+1)\frac{\lambda+\alpha+\mu}{\mu\alpha} + \frac{1}{\mu}\Big). \quad (9.2.43)$$

By solving $S_e(j) \geq 0$ we can obtain that the arriving customer prefers to enter if and only if $j \leq n_e$. □

9.2.3.2 Socially Optimal Strategy

Assume that all customers follow a threshold strategy n. Next, we consider the problem of social optimization. First, we consider the stationary distribution of the system $p_{i,j}$ with state space $S = \{0, 1\} \times \{0, 1, \ldots, n+1\}$.

Lemma 9.2.2 *In the observable $M/M/1$ retrial queue, if the customers follow the threshold-n strategy, the system is stable and the stationary distribution is given by*

$$p_{0,0} = A(n)\frac{\mu\alpha}{\lambda^2}, \quad (9.2.44)$$

$$p_{0,j} = A(n)\rho^{j-1}, \quad j = 1, 2, \ldots, n+1, \quad (9.2.45)$$

$$p_{1,j} = A(n)\frac{\alpha}{\lambda}\rho^j, \quad j = 0, 1, \ldots, n+1, \quad (9.2.46)$$

where

$$\rho = \frac{\lambda(\lambda+\mu)}{\mu\alpha}, \quad (9.2.47)$$

$$A(n) = \Big(\frac{\mu\alpha}{\lambda^2} + (1+\rho+\cdots+\rho^n) + \frac{\alpha}{\lambda}(1+\rho+\cdots+\rho^{n+1})\Big)^{-1}. \quad (9.2.48)$$

Next, we discuss the $\rho \neq 1$, $\rho = 1$ which can be treated similarly.

Theorem 9.2.6 *In the observable M/M/1 retrial queue, define the function*

$$g(x) = \frac{\alpha}{\lambda+\alpha}(x+1) + \frac{\lambda+\alpha\rho}{(\lambda+\alpha)(1-\rho)}\Big((x+1) - \rho\frac{1-\rho^{x+1}}{1-\rho}\Big) - 1. \quad (9.2.49)$$

If $g(0) > x_e$, the optimal strategy is balking. Otherwise, there exists a unique threshold joining strategy "enter the retrial orbit if there are at most n_{soc} customers in the orbit, whenever finding the server busy" that maximizes the social net benefit per time unit. The threshold n_{soc} is given from

$$n_{soc} = \lfloor x_{soc} \rfloor, \quad (9.2.50)$$

where x_{soc} is the unique nonnegative solution of the equation $g(x) = x_e$.

Proof When all customers follow the same threshold-n strategy, the effective arrival rate at the system is

$$\lambda^*(n) = \lambda(1 - p_{1,n+1}(n)) = \lambda - A(n)\alpha\rho^{n+1}. \tag{9.2.51}$$

The expected sojourn time of customers in system is

$$\mathbb{E}[S^*(n)] = \sum_{j=0}^{n+1} \frac{p_{1,j}(n)}{1 - p_{1,n+1}(n)} \frac{1}{\mu} + \sum_{j=0}^{n} \frac{p_{0,j}(n)}{1 - p_{1,n+1}(n)} \Big(\frac{1}{\mu} + (j+1)\frac{\lambda + \alpha + \mu}{\mu\alpha}\Big) \tag{9.2.52}$$

the expected social net benefit per time unit is

$$\begin{aligned} S_{soc}(n) &= \lambda^*(n)R - C\lambda^*(n)\mathbb{E}[S^*(n)] \\ &= C\frac{\lambda + \alpha + \mu}{\mu + \alpha}\Big(\lambda^*(n)(x_e + 1) - \lambda\sum_{j=0}^{n}(j+1)p_{1,j}(n)\Big). \end{aligned} \tag{9.2.53}$$

We will now show that $S_{soc}(n)$ is unimodal and attains its maximum at n_{soc}. We study the increments $S_{soc}(n) - S_{soc}(n-1)$.

$$S_{soc}(n) - S_{soc}(n-1) \geq 0 \Leftrightarrow g(n) \leq x_e, \tag{9.2.54}$$

where

$$g(n) = \frac{g_1(n)}{g_2(n)} - 1, \tag{9.2.55}$$

$$\begin{aligned} g_1(n) &= \lambda\Big(\sum_{j=0}^{n}(j+1)p_{1,j}(n) - \sum_{j=0}^{n-1}(j+1)p_{1,j}(n-1)\Big) \\ &= \frac{\alpha A(n)A(n-1)\rho^n}{\lambda^2(1-\rho)^2}\Big(\mu\alpha(n+1)(1-\rho) - \lambda^2 - \lambda\alpha\rho + \lambda^2\rho^{n+1} + \lambda\alpha\rho^{n+2}\Big), \end{aligned} \tag{9.2.56}$$

$$g_2(n) = \lambda^*(n) - \lambda^*(n-1) = \frac{\mu\alpha^2\rho^n}{\lambda^2}A(n)A(n-1). \tag{9.2.57}$$

The functions can be considerably simplified (9.2.49) can be obtained by using x to replace n. To prove now the unimodality of $S_{soc}(n)$, it is now sufficient to prove that $g(x)$ is increasing. Define the function

$$f(x) = g(x) - x. \tag{9.2.58}$$

Differentiating it, we have

$$\frac{\mathrm{d}}{\mathrm{d}x}f(x) = \frac{\rho}{1-\rho} + \frac{(\lambda+\alpha\rho)\rho^{x+2}\ln\rho}{(\lambda+\alpha)(1-\rho)^2} \geq 0, \quad x \geq 0. \tag{9.2.59}$$

Therefore, $f(x)$ and $g(x)$ are both increasing and $\lim_{x\to\infty} g(x) = \infty$. So there exists two cases: $g(0) > x_e$, then $g(n) > x_e$ for all $n \geq 0$; or there exists a unique solution n_{soc} such that $g(n_{soc}) \leq x_e < g(n_{soc}+1)$. In the former case, $S_{soc}(n)$ is decreasing and so it is social optimum for the customers that find a busy server to balk. In the latter case, in light of (9.2.54), we have that $n_{soc} = \lfloor x_{soc} \rfloor$ with x_{soc} being the unique solution of the equation $g(x) = x_e$. □

9.2.3.3 Strategy of Profit Maximization

We now consider the profit maximization problem. By imposing a nonnegative admission fee p.

Theorem 9.2.7 *Define function*

$$h(x) = x + \frac{\mu(1-\rho^{x+1})(1-\frac{\lambda+\alpha\rho}{\lambda+\alpha}\rho^{x+2})}{\lambda\rho^x(1-\rho)^2}. \tag{9.2.60}$$

If $h(0) > x_e$, then the strategy that maximizes the administrator's profit per time unit prescribes that all customers who find a busy server should balk. Otherwise, there exists a unique threshold joining strategy "enter the retrial orbit if there are at most n_{prof} customers in the orbit, whenever finding the server busy" that maximizes the administrator s profit per time unit. The threshold n_{prof} is given from

$$n_{prof} = \lfloor x_{prof} \rfloor, \tag{9.2.61}$$

where x_{soc} is the unique nonnegative solution of the equation $h(x) = x_e$.

Proof The administrator's profit per time unit is $S_{prof}(n) = \lambda^*(n)p(n)$, where $p(n)$ is the fee that he imposes on the customers to force them to follow a certain threshold n. We obtain

$$p(n) = R - C\Big((n+1)\frac{\lambda+\alpha+\mu}{\mu\alpha} + \frac{1}{\mu})\Big). \tag{9.2.62}$$

then

$$S_{prof}(n) = \lambda^*(n)C\frac{\lambda+\alpha+\mu}{\mu\alpha}(x_e - n). \tag{9.2.63}$$

We will now prove that the function $S_{prof}(n)$ is unimodal.

$$\frac{S_{prof}(n)}{S_{prof}(n-1)} \geq 1 \Leftrightarrow h(n) \leq x_e, \tag{9.2.64}$$

where $h(n) = n + \frac{\lambda^*(n-1)}{\lambda^*(n)-\lambda^*(n-1)}$. A proof similar to Theorem 9.2.6 leads to a conclusion. □

Finally, we proceed to the comparison of the optimal joining probabilities for the three different aspects of the problem.

Theorem 9.2.8 *The optimal joining thresholds n_e, n_{soc} and n_{prof} are ordered as*

$$n_{prof} \leq n_{soc} \leq n_e. \tag{9.2.65}$$

Proof We will first prove that $n_{soc} \leq n_e$. Note that $f(x)$ is increasing. Moreover, $f(0) = g(0) > 0$, so $f(x_{soc}) \geq 0$, that is, $g(x_{soc}) - x_{soc} \geq 0$. But, $g(x_{soc}) = x_e$ and we obtain $x_{soc} \leq x_e$. Taking floors yields $n_{soc} \leq n_e$.

To prove that $n_{prof} \leq n_{soc}$. When $n > n_e$, because of (9.2.63), we have $S_{prof}(n) < 0$, i.e., $n_{prof} \leq n_e$. Therefore, in the case where $n_{soc} = n_e$, we have immediately $n_{prof} \leq n_{soc}$. Suppose that $n_{soc} < n_e$, we have that $n_{prof} \leq n_{soc}$, that is, $S_{prof}(n)$ is decreasing in $\{n_{soc}, n_{soc}+1, \ldots, n_e-1\}$. From the unimodality of the function $S_{prof}(n)$, taking into account $g(n_{soc}) \leq x_e \leq g(n_{soc}+1)$ and $g(x)$ is increasing, we have that

$$g(n+1) \geq g(n_{soc}+1) > x_e, \quad n \geq n_{soc}. \tag{9.2.66}$$

To prove that $S_{prof}(n)$ is decreasing in $\{n_{soc}, n_{soc}+1, \ldots, n_e-1\}$, we have to prove in light of (9.2.64) that

$$h(n+1) \geq x_e, \quad n_{soc} \leq n < n_e. \tag{9.2.67}$$

Therefore, it suffices to prove that $h(n) \geq g(n)$, that is,

$$\begin{aligned} n + &\frac{\mu(1-\rho^{n+1})(1-\frac{\lambda+\alpha\rho}{\lambda+\alpha}\rho^{n+2})}{\lambda\rho^n(1-\rho)^2} \\ &\geq \frac{\alpha}{\lambda+\alpha}(n+1) + \frac{\lambda+\alpha\rho}{(\lambda+\alpha)(1-\rho)}\left((n+1) - \rho\frac{1-\rho^{n+1}}{1-\rho}\right) - 1. \end{aligned} \tag{9.2.68}$$

In case of $\rho < 1$, the above equation is written equivalently as

$$n+1 \leq \frac{1-\rho^{n+1}}{(1-\rho)\rho^{n+1}}\left\{\frac{\mu(1-\frac{\lambda+\alpha\rho}{\lambda+\alpha}\rho^{n+2})}{\lambda} + \frac{\lambda(\lambda+\alpha\rho)}{\mu\alpha}\rho^n\right\}.$$

Note that

$$\frac{1-\rho^{n+1}}{(1-\rho)\rho^{n+1}} = \sum_{j=0}^{n+1}\left(\frac{1}{\rho}\right)^j \geq n+1.$$

So, to prove (9.2.68), it suffices to prove that

$$\frac{\mu(1-\frac{\lambda+\alpha\rho}{\lambda+\alpha}\rho^{n+2})}{\lambda}+\frac{\lambda(\lambda+\alpha\rho)}{\mu\alpha}\rho^{n}\geq 1.$$

After some simplification, this is seen to be equivalent to

$$\frac{1}{\rho}+\frac{\lambda}{\rho\alpha}\Big(1-\frac{\lambda+\alpha\rho}{\lambda+\alpha}\rho^{n+2}\Big)>1,$$

which clearly holds for $\rho<1$. In the case $\rho>1$, the same line of argument with all the inequalities reversed is again valid. Therefore, the relationship $n_{prof}\leq n_{soc}$ is valid in all cases. □

References

Aguir, S., F. Karaesmen, O. Z. Akşin, and F. Chauvet. 2004. The impact of retrials on call center performance. *OR Spectrum* 26: 353–376.

Artalejo, Jesus R. 1995. A queueing system with returning customers and waiting line. *Operations Research Letters* 17: 191–199.

Artalejo, Jesus R., and Antonio Gómez-Corral. 1997. Steady state solution of a single-server queue with linear repeated requests. *Journal of Applied Probability* 34: 223–233.

Artalejo, Jesus R., and Antonio Gómez-Corral. 1998. Analysis of a stochastic clearing system with repeated attempts. *Stochastic Models* 14: 623–645.

Artalejo, Jesus R., and Antonio Gómez-Corral. 2008. *Retrial Queueing Systems: A Computational Approach*. Berlin: Springer.

Cui, Shiliang, Su. Xuanming, and Senthil Veeraraghavan. 2013. A model of rational retrials in queues. *Operations Research* 67 (6): 1699–1718.

Economou, Antonis, and Spyridoula Kanta. 2011. Equilibrium customer strategies and social-profit maximization in the single-server constant retrial queue. *Naval Research Logistics* 58: 107–122.

Elcan, A. 1994. Optimal customer return rate for an M/M/1 queueing system with retrials. *Probability in the Engineering and Informational Sciences* 8: 521–539.

Falin, G. I. 2008. The M/M/1 retrial queue with retrials due to server failures. *Queueing Systems* 58: 155–160.

Falin, G. I., and J. G. C. Templeton. 1997. *Retrial Queues*. London: Chapman & Hall.

Hassin, Refael, and Moshe Haviv. 1996. On optimal and equilibrium retrial rates in a queueing system. *Probability in the Engineering and Informational Sciences* 10: 223–227.

Kulkarni, V. G. 1983. A game theoretic model for two types of customers competing for service. *Operation Research Letters* 2: 119–122.

Kulkarni, V. G. 1983. On queueing systems with retrials. *Journal of Applied Probability* 20: 380–389.

Li, Kaili, and Jinting Wang. 2021. Equilibrium balking strategies in the single-server retrial queue with constant retrial rate and catastrophes. *Quality Technology and Quantitative Management* 18 (2): 156–178.

Wang, Zhongbin, and Jinting Wang. 2019. Information heterogeneity in a retrial queue: throughput and social welfare maximization. *Queueing Systems* 92 (1/2): 131–172.

Wang, Zhongbin, and Jinting Wang. 2021. Strategic priority-purchasing and joining rules in a retrial queue. *IMA Journal of Management Mathematics* 32 (2): 161–194.

Wang, Jinting, and F. Zhang. 2013. Strategic joining in M/M/1 retrial queues. *European Journal of Operational Research* 230: 76–87.

Zhang, Yu, and Jinting Wang. 2023. Managing retrial queueing systems with boundedly rational customers. *Journal of the Operational Research Society* 74 (3): 748–761.

Zhang, Yu., and Jinting Wang. 2022. Quality-speed trade-offs in customer-intensive services with boundedly rational customers and retrials. *Computers & Industrial Engineering* 167: 107983.

Zhang, F., Jinting Wang, and Bin Liu. 2012. On the optimal and equilibrium retrial rates in an unreliable retrial queue with vacations. *Journal of Industrial and Management Optimization* 8: 861–875.

Zhang, Yu., Jinting Wang, and Fang Wang. 2016. Equilibrium pricing strategies in retrial queueing systems with complementary service. *Applied Mathematical Modelling* 40: 5775–5792.

Chapter 10
Applications in Wireless Communication Systems

In the past decades, the development of communication and computer network technology has made the queueing performance analysis of communication system an important research direction, especially the application of retrial queueing model in wireless communication. With the popularization of computer technology, communication technology and multimedia, users can communicate with others or obtain information anytime and anywhere. In the process of transmitting and receiving information through the Internet, the equipment routing information about local area network (LAN) between computers plays an important role. The retrial queueing model is very suitable to describe the transmission mode of local area network. It is well known that in a LAN, many base stations are connected to a common network server (service stations). When a base station receives a message (a user arrives), it can observe the state of the service station. If the service station is idle, the message can be transmitted to the target base station through the service station. If the service station is not idle when a message arrives, then this information will be stored (can be regarded as being sent to a retrial space) for the next retransmission. In the first section of this chapter, we will consider the pricing strategy of the service provider (service station) and the maximization of the revenue of the social manager when the transmission rate of each base station reaches equilibrium. This part is mainly based on the work of Wang and Zhang (2015).

On the other hand, retrial queueing system also has important applications in cognitive radio communications. In recent years, a series of studies on radio networks have emphasized the technology of spectrum sharing (such as spectrum sensing, dynamic spectrum access protocol, resource allocation); For details, please refer to Niyato and Hossain (2008a, b), Elias et al. (2011), Hossain et al. (2009), Zhang et al. (2014, 2015a, b). In a cognitive radio network, there are two types of users: authorized user or primary user (PU), which has the right to occupy the spectrum at any time; And unauthorized users or secondary users (SUs). Secondary users can only obtain services when the spectrum is not occupied by authorized users. Therefore, cognitive radio networks can intelligently and dynamically access secondary users when

J. Wang, *Fundamentals of Queueing-Game Models*,
https://doi.org/10.1007/978-981-95-0261-5_10

the spectrum is idle, so as to make effective use of the spectrum. Some literatures (such as Jagannathan et al. 2012, Li and Han 2011) and Do et al. (2012) study the individually-optimal strategy and socially-optimal strategy of secondary users under different system information levels. However, these literatures assume that secondary users serve in the order of first-come, first-served (FCFS), without considering the decentralized dynamic spectrum access, so that the behavior of secondary users is ignored. Such restrictions cannot characterize the essential characteristics of behavioral decisions of secondary users. In addition, if it is assumed that the arriving secondary users are queued at the end of the queue according to their arrival time, it will have no negative external utility to the secondary users already in the system. In practice, however, the secondary users with blocked services do not serve according to the first-come, first-served rule, and a new secondary user usually affects other secondary customers in the system. Therefore, we consider establishing a retrial queueing system with failure and repair, in which secondary users retry independently to obtain services. The dedicated spectrum of primary users can be regarded as a server, in which primary users have higher priority than secondary users.

The second part of this chapter is adapted from Wang and Li (2016). Based on the concept of dynamic spectrum sharing, when the secondary user is a strategic customer, this part gives the equilibrium queueing strategy of the secondary user in Sect. 10.2 (refer to Hassin and Haviv (2003) and Stidham (2009) to match the application of queueing theory in this model). At the same time, we find the equilibrium joining strategy to maximize the social benefits of secondary users in the case of cooperation, and seek a pricing strategy that makes the social optimization consistent with the individual equilibrium.

10.1 LAN Application with Delayed Vacation

The random access protocol in WLAN is an important communication protocol at the application level. Its operation rules can be modeled and analyzed by the retrial queueing system. In order to connect the two nodes of the network, it is often necessary to establish a switching virtual channel to connect. One way to reduce the cost of switching virtual channels is to set an idle timer to reduce the number of setups. Specifically, after the server successfully transmits a message, it will be idle and detect whether new messages arrive or blocked messages try to retry. If no messages arrive during this period of time, the channel will be closed and the server enters the vacation state. If information arrives during this period, the server continues to remain available. In the relevant literature of queueing theory, this model is called delayed vacation model, and its application in communication networks has also been studied by many scholars, such as Hassan and Atiquzzaman (1997).

In addition, the terms “message generation interval, transmission time, backoff time (retransmission interval), overtime interval and additional reserved working time” in the communication protocol correspond to “arrival interval, service time, retry time, delay time and vacation time” in queueing theory. Therefore, in terms of queueing theory, we consider a single-server retrial queueing system without waiting space and with delayed vacation. The vacation policy here can reduce the operation cost of the server and it can also help the system running normally when the server is under repair to prevent the system from crashing. At the same time, such a vacation policy can also reduce the probability of information being blocked in the transmission process. Delayed vacation queueing mechanism is described in detail in the work of Frey and Takahashi (1999), Sakai et al. (1998) and Hassan and Atiquzzaman (1997).

This section is adapted from Wang and Zhang (2015). We introduce the basic models and related assumptions in 10.1.1. Then, in 10.1.2 and 10.1.3, we study the equilibrium joining strategy of customers and the revenue-maximization of the service provider. In 10.1.4, we consider the social welfare maximization problem. Finally, some numerical experiments are presented in 10.1.5.

10.1.1 Model Description

We consider a local area network (LAN) with a single Internet service provider shared by multiple base stations. At each base station, users can send a message (packet) through the channel. Before sending a message, we allow the base station to detect the channel state. If the channel is idle, the message can occupy the channel immediately and be transmitted to the destination terminal. On the contrary, if the channel is busy or unavailable, the packet will try to retransmit after a certain time. Such an interval is determined by the communication protocol, such as a backoff process. In practical communication systems, people usually use truncated binary exponential time to describe the retrial time of a packet.

Suppose a customer arrives at the system with a potential arrival rate of λ. If a customer finds that the server is idle upon arrival, he will occupy the server immediately and leave after receiving the service. Otherwise, if he finds that the server is unavailable (busy or on vacation), he will enter an unlimited retrial space, called orbit, and become a retrial customer. The retrial process of each retrial customer forms a Poisson process with an intensity of θ. For new arriving customers and retrial customers, the service time follows the exponential distribution with a parameter of μ. It is assumed that the reservation and vacation time follow the exponential distribution with parameters α and β, respectively.

We assume that customers are homogeneous. The admission fee for customers entering the system is p; After completing the service, each customer will be paid R unit. Meanwhile, during the waiting process of retial obit, they will spend the waiting cost of C unit time. At the time of arrival, the customer will be informed whether the server is available.

Denote by $I(t)$ the state of the server at time t. The events $I(t) = 0, 1$ or 2 correspond to, respectively, the states that the server is idle, busy or on vacation. Let $N(t)$ be the number of customers in the orbit at time t. It is clear that the stochastic process $\{(I(t), N(t)), t \geq 0\}$ is a two-dimensional continuous time Markov chain with state space $\{0, 1, 2\} \times \{0, 1, 2, \ldots\}$.

Assume that all customers follow a mixed strategy of "join with probability q while observing an unavailable server". So, the effective arrival rate is λq for states where the server is not available. The queueing system is stable if and only if $\lambda q < \mu$. We first consider the case that $\lambda q < \mu$ and let $p(i, j)$ be the steady-state probability of state (i, j). The balance equations are presented below for $j = 0, 1, 2, \ldots$.

$$(\lambda + \alpha + j\theta)p(0, j) = \mu p(1, j) + \beta p(2, j), \tag{10.1.1}$$

$$(\lambda q + \mu)p(1, j) = \lambda p(0, j) + (j + 1)\theta p(0, j + 1) + \lambda q p(1, j - 1), \tag{10.1.2}$$

$$(\lambda q + \beta)p(2, j) = \alpha p(0, j) + \lambda q p(2, j - 1), \tag{10.1.3}$$

where $p(1, -1) = p(2, -1) = 0$.

The above equations are to be solved by using the generating function technique. To this end, we define the following partial generating functions:

$$p_i(z) = \sum_{j=0}^{\infty} z^j p(i, j), \quad i = 0, 1, 2. \tag{10.1.4}$$

Theorem 10.1.1 *In the steady state, we consider the M/M/1 retrial queue with delayed vacations that all customers join with probability q when they arrive and find the server unavailable. The following results are obtained.*

(1) The probabilities that the server is idle, busy or on vacation are, respectively, given by

$$p_0(1) = \frac{\beta(\mu - \lambda q)}{\lambda\beta + \lambda q\alpha + (\alpha + \beta)(\mu - \lambda q)}, \tag{10.1.5}$$

$$p_1(1) = \frac{\lambda\beta + \lambda q\alpha}{\lambda\beta + \lambda q\alpha + (\alpha + \beta)(\mu - \lambda q)}, \tag{10.1.6}$$

$$p_2(1) = \frac{\alpha(\mu - \lambda q)}{\lambda\beta + \lambda q\alpha + (\alpha + \beta)(\mu - \lambda q)}. \tag{10.1.7}$$

(2) The mean numbers of customers in the orbit when the server is idle, busy or on vacation are, respectively, given by

$$N_0 = \frac{\lambda q(\alpha\mu + \lambda\beta)}{\theta[\lambda\beta + \lambda q\alpha + (\alpha + \beta)(\mu - \lambda q)]}, \tag{10.1.8}$$

$$N_1 = \frac{\lambda q[(\lambda\beta + \lambda q\alpha)(\alpha\mu + \lambda\beta + \theta\beta) + \lambda q\alpha\theta(\mu - \lambda q)]}{\theta\beta(\mu - \lambda q)[\lambda\beta + \lambda q\alpha + (\alpha + \beta)(\mu - \lambda q)]}, \tag{10.1.9}$$

$$N_2 = \frac{\lambda q\alpha[\alpha\mu + \lambda\beta + \theta(\mu - \lambda q)]}{\theta\beta[\lambda\beta + \lambda q\alpha + (\alpha + \beta)(\mu - \lambda q)]}. \tag{10.1.10}$$

Proof Multiplying Eqs. (10.1.1)–(10.1.3) by z^j, and summing up over all j, we can derive the following basic equations after some algebraic manipulations:

$$(\lambda+\alpha)p_0(z)+\theta z\frac{\mathrm{d}}{\mathrm{d}z}p_0(z)=\mu p_1(z)+\beta p_2(z), \tag{10.1.11}$$

$$(\lambda q(1-z)+\mu)p_1(z)=\lambda p_0(z)+\theta\frac{\mathrm{d}}{\mathrm{d}z}p_0(z), \tag{10.1.12}$$

$$(\lambda q(1-z)+\beta)p_2(z)=\alpha p_0(z). \tag{10.1.13}$$

Substituting (10.1.12) and (10.1.13) into (10.1.11), eliminating $\frac{\mathrm{d}}{\mathrm{d}z}p_0(z)$, we can obtain that

$$\lambda p_0(z)=(\mu-\lambda qz)p_1(z)-\lambda qp_2(z). \tag{10.1.14}$$

Inserting $z=1$ into (10.1.13) and (10.1.14), we have

$$\beta p_2(1)=\alpha p_0(1), \tag{10.1.15}$$

$$\lambda p_0(1)=(\mu-\lambda q)p_1(1)-\lambda qp_2(1). \tag{10.1.16}$$

Together with the normalizing condition:

$$\sum_{j=0}^{\infty}(p(0,j)+p(1,j)+p(2,j))=p_0(1)+p_1(1)+p_2(1)=1,$$

we can obtain (10.1.5)–(10.1.7).

By putting $z=1$ into (10.1.12) we obtain $N_0=\frac{\mathrm{d}}{\mathrm{d}z}p_0(z)\Big|_{z=1}$ given by (10.1.8), with the help of (10.1.5) and (10.1.6). In a similar approach, differentiating (10.1.13) and (10.1.14), and letting $z=1$, with the help of (10.1.7), (10.1.8) and (10.1.6), (10.1.8) (10.1.10), we can obtain (10.1.10) and (10.1.9). □

We now tag a customer in the orbit. Define $T(i,k)$ as the expected time until service begins for the tagged customer, given that k other customers are in the orbit and the server is in state i. We have the following lemma.

Lemma 10.1.1 *For $k\geq 0$, the expected waiting times $T(i,k)$ $i=0,1,2$, until service begins for a customer in the orbit in the M/M/1 retrial queue with delayed vacations in which all customers enter with probability q when they arrive finding the server unavailable, given that there are k other customers in the orbit, are given by*

$$T(0,k)=mk+w, \tag{10.1.17}$$

$$T(1,k)=mk+b, \tag{10.1.18}$$

$$T(2,k)=mk+v, \tag{10.1.19}$$

where

$$m = \frac{1}{2\mu - \lambda q}, \tag{10.1.20}$$

and

$$w = \frac{(\alpha+\beta)(2\mu-\lambda q) + 2\lambda\beta + \lambda q\alpha}{\theta\beta(2\mu-\lambda q)}, \tag{10.1.21}$$

$$b = \frac{2\beta(\lambda+\theta) + 2\mu(\alpha+\beta) - \lambda q\beta}{\theta\beta(2\mu-\lambda q)}, \tag{10.1.22}$$

$$v = \frac{2\lambda\beta + 2\mu(\alpha+\beta+\theta) - \lambda q\beta}{\theta\beta(2\mu-\lambda q)}. \tag{10.1.23}$$

The proof of this lemma can be conducted by following the proof of Lemma 2 in Wang and Zhang (2013) and therefore is omitted here.

Now we consider a customer who arrives to find the server unavailable, i.e., the server is either busy or on vacation. Due to the PASTA property, the probability that the customer observes j other customers in the orbit and the server is busy or on vacation equals the stationary probability that the server is busy or on vacation and j customers are in the orbit, which can be denoted by $p(1, j)$ or $p(2, j)$ respectively. If this customer decides to join the system, her expected waiting time in the orbit (from arrival to starting receiving service) is equal to $T(1, j)$ or $T(2, j)$. We can calculate the conditional expected time that a customer waits before beginning service if she decides to enter, given that the server is found unavailable upon arrival.

Theorem 10.1.2 *In steady state, we consider the M/M/1 retrial queue with delayed vacations that all customers join the system with probability q when they arrive finding the server unavailable, the conditional expected time that an arriving customer waits before beginning service if she decides to join, given that the server is found unavailable upon arrival, is*

$$T(q) = \frac{1}{\theta} + \frac{\alpha\mu}{\beta(\lambda\beta+\alpha\mu)} + \frac{\lambda\beta+\alpha\mu}{\theta\beta(\mu-\lambda q)} + \frac{\lambda\beta+\lambda q\alpha}{(\lambda\beta+\alpha\mu)(\mu-\lambda q)}. \tag{10.1.24}$$

Furthermore, $T(q)$ *is strictly increasing for* $q \in [0, 1]$.

Proof When the system is in the steady state, the conditional expected waiting time in the orbit for an arriving customer who finds the server unavailable an who decides to enter is

$$T(q) = \frac{\sum_{j=0}^{\infty} p(1,j)T(1,j) + \sum_{j=0}^{\infty} p(2,j)T(2,j)}{\sum_{j=0}^{\infty}[p(1,j)+p(2,j)]}$$
$$= \frac{m\sum_{j=0}^{\infty}[jp(1,j)+jp(2,j)] + b\sum_{j=0}^{\infty} p(1,j) + v\sum_{j=0}^{\infty} p(2,j)}{\sum_{j=0}^{\infty}[p(1,j)+p(2,j)]}$$
$$= \frac{m(N_1+N_2)+bp_1(1)+vp_2(1)}{p_1(1)+p_2(1)}, \qquad (10.1.25)$$

where m, b, v, $p_1(1)$, $p_2(1)$, N_1 and N_2 are given by (10.1.20), (10.1.22), (10.1.23), (10.1.6), (10.1.7), (10.1.9) and (10.1.10). Substituting all of them into (10.1.25), after some algebra, yields (10.1.24). Finally, the proof of the statement that $T(q)$ is strictly increasing for $q \in [0, 1]$ is elementary and therefore is omitted here. □

Based on the reward Ccost structure assumed in this chapter, the expected net benefit of an arriving customer if she finds the server unavailable and decides to enter is

$$S(q) = R - p - CT(q)$$
$$= R - p - C\Big(\frac{\alpha\mu}{\beta(\lambda\beta+\alpha\mu)} + \frac{(\lambda+\mu-\lambda q)\beta+\alpha\mu}{\theta\beta(\mu-\lambda q)} + \frac{\lambda\beta+\lambda q\alpha}{(\lambda\beta+\alpha\mu)(\mu-\lambda q)}\Big), \qquad (10.1.26)$$

which has a unique maximum

$$S(0) = R - p - C\Big(\frac{\lambda\beta+\alpha\mu+\beta\mu}{\theta\beta\mu} + \frac{\alpha\mu^2+\lambda\beta^2}{\beta\mu(\lambda\beta+\alpha\mu)}\Big), \qquad (10.1.27)$$

and a unique minimum

$$S(1) = R - p - C\Big(\frac{\mu(\alpha+\beta)}{\theta\beta(\mu-\lambda)} + \frac{\alpha\mu}{\beta(\lambda\beta+\alpha\mu)} + \frac{\lambda(\alpha+\beta)}{(\lambda\beta+\alpha\mu)(\mu-\lambda)}\Big). \qquad (10.1.28)$$

10.1.2 Nash Equilibrium Strategy

Until now, the equilibrium behaviour of the customers when they arrive and observe an unavailable server can be summarized in the following theorem.

Theorem 10.1.3 *In the M/M/1 retrial queue with delayed vacations and $\lambda < \mu$ for any given price $p < R$, a unique Nash equilibrium mixed strategy "join with probability q_e while observing an unavailable server" exists, where q_e is given by*

$$q_e = \begin{cases} 0, & \text{if } 0 < \frac{R-p}{C} < T(0), \\ \frac{1}{\lambda}\Big(\mu - \frac{\beta(\lambda+\theta)+\alpha\mu}{\theta\beta\Big(\frac{R-p}{C} - \frac{1}{\theta} - \frac{\alpha(\beta+\mu)}{\beta(\lambda\beta+\alpha\mu)}\Big)}\Big), & \text{if } T(0) \le \frac{R-p}{C} \le T(1), \\ 1, & \text{if } \frac{R-p}{C} > T(1), \end{cases} \qquad (10.1.29)$$

where

$$T(0) = \frac{\lambda\beta + \alpha\mu + \beta\mu}{\theta\beta\mu} + \frac{\alpha\mu^2 + \lambda\beta^2}{\beta\mu(\lambda\beta + \alpha\mu)}, \tag{10.1.30}$$

$$T(1) = \frac{\mu(\alpha + \beta)}{\theta\beta(\mu - \lambda)} + \frac{\alpha\mu}{\beta(\lambda\beta + \alpha\mu)} + \frac{\lambda(\alpha + \beta)}{(\lambda\beta + \alpha\mu)(\mu - \lambda)}. \tag{10.1.31}$$

Proof The conclusion follows from Theorem 10.1.2 and the fact that $S(q)$ is strictly decreasing for $q \in [0, 1]$. □

In Theorem 10.1.3 we assume that $\lambda < \mu$, while when $\lambda \geq \mu$, we can obtain similar results by a similar analysis.

Theorem 10.1.4 *In the M/M/1 retrial queue with delayed vacations and $\lambda \geq \mu$ for any given price $p < R$, a unique Nash equilibrium mixed strategy "join with probability q_e while observing an unavailable server" exists, where q_e is given by*

$$q_e = \begin{cases} 0, & \text{if } 0 < \frac{R-p}{C} < T(0), \\ \frac{1}{\lambda}\left(\mu - \frac{\beta(\lambda+\theta)+\alpha\mu}{\theta\beta\left(\frac{R-p}{C} - \frac{1}{\theta} - \frac{\alpha(\beta+\mu)}{\beta(\lambda\beta+\alpha\mu)}\right)}\right), & \text{if } \frac{R-p}{C} \geq T(0). \end{cases} \tag{10.1.32}$$

10.1.3 Server's Revenue Maximization Problem

We focus on the server's revenue maximization problem with the assumption that the server can set the service price itself in this subsection. From 10.1.2 we know that for any given price p, there exists an equilibrium strategy for the customers. It is also readily seen that the equilibrium joining probability q_e, given by (10.1.29) or (10.1.32), is a function of price p. With the full awareness of the demand behaviour of the customers, we now analyse the decision of the server that would like to make a profit by providing service to these customers. It is obviously that if the server charges a high-enough price, there should be only some fraction of the customers entering the system and purchasing service. Also, the server has the alternative of providing service to more customers by imposing a relatively low price. The price will be selected by the server to maximize its own expected profit. In other words, the posted price is the server's strategic variable.

Consequently, the server's problem is to maximize the following function for $0 < p < R$. That is,

$$f(p) = \lambda^*(p)p, \tag{10.1.33}$$

where $\lambda^*(p)$ is the mean effective arrival rate for the customers who decide to join the queueing system for a given price p and is given as follows:

$$\lambda^*(p) = \lambda p_0(1) + \lambda q_e(p)(1 - p_0(1)) = \lambda\mu\Big(\frac{\beta + \alpha q_e(p)}{\lambda\beta + \alpha\mu + \beta(\mu - \lambda q_e(p))}\Big). \tag{10.1.34}$$

Similar to the previous analysis, we consider two cases that are defined by $\lambda < \mu$ and $\lambda \geq \mu$. We study them separately in the sequel.

We consider the first case with the condition $\lambda < \mu$. In this case $q_e(p)$ is given by (10.1.29).

If $R < CT(0)$, the expected waiting cost of an arriving customer who finds the server unavailable and decides to join the queueing system without customers in the orbit exceeds the reward she would gain from service, and therefore the customer never enters the system even if there is no price imposed on her. Thus, the customers enter the system only when they arrive finding an idle server. Inserting $q_e(p) = 0$ into (10.1.34) and (10.1.33), we can obtain the revenue function of the server which can be written as

$$f(p) = \frac{\lambda\beta\mu}{\alpha\mu + \lambda\beta + \beta\mu} p. \tag{10.1.35}$$

Note that the optimal price p_l will be determined as the value approximate to R. Next we analyse the situation when $R \geq CT(0)$ and have the following result.

Theorem 10.1.5 *The objective function of the server $f(p)$ is concave for $p \in ((R - CT(1))^+, R - CT(0)]$, where $x^+ = max\{0, x\}$.*

Proof If follows from the second branch of (10.1.29) that when $max\{0, R - CT(1)\} < p \leq R - CT(0)$, $q_e(p)$ is given by

$$q_e(p) = \frac{1}{\lambda}\left(\mu - \frac{\beta(\lambda + \theta) + \alpha\mu}{\theta\beta\left(\frac{R-p}{C} - \frac{1}{\theta} - \frac{\alpha(\beta+\mu)}{\beta(\lambda\beta+\alpha\mu)}\right)}\right), \tag{10.1.36}$$

and is decreasing and concave in p since

$$\frac{\mathrm{d}q_e(p)}{\mathrm{d}p} = -\frac{\beta(\lambda + \theta) + \alpha\mu}{\lambda\beta\theta C}\Big(\frac{R-p}{C} - \frac{1}{\theta} - \frac{\alpha(\beta + \mu)}{\beta(\lambda\beta + \alpha\mu)}\Big)^{-2} < 0, \tag{10.1.37}$$

$$\frac{\mathrm{d}^2 q_e(p)}{\mathrm{d}p^2} = -\frac{2[\beta(\lambda + \theta) + \alpha\mu]}{\lambda\beta\theta C^2}\Big(\frac{R-p}{C} - \frac{1}{\theta} - \frac{\alpha(\beta + \mu)}{\beta(\lambda\beta + \alpha\mu)}\Big)^{-3} < 0. \tag{10.1.38}$$

Taking the first and second order of $\lambda^*(p)$ given by (10.1.34), we obtain the following results.

$$\frac{\mathrm{d}\lambda^*(p)}{\mathrm{d}p} = \frac{\lambda\mu(\alpha + \beta)(\lambda\beta + \alpha\mu)}{[\lambda\beta + \alpha\mu + \beta(\mu - \lambda q_e(p))]^2}\frac{\mathrm{d}q_e(p)}{\mathrm{d}p} < 0, \tag{10.1.39}$$

$$\frac{\mathrm{d}^2\lambda^*(p)}{\mathrm{d}p^2} = \frac{\lambda\mu(\alpha + \beta)(\lambda\beta + \alpha\mu)^2}{[\lambda\beta + \alpha\mu + \beta(\mu - \lambda q_e(p))]^3}\frac{\mathrm{d}^2 q_e(p)}{\mathrm{d}p^2} < 0. \tag{10.1.40}$$

As such, the second-order condition for the server is

$$\frac{d^2 f(p)}{dp^2} = p\frac{d^2\lambda^*(p)}{dp^2} + 2\frac{d\lambda^*(p)}{dp} < 0, \quad (10.1.41)$$

which shows that $f(p)$ is concave. □

When $p \in (R - CT(0), R)$, $q_e(p) = 0$, and $f(p)$ given by (10.1.35) is maximized at p_l. We next discuss the situations when $p \in (0, R - CT(0)]$, which contains two cases.

(1) If $R - CT(1) \leq 0$, we consider the case when $p \in [0, R - CT(0)]$. It is not difficult to see that when $p = 0$, $f(p) = 0$, and thus it is not the optimal price. From 10.1.5, $f(p)$ is concave for $p \in (0, R - CT(0)]$ and thus there exists at most one stationary point that satisfies the first-order condition for optimality

$$\lambda^*(p) + p\frac{d\lambda^*(p)}{dp} = 0. \quad (10.1.42)$$

We denote this maximal point by p_m. In this case, the optimal price may either be p_m, or p_l. Which value of price the server will choose depends on the comparison between $f(p_l)$ and $f(p_m)$.

(2) If $R - CT(1) > 0$, when $0 < p \leq R - CT(1)$, $q_e(p) = 1$. Inserting $q_e(p) = 1$ into (10.1.33), the revenue function of the server is

$$f(p) = \lambda p. \quad (10.1.43)$$

And the optimal price is clearly attained at $R - CT(1)$. On the other hand, it follows from 10.1.5 that $f(p)$ is concave for $p \in (R - CT(1), R - CT(0)]$, and thus there exists at most one stationary point that satisfies Eq. (10.1.42). We denote this maximal point by p_h if there exists. Hence in this case, the optimal price for the server is the corresponding price of the maximum of $f(p_l)$, $f(R - CT(1))$ and $f(p_h)$.

In summary, the above results for the case $\lambda < \mu$ are given in the following theorem.

Theorem 10.1.6 *In the M/M/1 retrial queue with delayed vacations and $\lambda^*(p) < \mu$, the optimal price p_e that maximizes the profit of the server can be determined with the help of the following procedure.*

(1) If $R < CT(0)$, the optimal price of the server is p_l which is approximate to R;
(2) If $CT(0) \leq R \leq CT(1)$, solve Eq. (10.1.42) for $p \in (0, R - CT(0)]$ in which $\lambda^*(p)$ and $q_e(p)$ are given by (10.1.34) and (10.1.36), respectively, and denote the solution by p_m if there exists. Then the optimal price of the server is found to be the corresponding price of $p_e = \arg\max\{f(p), p \in \{p_l, p_m\}\}$;

(3) *If* $R > CT(1)$, *denoted by* p_h *the solution to Eq. (10.1.42) for* $p \in (R - CT(1), R - CT(0)]$ *if there exists, where* $\lambda^*(p)$ *and* $q_e(p)$ *are given by (10.1.34) and (10.1.36) respectively. The optimal price of the server is* $p_e = \arg\max\{f(p), p \in \{p_l, R - CT(1), p_h\}\}$,

where $T(0)$ *and* $T(1)$ *are provided by (10.1.30) and (10.1.31).*

In the same manner, for the case $\lambda \geq \mu$, an analogous result for determining the optimal price of the server is given as follows. The proof is omitted here for the similarity.

Theorem 10.1.7 *In the M/M/1 retrial queue with delayed vacations and* $\lambda \geq \mu$, *the optimal price* p_e *that maximizes the profit of the server can be determined with the help of the following procedure*

(1) *If* $R < CT(0)$, *the optimal price is* p_l, *which is approximate to* R;
(2) *If* $R \geq CT(0)$, *solve Eq. (10.1.42) for* $p \in (0, R - CT(0)]$ *in which* $\lambda^*(p)$ *and* $q_e(p)$ *are given by (10.1.34) and (10.1.36) respectively, and denote the solution by* p_s *if there exists. Then the optimal price of the server is* $p_e = \arg\max\{f(p), p \in \{p_l, p_s\}\}$,

where $T(0)$ *is given by (10.1.30).*

10.1.4 Socially Optimal Strategy

By definition, social welfare is the sum of the welfare of the server and customers. For a given price p and and joining probability q, the customers' surplus per time unit is $CS = \lambda p_0(1)(R - p) + \lambda q(p_1(1) + p_2(1))(R - p - CT(q))$. Meanwhile, the server's surplus per time unit is $SS = (\lambda p_0(1) + \lambda q(p_1(1) + p_2(1)))p$. Therefore, social welfare is

$$
\begin{aligned}
SW &= CS + SS \\
&= (\lambda p_0(1) + \lambda q(p_1(1) + p_2(1)))R - C\lambda q(p_1(1) + p_2(1))T(q) \\
&= \frac{(\lambda\beta + \lambda q\alpha)\mu}{\lambda\beta + \lambda q\alpha + (\alpha+\beta)(\mu - \lambda q)}R - \frac{\lambda q[(\alpha+\beta)(\alpha\mu + \lambda\beta) + \alpha\mu\theta]}{\theta\beta[\lambda\beta + \lambda q\alpha + (\alpha+\beta)(\mu - \lambda q)]}C \\
&\quad - \frac{\lambda q(\lambda\beta + \lambda q\alpha)(\alpha\mu + \lambda\beta + \theta\beta)}{\theta\beta(\mu - \lambda q)[\lambda\beta + \lambda q\alpha + (\alpha+\beta)(\mu - \lambda q)]}C. \qquad (10.1.44)
\end{aligned}
$$

We observe that social welfare is a function of q. When a price is charged, customers choose a corresponding optimal joining strategy that maximizes their own utility. The social planner's goal is to motivate the customers to adopt the socially optimal strategy. We first consider the problem that is to maximize SW over q, i.e., to find the socially optimal joining probability q. Next we will prove that there exists a unique socially optimal balking strategy.

Theorem 10.1.8 *In the M/M/1 retrial queue with delayed vacations and* $\lambda < \mu$, *a unique mixed strategy "enter with probability* q^* *while observing an unavailable server" that maximizes the social net benefit per time unit exists and is given by*

$$q^* = \begin{cases} 0, & \text{if } 0 < \frac{R}{C} < W_1, \\ q^{**}, & \text{if } W_1 \le \frac{R}{C} \le W_2, \\ 1, & \text{if } \frac{R}{C} > W_2, \end{cases} \tag{10.1.45}$$

where

$$q^{**} = \frac{\mu\beta A - \beta B - \sqrt{\beta^2 B^2 + AB\beta(\lambda\beta + \alpha\mu)}}{\lambda\beta A}, \tag{10.1.46}$$

$$W_1 = \frac{(\lambda\beta + \alpha\mu + \beta\mu)[(\lambda\beta + \alpha\mu)(\lambda\beta + \alpha\mu + \beta\mu + \theta\beta) + \alpha\mu\theta(\beta - \mu)]}{\theta\beta\mu^2(\alpha + \beta)(\lambda\beta + \alpha\mu)}, \tag{10.1.47}$$

$$W_2 = \frac{\mu(\lambda\beta + \alpha\mu + \theta\beta)[\alpha\mu^2(\alpha + 2\beta) + \lambda\beta^2(2\mu - \lambda)]}{\theta\beta\mu(\alpha + \beta)(\lambda\beta + \alpha\mu)(\mu - \lambda)^2} - \frac{(\mu - \lambda)^2[\beta^2(\theta - \mu)(\lambda\beta + \alpha\mu) - \alpha\theta(\beta - \mu)(\lambda\beta + \alpha\mu + \beta\mu)]}{\theta\beta\mu(\alpha + \beta)(\lambda\beta + \alpha\mu)(\mu - \lambda)^2}, \tag{10.1.48}$$

and

$$A = \theta\beta\mu(\alpha + \beta)(\lambda\beta + \alpha\mu)R + \beta^2(\lambda\beta + \alpha\mu)(\theta - \mu)C - \alpha\theta(\beta - \mu)(\lambda\beta + \alpha\mu + \beta\mu)C, \tag{10.1.49}$$

$$B = \beta\mu(\lambda\beta + \alpha\mu)(\lambda\beta + \alpha\mu + \theta\beta)C. \tag{10.1.50}$$

Proof We omit the proof for the sake of brevity. The results can be proved by following the steps of the proof of Theorem 4 in Wang and Zhang (2013). □

For the case $\lambda \ge \mu$, the analogue of the above Theorem is presented below.

Theorem 10.1.9 *In the M/M/1 retrial queue with delayed vacations and* $\lambda \ge \mu$, *a unique socially optimal strategy "enter with probability* q^* *while observing an unavailable server" exists and is given as follows.*

$$q^* = \begin{cases} 0, & \text{if } 0 < \frac{R}{C} < W_1; \\ q^{**}, & \text{if } \frac{R}{C} > W_1; \end{cases} \tag{10.1.51}$$

where q^{**} *and* W_1 *are given by (10.1.46) and (10.1.47), respectively.*

The last problem for social planner is how to choose a suitable price for service to induce selfish customers to behave in a socially optimal way, i.e., to eliminate the difference between the individually and the socially optimal joining rates. First, we analyse the scenario that $\lambda < \mu$ and have the following theorem.

Theorem 10.1.10 *In the M/M/1 retrial queue with delayed vacations and* $\lambda < \mu$, *the socially optimal price* p^* *that maximizes the social welfare per time unit is given by*

$$p^* = \begin{cases} C(W_1 - T(0)), & \text{if } 0 < \frac{R}{C} < W_1, \\ R - C\Big(\frac{1}{\theta} + \frac{\alpha(\mu-\beta)}{\beta(\lambda\beta+\alpha\mu)} + \frac{\beta(\lambda+\theta)+\alpha\mu}{\theta\beta(\mu-\lambda q^{**})}\Big), & \text{if } W_1 \le \frac{R}{C} \le W_2, \\ C(W_2 - T(1)), & \text{if } \frac{R}{C} > W_2, \end{cases} \quad (10.1.52)$$

where $T(0)$, $T(1)$, W_1, W_2 *and* q^{**} *are given by (10.1.30), (10.1.31), (10.1.47), (10.1.48) and (10.1.46).*

Proof If $\frac{R}{C} < W_1$, the socially optimal joining probability is 0. Let $p = C(W_1 - T(0))$, then we have $\frac{R-p}{C} < T(0)$, which is the condition that the equilibrium joining probability is 0. So $p = C(W_1 - T(0))$ is the optimal price that makes the equilibrium joining probability coincide with the socially optimal joining probability, which gives the first branch of (10.1.52).

In a similar way, we can obtain the third branch of (10.1.52).

When $\frac{R}{C} \in [W_1, W_2]$, we know from (10.1.45) that the socially optimal joining probability is q^{**}. Let $q^{**} = \frac{1}{\lambda}\Bigg(\mu - \frac{\beta(\lambda+\theta)+\alpha\mu}{\theta\beta\Big(\frac{R-p}{C} - \frac{1}{\theta} - \frac{\alpha(\mu-\beta)}{\beta(\lambda\beta+\alpha\mu)}\Big)}\Bigg)$, $p = R - C\Big(\frac{1}{\theta} + \frac{\alpha(\mu-\beta)}{\beta(\lambda\beta+\alpha\mu)} + \frac{\beta(\lambda+\theta)+\alpha\mu}{\theta\beta(\mu-\lambda q^{**})}\Big)$. If such a price is charged, $\frac{R-p}{C} \in [T(0), T(1)]$. In virtue of the second branch of (10.1.29), the resulting equilibrium joining probability is reduced to q^{**}. Thus, the equilibrium joining probability can coincide with the socially optimal one by charging each entering customer a service price which is given in the second branch of (10.1.52). □

For the opposite scenario $\lambda \ge \mu$, the socially optimal price can also be obtained.

Theorem 10.1.11 *In the M/M/1 retrial queue with delayed vacations and* $\lambda \ge \mu$, *the socially optimal price* p^* *is given as*

$$p^* = \begin{cases} C(W_1 - T(0)), & \text{if } 0 < \frac{R}{C} < W_1, \\ R - C\Big(\frac{1}{\theta} + \frac{\alpha(\mu-\beta)}{\beta(\lambda\beta+\alpha\mu)} + \frac{\beta(\lambda+\theta)+\alpha\mu}{\theta\beta(\mu-\lambda q^{**})}\Big), & \text{if } \frac{R}{C} \ge W_1. \end{cases} \quad (10.1.53)$$

10.1.5 Numerical Examples

In this section, we investigate the effects of the parameters on the pricing of the service provider and the system administrator via numerical experiments. Specifically, we explore the sensitivity of the revenue-maximizing as well as the socially optimal prices with respect to some main indicators of the system.

We observe in Fig. 10.1 p_e is increasing and p^* is non-increasing with respect to service rate μ. This is because when μ increases, the service provider can serve more

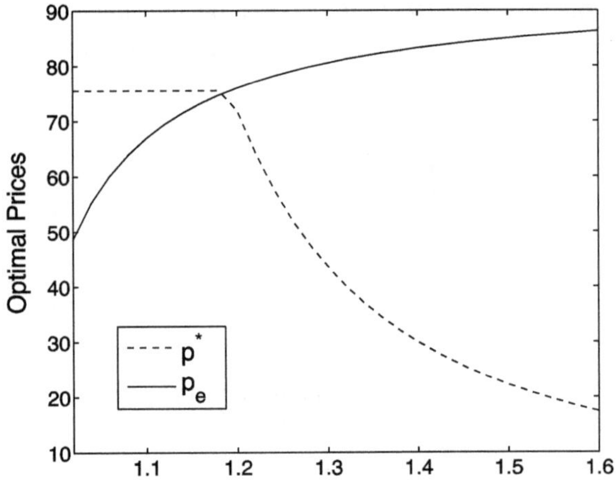

Fig. 10.1 Revenue-maximizing and socially optimal prices versus μ for $\lambda = 0.9, \theta = 0.3, \alpha$=0.2, $\beta = 0.4, R = 100, C = 1$

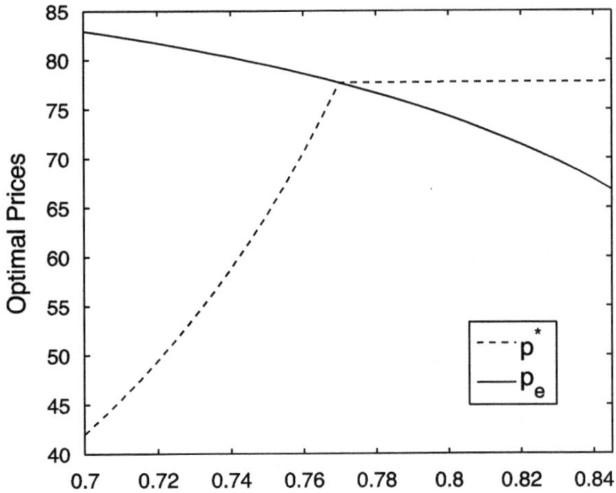

Fig. 10.2 Revenue-maximizing and socially optimal prices versus λ for $\mu = 1, \theta = 0.3, \alpha$=0.2, $\beta = 0.4, R = 100, C = 1$

customers per time unit and thus charges a higher price, while the social planner may reduce the price to induce more customers to enter the system and to compensate for the extra waiting cost incurred by added joining customers. Besides, when μ is small, the service provider may set a low price (lower than the socially optimal price) to attract customers to enter the system for receiving the service and the afford the service fee. In contrast, when μ is large enough, there are many customers who are willing to receive the service. Therefore, the service provider can set a high price (higher than the socially optimal price) to maximize its revenue.

Figure 10.2 illustrates the effect of arrival rate on pricing strategies of the service provider and the administrator. We note that p_e is decreasing and p^* is non-decreasing in λ. The reason is that when λ increases, arriving customers may expect that the orbit is congested and are reluctant to join. The service provider has to set a lower price (which is even lower than the socially optimal one) when λ is large enough

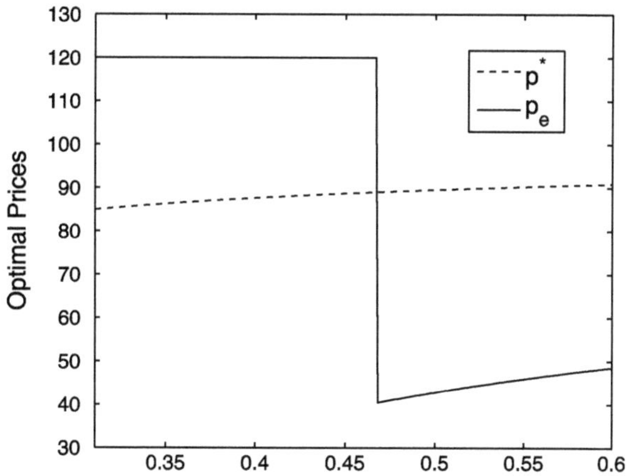

Fig. 10.3 Revenue-maximizing and socially optimal prices versus β for $\lambda = 0.9, \mu = 1, \theta = 0.3, \alpha=0.2, R = 120, C = 1$

to retain more potential customers who may pay the service fee. However, with the increase of λ, to avoid the system getting more congested, the administrator will impose a relatively higher price for admission to join the system.

As for β, when it increases, the mean vacation time decreases, which reduces the waiting time of customers in the orbit and leads to a less waiting cost. Both the service provider and the administrator will benefit from setting a higher price. As is seen in Fig. 10.3, p_e is piecewise non-decreasing and p^* is increasing in β. When β is small, the service provider can charge a price which is approximate to the reward of customers for completing service. In contrast, when β is large enough, the mean vacation time becomes short, the service provider may set a price which is lower than the socially optimal price to further increase the number of customers who pay the service fee per time unit in order to maximize its own revenue. We also note that p_e is not continuous as we vary continuously the value of β. The phenomenon that the plot of p_e contains a sudden shift can be interpreted as follows. Due to the fact that the inequality $R > CT(1)$ holds for all β that varies from 0.3 to 0.6, the optimal price of the server p_e is the corresponding price of $\max\{f(p_l), f(R - CT(1)), f(p_h)\}$. When β starts to increase from 0.3, the optimal price is $p_e = p_l$, which is approximate to R. When β continues to increase to some value between 0.45 and 0.5, the optimal price pe changes suddenly from p_l to p_h, which is smaller than p_l. Therefore, there is a sudden change and p_e is not continuous at this point.

Figure 10.4 depicts the revenue-maximizing as well as the socially optimal prices with respect to α. It is observed that when α increases, the reserved time of the server after completing the service of a certain customer becomes shorter, due to which the customers in the system often compete for service with each other more fiercely and have less chance to receive their required service. As a result, arriving customers may have less incentive to join the system. In order to attract more customers to join the system, both the service provider and the administrator have to set a lower price to reduce the total cost of customers, which is regarded as the condition that both

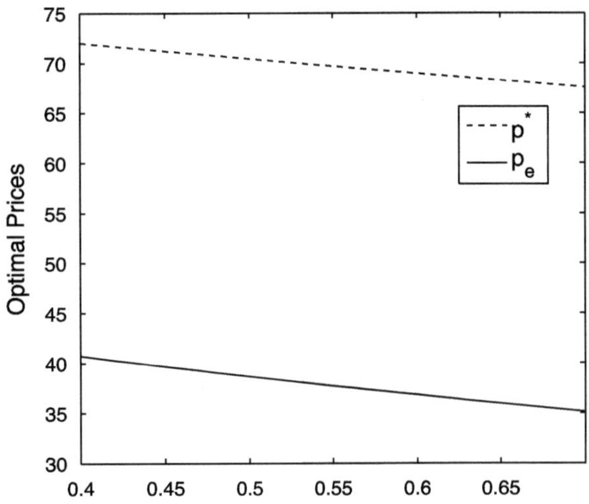

Fig. 10.4 Revenue-maximizing and socially optimal prices versus α for $\lambda = 0.9$, $\mu = 1$, $\theta = 0.3$, $\beta = 0.4$, $R = 100$, $C = 1$

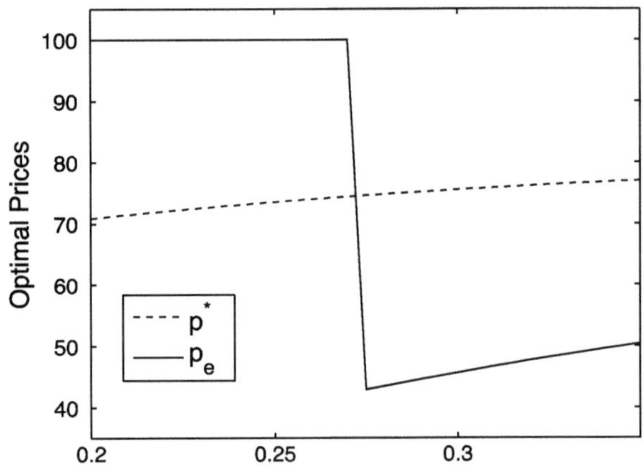

Fig. 10.5 The joining probability of customers versus θ for $\lambda = 0.9$, $\mu = 1$, $\alpha = 0.2$, $\beta = 0.4$, $R = 100$, $C = 1$

revenue-maximizing and the socially optimal prices are decreasing as α increases. This explains the monotonically decreasing of two curves in Fig. 10.4.

Similar to Fig. 10.3, in Fig. 10.5 it is readily observe that p_e is piecewise non-decreasing and p^* is increasing in θ. This is because the fact that more frequent retrials will reduce the waiting time of customers in the orbit; as a result, the customers incur a less waiting cost as well as a less total cost. Which means, both the service provider and the administrator will benefit from setting a higher price. When θ is small, the optimal price charged by the service provider is approximate to the reward for completing service. When θ is large enough, the customers who retry more frequently have more chances to receive the service, which leads to a less waiting time. The service provider may set a price which is lower than the socially optimal one to further attract larger number of customers who enter the system and pay the service

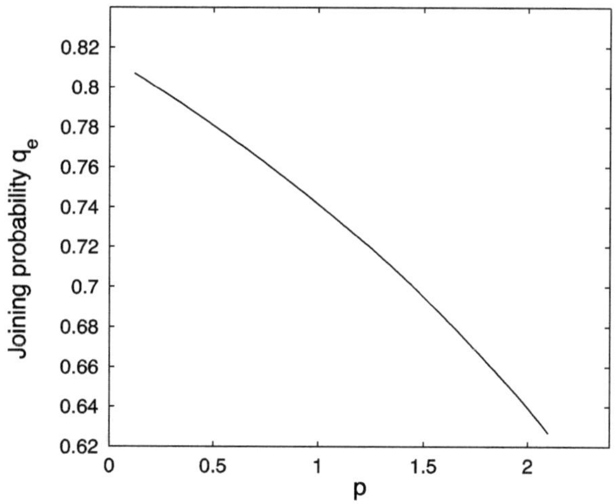

Fig. 10.6 The joining probability of customers versus p for $\lambda = 0.8$, $\mu = 1$, $\alpha = 0.1$, $\theta = 0.7$, $\beta = 0.5$, $R = 9$, $C = 1$

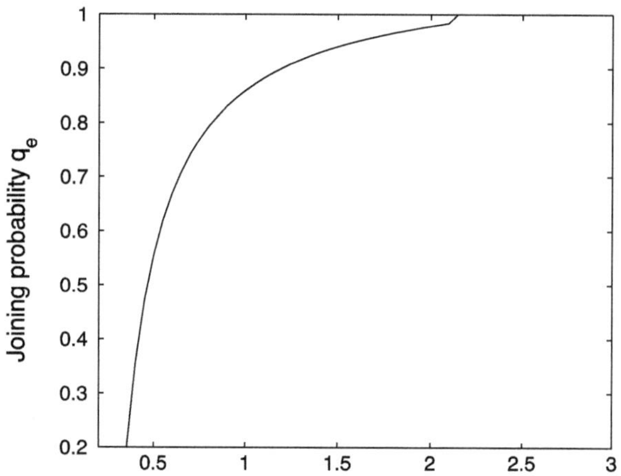

Fig. 10.7 The joining probability of customers versus θ for $\lambda = 0.8$, $\mu = 1$, $\alpha = 0.1$, $\beta = 0.5$, $R = 9$, $C = 1$

fee so as to maximize its own revenue. The reason why the optimal price pe changes suddenly can be similarly explained as in Fig. 10.3.

Figure 10.6 depicts the joining probability of a strategic customer with respect to the varying values of p. It is readily seen that the joining probability drops almost linearly when the price increases, as is to be expected, whereas in Fig. 10.7 we display the joining probability of a strategic customer as a function of retrial rate θ. It is observed that when θ increases, the joining probability q_e for a strategic customer also increases. When the value of q_e reaches a threshold, the joining probability approaches 1 quickly, due to the fact that when the retrial rate becomes too large, the retrial queueing system behaves as a traditional Markovian queue with FCFS discipline. Therefore, the joining probability increases correspondingly.

10.2 Application: Cognitive Radio Networks

We consider another application of queueing game on the cognitive radio (CR) communication systems in this section. It is well-known that the radio frequency spectrum is limited in wireless communication applications. To improve utilization of this scarce resource, Mitola (2001) proposed the CR technology as an effective approach while simultaneously retain the primary communication network infrastructure. There are two types of users in the CR systems, namely, primary users (PUs) or licensed users owning priority to access and use spectrums at all times, and secondary users (SUs) or unlicensed users that access spectrums opportunistically without interfering with PUs. Accordingly, when spectrums are unoccupied by PUs, CR technology enables SUs to use opportunistically these spectrums in an intelligent, dynamic and efficient way. Extensive research on CR systems has elaborated this spectrum sharing mechanism, including spectrum sensing, dynamic spectrum access protocol, resource allocation. Interested readers are referred to Niyato and Hossain (2008a, b), Elias et al. (2011), Hossain et al. (2009), Zhang et al. (2014, 2015a, b) for more detailed discussions.

In dynamic spectrum sharing framework, opportunistic spectrum access (OSA) is categorized as two modes: centralized and decentralized operating cognitive medium access control (MAC) protocols. In the former case a central controller determines the spectrum access for all SUs; while in the latter case, a secondary user can decide to join or not to join the spectrum autonomously. In a decentralized system the joining behavior of SUs have important influence on the system performance and consequently the decision-making of service providers. The underlying game-theoretic analyses on the users behavior match consistently to the rational queueing studies in the queueing literature (see Hassin and Haviv 2003 and Stidham 2009). To be specific, in the decentralized dynamic spectrum access (DSA), secondary users are allowed to make their own decisions in a non-cooperative way that can be regarded as a game among these secondary users. The central issue in this case is to find the Nash equilibrium joining strategy.

In contrast, in centralized DSA case, the core problem is to identify the cooperative strategy operated by a central planer maximizing the overall welfare of the system. Although many studies including Jagannathan et al. (2012), Li et al. (2011) and Do et al. (2012) have addressed this problem by investigating social optimization strategy as well as individual optimal strategy of SUs with different information scenarios by virtue of game-theoretic method, the decentralized behavior of SUs has been partially characterized because in these studies secondary users are assumed to form a queue in CR environment and therefore only the head of the queue can sense idle channels in an FCFS fashion. Evidently, this constraint does not essentially describe the competition and congestion among secondary users. In fact, in a wireless communication system with decentralized cognitive MAC protocol, the arriving SUs compete for access with random service discipline. In the aforementioned studies, however, an arriving SU is assumed to be placed at the end of the queue, and this assumption ignores negative externalities (brought by the new arriving SU) imposed on those existed SUs in the system. In this section the noncooperative and cooperative joining strategies in CR systems with random access is considered and several key results are obtained. Also, the relevant discussion can be found in Wang and Li (2016).

10.2.1 Problem Formulation

A CR system with a single authorized spectrum is considered, in which the spectrum (or called band) is regarded as a single server serving a single PU and SUs. As mentioned, secondary users opportunistically get access to the spectrum when it is not used by the PU or other SU. That is, if the spectrum is idle, SUs are allowed to sense and get use of the spectrum. At the same time, the spectrum remains transparent to the primary user who owns the PU priority. Once the PU requires service, the spectrum stops serving an SU who occupies the spectrum if any and begins to serve the PU immediately. The interrupted SU is marked as a special user waiting for the server's availability and will occupy the spectrum once the primary user completes service. Accordingly, the spectrum has four states: (1) idle, (2) occupied by an SU, (3) occupied by PU along with one preempted SU, and (4) occupied by PU without a preempted SU.

We assume that the service time of the PU is random and exponentially distributed with rate β. The inter-arrival times of the PU's tasks are exponentially distributed with rate α. SUs arrive according to a Poisson process with arrival rate λ. The SUs's service times are exponentially distributed, with a rate μ, which is only served with the absence of PU.

In the CR system each SU can sense whether the spectrum is idle or not. However, the SU does not know whether the spectrum is used by the PU or an SU when the spectrum is sensed to be unavailable. Also, the SU does not know how many other SUs are presently attempting to use the spectrum. Every SU must decide to join or not to join the spectrum according to the available spectrum's information, i.e., they decide to balk (leave the system) or to join the system and use the spectrum eventually. In addition, the decisions of SUs are irrevocable, retrials of balking customers and reneging of joining customers are not allowed. An SU deciding to join the system immediately occupies the spectrum if the spectrum is idle; otherwise, if the spectrum is found busy, this joining SU is directed into a virtual space (orbit) with an infinite capacity and becomes a repeated user. In the retrial orbit, every SU generates a Poisson flow of retrials with rate θ, independent of other SUs' retrials, until it gets access and completes service. After which, the SU leaves and has no further effects on the system. The inter-arrival times of SUs, service times of SUs, SUs' retrial times, PU's inter-arrival times and service times are assumed to be mutually independent.

Once an SU joins the system, waiting cost incurs continuously which is accumulated from arrival in to leaving the system. Denote by C the cost per time unit for the stay of SUs in the system. When the SU completes its job, it receives a service reward of R. By Wang et al. (2001), the so-called "generalized service time" of an SU, which equals to the length of stay time of the SU in the system, is given by $\frac{1}{\mu}(1+\frac{\alpha}{\beta})$; Therefore, the cost incurred during the generalized time is given by $\frac{C}{\mu}(1+\frac{\alpha}{\beta})$, which is independent of SUs' strategy, and one can subtract it from service reward R. Accordingly, without loss of generality, it is assumed throughout that the waiting costs for the SUs are incurred only for their stay time in the orbit. For convenience, we define $\rho_1 = \frac{\lambda}{\mu}$ and $\rho_2 = \frac{\alpha}{\beta}$ for the traffic intensity of SUs and PU respectively. Also, denote by $\rho = \rho_1(1+\rho_2)$, $\rho_\theta = \frac{\theta}{\mu}$, and $\bar{a} \equiv 1-a$ for any real number $a \in [0, 1]$.

Throughout the section, we assume that all SUs are strategic and risk neutral. Two scenarios are studied: (1) SUs are non-cooperative, i.e., SUs wish to maximize their own expected benefit or to choose the Nash equilibrium joining strategy. (2) cooperative case, i.e. a secondary user will attempt to maximize the total expected benefit of all SUs. By assumptions, there is no cost during the period since SUs begin to access the spectrum, which follows that, if the spectrum is found idle, the arriving SU prefers to join and occupies the spectrum immediately, regardless of non-cooperative or cooperative cases, because the arriving SU has a definitely positive net benefit R. Thus, the act of joining is a dominant strategy for such an SU. This strategy does not need to take into account the strategies of any other SUs. Due to this fact, in the sequel, we focus on the behavior of SUs that find the spectrum unavailable upon arrival. Evidently, there are two pure strategies: to join or to balk, in the present formulation. In addition to the two pure strategies, we can derive a mixed strategy that is specified by the joining probability of an arriving SU finding the spectrum unavailable.

The spectrum's state is denoted by $I(t)$ at time t, and $I(t) = 0, 1, 2$ or 3 denote the following events: (1) the spectrum is idle, (2) occupied by an SU, (3) occupied by PU with one interrupted SU, or 4) occupied by the PU with no interrupted SU, respectively. Denote by $N(t)$ the SUs' number in the orbit at time t. The stochastic process $\{(I(t), N(t)), t \geq 0\}$ is obviously a two-dimensional Markov chain with state space $\{0, 1, 2, 3\} \times \{0, 1, 2, \ldots\}$. In this section we assume that all SUs are indistinguishable, they follow a mixed strategy "join with probability q while observing an unavailable spectrum and balk with probability $1-q$". Accordingly, we know that the effective arrival rate is λq for states that spectrum is not idle. By a similar argument as in Kulkarni and Choi (1990), it can be shown that the system is stable if and only if $q\rho < 1$.

In the steady-state, we denote by $p(i, j)$ the steady-state probability of state (i, j). The balance equations of the system are given below ($j = 0, 1, 2, \ldots$).

$$(\lambda + \alpha + j\theta)p(0, j) = \mu p(1, j) + \beta p(3, j), \tag{10.2.1}$$

$$\begin{aligned}(\lambda q + \alpha + \mu)p(1, j) &= \lambda p(0, j) + \lambda q p(1, j-1) \\ &\quad + (j+1)\theta p(0, j+1) + \beta p(2, j),\end{aligned} \tag{10.2.2}$$

$$(\lambda q + \beta)p(2, j) = \lambda q p(2, j-1) + \alpha p(1, j), \tag{10.2.3}$$

$$(\lambda q + \beta)p(3, j) = \lambda q p(3, j-1) + \alpha p(0, j), \tag{10.2.4}$$

where $p(1, -1) = p(2, -1) = p(3, -1) = 0$.

By using generating function method the above equations can be solved. Define:

$$p_i(z) = \sum_{j=0}^{\infty} z^j p(i, j), \quad i = 0, 1, 2, 3. \tag{10.2.5}$$

Theorem 10.2.1 *In the steady state, all SUs join the system with probability q upon their arrival finding the spectrum unavailable, the following results hold:*

(1) The steady-state probabilities that the spectrum is idle, used by an SU, used by the PU with a preempted SU, or used by the PU with no preempted SU, are respectively presented by

$$p_0(1) = \frac{1 - q\rho}{(1 + \rho_2)(1 + \bar{q}\rho_1)}, \tag{10.2.6}$$

$$p_1(1) = \frac{\rho_1(1 + q\rho_2)}{(1 + \rho_2)(1 + \bar{q}\rho_1)}, \tag{10.2.7}$$

$$p_2(1) = \frac{\rho_1\rho_2(1 + q\rho_2)}{(1 + \rho_2)(1 + \bar{q}\rho_1)}, \tag{10.2.8}$$

$$p_3(1) = \frac{\rho_2(1 - q\rho)}{(1 + \rho_2)(1 + \bar{q}\rho_1)}. \tag{10.2.9}$$

(2) In the steady-state, the expected numbers of SUs in the orbit of following cases: 1) when the spectrum is idle, 2) the spectrum is used by an SU, 3) the spectrum is used by the PU with one preempted SU, or 4) the spectrum is used by the PU with no preempted SU, are respectively obtained and given by

$$N_0 = \frac{\rho_1 q(\rho + \rho_2)}{\rho_\theta(1 + \rho_2)(1 + \bar{q}\rho_1)}, \tag{10.2.10}$$

$$N_1 = \frac{q\rho_1^2(1 + \rho_2 q)[(\rho_1 + \rho_\theta)(1 + \rho_2) + \rho_2]}{\rho_\theta(1 + \rho_2)(1 + \bar{q}\rho_1)(1 - q\rho)} + \frac{\lambda q^2\rho_1\rho_2}{\beta(1 + \rho_2)(1 - q\rho)}, \tag{10.2.11}$$

$$N_2 = \frac{\lambda q\alpha\rho_1[q\rho_2(1 + \bar{q}\rho_1) + (1 + q\rho_2)(1 - q\rho)]}{(1 + \rho_2)(1 + \bar{q}\rho_1)(1 - q\rho)} + \frac{q\alpha\beta\rho_1^2(1 + q\rho_2)[\rho_2 + \rho + \rho_\theta(1 + \rho_2)]}{\rho_\theta(1 + \rho_2)(1 + \bar{q}\rho_1)(1 - q\rho)}, \tag{10.2.12}$$

$$N_3 = \frac{q\rho_1\rho_2[\beta(\rho + \rho_2) + \theta(1 - q\rho)]}{\rho_\theta\beta(1 + \rho_2)(1 + \bar{q}\rho_1)}. \tag{10.2.13}$$

Proof We first multiply Eqs. (10.2.1)–(10.2.4) by z^j and then sum up over all j, the following basic equations can be derived after algebraic manipulations:

$$(\lambda + \alpha)p_0(z) + \theta z p_0'(z) = \mu p_1(z) + \beta p_3(z), \tag{10.2.14}$$

$$(\lambda q(1 - z) + \alpha + \mu)p_1(z) = \lambda p_0(z) + \theta p_0'(z) + \beta p_2(z), \tag{10.2.15}$$

$$(\lambda q(1 - z) + \beta)p_2(z) = \alpha p_1(z), \tag{10.2.16}$$

$$(\lambda q(1 - z) + \beta)p_3(z) = \alpha p_0(z), \tag{10.2.17}$$

After eliminating $p_1(z)$, $p_2(z)$ and $p_3(z)$ in Eqs. (10.2.14)–(10.2.17), we obtain $p_0'(z)$.

$$\begin{aligned} p_0'(z) = &\frac{\lambda q p_0(z)}{\theta(\lambda q(1-z)+\beta)((\mu-\lambda qz)(\lambda q(1-z)+\beta)-\lambda q\alpha z)} \\ &\times\Big(\lambda(\lambda q(1-z)+\alpha+\beta)(\lambda q(1-z)+\beta) \\ &+\alpha((\lambda q(1-z)+\beta)(\lambda q(1-z)+\alpha+\mu)-\alpha\beta)\Big). \end{aligned} \tag{10.2.18}$$

Inserting $z = 1$ in the above equation, we have

$$p_0'(1) = \frac{\lambda q(\rho+\rho_2)}{\theta(1-q\rho)} p_0(1). \tag{10.2.19}$$

We differentiate (10.2.18), and set $z = 1$ in the result, which yields

$$\begin{aligned} p_0''(1) = &\frac{1}{\theta\beta(1-q\rho)} \times \Big\{[-\lambda^2 q^2(\rho_1(2+\rho_2)+\alpha(1+\rho_2)/\mu+\rho_2)]p_0(1) \\ &+ [\lambda q(\beta(\rho+\rho_2)+\theta(\alpha+\beta+2\mu)/\mu-\theta\rho_1 q(2+\rho_2))]p_0'(1)\Big\}. \end{aligned} \tag{10.2.20}$$

Next, we let $z = 1$ in the Eqs. (10.2.14)–(10.2.17), and take into account (10.2.19) along with the following normalizing condition

$$\sum_{j=0}^{\infty}(p(0,j)+p(1,j)+p(2,j)+p(3,j)) = p_0(1)+p_1(1)+p_2(1)+p_3(1) = 1,$$

which arrives the results (10.2.6)–(10.2.9) after some tedious algebra.

We first substitute (10.2.6) into (10.2.19), which arrives at $N_0 = p_0'(1)$ which is given by (10.2.10). Then we differentiate (10.2.17) and evaluate it at the point $z = 1$, which yields $N_3 = p_3'(1)$ that is given by (10.2.13) by substituting (10.2.9) and (10.2.10). Similarly, we differentiate (10.2.14) and (10.2.16), and let $z = 1$, by virtue of (10.2.6), (10.2.10), (10.2.13), (10.2.20), (10.2.8) and (10.2.11), the results of (10.2.11) and (10.2.12) can be obtained, respectively. □

It is remarked that in item 1) of Theorem 10.2.1, each probability for those four cases is independent of parameter θ. This fact is in line with known results in the literature of retrial queues. To be specific, the steady-state distribution of the server's state (in other words, the stationary distribution of the states of spectrum) does not depend on the retrial rete. This fact has been explained (see the context on pages 12–13) in the monograph by Falin and Templeton (1997).

10.2.2 Non-cooperative Solution

We consider noncooperative scenario in this subsection. In this situation all SUs behave in a selfish manner, which means that the goal of each secondary user is to maximize his or her own expected benefit. Upon arrival, each SU decides whether to join the system or not based on a linear cost function in which the SU has to evaluate the expected waiting cost when finding the spectrum unavailable against the received reward when the service is completed. To be specific, each SU strictly prefers to join the system if the reward exceeds the expected waiting cost in the orbit, and is indifferent between joining and balking if the reward for service equals to the waiting cost. In this interactive system each SU's behavior affects other SUs' behavior and decision, as well as the CR system's performance. In this sense, the CR system can be modeled a symmetric non-cooperative game among secondary users and the core is to obtain the Nash equilibrium solution in a competitive environment.

At first we need to derive expressions for the conditional expected waiting time of an arriving secondary user, given that this SU senses upon arrival that the spectrum is unavailable and makes a decision to join the orbit. The following theorem gives such a solution.

Theorem 10.2.2 *In the steady state, the conditional expected waiting time of an arriving secondary user, given that the SU senses upon arrival that the spectrum unavailable and makes a decision to join the orbit with probability q, is given by*

$$T(q) = \frac{(1+\rho_2)(1+\bar{q}\rho_1)}{\theta(1-q\rho)} + \frac{\rho_2(1+\bar{q}\rho_1)}{\beta(1-q\rho)(\rho+\rho_2)} + \frac{\rho_1(1+\rho_2)^2(1+q\rho_2)}{\mu(1-q\rho)(\rho+\rho_2)}. \quad (10.2.21)$$

Additionally, $T(q)$ is strictly increasing for q.

Proof First, we know the stationary probability that the secondary user senses the PU band is used by a secondary user, or used by the PU with one preempted secondary user, or used by the PU with no preempted secondary user, along with that there are j other SUs in the orbit, coincides with the stationary probability $p(1, j)$, $p(2, j)$, or $p(3, j)$ in virtue of the PASTA property. Whereupon, the total arrival rate of SUs in the orbit is

$$\lambda_{ret} = \lambda q(p_1(1) + p_2(1) + p_3(1)) = \frac{\lambda q(\rho+\rho_2)}{(1+\rho_2)(1+\bar{q}\rho_1)}. \quad (10.2.22)$$

Hence, the mean number of secondary users in the orbit is given by

$$\begin{aligned} N &= N_0 + N_1 + N_2 + N_3 \\ &= \frac{\lambda q(\rho+\rho_2)}{\theta(1-q\rho)} + \frac{\lambda q\rho_2}{\beta(1+\rho_2)(1-q\rho)} + \frac{q\rho_1\rho(1+q\rho_2)}{(1+\bar{q}\rho_1)(1-q\rho)}. \end{aligned} \quad (10.2.23)$$

By virtue of Little's formula, we obtain the mean waiting time of a secondary user that joins the orbit as

$$T(q) = \frac{N}{\lambda_{ret}} = \frac{(1+\rho_2)(1+\bar{q}\rho_1)}{\theta(1-q\rho)} + \frac{\rho_2(1+\bar{q}\rho_1)}{\beta(1-q\rho)(\rho+\rho_2)} + \frac{\rho_1(1+\rho_2)^2(1+q\rho_2)}{\mu(1-q\rho)(\rho+\rho_2)}. \quad (10.2.24)$$

It is readily seen that $T(q)$ is strictly increasing for q and therefore we complete the proof. □

By definition, an arriving secondary user's expected net benefit given that the SU finds the spectrum unavailable and chooses to join is

$$\begin{aligned} S(q) &= R - CT(q) \\ &= R - C\Big[\frac{(1+\rho_2)(1+\bar{q}\rho_1)}{\theta(1-q\rho)} + \frac{\rho_2(1+\bar{q}\rho_1)}{\beta(1-q\rho)(\rho+\rho_2)} + \frac{\rho_1(1+\rho_2)^2(1+q\rho_2)}{\mu(1-q\rho)(\rho+\rho_2)}\Big]. \end{aligned} \quad (10.2.25)$$

Moreover, by Theorem 10.2.2, it can be proved that $S(q)$ is strictly decreasing for q. We proceed to consider the SUs' equilibrium behavior when they arrive and sense the spectrum unavailable.

The stability condition $q\rho < 1$ implies that $\rho < 1$ or $\rho \geq 1$. For this reason, we consider the mixed strategy in two cases, i.e., $\rho < 1$ and $\rho \geq 1$, and the results are presented in Theorem 10.2.3 and Theorem 10.2.4 respectively.

Theorem 10.2.3 *In the case of $\rho < 1$, a unique Nash equilibrium mixed strategy "join the system with probability q_e while sensing the spectrum unavailable" exists, where the equilibrium probability q_e is given by*

$$q_e = \begin{cases} 0, & if\ \frac{R}{C} < T(0), \\ q_{ee}, & if\ T(0) \leq \frac{R}{C} \leq T(1), \\ 1, & if\ \frac{R}{C} > T(1), \end{cases} \quad (10.2.26)$$

where

$$q_{ee} = \frac{\Big\{R\theta(\rho+\rho_2) - C\rho_\theta\rho(1+\rho_2) - C(1+\rho_1)(\rho(1+\rho_2) + \rho_2(1+\rho_2+\frac{\theta}{\beta}))\Big\}}{\Big\{R\theta\rho_1(1+\rho_2)(\rho+\rho_2) + C\rho_\theta\rho\rho_2(1+\rho_2) - C\rho_1(\rho(1+\rho_2) + \rho_2(1+\rho_2+\frac{\theta}{\beta}))\Big\}}, \quad (10.2.27)$$

$$T(0) = \frac{(1+\rho_1)(1+\rho_2)}{\theta} + \frac{\mu\rho_2(1+\rho_1) + \beta\rho(1+\rho_2)}{\beta\mu(\rho+\rho_2)}, \quad (10.2.28)$$

$$T(1) = \frac{1+\rho_2}{\theta(1-\rho)} + \frac{\rho\beta(1+\rho_2)^2 + \mu\rho_2}{\mu\beta(1-\rho)(\rho+\rho_2)}. \quad (10.2.29)$$

Proof It is evident that, when $\mu\beta > \lambda(\alpha+\beta)$, $S(q)$ is strictly decreasing for $q \in [0, 1]$, and has a unique maximum

$$S(0) = R - CT(0) = R - C\Big(\frac{(1+\rho_1)(1+\rho_2)}{\theta} + \frac{\mu\rho_2(1+\rho_1) + \beta\rho(1+\rho_2)}{\beta\mu(\rho+\rho_2)}\Big), \quad (10.2.30)$$

and a unique minimum

$$S(1) = R - CT(1) = R - C\Big(\frac{1+\rho_2}{\theta(1-\rho)} + \frac{\rho\beta(1+\rho_2)^2 + \mu\rho_2}{\mu\beta(1-\rho)(\rho+\rho_2)}\Big). \tag{10.2.31}$$

As a consequence, when $\frac{R}{C} < T(0)$, we know that $S(q)$ is negative for every q, and therefore the best response of an arriving secondary user is balking and the unique equilibrium point is $q_e = 0$, which attains the first sector of (10.2.26).

Similarly, when $T(0) \le \frac{R}{C} \le T(1)$, we know that there exists a unique solution for $S(q) = 0$. The solution lies in [0, 1], and therefore gets the second sector of (10.2.26).

When $\frac{R}{C} > T(1)$, for every q an arriving secondary user joining the orbit gets a positive benefit. That is, the best response of SUs is 1. Consequently, the choice of joining is the unique Nash equilibrium (NE) strategy. The third sector of (10.2.26) is attained. □

On the other hand, we can obtain the following results when $\rho \ge 1$.

Theorem 10.2.4 *In the case of $\rho \ge 1$, for SUs, a unique Nash equilibrium mixed strategy "join with probability q_e while sensing the spectrum unavailable" exists, in which q_e is given by*

$$q_e = \begin{cases} 0, & \text{if } \frac{R}{C} < T(0), \\ q_{ee}, & \text{if } \frac{R}{C} \ge T(0), \end{cases} \tag{10.2.32}$$

where the values of q_{ee} and $T(0)$ are given by (10.2.27) and (10.2.28), respectively.

10.2.3 Cooperative Solution

In this subsection we investigate the optimal joining strategy of secondary users from a social point of view. In this situation the goal is to maximizes the SUs' overall welfare per time unit. It is noticed that if SUs are non-cooperative, they competitively get access the spectrum. By contrast, the CR system's performance can be improved if the secondary users cooperate with each other to attain optimal spectrum access.

When all SUs follow a mixed policy "join the system with probability q while sensing an unavailable spectrum upon arrival", the social benefit per time unit is defined as

$$S_{soc}(q) = \lambda^* R - CN, \tag{10.2.33}$$

where λ^* represents the mean arrival rate of SUs joining the CR system and N denotes the expected number of secondary users in the orbit that is given by (10.2.23). By virtue of (10.2.6)–(10.2.9) and the PASTA property, we get

$$\lambda^* = \lambda p_0(1) + \lambda q(p_1(1) + p_2(1) + p_3(1)) = \frac{\mu\rho_1(1+q\rho_2)}{(1+\rho_2)(1+\bar{q}\rho_1)}. \tag{10.2.34}$$

We substitute (10.2.23) and (10.2.34) into (10.2.33), and then obtain that

$$S_{soc}(q) = \frac{R\mu\rho_1(1+q\rho_2)}{(1+\rho_2)(1+\bar{q}\rho_1)} - \frac{C\lambda q}{\theta\beta(1+\rho_2)(1+\bar{q}\rho_1)(1-q\rho)} \times\Big(\beta(1+\rho_2)(1+\bar{q}\rho_1)(\rho+\rho_2)+\theta\rho_2(1+\bar{q}\rho_1)+\beta\rho_\theta\rho(1+\rho_2)(1+q\rho_2)\Big). \quad (10.2.35)$$

We proceed to consider the cooperative strategy that maximizes the social benefit over q. In other words, we aim to find the socially optimal joining probability q^* of secondary users when they arrive at the system and find the spectrum unavailable. The corresponding results are summarized in the following Theorems 10.2.5 and Theorem 10.2.6.

Theorem 10.2.5 *In the considered CR system, when $\rho < 1$, there exists a unique mixed strategy "join the system with probability q^* while sensing an unavailable spectrum" that maximizes the social net benefit per time. It is given by*

$$q^* = \begin{cases} 0, & \text{if } \frac{R}{C} < W_1, \\ \frac{1}{\rho}\left[1 - \frac{x_2(1+\rho_2)}{\mu}\right], & \text{if } W_1 \le \frac{R}{C} \le W_2, \\ 1, & \text{if } \frac{R}{C} > W_2, \end{cases} \quad (10.2.36)$$

where

$$x_2 = \frac{-B+\sqrt{B^2 - AB(\lambda + \frac{\alpha\mu}{\alpha+\beta})}}{A}, \quad (10.2.37)$$

$$A = (\lambda(\alpha+\beta)+\alpha\mu)(C\theta(\alpha+\beta)^2 - C\beta^2\mu + R\theta\beta\mu(\alpha+\beta)) - \frac{C\alpha\mu\theta(\beta\mu-(\alpha+\beta)^2)}{1+\rho_2}, \quad (10.2.38)$$

$$B = -\frac{C\beta\mu(\lambda(\alpha+\beta)+\alpha\mu)}{1+\rho_2} \times \left((\lambda+\theta)(\alpha+\beta)+\alpha\mu+\frac{\alpha\mu\theta}{\alpha+\beta}\right) < 0, \quad (10.2.39)$$

$$W_1 = \frac{1}{\theta}(1+\rho_1)^2(1+\rho_2) + \frac{\rho_2(1+\rho_1)^2}{\beta(\rho+\rho_2)} + \frac{\rho(1+\rho_1)(1+\rho_2)}{\mu(\rho+\rho_2)}, \quad (10.2.40)$$

$$W_2 = \frac{(1+\rho_2)(\mu+\rho\theta)}{\mu\theta(1-\rho)^2} + \frac{\rho_2}{\beta(1-\rho)^2(\rho+\rho_2)} + \frac{\rho(1+\rho_2)(1+\rho+\rho_2)}{\mu(1-\rho)(\rho+\rho_2)}. \quad (10.2.41)$$

Proof By (10.2.35), when all SUs follow a mixed policy "join the system with probability q while sensing an unavailable spectrum", we have known the social benefit per time unit. Let $x = \frac{\mu\beta}{\alpha+\beta} - \lambda q$, the result of (10.2.35) can be arranged as a function of x:

$$f(x) = \frac{C(\rho+\rho_2)}{\mu\theta(1+\rho_2)} + \frac{C\rho_2(1-(1+\rho_2)^2)}{(1+\rho_2)^2} - \frac{R\mu\rho_2}{1+\rho_2} + \frac{Ax+B}{\theta\beta(\alpha+\beta)x((\alpha+\beta)x+\lambda(\alpha+\beta)+\alpha\mu)}, \quad (10.2.42)$$

where the expressions of A and B are given by (10.2.38) and (10.2.39), respectively.

In light of this, the original problem of maximizing $S_{soc}(q)$ for $q \in [0, 1]$ is apparently equivalent to the new problem of maximizing $f(x)$ for $x \in \left[\frac{\mu(1-\rho)}{1+\rho_2}, \frac{\mu}{1+\rho_2}\right]$. We first differentiate $f(x)$ with respect to x and yield

$$f'(x) = -\frac{A(\alpha+\beta)x^2 + 2B(\alpha+\beta)x + B(\lambda(\alpha+\beta)+\alpha\mu)}{\theta\beta(\alpha+\beta)x^2((\alpha+\beta)x + \lambda(\alpha+\beta)+\alpha\mu)^2}. \quad (10.2.43)$$

Henceforth, we consider two cases:

(1) When

$$\frac{R}{C} \le \frac{\rho_2(\mu - \beta(1+\rho_2)^2)}{\mu\beta(1+\rho_2)^2(\rho+\rho_2)} + \frac{1-\rho_\theta(1+\rho_2)^2}{\theta(1+\rho_2)} \Leftrightarrow A \le 0.$$

In this case, it is observed that $f'(x) > 0$ for any $x \in \left[\frac{\mu(1-\rho)}{1+\rho_2}, \frac{\mu}{1+\rho_2}\right]$. This implies that $f(x)$ is strictly increasing in x and the unique maximum is achieved at $x = \frac{\mu}{1+\rho_2}$. Consequently, the best response of arriving secondary users that find the spectrum unavailable will join the orbit with probability $q^* = 0$.

(2) When

$$\frac{R}{C} > \frac{\rho_2(\mu - \beta(1+\rho_2)^2)}{\mu\beta(1+\rho_2)^2(\rho+\rho_2)} + \frac{1-\rho_\theta(1+\rho_2)^2}{\theta(1+\rho_2)} \Leftrightarrow A > 0.$$

It should be noted that in this case, the function $f(x)$ may not be always monotone in x for $x \in \left[\frac{\mu(1-\rho)}{1+\rho_2}, \frac{\mu}{1+\rho_2}\right]$. Further, solving the equation $f'(x) = 0$ we can obtain two roots of the equation as follows.

$$x_1 = \frac{-B - \sqrt{B^2 - AB(\lambda + \frac{\alpha\mu}{\alpha+\beta})}}{A} < 0, \quad (10.2.44)$$

$$x_2 = \frac{-B + \sqrt{B^2 - AB(\lambda + \frac{\alpha\mu}{\alpha+\beta})}}{A} > 0. \quad (10.2.45)$$

As a consequence, the function $f(x)$ is decreasing in $(-\infty, x_1)$ and $(x_2, +\infty)$ respectively, and it is increasing in $[x_1, x_2]$.

Accordingly, we discuss the following three subcases.

Subcase (2.1): When $\frac{\mu}{1+\rho_2} < x_2 \Leftrightarrow \frac{R}{C} < W_1$. In this subcase, the function $f(x)$ is increasing for $x \in \left[\frac{\mu(1-\rho)}{1+\rho_2}, \frac{\mu}{1+\rho_2}\right]$, and the unique maximum is attained at $x = \frac{\mu}{1+\rho_2}$, i.e., $q^* = 0$.

Subcase (2.2): When $\frac{\mu(1-\rho)}{1+\rho_2} \le x_2 \le \frac{\mu}{1+\rho_2} \Leftrightarrow W_1 \le \frac{R}{C} \le W_2$. In this subcase, the function $f(x)$ is unimodal for $x \in \left[\frac{\mu(1-\rho)}{1+\rho_2}, \frac{\mu}{1+\rho_2}\right]$. More specifically, the function is increasing for $x \in [\frac{\mu(1-\rho)}{1+\rho_2}, x_2]$ and decreasing for $x \in [x_2, \frac{\mu}{1+\rho_2}]$. Whereupon, the optimal solution is attained at $x = x_2$, i.e., $q^* = \frac{1}{\rho}\left[1 - \frac{x_2(1+\rho_2)}{\mu}\right]$. To summarize, the optimal strategy of secondary users when they find the spectrum unavailable is choosing the strategy "to join the system with probability $q^* = \frac{1}{\rho}\left[1 - \frac{x_2(1+\rho_2)}{\mu}\right]$".

Subcase (2.3): When $\frac{\mu(1-\rho)}{1+\rho_2} > x_2 \Leftrightarrow \frac{R}{C} > W_2$. In this subcase, the function $f(x)$ is decreasing for $x \in \left[\frac{\mu(1-\rho)}{1+\rho_2}, \frac{\mu}{1+\rho_2}\right]$, and therefore the function takes its maximum at $x = \frac{\mu(1-\rho)}{1+\rho_2}$, i.e., the best response of SUs is $q^* = 1$.

Based on the above discussions, we conclude by combining above results in two cases and by noting

$$\frac{\rho_2(\mu - \beta(1+\rho_2)^2)}{\mu\beta(1+\rho_2)^2(\rho+\rho_2)} + \frac{1 - \rho_\theta(1+\rho_2)^2}{\theta(1+\rho_2)} < W_1,$$

the socially optimal probability q^* in the theorem statement can be attained. □

For the case $\rho \geq 1$, in a parallel way and by using a similar argument, the corresponding results can be obtained and are given in the following theorem.

Theorem 10.2.6 *In the considered CR system with $\rho \geq 1$, there exists a unique mixed strategy "join the system with probability q^* while sensing an unavailable spectrum" that maximizes the social net benefit per time unit and it is given by*

$$q^* = \begin{cases} 0, & \text{if } \frac{R}{C} < W_1, \\ \frac{1}{\rho}\left[1 - \frac{x_2(1+\rho_2)}{\mu}\right], & \text{if } \frac{R}{C} \geq W_1, \end{cases} \tag{10.2.46}$$

where the values of x_2 and W_1 are provided by (10.2.37) and (10.2.40), respectively.

10.2.4 Admission Fee

As of now, the Nash equilibrium strategy and the socially optimal strategy of SUs have been obtained when they behave in a non-cooperative or a cooperative way. It can be easily see that the individual's decision (Nash equilibrium solution) deviates from the socially preferred one. In other words, attention of narrow self-interest usually does not give rise to overall optimality. In reality, however, for the sake of efficient and fair management, the Nash equilibrium joining probability q_e is expected to coincide with the socially optimal joining probability q^*.

To lead individually optimizing SUs to choose strategies in a socially optimal way, it is necessary to compare q_e and q^* at first.

Theorem 10.2.7 *For the considered CR system, the equilibrium joining probability* q_e *is not smaller than the socially optimal joining probability* q^*, *i.e.,*

$$q^* \leq q_e. \tag{10.2.47}$$

Proof In principle, there are two cases should be addressed with $\rho < 1$ and $\rho \geq 1$. Here we only consider the case $\rho < 1$. For the case $\rho \geq 1$, the result holds by a similar way.

Next, according to (10.2.26) and (10.2.36), it can be proved that $T(0) < W_1$ and $T(1) < W_2$. Hence, we only need to compare $T(1)$ and W_1 in the following two situations.

(1) If $W_1 > T(1)$, we will have the following five subcases.

−− **Subcase (1a)**: If $\frac{R}{C} < T(0)$, then we have $q^* = q_e = 0$.

−− **Subcase (1b)**: If $T(0) \leq \frac{R}{C} \leq T(1)$, then we get $q^* = 0$ and $q_e = q_{ee} \in [0, 1]$. Therefore, the inequality $q^* \leq q_e$ holds.

−− **Subcase (1c)**: If $T(1) < \frac{R}{C} < W_1$, then we get $q^* = 0$ and $q_e = 1$.

−− **Subcase (1d)**: If $W_1 \leq \frac{R}{C} \leq W_2$, then we get $q^* = \frac{1}{\rho}\left[1 - \frac{x_2(1+\rho_2)}{\mu}\right] \in [0, 1]$ and $q_e = 1$, and obviously we have $q^* \leq q_e$.

−− **Subcase (1e)**: If $\frac{R}{C} > W_2$, then $q^* = q_e = 1$.

(2) If $W_1 \leq T(1)$, there exists five subcases to be discussed.

−− **Subcase (2a)**: If $\frac{R}{C} < T(0)$, then we get that $q^* = q_e = 0$.

−− **Subcase (2b)**: If $T(0) \leq \frac{R}{C} \leq W_1$, then we get $q^* = 0$ and $q_e = q_{ee} \in [0, 1]$. Hence, the inequality $q^* \leq q_e$ holds.

−− **Subcase (2c)**: If $W_1 < \frac{R}{C} \leq T(1)$, then we get $q^* = \frac{1}{\rho}\left[1 - \frac{x_2(1+\rho_2)}{\mu}\right]$ and $q_e = q_{ee}$. After some algebra, we obtain that $q^* < q_e$.

−− **Subcase (2d)**: If $T(1) < \frac{R}{C} \leq W_2$, then we get $q^* = \frac{1}{\rho}\left[1 - \frac{x_2(1+\rho_2)}{\mu}\right] \in [0, 1]$ and $q_e = 1$, and obviously we get that $q^* \leq q_e$.

−− **Subcase (2e)**: If $\frac{R}{C} > W_2$, then $q^* = q_e = 1$.

To summarize, the inequality of (10.2.47) holds for all cases. □

It is noted by Theorem 10.2.7 we know that the equilibrium joining probability is greater than the socially optimal joining probability when second users find the spectrum unavailable upon arrival. Intuitively, if a secondary user has the right to make decision autonomously and independently without considering other secondary users' benefit, the secondary user have more incentive to join the orbit than if it is in a cooperative environment. From the economic viewpoint, when an arriving SU decides to enter the orbit, its decision and behavior (joining the orbit) increases the delay for other SUs in the orbit. Such an effect is called "negative externality" in economics. Theorem 10.2.7 reveals a fact: the SUs that independently maximize their own benefit ignore those negative externalities that they impose on the other SUs in the orbit and later arrivals, and thus they tend to overuse the system. Nevertheless, these externalities are taken into account when all secondary users cooperate with each other and aim to maximize the sum of benefits from service minus the waiting costs per time unit. As a consequence, in the cooperative situation, fewer customers would join the system, inducing a less congested system. The resulting individual equilibrium joining probability is therefore larger than the socially optimal one.

In order to utilize the spectrum more efficiently and to eliminate the difference between the individual equilibrium and the socially optimal strategies, the system administrator may impose a positive admission fee p for the secondary users that choose to join the orbit (i.e., the reward for secondary users joining the orbit will be decreased to $R - p$ from the original R in the non-cooperative scenario). In this way, self-interested secondary users are expected to reduce their effective arrival rate.

In the sequel we consider the scenario of $\rho < 1$ and present the following Theorem 10.2.8.

Theorem 10.2.8 *In the considered CR network with* $\rho < 1$, *there exists an admission fee* $p > 0$ *such that if it is imposed for each SU that joins the orbit, then the resulting equilibrium joining probability by replacing* R *with* $R - p$ *in Eq. (10.2.25), will coincide with the socially optimal one in Eq. (10.2.36), and is given by*

$$p^* = \begin{cases} C(W_1 - T(0)), & \text{if } \frac{R}{C} < W_1, \\ p^{**}, & \text{if } W_1 \le \frac{R}{C} \le W_2, \\ C(W_2 - T(1)), & \text{if } \frac{R}{C} > W_2, \end{cases} \tag{10.2.48}$$

where

$$p^{**} = R - \frac{C\rho_2 + C(1+\rho_2)(\rho_1 - \rho_2 x_2/\mu)}{(\rho+\rho_2)x_2} - \frac{C\left(\lambda\rho_2 + \rho(\lambda + x_2)\right)}{\theta(\rho+\rho_2)x_2} - \frac{C\rho_2(1+\rho_2+\frac{\theta}{\beta})(\frac{\rho_2\mu}{1+\rho_2} + \lambda + x_2)}{\theta(1+\rho_2)(\rho+\rho_2)x_2}, \tag{10.2.49}$$

and the values of $T(0)$, $T(1)$, W_1, W_2 *along with* x_2 *are given by Eqs. (10.2.28), (10.2.29), (10.2.40), (10.2.41) and (10.2.37).*

Proof It is noticed when the reward for secondary users that join the orbit is reduced from R to $R - p$, the social objective function is the same as Eq. (10.2.35) since the admission fee p is not included in the social objective function with a transfer of income from one social group (SUs) to another (the administrator of the system). Thus, the resulting socially optimal joining probability remains the same as Eq. (10.2.36). However, the equilibrium joining probability is changed to

$$q_e = \begin{cases} 0, & \text{if } 0 < \frac{R-p}{C} < T(0), \\ q_{ee}, & \text{if } T(0) \le \frac{R-p}{C} \le T(1), \\ 1, & \text{if } \frac{R-p}{C} > T(1), \end{cases} \tag{10.2.50}$$

Accordingly, we discuss three cases as follows.

- For the case of $\frac{R}{C} < W_1$, the socially optimal joining probability $q^* = 0$. Letting $p = C(W_1 - T(0))$, then we have $\frac{R-p}{C} < T(0)$, which is the condition that the equilibrium joining probability is 0. As a consequence, we conclude that $p =$

$C(W_1 - T(0))$ is the appropriate admission fee that leads the equilibrium joining probability coincide with the socially optimal one. This gives the first part of (10.2.48).

- By a similar argument, we obtain the third part of (10.2.48).
- For the case $W_1 \leq \frac{R}{C} \leq W_2$, the socially optimal solution is $q^* = \frac{1}{\rho}\left[1 - \frac{x_2(1+\rho_2)}{\mu}\right]$. Letting $q_{ee} = \frac{1}{\rho}\left[1 - \frac{x_2(1+\rho_2)}{\mu}\right]$, then we get $p = p^{**}$ that is given by (10.2.49). It is readily seen that if such a fee is charged, $T(0) \leq \frac{R-p}{C} \leq T(1)$. Due to the second branch of (10.2.50) the resulting equilibrium joining probability is reduced to $\frac{1}{\rho}\left[1 - \frac{x_2(1+\rho_2)}{\mu}\right]$. Consequently, the equilibrium joining probability coincides with the socially optimal joining probability by imposing for each secondary user that joins the orbit. Thus the second part of (10.2.48) is attained.

□

On the other hand, for the scenario with $\rho \geq 1$, the admission fee that induces the equilibrium and the socially optimal solutions coincide is presented below by following a similar argument.

Theorem 10.2.9 *For the considered CR system with $\mu\beta \leq \lambda(\alpha + \beta)$, there exists an admission fee $p > 0$ such that if it is imposed for each secondary user that joins the orbit, then the resulting equilibrium solution coincides with the socially optimal one, and is given by*

$$p^* = \begin{cases} C(W_1 - T(0)), & \text{if } \frac{R}{C} < W_1, \\ p^{**}, & \text{if } \frac{R}{C} \geq W_1, \end{cases} \tag{10.2.51}$$

where the values of p^{**}, $T(1)$, W_1 and x_2 are provided by (10.2.49), (10.2.29), (10.2.40) and (10.2.37).

10.3 Summary

In this chapter we focused on two applications in wireless LAN and cognitive radio networks by using queueing game approach. Nowadays with rapid development of modern information technology, cloud computing, edge computing, fog computing, among others, are playing an increasingly important role in many fields. Conflicts between the individually optimal behavior and socially optimal solution become ubiquitous and are increasing sharply. In this means, game theory in general and queueing game in particular play an increasingly important role in analysing these conflicts and revealing the underlying mechanism with resource allocation and/or congestion. We mention in the cloud computing scenario there are several publications in the queueing game literature, see for example, Zhu et al. (2023), Zhu et al. (2024), among others.

References

Do, C., N. Tran, M. V. Nguyen, C. S. Hong, and S. Lee. 2012. Social optimization strategy in unobserved queueing systems in cognitive radio networks. *IEEE Communications Letters* 16: 1944–1947.

Elias, J., F. Martignon, A. Capone, and Eitan Altman. 2011. Non-cooperative spectrum access in cognitive radio networks: A game theoretical model. *Computer Networks* 55: 3832–3846.

Falin, G. I., and J. G. C. Templeton. 1997. *Retrial Queues*. London: Chapman & Hall.

Frey, A., and Y. Takahashi. 1999. An $M^X/GI/1/N$ queue with close-down and vacation times. *Journal of Applied Mathematics and Stochastic Analysis* 12: 63–83.

Hassan, M., and M. Atiquzzaman. 1997. A delayed vacation model of an M/G/1 queue with setup time and its application to SVCC-based ATM networks. *IEICE Transactions on Communications* E802B: 317–323.

Hassin, Refael, and Moshe Haviv. 2003. *To Queue or Not to Queue: Equilibrium Behavior in Queueing Systems*. Boston: Kluwer Academic Publishers.

Hossain, E., D. Niyato, and Zhu Han. 2009. *Dynamic Spectrum Access and Management in Cognitive Radio Networks*. Cambridge: Cambridge University Press.

Jagannathan, K., I. Menache, E. Modiano, and G. Zussman. 2012. Non-cooperative spectrum access-the dedicated versus free spectrum choice. *IEEE Journal on Selected Areas in Communications* 30: 2251–2261.

Kulkarni, V. G., and B. D. Choi. 1990. Retrial queues with server subject to breakdowns and repairs. *Queueing Systems* 7: 191–208.

Li, H., and Zhu Han. 2011. Socially optimal queuing control in CR networks subject to service interruptions: To queue or not to queue? *IEEE Transactions on Wireless Communications* 10: 1656–1666.

Mitola, J. III. 2001. Cognitive radio for flexible mobile multimedia communications. *Mobile Networks and Applications* 6: 435–441.

Niyato, D., and E. Hossain. 2008. Competitive spectrum sharing in cognitive radio networks: A dynamic game approach. *IEEE Transactions on Wireless Communications* 7: 2651–2660.

Niyato, D., and E. Hossain. 2008. Competitive pricing for spectrum sharing in cognitive radio networks: Dynamic game, inefficiency of Nash equilibrium, and collusion. *IEEE Journal on Selected Areas in Communications* 26: 192–202.

Sakai, Y., Y. Takahashi, Y. Takahashi, and T. Hasegawa. 1998. A composite queue with vacation/set-up/close-down times.

Stidham, S., Jr. 2009. *Optimal Design of Queueing Systems*. Taylor and Francis, Boca Raton: CRC Press.

Wang, Jinting, and Wei Wayne Li. 2016. Non-cooperative and cooperative joining strategies in cognitive radio networks with random access. *IEEE Transactions on Vehicular Technology* 65 (7): 5624–5636.

Wang, Jinting, and F. Zhang. 2013. Strategic joining in M/M/1 retrial queues. *European Journal of Operational Research* 230: 76–87.

Wang, Jinting, and Feng Zhang. 2015. Monopoly pricing in a retrial queue with delayed vacations for local area network applications. *IMA Journal of Management Mathematics* 27 (2): 315–334.

Wang, J., J. Cao, and Q. Li. 2001. Reliability analysis of the retrial queue with server breakdowns and repairs. *Queueing Systems* 38: 63–380.

Zhang, Haijun, Chunxiao Jiang, Norman C. Beaulieu, Xiaoli Chu, Xianbin Wang, and Tony Q.S.. Quek. 2015b. Resource allocation for cognitive small cell networks: A cooperative bargaining game theoretic approach. *IEEE Transactions on Wireless Communications* 14 (6): 3481–3493.

Zhang, Haijun, Chunxiao Jiang, Xiaotao Mao, and Hsiao-Hwa. Chen. 2015a. Interference-limit resource optimization in cognitive femtocells with fairness and imperfect spectrum sensing. *IEEE Transactions on Vehicular Technology* 65 (3): 1761–1771.

Zhang, Haijun, Chunxiao Jiang, Norman C. Beaulieu, Xiaoli Chu, Xiangming Wen, and Meixia Tao. 2014. Resource allocation in spectrum-sharing OFDMA femtocells with heterogeneous services. *IEEE Transactions on Communications* 62 (7): 2366–2377.

Zhu, Sheng, Jinting Wang, and Wei Wayne Li. 2023. Cloud or in-house service? Strategic joining and social optimality in hybrid service systems with retrial orbit. *IEEE Systems Journal* 17 (3): 3810–3821.

Zhu, Sheng, Jinting Wang, and Wei Wayne Li. 2024. Should opportunists be encouraged? Optimal decisions in hybrid cloud service systems. *IEEE Transactions on Network and Service Management* 21 (3): 3330–3343.

Chapter 11
Applications in Service-Inventory Systems

In today's highly competitive market of consumer products, customers have the opportunity to compare and choose among the products of a considerable number of companies. Considering the relatively high waiting cost for products and close substitutes among competitors, customer always faces one question whether to join and wait in a selling system (or online shop) upon arrival or to balk the system for another one. By a certain reward-cost function which reflects his reward after service and the inevitable sojourn cost, he seeks the individually optimal strategy to maximize his benefit against other customers who have the same benefit function. The individually optimal strategy is also called individual equilibrium strategy. Such self-organized or endogenic equilibrium unveils differences in individually-optimal and socially-optimal behavior and thus be used in modelling various issues arising from operations management and service pricing, to mention only a few. Interested readers can refer to recent works on this topic, see Wang et al. (2019, 2021a, b, 2024), Wang and Sun (2022) and the references therein.

The work on equilibrium analysis of queueing systems dates back to the seminal study of Naor (1969) who firstly study the queueing system from the game-theoretic point of view. He considered an M/M/1 queueing system where the customers are informed the queue length along with server's status and obtained customers' individually optimal strategy. On the other hand, customers in the system can be regarded as a whole and there exists an overall optimal strategy. It was showed that the individually optimal strategy is different from the overall optimal strategy. Later, Edelson and Hilderbrand (1975) extend Naor's model by assuming that customers can not observe the length of queue upon their arrivals. Since then, there is a large amount of work analyzing the strategic behavior of customers, such as unreliable queues (see Economou et al. 2008; Wang and Zhang 2011; Li et al. 2014); retrial queues (see, for example, Elcan 1994; Hassin and Haviv 1996; Economou et al. 2011; Wang and Zhang 2013); clearing queueing systems, see Boudali and Economou (2012), Economou et al. (2013); vacation queues, see Guo et al. (2011), Zhang et al. (2012),

J. Wang, *Fundamentals of Queueing-Game Models*,
https://doi.org/10.1007/978-981-95-0261-5_11

Guo et al. (2012), Liu et al. (2012), Guo et al. (2013); egalitarian processor-sharing queue, see Yu et al. (2014), among others.

To induce the selfish customer to behave in a social way, Naor (1969) also study the static price which is levied on the customers joining the queue. Edelson and Hilderbrand (1975) show that the monopoly equilibrium toll also maximizes the social welfare function in Naor's model when the customers make decisions before observing the state of the system. Mendelson and Whang (1990) derive an incentive-compatible pricing mechanism with multiple classes of customers which can maximize the expected net benefit of the system based on the decentralized individual behavior. Chen and Frank (2001) present state dependent pricing in an M/M/1 model where the firm adjusts the price according to the length of the queue. Facing homogeneous customers, the firm's pricing rule also maximizes social welfare. While for heterogenous customers, this equivalence does not hold. Pricing mechanism is also essential in other fields. For example, Chen and Wan (2003) consider a competition of two make-to-order firms; Chen and Frank (2004) formulate the pricing and service rate decision of a monopolist in a short or long run; Chen et al. (2015) investigate two compensation mechanisms when a stockout occurs; Wang and Zhang (2015) study an M/M/1 retrial queue for local area network applications. More comprehensive surveys can be found in the excellent monographs by Hassin and Haviv (2003), Stidham (2009), Hassin (2016).

Recently, the integrated service-inventory systems in which the service process is often connected with inventory management have been studied extensively. Berman and Kim (1999) investigate a service-inventory system where demands arrive the system according to a Poisson process and the service times are exponentially distributed. Under the assumptions that the capacity is finite or infinite, the authors characterized the optimal order policy for the service facility. Later on, Berman and Sapna (2000) generalize this model to the M/G/1 queue in which the reorder point is zero and yielded the closed form expressions of the stationary distribution and performance measures. Schwarz et al. (2006) establish the explicit formulas of stationary distribution of joint queue length and inventory process for an M/M/1 service-inventory system under different kinds of order policies and lost sales. Krishnamoorthy et al. (2012) consider a single server service-inventory system where the service may be interrupted, in which the inventory is replenished according to (s, S)-policy and customers retry for the service if the server is busy. Through the matrix analytic method the authors obtained steady state distributions and some performance measures. Recently, Baek and Moon (2014) study an M/M/1 service-inventory system. They derive the stationary joint distribution of the queue length and the on-hand items in product form. It should be noted that, different from aforementioned works, in Baek and Moon (2014) inventory is replenished by an external order under (r, Q)-policy or an internal production.

Starting with the work of Whitin (1955) who investigates the newsvendor problem with price dependent demand, the academic community has gradually recognized its importance and appeared many models that coordinate pricing and inventory control. Chen and Levi (2004), Chen and Levi (2006) present a finite horizon, periodic review model and an infinite horizon, continuous review model respectively. Assuming all shortages are backlogged, they established an optimal inventory policy along with pricing strategy. Beside this, Chen, Ray and Song (2006), Huh and Janakiraman (2008) consider lost sales model and also obtain joint inventory and pricing control solution. But they all assume the lead time is zero. For nonzero lead time and lost sales model, it is an intractable problem to establish joint optimal control. For a review, the reader is referred to Bijvank and Vis (2011). In this chapter, we focus on the individual equilibrium strategy, socially optimal strategy and optimal pricing separately instead of joint optimal strategy.

Upon publication of Wang et al. (2017), there is no work analyzing the equilibrium behavior of customers along with the pricing mechanism in an M/M/1 service-inventory system. The first part of this chapter is adapted from Wang et al. (2017), we first investigate the individual equilibrium strategy and the socially optimal strategy along with the optimal price in consideration of customer-choice behavior. Interestingly, it is found that the individual equilibrium joining probability (or threshold) is not always greater than the socially optimal one. Based on these results, the first part analyzes the revenue function to maximize the server's benefit. In the fully unobservable case we can find all possible optimal prices levied on the customers, while in the partially observable case we numerically obtain the optimal price because the solution of individually equilibrium threshold is implicit. It is revealed that providing more transparent information to the customers may not be more profitable to the server.

The structure of the first part is as follows. The model description is given in Sect. 11.1. Section 11.2 presents the individually and socially optimal strategies in the fully unobservable case. The optimal price imposed by the server is obtained. Section 11.3 devotes to the partially observable case. Numerical experiments are carried out in Sect. 11.4 and conclusions are summarized in Sect. 11.5.

The second part of this chapter is mainly based on Zhang and Wang (2019) in which the authors study the optimal inventory threshold for a dynamic service make-to-stock system with strategic customers. The organization of the second part is illustrated in Sect. 11.6.

11.1 Model Description

We consider a service-inventory system in which all the customers arrive according to a Poisson process with rate λ and the service times are exponentially distributed with rate μ. At the instant of service completion each customer takes one item from the inventory if any. This process is different from the Make-to-Stock queue, for example Chen et al. (2015), where the customer does not need any service time and takes one item immediately if the inventory is not empty. In our work, for ease of understanding, the service time can be regarded as time-interval of taking delivery of item from the inventory. After this, customer completes the service and leaves the system immediately.

The inventory status is a system parameter taking value from 0 to Q (Q is the capacity of the inventory), which is controlled by a predetermined continuous review policy. When the number of items decreases to zero, it takes an exponentially distributed lead time with parameter θ for the replenishment to arrive. For simplicity, the items are added to the inventory and put on the shelf upon arrival.

System manager publishes real-time system information to customers all the time, more specifically: (1) all customers are informed whether the inventory is empty or not; (2) if the inventory is not empty, the manager may disclose the number of items in inventory; (3) once the inventory is empty, a replenishment order will be sent immediately and new customers will be rejected during the lead time. Remarkably any customer, even if he is at the head of the queue, would not be served until the replenishment is finished.

Our model setting is practical and reasonable, for example, it is suitable for the operation management of a bakeshop. Considering the cookie's shelf life, shop manager always provides Q cookies everyday. These cookies will be sold out or not depending on the demands in random. In reality, stockout always arises in retail stores or supermarkets and when some commodities are out of stock, the sellers often remind customers by putting a tag on the shelf. Therefore, arriving customers have options to choose another similar one or go to another store, which can be regarded as lost sales (refer to the monograph by Zipkin 2000). Hence we consider the profit loss as a part of our objective function.

If the inventory is not empty upon arrival, customers must make a decision whether to join or to balk the system based on the information available. However, it is unsuitable for him to make decision just according to the number of customers before him and the number of items on shelf. The reason is that the inventory may be empty before he could receive the service. The graphical description of this system is presented in Fig. 11.1.

The state of system at time t is characterized by the pair $(J(t), K(t))$, where $J(t)$ denotes the number of customers in system and $K(t)$ denotes the number of items on shelf. It is clear that the process $\{(J(t), K(t)),\ t \geq 0\}$ is a continuous time

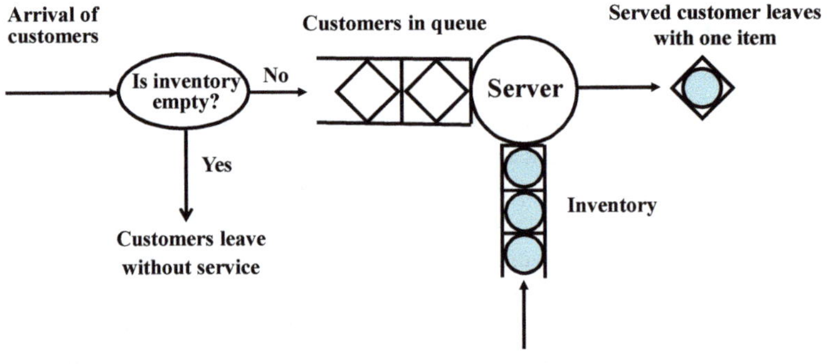

Fig. 11.1 The service-inventory system

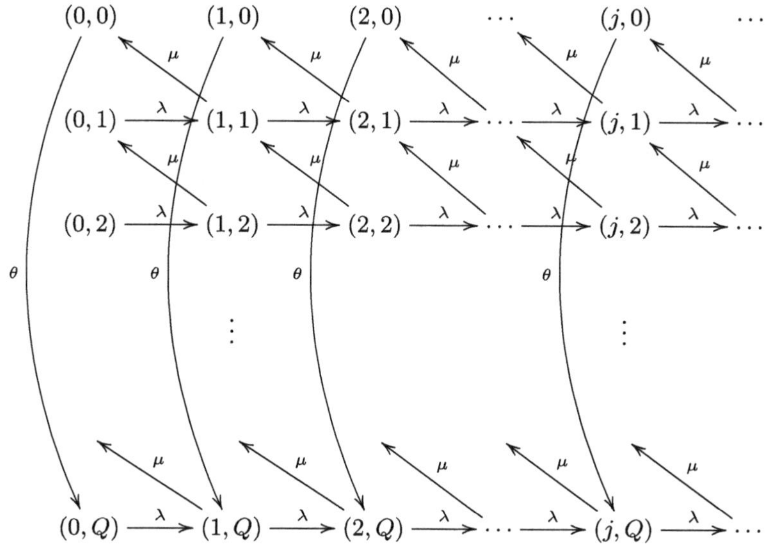

Fig. 11.2 Transition rate diagram of the service-inventory system

Markov chain with state space $\Omega = \{(j, k) | \ j \geq 0, \ k = 0, 1, \ldots, Q\}$. The transition rate diagram is shown in Fig. 11.2.

It is assumed that each customer receives a reward of R units after service completion, and he has to pay a price p $(0 \leq p < R)$ for an item. A waiting cost of h units per time unit is continuously accumulated from the time that a joining customer enters the system until he leaves the system after service, including lead times if any. Customers must make a "to purchase or not to purchase" decision based on their perception of the system's long-run performance measure, i.e., the expected waiting time. They are risk neutral, i.e., they maximize the expected value of their net benefit. A decision to join is irrevocable, which means that retrials of balking customer and reneging are not allowed.

In this work we mainly investigate the following two cases: (1) Fully unobservable case: Customers can not observe $J(t)$ and $K(t)$. (2) Partially observable case: Customers can observe $J(t)$, but can not observe $K(t)$.

11.2 Fully Unobservable Case

In this case, customer can observe neither the number of customers in system nor the number of items on shelf upon arrival. Since all customers in system are homogeneous, we suppose that all the customers follow a mixed strategy "enter with probability q and balk with probability $1 - q$ while the inventory is not empty". The transition diagram is shown in Fig. 11.3.

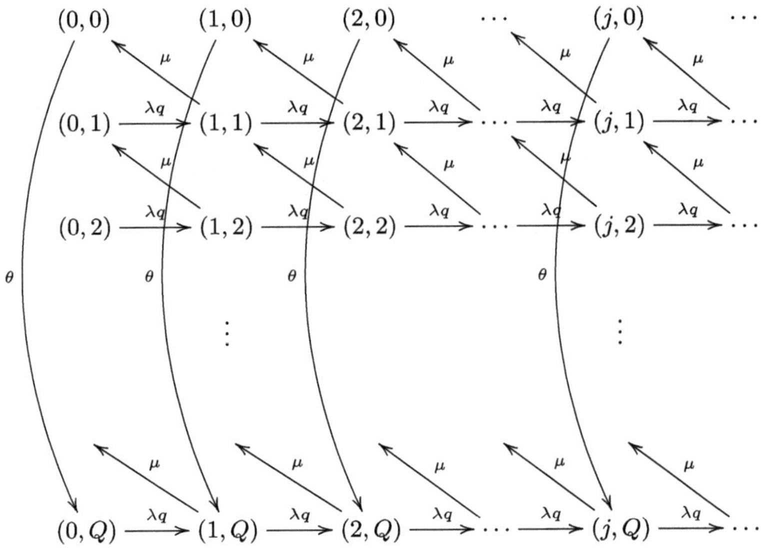

Fig. 11.3 The service-inventory system in the fully unobservable case

We focus on the following three issues: (1) To analyze the individual behavior of customer who is selfish. (2) To derive the optimal price levied on the customers to maximize the server's revenue. (3) To derive the socially optimal strategy. It is shown that the individually optimal strategy and the socially optimal strategy are all weakly dominant strategies. In general, the individual equilibrium strategy does not coincide with the socially optimal strategy.

In the sequel, we give the effective arrival rate λ_A, the expected number of customers in the system $\mathbb{E}[N]$, the expected sojourn time in the system $\mathbb{E}[W]$ by the steady state probabilities $p(j,k)$ of the system. It is evident that the system is stable if and only if $\lambda q < \mu$. For simplicity, we only analyze the case $\lambda < \mu$ and the corresponding results for the case $\lambda \geq \mu$ are summarized as remarks.

Proposition 11.2.1 *In the fully unobservable case of the M/M/1 service-inventory system, the steady state probabilities* $\{p(j,k) : (j,k) \in \Omega\}$ *are given as follows:*

$$p(0,0) = \left(1 - \frac{\lambda q}{\mu}\right) \frac{\lambda q}{\lambda q + Q\theta}, \tag{11.2.1}$$

$$p(j,k) = \left(\frac{\lambda q}{\mu}\right)^j \frac{\theta}{\lambda q} p(0,0), \quad j \geq 0, \quad k = 1, 2, \ldots, \ Q, \tag{11.2.2}$$

$$p(j,0) = \left(\frac{\lambda q}{\mu}\right)^j p(0,0), \quad j \geq 1, \tag{11.2.3}$$

$$p(j) = \sum_{k=0}^{Q} p(j,k) = \left(1 - \frac{\lambda q}{\mu}\right)\left(\frac{\lambda q}{\mu}\right)^j, \quad j \geq 0. \tag{11.2.4}$$

Proof The argument is similar to that of Theorem 2.5 in Schwarz et al. (2006) and we omit it. □

Using the steady state probabilities (11.2.1)–(11.2.4) and Little's Law, we have

$$\lambda_A = \lambda q \left[1 - \sum_{j=0}^{\infty} p(j,0)\right] = \frac{\lambda q Q\theta}{\lambda q + Q\theta}, \tag{11.2.5}$$

$$\mathbb{E}[N] = \sum_{j=1}^{\infty} j p(j) = \frac{\lambda q}{\mu - \lambda q}, \tag{11.2.6}$$

$$\mathbb{E}[W] = \frac{\mathbb{E}[N]}{\lambda_A} = \frac{\lambda q + Q\theta}{Q\theta(\mu - \lambda q)}. \tag{11.2.7}$$

Remark 11.2.1 We observe that the Eqs. (11.2.4) and (11.2.6) are the same as the formulae of the M/M/1 queue shown in Section 2.2 of Shortle et al. (2018). However, the expected sojourn time $\mathbb{E}[W]$ is not the case because of the effective arrival rate λ_A. When we let $Q \to \infty$ or $\theta \to \infty$, which means the number of items per replenishment order is infinite or the lead time approaches to zero, this model behaves like the M/M/1 queue in Shortle et al. (2018). Similarly, so do the Eqs. (11.2.5)–(11.2.7).

11.2.1 Individual Equilibrium Strategy

Theorem 11.2.1 *In the fully unobservable case of the M/M/1 service-inventory system, a unique mixed strategy 'enter with probability q_e' exists and it is given by*

$$q_e = \begin{cases} 0, & \text{if } R - p \le \frac{h}{\mu}, \\ q_e^\star, & \text{if } \frac{h}{\mu} < R - p < h\left[\frac{1}{\mu-\lambda} + \frac{\lambda}{Q\theta(\mu-\lambda)}\right], \\ 1, & \text{if } R - p \ge h\left[\frac{1}{\mu-\lambda} + \frac{\lambda}{Q\theta(\mu-\lambda)}\right], \end{cases} \tag{11.2.8}$$

where $q_e^\star = \frac{Q\theta[(R-p)\mu - h]}{\lambda[(R-p)Q\theta + h]}$.

Proof When a tagged customer decides to enter the system at his arrival instant, the expected net benefit is

$$I_{fu}(q) = R - p - h\mathbb{E}[W] = R - p - h\left[\frac{\lambda q + Q\theta}{Q\theta(\mu - \lambda q)}\right],$$

and it is easy to show that $I_{fu}(q)$ is strictly decreasing in $q \in [0, 1]$.

- If $I_{fu}(0) = R - p - \frac{h}{\mu} \le 0$, the maximum benefit is non-positive. Customers do not enter the system even if there is no customer in front of him and the inventory is not empty. The first part of (11.2.8) is proved.

- If $I_{fu}(1) = R - p - h\left[\frac{1}{\mu-\lambda} + \frac{\lambda}{Q\theta(\mu-\lambda)}\right] \geq 0$ or $R - p \geq h\left[\frac{1}{\mu-\lambda} + \frac{\lambda}{Q\theta(\mu-\lambda)}\right]$, customers prefer to join the system because the minimal benefit is non-negative. The last part of (11.2.8) is at hand.
- If $\frac{h}{\mu} < R - p < h\left[\frac{1}{\mu-\lambda} + \frac{\lambda}{Q\theta(\mu-\lambda)}\right]$, there exists a unique $q_e^\star \in (0, 1)$ such that $I_{fu}(q) = 0$. In this case, the tagged customer is indifferent between joining and balking the system.

This completes the proof. □

From Theorem 11.2.1 we know that $I_{fu}(q)$ is a decreasing function. When the joining probability q adopted by other customers is smaller than q_e, the expected net benefit of an arriving customer is positive if he chooses to enter the system, thus the unique best response is 1. Conversely, the unique best response is 0 if $q > q_e$ because that the expected net benefit is negative. If $q = q_e$, every strategy is the best response since the expected net benefit is always zero. This behavior illustrates a situation that an individual's best response is a decreasing function of the strategy selected by the other customers. Therefore, we have an 'avoid the crowd' (ATC) situation. The rationale behind this terminology is that the higher the value selected by the others, the lower one's best response becomes.

Remark 11.2.2 When $\lambda \geq \mu$, the result in Theorem 11.2.1 also holds by following a similar analysis and the unique mixed strategy 'enter with probability q_e' is given by

$$q_e = \begin{cases} 0, \text{ if } & R - p \leq \frac{h}{\mu}, \\ q_e^\star, \text{ if } & \frac{h}{\mu} < R - p. \end{cases}$$

11.2.2 Revenue Maximization

In Sect. 11.2.1, Eq. (11.2.8) shows that there exists an equilibrium strategy q_e for customers, which depends on the value of p. It is readily seen that q_e is strictly decreasing in p, which means that a large proportion of customers will choose to enter the system for service if a lower p is levied. In this subsection, we will seek an optimal price p to maximize the server's revenue.

According to the description of the system, the replenishment order would be triggered only when inventory is empty. Using Eqs. (11.2.1)–(11.2.3), the effective replenishment order rate θ_{Re} per time unit can be obtained as

$$\theta_{Re} = \theta \sum_{j=0}^{\infty} p(j, 0) = \frac{\lambda q \theta}{\lambda q + Q\theta}, \tag{11.2.9}$$

the expected number of items in inventory is

$$Q_I = \sum_{k=1}^{Q} k \sum_{j=0}^{\infty} p(j,k) = \frac{Q\theta(Q+1)}{2(\lambda q + Q\theta)}, \tag{11.2.10}$$

and the loss rate λ_L of customers per time unit can be solved by

$$\lambda_L = \lambda q - \lambda_A = \frac{(\lambda q)^2}{\lambda q + Q\theta}.$$

So the revenue function of the sever is given by

$$\begin{aligned} U_{fu}(p) = & \; p \cdot \lambda_A - c \cdot \theta_{Re} - f \cdot Q_I - l \cdot \lambda_L = (p+l)\lambda_A - c\theta_{Re} - fQ_I - l\lambda q \\ = & \; (p+l)\frac{\lambda q Q\theta}{\lambda q + Q\theta} - c\frac{\lambda q \theta}{\lambda q + Q\theta} - f\frac{Q\theta(Q+1)}{2(\lambda q + Q\theta)} - l\lambda q, \end{aligned} \tag{11.2.11}$$

where c is the cost for every replenishment order, f is the holding cost for one item per time unit, l is the loss cost of each lost customer and $q \triangleq q_e(p)$ defined as a function on p is the equilibrium joining probability in Eq. (11.2.8). Let $p_1 \triangleq R - h[\frac{1}{\mu-\lambda} + \frac{\lambda}{Q\theta(\mu-\lambda)}]$, $p_2 \triangleq R - \frac{h}{\mu}$, and let $p_u^{\star} \in \big(\max\{0, p_1\}, p_2\big)$ be the solution of $\frac{dU_{fu}(p)}{dp} = 0$ if it exists. And we have $p_1 < p_2$ from Theorem 11.2.1.

With full awareness of customers' strategy behavior, we set p as the decision variable and seek an optimal price to maximize the revenue function. If $p \in (p_2, R)$, we have $q_e(p) = 0$ from Eq. (11.2.8). In this case, customers never enter the system and the revenue function degenerates to $-\frac{f(Q+1)}{2}$, which means the expected holding cost of all the items per time unit. Obviously, the server does not expect this situation.

Next, we discuss the situation $p \in [0, p_2]$ in two cases.

Case 1: For $p_1 > 0$. In this case, (i) When $0 \le p \le p_1$, $q_e(p) = 1$ and it is evident from Eq. (11.2.11) that the optimal price p_u is p_1. (ii) When $p_1 \le p \le p_2$, $q_e(p) = \frac{Q\theta[(R-p)\mu-h]}{\lambda[(R-p)Q\theta+h]}$. Considering the continuity of $U_{fu}(p)$, the value of p_u can be chosen from p_1, p_2, and $p_u^{\star}$ if they exist. To summarize, the optimal price p_u can be chosen from p_1, p_2, and $p_u^{\star}$.

Case 2: For $p_1 \le 0$. In this case, when $0 \le p \le p_2$, we have $q_e(p) = \frac{Q\theta[(R-p)\mu-h]}{\lambda[(R-p)Q\theta+h]}$. Similar to the case 1, considering the continuity of $U_{fu}(p)$, the optimal price p_u can be chosen from 0, p_2, and $p_u^{\star}$ if they exist.

The main results based on the above analysis are summarized in the following theorem.

Theorem 11.2.2 *In the fully unobservable case of the M/M/1 service-inventory system, the optimal price levied on the customer is* $p_u = \arg\max\Big\{U_{fu}(p), p \in \big\{\max\{p_1, 0\},\ p_2,\ p_u^{\star}\big\}\Big\}$ *which can maximize the server's revenue* $U_{fu}(p)$ *in* $p \in [0, p_2]$.

Remark 11.2.3 When $\lambda \geq \mu$, the optimal price levied on the customer that maximizes the server's revenue $U_{fu}(p)$ in $p \in [0, p_2]$ is $p_u = \arg\max\left\{U_{fu}(p), p \in \{0,\ p_2,\ p_u^{\star}\}\right\}$.

11.2.3 Socially Optimal Strategy

Now we proceed to the socially optimal strategy. For a given price p and a joining probability q, the surplus of all customers is $S_1 = \lambda_A(R - p) - h\mathbb{E}[N]$ and the server revenue is $S_2 = p\lambda_A - c\theta_{Re} - fQ_I - l\lambda_L$. Considering Eqs. (11.2.5)–(11.2.6) and (11.2.9)–(11.2.10), we have the following expected social welfare per time unit:

$$\begin{aligned} S_{fu}(q) &= S_1 + S_2 = (R+l)\lambda_A - hE[N] - c\theta_{Re} - fQ_I - l\lambda q \\ &= (R+l)\frac{\lambda q Q\theta}{\lambda q + Q\theta} - h\frac{\lambda q}{\mu - \lambda q} - c\frac{\lambda q\theta}{\lambda q + Q\theta} - f\frac{Q\theta(Q+1)}{2(\lambda q + Q\theta)} - l\lambda q \end{aligned} \tag{11.2.12}$$

Theorem 11.2.3 *In the fully unobservable case of the M/M/1 service-inventory system, a unique mixed strategy 'enter with probability q_s' which maximizes the social welfare per time unit exists and the probability is given by*

$$q_s = \begin{cases} 0, & \text{if } R \leq \frac{h}{\mu} + \frac{c}{Q} - \frac{f(Q+1)}{2Q\theta}, \\ \min\{1, q_s^{\star}\}, & \text{if } \frac{h}{\mu} + \frac{c}{Q} - \frac{f(Q+1)}{2Q\theta} < R, \end{cases}$$

where $q_s^{\star}$ satisfies $\frac{dS_{fu}(q)}{dq} = 0$.

Proof According to the Eq. (11.2.12), we have

$$\frac{dS_{fu}(q)}{dq} = \frac{(R+l)\lambda(Q\theta)^2}{(\lambda q + Q\theta)^2} - \frac{h\lambda\mu}{(\mu - \lambda q)^2} - \frac{c\lambda Q\theta^2}{(\lambda q + Q\theta)^2} + \frac{f\lambda Q\theta(Q+1)}{2(\lambda q + Q\theta)^2} - l\lambda$$

and

$$\frac{d^2S_{fu}(q)}{dq^2} = -\frac{2(R+l)(\lambda Q\theta)^2}{(\lambda q + Q\theta)^3} - \frac{2h\lambda^2\mu}{(\mu - \lambda q)^3} + \frac{2cQ(\lambda\theta)^2}{(\lambda q + Q\theta)^3} - \frac{f\lambda^2 Q\theta(Q+1)}{(\lambda q + Q\theta)^3}.$$

Case 1: If $R < \frac{c}{Q} - \frac{f(Q+1)}{2Q\theta} - l$, then $\frac{dS_{fu}(q)}{dq} < 0$ and $S_{fu}(q)$ reaches its maximum at $q_s = 0$.

Case 2: If $R \geq \frac{c}{Q} - \frac{f(Q+1)}{2Q\theta} - l$, then $\frac{d^2S_{fu}(q)}{dq^2} < 0$ and moreover $S_{fu}(q)$ is strictly concave in $q \in [0, 1]$;

Case 2(i): If $\frac{c}{Q} - \frac{f(Q+1)}{2Q\theta} - l \leq R \leq \frac{h}{\mu} + \frac{c}{Q} - \frac{f(Q+1)}{2Q\theta}$, then $\frac{dS_{fu}(q)}{dq}|_{q=0} = \lambda(R + l - \frac{h}{\mu} - \frac{c}{Q} + \frac{f(Q+1)}{2Q\theta}) \leq 0$ and $S_{fu}(q)$ also reaches its maximum at $q_s = 0$;

Case 2(ii): If $R > \frac{h}{\mu} + \frac{c}{Q} - \frac{f(Q+1)}{2Q\theta}$, then $\frac{dS_{fu}(q)}{dq}|_{q=0} > 0$ and denote $q_s^\star$ be the unique solution of $\frac{dS_{fu}(q)}{dq} = 0$. Hence $S_{fu}(q)$ obtains its maximum at $q_s = \min\{1, q_s^\star\}$. This completes the proof. □

11.3 Partially Observable Case

In this case customers can not observe the number of items, but they can observe the number of customers in the system upon arrivals. We seek for the individual equilibrium and socially optimal strategies within the class of pure threshold strategies. Due to the implicit expression of individual equilibrium strategy, we will numerically investigate optimal revenue price p_o of server like in Sect. 11.2.2.

Given a threshold n, the transition rate diagram is depicted in Fig. 11.4. To ensure a customer enter the system when there is no customer in front of him and the inventory is not empty, the condition

$$R - p - \frac{h}{\mu} > 0 \tag{11.3.1}$$

is necessary.

11.3.1 Individual Equilibrium Strategy

In this subsection, we derive the individual equilibrium threshold n_e of customers. To this end, we present the steady state probabilities of the system in the following Proposition. By replacing n with n_e in Fig. 11.4, the corresponding transition rate diagram on threshold n_e is at hand and is omitted here.

Proposition 11.3.1 *In the partially observable M/M/1 service-inventory system, where a customer enters the system according to a pure threshold strategy 'while arriving at time t, observes the number of customers $J(t)$ ahead of him, enters if $J(t) \le n_e$ and balks otherwise', the steady state probabilities $\big(p(j,k) : (j,k) \in \Omega_{n_e} = \big\{\{0, 1, \ldots, n_e + 1\} \times \{0, 1, \ldots, Q\}\backslash(n_e + 1, 0)\big\}\big)$ on threshold n_e are given as follows:*

$$p_{n_e}(0,0) = \left\{\frac{\mu}{\mu-\lambda}(1+\frac{Q\theta}{\lambda}) - (\frac{\mu+Q\theta}{\mu-\lambda})(\frac{\lambda}{\mu})^{n_e+1}\right\}^{-1}, \tag{11.3.2}$$

$$p_{n_e}(j,k) = \left(\frac{\lambda}{\mu}\right)^j \frac{\theta}{\lambda} p_{n_e}(0,0), \quad j = 0, 1, \ldots, n_e + 1, \ k = 1, 2, \ldots, Q, \tag{11.3.3}$$

$$p_{n_e}(j,0) = \left(\frac{\lambda}{\mu}\right)^j p_{n_e}(0,0), \quad j = 1, 2, \ldots, n_e.$$

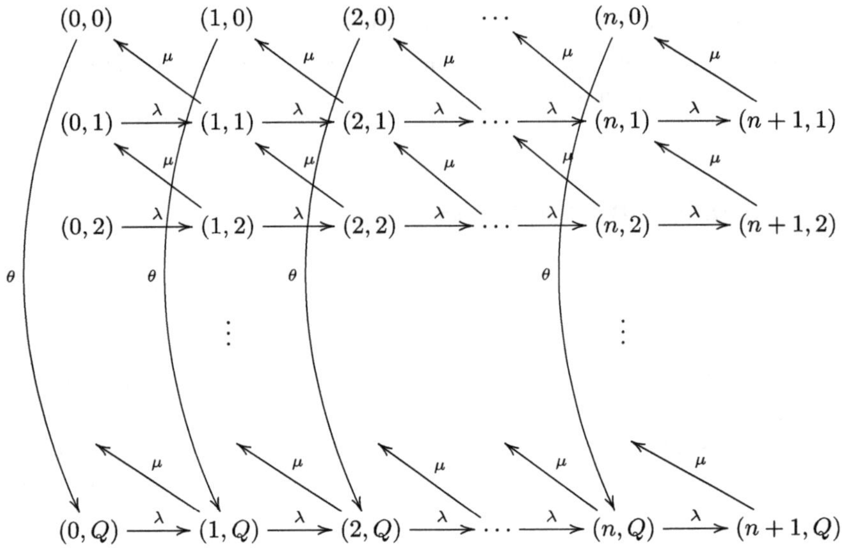

Fig. 11.4 The service-inventory system in the partially observable case

Proof The proof is similar to the one in Theorem 6.2 of (2006) and we omit it here. □

Based on Proposition 11.3.1, we now establish the expected sojourn time in the system of a tagged customer who decides to join the system after observing j customers ahead of him upon arrival.

Lemma 11.3.1 *In the partially observable M/M/1 service-inventory system, where a customer enters the system according to a pure threshold strategy "while arriving at time t, observes $J(t)$; enters if $J(t) \leq n_e$ and balks otherwise", the expected sojourn time of a customer who observes j customers ahead of him upon arrival and decides to enter is given by*

$$T(j) = \frac{1}{Q}\sum_{k=1}^{Q}\Big(\frac{j+1}{\mu} + \lceil\frac{j+1-k}{Q}\rceil \cdot \frac{1}{\theta}\Big), \tag{11.3.4}$$

for $j = 0, 1, \ldots, n_e$. Furthermore, $T(j)$ is a strictly increasing function in j.

Proof First, we denote $\Pr(K(t) = k|J(t) = j)$ by the probability that the number of items is k when the tagged customer observes j customers in the system at arrival instant t. Conditioning on the number of items that he finds upon arrival, we have

$$T(j) = \sum_{k=1}^{Q} T(j,k)\Pr(K(t) = k|J(t) = j), \tag{11.3.5}$$

where

$$T(j,k) = \frac{j+1}{\mu} + \left\lceil \frac{j+1-k}{Q} \right\rceil \cdot \frac{1}{\theta} \tag{11.3.6}$$

is the expected sojourn time given that he finds the system at state (j, k) upon arrival. Due to the PASTA property with Eq. (11.3.3), we have

$$\Pr(K(t) = k | J(t) = j) = \frac{p_{n_e}(j,k)}{\sum_{k=1}^{Q} p_{n_e}(j,k)} = \frac{1}{Q} \tag{11.3.7}$$

for $k = 1, 2, \ldots, Q$ and $j = 0, 1, \ldots, n_e$. Inserting Eqs. (11.3.6) and (11.3.7) into (11.3.5), we have the result in Eq. (11.3.4).

Next, we have

$$T(j+1) - T(j) = \frac{1}{Q} \sum_{k=1}^{Q} \left\{ \frac{1}{\mu} + \left(\lceil \frac{j+2-k}{Q} \rceil - \lceil \frac{j+1-k}{Q} \rceil \right) \cdot \frac{1}{\theta} \right\} > 0,$$

which shows that $T(j)$ is a strictly increasing function in j. This completes the proof. □

Theorem 11.3.1 *In the partially observable M/M/1 service-inventory system, there exists a unique individual equilibrium threshold* n_e.

Proof With the help of Eq. (11.3.4), the net benefit of a customer who observes j customers ahead of him and decides to enter is given by

$$I_{ao}(j) = R - p - hT(j) = R - p - \frac{h}{Q} \sum_{k=1}^{Q} \left(\frac{j+1}{\mu} + \lceil \frac{j+1-k}{Q} \rceil \cdot \frac{1}{\theta} \right). \tag{11.3.8}$$

It is obvious that $I_{ao}(j)$ is strictly decreasing in j from Lemma 11.3.1 and Eq. (11.3.8). Considering $I_{ao}(0) = R - p - \frac{h}{\mu} > 0$ and $\lim_{j\to\infty} I_{ao}(j) = -\infty$, there exists a unique real number x_e such that $I_{ao}(x_e) = 0$. Denote $n_e = \lfloor x_e \rfloor$ which satisfies $I_{ao}(n_e) \geq 0$ and $I_{ao}(n_e + 1) < 0$. This completes the proof. □

Given the individually optimal threshold n_e defined above, similar to Sect. 11.2.2, the server seeks an optimal price to maximize the following revenue function

$$U_{ao}(p) = \lambda p \sum_{j=0}^{n_e} p_{n_e}(j) - c\theta \sum_{j=0}^{n_e} p_{n_e}(j,0) - f \sum_{k=1}^{Q} k \sum_{j=0}^{n_e+1} p_{n_e}(j,k) - l\lambda \sum_{j=0}^{n_e} p_{n_e}(j,0),$$

where

$$p_{n_e}(j) = \sum_{k=1}^{Q} p_{n_e}(j, k),$$

the parameters $c,\ f,\ l$ have been defined in Sect. 11.2.2, $\lambda p \sum_{j=0}^{n_e} p_{n_e}(j)$ is the expected net benefit of server, $c\theta \sum_{j=0}^{n_e} p_{n_e}(j, 0)$ is the replenishment order cost, $f \sum_{k=1}^{Q} k \sum_{j=0}^{n_e+1} p_{n_e}(j, k)$ is the holding cost of all items and $l\lambda \sum_{j=0}^{n_e} p_{n_e}(j, 0)$ is the loss cost of lost customers. Since the explicit expression of n_e is not easy to obtain, we manage to find the optimal price p_o numerically to maximize the server's revenue by using Particle Swarm Optimization (PSO) algorithm (refer to Kennedy and Eberhart 1995 and Carlisle and Dozier 2001, Sect. 5).

11.3.2 Socially Optimal Strategy

To maximize the social welfare, the number of customers $J(t)$ in system should be controlled to obtain the social optimum. For a pre-specified threshold n, an arriving customer is permitted to enter the system when $J(t) \le n$ and is rejected otherwise. Similar to Proposition 11.3.1, the steady state probabilities on threshold n are given by

$$p_n(0,0) = \left\{\frac{\mu}{\mu-\lambda}(1+\frac{Q\theta}{\lambda}) - (\frac{\mu+Q\theta}{\mu-\lambda})(\frac{\lambda}{\mu})^{n+1}\right\}^{-1}, \tag{11.3.9}$$

$$p_n(j,k) = \left(\frac{\lambda}{\mu}\right)^{j}\frac{\theta}{\lambda}p_n(0,0),\quad j=0,1,\ldots,n+1,\ k=1,2,\ldots,\ Q, \tag{11.3.10}$$

$$p_n(j,0) = \left(\frac{\lambda}{\mu}\right)^{j} p_n(0,0),\quad j=1,2,\ldots,n, \tag{11.3.11}$$

$$p_n(j) = \sum_{k=1}^{Q} p_n(j,k). \tag{11.3.12}$$

Considering the surplus of all customers

$$F_{ao}(n) = \sum_{j=0}^{n} p_n(j)\lambda[R - p - hT(j)] \tag{11.3.13}$$

and the revenue of the server $\lambda p \sum_{j=0}^{n} p_n(j) - c\theta \sum_{j=0}^{n} p_n(j,0) - f \sum_{k=1}^{Q} k \sum_{j=0}^{n+1} p_n(j,k) - l\lambda \sum_{j=0}^{n} p_n(j,0)$, the socially optimal threshold n_s is determined by the following social welfare function per time unit

$$
\begin{aligned}
&S_{ao}(n)\\
&=\sum_{j=0}^{n} p_n(j)\lambda[R-p-hT(j)]+\lambda p\sum_{j=0}^{n} p_n(j)\\
&\quad-(c\theta+l\lambda)\sum_{j=0}^{n} p_n(j,0)-f\sum_{k=1}^{Q} k\sum_{j=0}^{n+1} p_n(j,k)\\
&=\sum_{j=0}^{n} p_n(j)\lambda[R-hT(j)]-\frac{(c\theta+l\lambda)\mu}{\mu-\lambda}\left[1-\left(\frac{\lambda}{\mu}\right)^{n+1}\right]p_n(0,0)\\
&\quad-\frac{f\mu\theta Q(Q+1)}{2\lambda(\mu-\lambda)}\left[1-\left(\frac{\lambda}{\mu}\right)^{n+2}\right]p_n(0,0)\\
&=S_1(n)+S_2(n),
\end{aligned}
$$

where

$$
\begin{aligned}
S_1(n) &= \sum_{j=0}^{n} p_n(j)\lambda[R-hT(j)],\\
S_2(n) &= -\frac{(c\theta+l\lambda)\mu}{\mu-\lambda}\left[1-\left(\frac{\lambda}{\mu}\right)^{n+1}\right]p_n(0,0)\\
&\quad-\frac{f\mu\theta Q(Q+1)}{2\lambda(\mu-\lambda)}\left[1-\left(\frac{\lambda}{\mu}\right)^{n+2}\right]p_n(0,0).
\end{aligned}
$$

Next we determine the maximum point n_s of $S_{ao}(n)$. To this end, we show that $S_1(n)$ is discretely unimodal function and $S_2(n)$ captures monotonicity along with concavity in n under some condition separately.

Following the technique in Knudsen (1972) or Guo and Hassin (2011), we first prove that $F_{ao}(n)$ is discretely unimodal function in n, reach its maximum at n_o and then establish the relationship between n_e and n_o. Due to the characteristic of two functions $S_1(n)$ and $F_{ao}(n)$, it is ready to obtain the similar quality of $S_1(n)$.

For a discretely unimodal function, a local maximum is a global maximum. Using the recursive method, we successively evaluate the function value $S_1(n)$ with discretely variable n until the maximum point $\underline{n_s}$ is achieved, i.e., we need to find the $\underline{n_s}$ such that the following inequalities hold:

$$
S_1(\underline{n_s}-1)\le S_1(\underline{n_s}) \quad \text{and} \quad S_1(\underline{n_s})\ge S_1(\underline{n_s}+1).
$$

Theorem 11.3.2 *(1) $F_{ao}(n)$ is discretely unimodal and it has a unique maximum point n_o, furthermore, $n_o \le n_e$;*

(2) $S_1(n)$ is discretely unimodal and it has a unique maximum point $\underline{n_s}$, furthermore, $\underline{n_s} \le n_e$.

The proof of Theorem 11.3.2 is omitted here, interested readers may refer to Wang and Zhang (2017) for details. Next, according to the presentation of $S_2(n)$, the following result is reachable.

Lemma 11.3.2 *(1) $S_2(n)$ is strictly increasing (decreasing) in n if and only if $\frac{f(Q+1)}{2} > (<)\ c\theta + l\lambda$;*

(2) $S_2(n)$ is strictly concave (convex) function in n if and only if $\frac{f(Q+1)}{2} > (<)\ c\theta + l\lambda$.

Proof (1) Based on the definitions of $S_2(n)$ and $p_n(0,0)$, we have

$$\begin{aligned} & S_2(n+1) - S_2(n) \\ &= \frac{[\frac{f(Q+1)}{2} - (c\theta + l\lambda)]\theta Q\lambda^{n+2}\mu^{n+2}(\lambda-\mu)^2}{[(\lambda+\theta Q)\mu^{n+2} - (\mu+\theta Q)\lambda^{n+2}][(\lambda+\theta Q)\mu^{n+3} - (\mu+\theta Q)\lambda^{n+3}]}. \end{aligned} \tag{11.3.14}$$

Obviously the denominator of Eq. (11.3.14) is positive, hence $S_2(n)$ is strictly increasing (decreasing) in n if and only if $\frac{f(Q+1)}{2} > (<)\ c\theta + l\lambda$.

(2)

$$\begin{aligned} & S_2(n+2) - 2S_2(n+1) + S_2(n) \\ &= \frac{[\frac{f(Q+1)}{2} - (c\theta + l\lambda)](\lambda-\mu)^3\theta Q\lambda^{n+2}\mu^{n+2}[(\lambda+\theta Q)\mu^{n+3} + (\mu+\theta Q)\lambda^{n+3}]}{[(\lambda+\theta Q)\mu^{n+2} - (\mu+\theta Q)\lambda^{n+2}][(\lambda+\theta Q)\mu^{n+3} - (\mu+\theta Q)\lambda^{n+3}][(\lambda+\theta Q)\mu^{n+4} - (\mu+\theta Q)\lambda^{n+4}]}. \end{aligned}$$

Therefore $S_2(n+2) - 2S_2(n+1) + S_2(n) < (>)\ 0 \Leftrightarrow \frac{f(Q+1)}{2} > (<)\ c\theta + l\lambda$, i.e., $S_2(n)$ is strictly concave (convex) in n if and only if $\frac{f(Q+1)}{2} > (<)\ c\theta + l\lambda$. This completes the proof. □

Since $S_1(n)$ is discretely unimodal, it reaches its maximum point at $\underline{n_s}$. While $S_2(n)$ is strictly increasing and concave in n if and only if $\frac{f(Q+1)}{2} > c\theta + l\lambda$, it is easy to verify the following conclusion.

Theorem 11.3.3 *If the expected social welfare $S_{ao}(n)$ has the maximum point n_s, then $n_s \geq \underline{n_s}$ when $\frac{f(Q+1)}{2} > c\theta + l\lambda$.*

11.4 Numerical Experiments

In this section we show the tendency of joining probabilities q_e, q_s, the thresholds n_e, n_s, the optimal prices p_u and p_o, the social optimums $S_{fu}(q_s)$, $S_{ao}(n_s)$ and the revenue maximums $U_{fu}(p_u)$, $U_{ao}(p_o)$ with respect to the arrival rate λ, the service rate μ, the customer reward R and the waiting cost h in fully unobservable case and in partially observable case.

The primary parameters along with their definitions and measurement units are summarized in Table 11.1. In the numerical examples, some parameters are fixed with values as follows: $\theta = 0.3$, $c = 2$, $f = 0.01$, $l = 0.05$, $Q = 100$. Other parameters' values and the tendency of performance measures can be seen in Figs. 11.5, 11.6, 11.7, 11.8 and 11.9. In what follows, the units of all parameters involved have been omitted for simplicity.

Table 11.1 Definition and measurement unit of each variable, parameter and symbol

	Definition	Measurement unit
R	Reward after service per customer	\$/per customer
$\lambda(\lambda_A, \lambda_L)$	The expected(effective, loss) arrival rate of customers per time unit	Customer number/per time unit
μ	Service rate of server per time unit	Customer number/per time unit
$\theta(\theta_{Re})$	The expected(effective) replenishment order rate per time unit	Order number/per time unit
h	Waiting cost per customer per time unit	\$/(per customer·per time unit)
c	Cost per replenishment order	\$/per order
f	Holding cost per item per time unit	\$/(per item·per time unit)
l	Loss cost per lost customer	\$/per customer
p	Price per item in inventory	\$/per item
Q	Number of items per replenishment order	Item number/per order
$E[N]$	The expected number of customers in the system	Customer number
$E[W]$	The expected sojourn time in the system	Time length
Q_I	The expected number of items in inventory	Item number

Firstly, Figs. 11.5 and 11.6 show us the tendency of $q_e,\ q_s$ and $n_e,\ n_s$ with respect to $\lambda,\ \mu,\ R,\ h$. Firstly, less customers are willing to join the queue for a greater value of λ, so q_e and n_e are decreasing in Figs. 11.5a and 11.6a. Meanwhile, to avoid the excessive congestion and obtain the social optimum, q_s and n_s are also decreasing in λ. The same situation can be seen in Figs. 11.5d and 11.6d. Since the service level can be improved by adjusting μ, we find $q_e,\ q_s$ and $n_e,\ n_s$ are increasing in μ from the Figs. 11.5b and 11.6b. Similar phenomena can also be found in Figs. 11.5c and 11.6c due to the greater reward R. In these cases, regardless of the number of customers in system is given or not, the joining probabilities and the thresholds have the same tendency.

Secondly, based on different starting point, individually optimal customers always consider their own benefits and ignore the external effect incurred on the others. Therefore individual equilibrium joining probability (or threshold) should be no less than social optimum one. This tendency can be seen in Figs. 11.5a, c, d and 11.6b, d. However, in our model, we can also find $q_s > q_e$ when $\mu = 0.6$ in Fig. 11.5b, $n_s > n_e$ when $\lambda = 0.1,\ 0.3$ in Fig. 11.6a or $R = 1.5$ in Fig. 11.6c. This is different

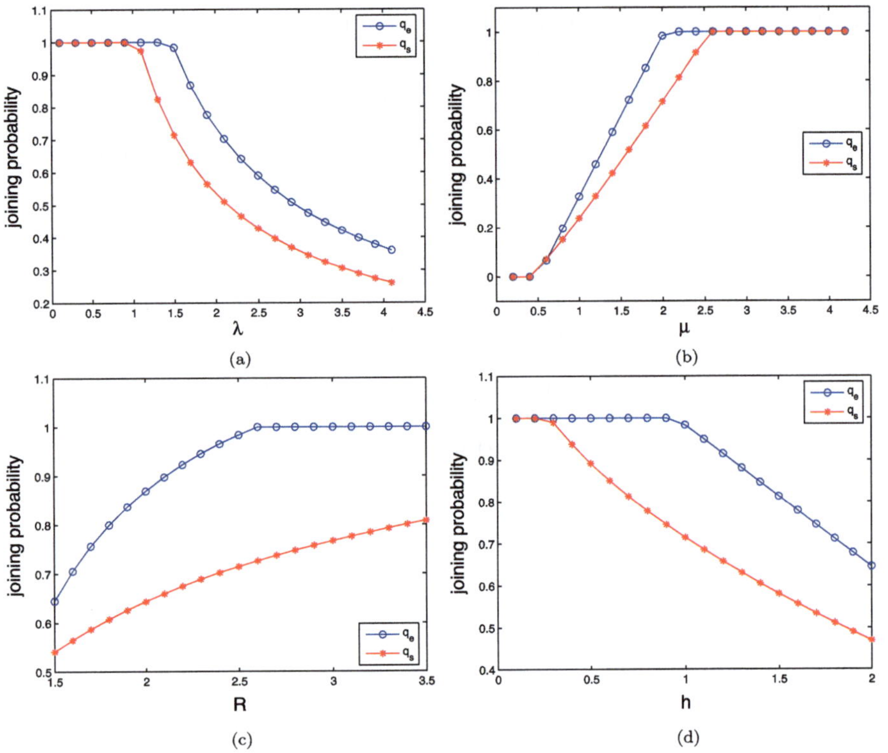

Fig. 11.5 The joining probabilities $(q_e,\ q_s)$ versus (λ, μ, R, h) separately where $p = 0.5$, $(\mu, R, h) = (2, 2.5, 1)$ for figure (**a**), $(\lambda, R, h) = (1.5, 2.5, 1)$ for figure (**b**), $(\lambda, \mu, h) = (1.5, 2, 1)$ for figure (**c**), $(\lambda, \mu, R) = (1.5, 2, 2.5)$ for figure (**d**)

from the normal queue model discussed in, for example, Naor (1969), because the service process is linked with inventory management problem. Moreover, to obtain positive social optimum, from Fig. 11.7a, c the case $\lambda = 0.3$ or $R = 1.5$ is more preferable.

Thirdly, we investigate the social welfare which focuses on all the customers and considers their benefits as a whole. From Fig. 11.7a, b, d we find the social optimums in the fully unobservable case and partially observable case have no obvious difference when λ and h are smaller or μ is bigger. For example, as μ increases, social optimum is consistent between these two cases. From this perspective, concealing the information may save some kind of maintenance cost which is difficult to be computed. As the system becomes congested or the waiting cost increases, from Fig. 11.7a, d we find the social optimum in partially observable case is always greater than that in fully unobservable case and the gap is growing. At this moment, disclosing the number of customers in system to the arriving customers is particularly

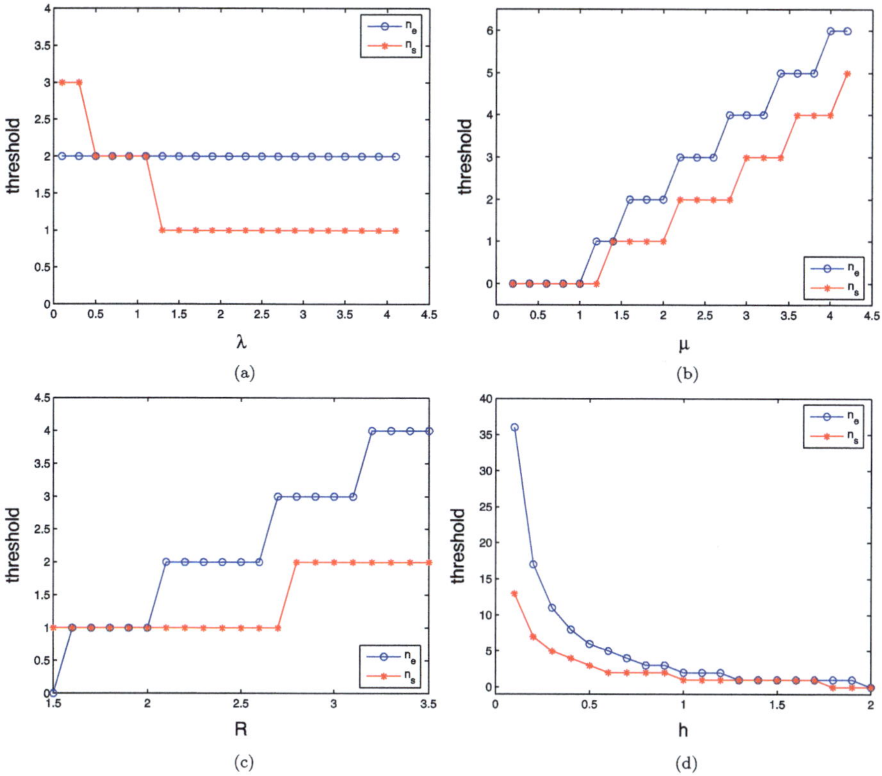

Fig. 11.6 The thresholds $(n_e,\ n_s)$ versus (λ, μ, R, h) separately where $p = 0.5$, $(\mu, R, h) = (2, 2.5, 1)$ for figure (**a**), $(\lambda, R, h) = (1.5, 2.5, 1)$ for figure (**b**), $(\lambda, \mu, h) = (1.5, 2, 1)$ for figure (**c**), $(\lambda, \mu, R) = (1.5, 2, 2.5)$ for figure (**d**)

important for the social welfare. Of course, this control policy is always appropriate to system manager. For convenience, the manager can publish real-time system information to customers all the time.

Finally, we illustrate the tendency of optimal prices p_u and p_o, revenue maximums $U_{fu}(p_u)$ and $U_{ao}(p_o)$. If the condition (11.3.1) does not hold, we set the optimal price to zero. On the one hand, the optimal prices p_u and p_o which capture the maximum of the revenue function mainly have the same monotonicity as the revenue functions. These tendencies can be seen in Figs. 11.8 and 11.9. On the other hand, the tendency of the revenue function is quite different from usual situation in which the social welfare with more transparent information is better.

Figure 11.8a displays $U_{fu}(p_u) \geq U_{ao}(p_o)$ for smaller λ and then $U_{fu}(p_u) < U_{ao}(p_o)$ in other situations. That is to say, to obtain more revenue, the server should conceal the system information to the arriving customers when the arrival rate is smaller and then disclose it when the arrival rate is greater. Since the arrival rate and the waiting cost have the same effect on the customer behavior, the same situation

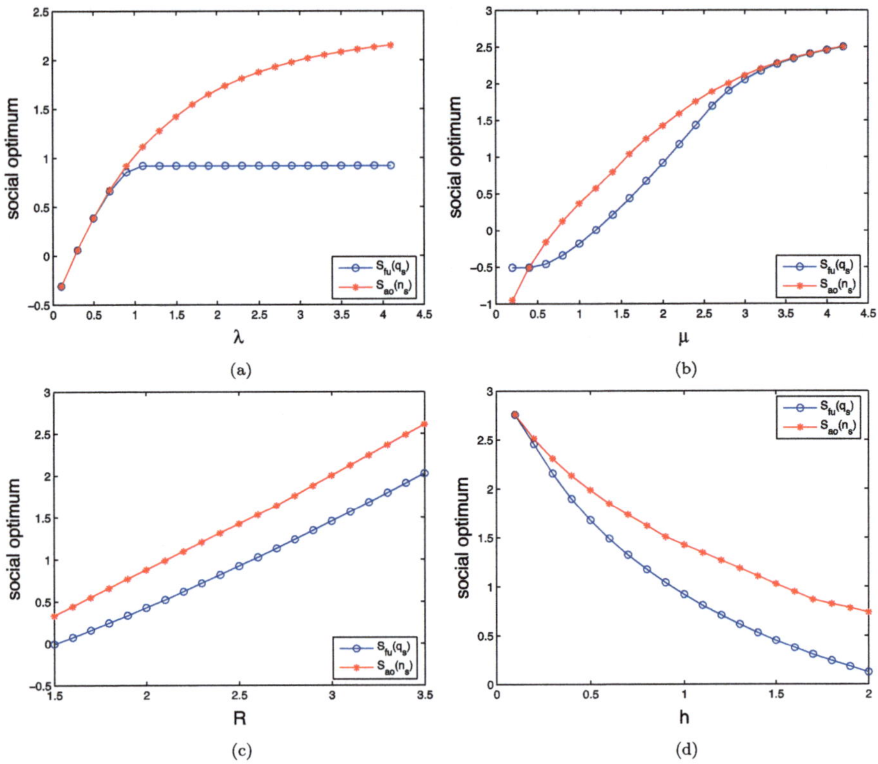

Fig. 11.7 The social optimums ($S_{fu}(q_s)$, $S_{ao}(n_s)$) versus (λ, μ, R, h) separately where $p = 0.5$, $(\mu, R, h) = (2, 2.5, 1)$ for figure (**a**), $(\lambda, R, h) = (1.5, 2.5, 1)$ for figure (**b**), $(\lambda, \mu, h) = (1.5, 2, 1)$ for figure (**c**), $(\lambda, \mu, R) = (1.5, 2, 2.5)$ for figure (**d**)

can be seen in Fig. 11.8d. However, Fig. 11.8b illustrates opposite conclusion where the server should disclose the system information to the arriving customers when the service rate is smaller and then conceal otherwise. Therefore, which case is beneficial to the server, whether or not disclosing the number of customers in system to the arriving customers is closely linked with different parameters.

Since p_u is determined by p_1, p_2, $p_u^{\star}$, we find it has a frequent fluctuation with parameter variation in Fig. 11.9. However, in the fully unobservable case, we realize that the revenue maximum $U_{fu}(p_u)$ is equal to social optimum $S_{fu}(q_s)$ in most cases, which can be seen in Figs. 11.7 and 11.8. This phenomenon is same as the conclusion in Edelson and Hilderbrand (1975) because these two models are extremely similar when the number of items in inventory is unobservable. In this situation, social welfare and server revenue can reach an agreement through imposing a price p.

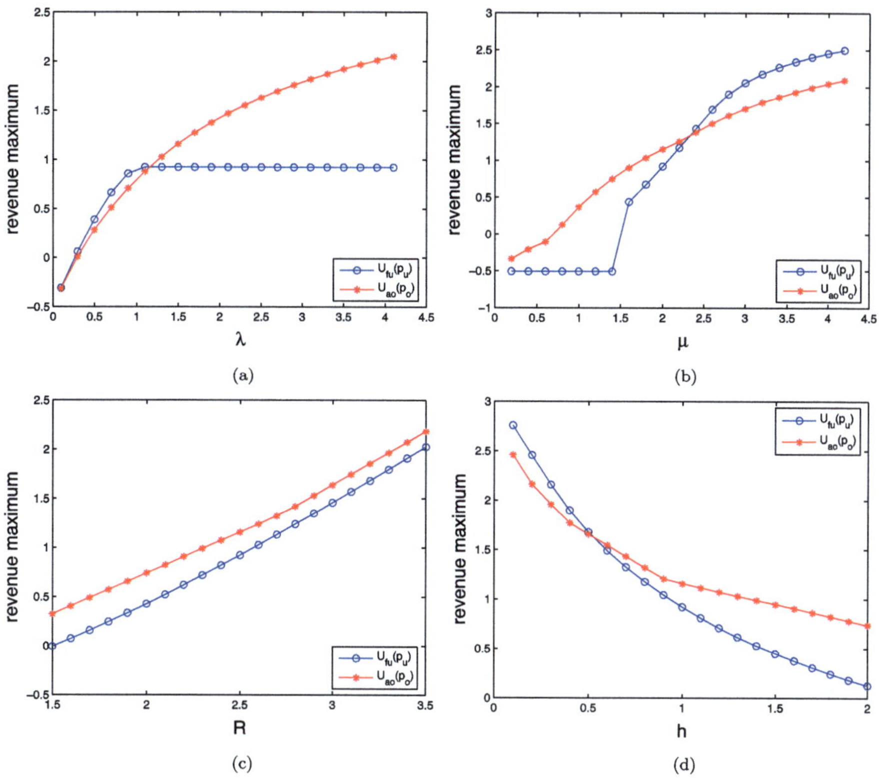

Fig. 11.8 The revenue maximums $(U_{fu}(p_u),\ U_{ao}(p_o))$ versus (λ, μ, R, h) separately where $(\mu, R, h) = (2, 2.5, 1)$ for figure (**a**), $(\lambda, R, h) = (1.5, 2.5, 1)$ for figure (**b**), $(\lambda, \mu, h) = (1.5, 2, 1)$ for figure (**c**), $(\lambda, \mu, R) = (1.5, 2, 2.5)$ for figure (**d**)

11.5 Discussions

In this section we studied a single-server service-inventory system with lost sales. Customers arrive to the system according to a Poisson process and they are not permitted to join the system if the number of items drops to zero. The service times are exponentially distributed and each customer takes exactly one item from the inventory after service. A continuous review policy is adopted to replenish the inventory. In the fully unobservable case and partially observable case, we obtained the individually optimal strategy and socially optimal strategy. Numerical experiments show that the former can be less than the latter which is different from the results available in the literature. Moreover, comparing the social optimum in these two cases, it is definitely beneficial to the social welfare when more transparent information is available to customers. But, after introducing the optimal price that maximizes the server revenue, the revenue maximum in partially observable case is not always greater than that in the fully unobservable case. The managerial insight is that providing more system

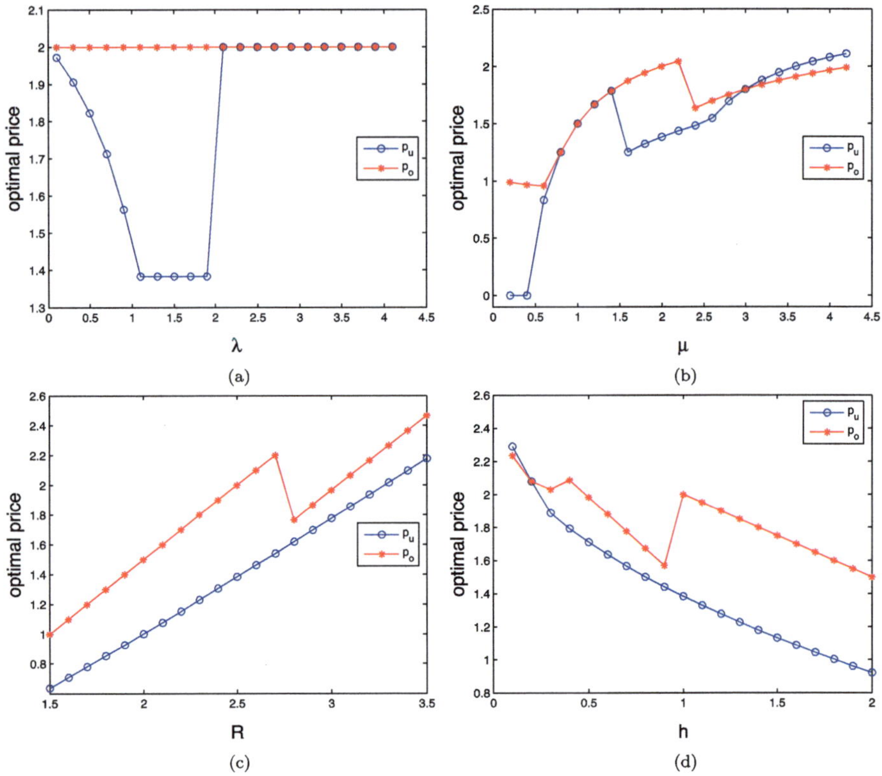

Fig. 11.9 The optimal prices $(p_u,\ p_o)$ versus (λ, μ, R, h) separately where $(\mu, R, h) = (2, 2.5, 1)$ for figure (**a**), $(\lambda, R, h) = (1.5, 2.5, 1)$ for figure (**b**), $(\lambda, \mu, h) = (1.5, 2, 1)$ for figure (**c**), $(\lambda, \mu, R) = (1.5, 2, 2.5)$ for figure (**d**)

information to customers may not be beneficial to the server. Therefore, distinct policy should be adopted aiming at social welfare and server revenue. Beside this, in the fully unobservable case, numerical results illustrate an reasonable phenomenon that the revenue maximum is equal to social optimum in most cases.

11.6 Optimal Inventory Threshold for a Make-to-Stock System

In this section we consider a Make-to-Stock (MTS) system with a single production facility and one type product, where the potential customers arrive according to a Poisson process and the service time follows an exponential distribution. To balance quality of service and related holding cost of products, the production manager adopts a dynamic production policy. If there is no customer waiting in the system, instead of shutting down, the system operates with low production rate until the number of products reaches to a certain threshold of inventory. When the inventory is empty and a new customer emerges, the system switches to high production rate where the switching time is assumed to be exponentially distributed. Arriving customers are delay-sensitive and strategic in the sense that if they encounter a stock out upon arrivals, they make decisions on whether to wait for product or leave without purchase based on their utility and the system information, i.e, whether the number of products is observable to them or not.

Such a Make-to-Stock inventory system is suitable for the product of expensive equipment with high holding cost and relatively low demand such as aircraft equipment, see Li et al. (2016). On the other hand, once the inventory of the particular product is empty, customers have options to choose another competitive brand for substitutes which leads to losses of customers or lost sales (see Zipkin 2000). In this situation, manager has to adopt some production/selling mechanisms to retain consumers, such as actions of discount, priority auction, or adjustment of the service rate. In this section we focus on a situation in which the production rate is adjustable and how the manager designs the inventory threshold according to the strategic behavior of customers.

There are two streams of research relevant to the analysis undertaken in this section, the first one concerned with Make-to-Stock production systems and the second one with queueing games with strategic customers. We first provide an overview of literature related to Make-to-Stock production systems. Gavish and Graves (1980) consider an M/D/1 production/inventory model with one type product. Through a two-critical-number continuous review policy, they propose an efficient search algorithm to find the optimal policy parameter values. Gavish and Graves (1981) and Srinivasan et al. (1991) extend the model in Gavish and Graves (1980) to the M/G/1 production/inventory system and obtain similar conclusions. Different from Poisson demand process in Gavish and Graves (1980, 1981), Graves and Keilson (1981) use the compensation method and they obtain the similar conclusions based on a Poisson process with demand size being an exponential random variable. Later, many works focused on the multiple product cases. Zheng and Zipkin (1990) assume that there are two distinct products and supplies of them are obtained through a common production facility. Through two information-control structures, the entire system is controlled and then compared by deciding when and which product to produce. In Veatch and Wein (1996) and Wein (1992), the authors extend the above work to K different classes of jobs with different production times. Other references focusing on multiple products can refer to Ha (1997), Bertsimas and Paschalidis (2001), Liu et al. (2012) and the references therein.

The aforementioned works in the first stream assume exogenously given demand processes, and there are also some articles dealing with delay-sensitive customers as in our work. Li (1992) first considers production/inventory policy for a single firm which produces a single product by taking the impatient customers into account. Similar to "newsboy" formulas, the explicit optimal policy is derived and then the model is extended to an $n-$firm market game due to delivery-time competition. Economopoulos et al. (2011) studied an M/M/1 MTS system with base stock and base backlog. During stockout periods, there exist three kinds of loss sales: balking, rejection and reneging. Using a finite capacity M/M/1 queueing model, the authors determine the optimal base stock and base backlog which maximize the mean profit. Li and Jiang (2013) analyze the product-service system with finite numbers of states. Being aware of the base stock strategy, impatient customers arrive to the system according to a Poisson process and balking probability is affected by the number of customers in queue. Although they choose to join the system, customers have the option to renege if the waiting time is beyond their upper boundary. Chen et al. (2015) present two compensation mechanisms, uniform compensation and priority auction, to impatient customers in an M/M/1 Make-to-Stock system when a stockout occurs. Li et al. (2016) consider a Make-to-Stock production system, derive the equilibrium customer purchasing strategy and minimized the expected cost rate function through a two-critical-number control policy.

As for the second stream of research, the strategic behavior of customers has been well studied in classic queueing systems where inventory is not considered. In these settings, individual customer will be caught in a dilemma when he decides whether to enter the system or not upon arrival according to a certain reward-cost structure. In this work, all the customers are assumed to be homogeneous and each one seeks the individual optimal strategy which can maximize his benefit against other customers who have the same benefit function. Therefore, the individual optimal strategy is symmetric strategy. Once the optimum is reached, no customer has the incentive to deviate unilaterally from this strategy assuming that all other customers make no change. So the individual optimal strategy is also called (Nash) equilibrium strategy (see Hassin and Haviv 2003 for details).

The seminal work focusing on equilibrium strategy analysis from the economic aspect dates back to the study of Naor (1969). Naor considers an M/M/1 queueing system where customers are informed the queue length, and solved the customers' individual equilibrium strategies. Later, Edelson and Hilderbrand (1975) extend the results in Naor (1969), where customers can not observe the queue length upon arrivals.

In the following, we include some of the works on queueing system with setup times or vacation policy. Burnetas and Economou (2007) consider an M/M/1 queueing system with setup times. Whenever this system becomes empty, the server is turned off. And if a customer arrives to an empty system, the server begins to set up. Under four different information levels, they derive equilibrium strategies for the customers. Economou et al. (2011) extend the results in Burnetas and Economou (2007) from exponentially distributed service and vacation times to general distribution. Given the server's state or not, the authors investigated customers' equilibrium strategies along with social optimal strategies, and then made some comparison between them. Zhang et al. (2013) generalize the result inBurnetas and Economou (2007) to the queueing systems with working vacation where the server operates with a lower service rate rather than completely stops during the vacation period. Liu et al. (2012) study the equilibrium behavior of customers in continuous/discrete time Markovian queueing systems under single vacation policy and observable cases. Ma et al. (2012) establish customers' equilibrium behavior in Geo/Geo/1 queueing system with multiple vacation policy and four different information levels. See also Guo and Hassin (2011), Li et al. (2016), among others. As for the main models and detailed results, interested readers are referred to the excellent monographs of Hassin and Haviv (2003), Hassin (2016) and references therein.

So far, only very few works have been carried out in the literature on the strategic customer behavior in Make-to-Stock system, and to the best of our knowledge, such a dynamic service policy in a Make-to-Stock system with strategic customers has not been studied in the literature. The most relevant work to our model is Li et al. (2016), in which a Make-to-Stock production system with a two-critical-number control policy is investigated. The authors derive the optimal strategy minimizing the expected cost rate function by considering the customer equilibrium behavior. However, the production rate is kept constant in their model.

In reality, a production manager can adjust production rate to balance quality of service and related production and holding cost of products. Many real life manufacture factories can be accommodated by our model. When orders come, manager should increase the production rate to avoid the excessive time delay. For example, there are a lot of demands for ice-cream in summer time, hence, an ice-cream factory would increase its workforce to cope with that demand. As demand decreases, manager should adjust the production rate and store a relative small amount of products in response to low demand rate considering its deterioration. Therefore the main task of inventory management system is to choose the appropriate inventory threshold for improvement of economic benefits.

The results in this section offer managerial guidelines for production managers and the main contributions can be summarized as follows. First, an M/M/1 Make-to-Stock system is investigated with dynamic service in which the customers are strategic and delay-sensitive. Second, based on the strategic behavior of customers, this section establishes an expected cost function of the system built on the performance measures. The objective is to seek the optimal inventory threshold minimizing the cost. It is formulated as a Stackelberg game between production manager and customers, where the former is the game leader. Third, by considering strategic customers who make decentralized decisions on joining the queue for purchase or balking, we gain new insights for operating the production system through adjustable production rate. When the number of products in inventory is observable, we obtain some kind of convexity of cost function and find an interval which contains the optimal threshold. When the number of products in inventory is unobservable, we can obtain the optimal threshold numerically. Comparing these two cases, it shows that the system would serve all the customer when they are patient enough, and therefore the two cases are entirely identical. With increasing of customers' delay sensitivity, the numerical results demonstrate that the expected cost function in observable case is not greater than that of unobservable case. Once the customers' delay sensitivity exceeds a certain level, the holding cost plays a dominant role. That is to say, Make-to-Stock policy is disadvantageous to the system.

To this end, we first investigate the individual optimum and then explore the system's cost optimum under the two information levels. This section is organized as follows. The model description is given in Sect. 11.6.1. Section 11.6.2 presents individual equilibrium strategy and the system's optimal threshold in observable and unobservable cases. Finally, Sects. 11.7 and 11.8 give some numerical experiments and conclusions, respectively. Some auxiliary proposition and proofs are presented in Appendix.

11.6.1 Model Description

In this paper we consider an M/M/1 make-to-stock system with a single production facility (server) and a single product. All the customers arrive according to a Poisson process with rate λ and the service discipline is first-come, first-served (FCFS). We assume that the production times are all exponentially distributed with low rate μ_l and high rate μ_h, respectively. If the inventory is empty (no product in inventory), a customer has to wait for an accomplishment of one product. And if the inventory is nonempty, he can take one product away without any delay. To serve for the potential customers in seconds, when the number of customers is zero, the server is operated with a low production rate μ_l until the inventory level reaches to the upper threshold S. Once the number of customers is nonzero, the server is switched to high production rate μ_h to shorten the waiting time of customers. Here the switching times are also assumed to be exponentially distributed with rate α. Without loss of generality, the inter-arrival times, the production times and the switching times are mutually independent.

The system state is specified at time t by the pair $(I(t), K(t))$. If $I(t) > 0$, $I(t)$ denotes the number of customers in system(implies the inventory is empty); if $I(t) \leq 0$, $-I(t)$ denotes the number of products in inventory. And $K(t)$ denotes the state of server, concretely, $K(t) = 2$ means the server is operated with low production rate μ_l, $K(t) = 1$ means the server is operated with high production rate μ_h and $K(t) = 0$ means the server is in switching state and does not offer product any more. Although the server is operated with high production rate in state 1, the inventory is always empty because the finished products are taken away instantaneously by the waiting customers.

It is clear that the process $\{(I(t), K(t)),\ t \geq 0\}$ is a continuous time Markov chain with state space $\Omega = \{(i, 2), -S \leq i \leq 0;\ (i, 0),\ i \geq 1;\ (i, 1),\ i \geq 1\}$. The transition rate diagram is shown in Fig. 11.10. Note that the related results when $\lambda = \mu_l$ can be regarded as the limitation of the case $\lambda \neq \mu_l$, and moreover our work is same as the model in Burnetas and Economou (2007) if $S = 0$ (i.e. Make-to-Order system). Therefore we consider $S \geq 1$ and $\lambda \neq \mu_l$ in the rest of the paper.

Once served, customer will receive a reward of R units and meanwhile, a waiting cost of c units per unit time is continuously accumulated from the time that he enter the system to the time he completes the service. In order to maximize his expected net benefit, the customer decides whether to join or balk the system upon arrival based on the given information. Once his action is chosen, it is irrevocable in the sense that neither retrial of balking customer nor reneging of entering customer are

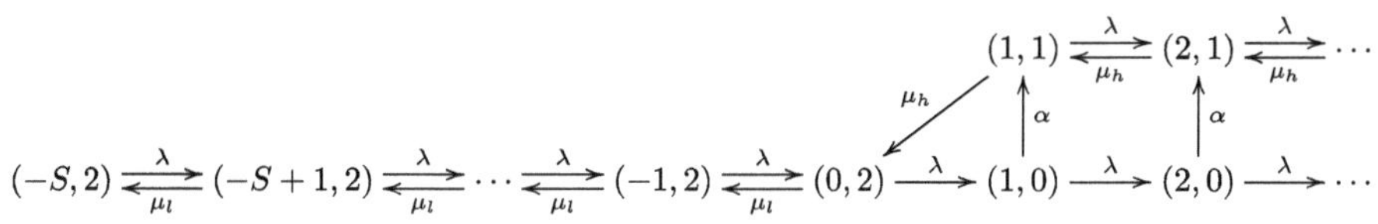

Fig. 11.10 Transition rate diagram of the system

allowed. According to the information available to the customers upon arrivals, we mainly investigate the following two cases in this paper:

- Observable inventory case: Customers can observe the number of products in inventory.
- Unobservable inventory case: Customers can not observe the number of products in inventory.

11.6.2 Main Results

In this section, we present individual optimum and cost optimum in two cases, i.e., observable inventory and unobservable inventory cases. According to the information level, we first seek for customer's individual optimal strategy(Nash equilibrium strategy) which can maximize his own benefits. And then we form a Stackelberg game between the production manager and customers, where the former is leader. Based on customers' behavior, the leader investigates the optimal inventory threshold that can minimize the expected cost function. And in this situation the customers also can reach to their individual optimum.

Through comparing the reward with the sojourn cost, customer may apply pure strategy(with joining probability $q = 0$ or $q = 1$) or mixed strategy(with joining probability $q \in [0, 1]$) to make his decision. In this work, we investigate the mixed strategy and derive the corresponding equilibrium joining probability.

The primary notations along with their definitions in this work are summarized in Table 11.2, and some performance measures used in this work are summarized in Table 11.3.

11.6.3 Observable Inventory Case

In this section, all the customers can observe the number of products. If the inventory is nonempty which implies the server is in state 2, a customer would join the system definitely, and then take one product away from the inventory without any delay. Once the inventory is empty, he can not determine the server's state (0 or 1) and the number of customers in system. Thus we just focus on his strategic behavior when

Table 11.2 Notations and definitions

Notation	Definition	Notation	Definition
λ	Arrival rate of customers per unit time	$\mu_l(\mu_h)$	Low(high) production rate of server per unit time
S	Inventory threshold	α	Switching rate between high and low rate per unit time
R	Reward after service per customer	c	Waiting cost per customer per unit time
f	Holding cost per product per unit time	q	Joining probability of customers

Table 11.3 The performance measures

Definition	Observable inventory case	Unobservable inventory case
Steady state probabilities	$p_{ob}(i, k)$	$p_{un}(i, k)$
Expected sojourn time of customer	$\mathbb{E}[W_e]$	$\mathbb{E}[W_{un}]$
Equilibrium joining probability of customer	q_{ob}	q_{un}
Expected net benefit of customer	$I_{ob}(q)$	$I_{un}(q)$
Expected number of customers in system	$\mathbb{E}[N_{ob}]$	$\mathbb{E}[N_{un}]$
Expected number of products in inventory	$\mathbb{E}[I_{ob}]$	$\mathbb{E}[I_{un}]$
Expected cost function of system	$F_{ob}(S)$	$F_{un}(S)$

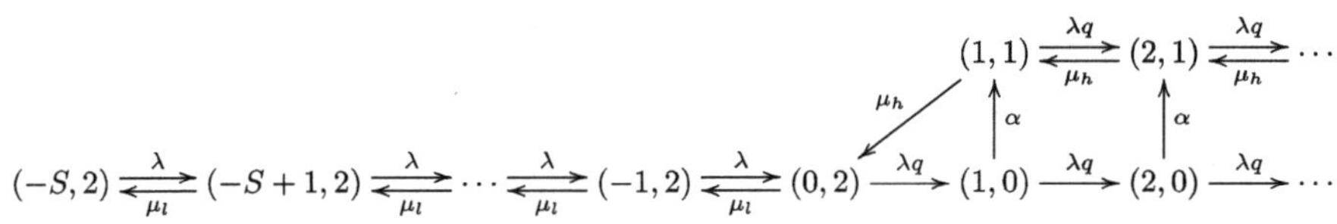

Fig. 11.11 Transition rate diagram of the system for observable inventory case

he finds the inventory empty upon arrival. Since all the customers are assumed to be homogeneous, we only consider the following mixed strategy: "customer joins the system with probability q, balks with probability $1 - q$". The corresponding transition rate diagram is shown in Fig. 11.11.

To discuss this situation, the steady state probabilities of the system are necessary and the preliminary results are summarized in the following lemma. It is readily proved that the system is stable if and only if $\lambda q < \mu_h$. For simplicity, we mainly investigate the results within the case $\lambda < \mu_h$ and just present the remainder results for the case $\lambda \geq \mu_h$. For brevity we consider only the typical case where $\lambda q + \alpha \neq \mu_h$. In what follows, denote $\varphi = \frac{\mu_l}{\lambda q}$, $\psi = \frac{\mu_h(\lambda q+\alpha)}{\alpha(\mu_h-\lambda q)}$.

Lemma 11.6.1 *In the observable inventory case, the steady state probabilities* $\{p_{ob}(i,k) : (i,k) \in \Omega\}$ *are given as follows:*

$$p_{ob}(0,2) = \left\{\frac{\mu_l[(\frac{\lambda}{\mu_l})^{-S} - 1]}{\mu_l - \lambda} + \frac{\mu_h(\lambda q + \alpha)}{\alpha(\mu_h - \lambda q)}\right\}^{-1}, \tag{11.6.1}$$

$$p_{ob}(i,2) = \left(\frac{\lambda}{\mu_l}\right)^i p_{ob}(0,2),\ \ i = -S, \ldots, -1, \tag{11.6.2}$$

$$p_{ob}(i,0) = \left(\frac{\lambda q}{\lambda q + \alpha}\right)^i p_{ob}(0,2),\ \ i = 1, 2, \ldots, \tag{11.6.3}$$

$$p_{ob}(i,1) = \frac{\lambda q + \alpha}{\mu_h - \lambda q - \alpha}\left[\left(\frac{\lambda q}{\lambda q + \alpha}\right)^i - \left(\frac{\lambda q}{\mu_h}\right)^i\right] p_{ob}(0,2),\ \ i = 1, 2, \ldots. \tag{11.6.4}$$

11.6.3.1 Individual Optimum

If the inventory is non-empty, the customer would enter the system and take one product away without hesitating. And if the inventory is empty, the customer would join the queue with probability through comparing the reward with the expected waiting cost based on the available information. According to the result above, we first obtain the expected sojourn time $\mathbb{E}[W_e]$ of an arriving customer when he finds the inventory empty and decides to join the system and then derive the corresponding equilibrium joining probability q_{ob}.

Theorem 11.6.1 *The expected sojourn time of an arriving customer when he find the inventory is empty and decides to join the system is*

$$\mathbb{E}[W_e] = \frac{\mu_h + \alpha - \lambda q}{\alpha(\mu_h - \lambda q)}. \tag{11.6.5}$$

Proof If the system is in state $(i, 0)$, $i \geq 1$ upon arrival, the sojourn time of an arriving customer is $\frac{1}{\alpha} + \frac{i+1}{\mu_h}$. If the system is in state $(i, 1)$, $i \geq 1$ upon arrival, the sojourn time of an arriving customer is $\frac{i+1}{\mu_h}$. If the system is in state $(0, 2)$, the sojourn time of an arriving customer is $\frac{1}{\alpha} + \frac{1}{\mu_h}$. Thus, the expected sojourn time of an arriving customer is

$$\begin{aligned}
\mathbb{E}[W_e] =& \frac{1}{1 - \sum_{i=-S}^{-1} p_{ob}(i,2)}\Big\{\sum_{i=1}^{\infty} p_{ob}(i,0)\left(\frac{1}{\alpha} + \frac{i+1}{\mu_h}\right) + \sum_{i=1}^{\infty} p_{ob}(i,1)\frac{i+1}{\mu_h} + p_{ob}(0,2)\left(\frac{1}{\alpha} + \frac{1}{\mu_h}\right)\Big\} \\
=& \frac{p_{ob}(0,2)}{1 - \sum\limits_{i=-S}^{-1} p_{ob}(i,2)}\Big\{\sum_{i=0}^{\infty}\left(\frac{\lambda q}{\lambda q + \alpha}\right)^i\left(\frac{1}{\alpha} + \frac{i+1}{\mu_h}\right) + \sum_{i=1}^{\infty}\frac{\lambda q + \alpha}{\mu_h - \lambda q - \alpha}\left[\left(\frac{\lambda q}{\lambda q + \alpha}\right)^i - \left(\frac{\lambda q}{\mu_h}\right)^i\right]\frac{i+1}{\mu_h}\Big\}
\end{aligned}$$

$$=\frac{\alpha(\mu_h-\lambda q)}{\mu_h(\lambda q+\alpha)}\Big\{\frac{1}{\alpha}\sum_{i=0}^{\infty}\left(\frac{\lambda q}{\lambda q+\alpha}\right)^i+\frac{1}{\mu_h}\sum_{i=0}^{\infty}(i+1)\left(\frac{\lambda q}{\lambda q+\alpha}\right)^i$$

$$+\frac{\lambda q+\alpha}{\mu_h(\mu_h-\lambda q-\alpha)}\sum_{i=1}^{\infty}(i+1)\left[\left(\frac{\lambda q}{\lambda q+\alpha}\right)^i-\left(\frac{\lambda q}{\mu_h}\right)^i\right]\Big\}.$$

Note that $\sum_{i=0}^{\infty}(i+1)x^i=\frac{1}{(1-x)^2}$ and $\sum\limits_{i=1}^{\infty}(i+1)x^i=\frac{1}{(1-x)^2}-1$, then

$$\begin{aligned}\mathbb{E}[W_e]=&\frac{\alpha(\mu_h-\lambda q)}{\mu_h(\lambda q+\alpha)}\left\{\frac{\lambda q+\alpha}{\alpha^2}+\frac{(\lambda q+\alpha)^2}{\mu_h\alpha^2}+\frac{\lambda q+\alpha}{\mu_h(\mu_h-\lambda q-\alpha)}\left[\frac{(\lambda q+\alpha)^2}{\alpha^2}-\frac{\mu_h^2}{(\mu_h-\lambda q)^2}\right]\right\}\\ =&\frac{\alpha(\mu_h-\lambda q)}{\mu_h(\lambda q+\alpha)}\left\{\frac{(\lambda q+\alpha)(\mu_h+\lambda q+\alpha)}{\mu_h\alpha^2}+\frac{\lambda q(\lambda q+\alpha)[(\lambda q+\alpha)(\mu_h-\lambda q)+\alpha\mu_h]}{\mu_h\alpha^2(\mu_h-\lambda q)^2}\right\}\\ =&\frac{\alpha(\mu_h-\lambda q)}{\mu_h(\lambda q+\alpha)}\cdot\frac{\mu_h(\lambda q+\alpha)(\mu_h+\alpha-\lambda q)}{\alpha^2(\mu_h-\lambda q)^2}\\ =&\frac{\mu_h+\alpha-\lambda q}{\alpha(\mu_h-\lambda q)},\end{aligned}$$

which also can be found in Theorem 5 of Burnetas and Economou (2007).

This completes the proof. □

In the above theorem, using total probability formula, we get the conditional expected sojourn time of an arriving customer who finds the inventory is empty and decides to join the system. This performance is independent of the others customer who find the inventory is nonempty because he does not need any waiting time in this situation. Therefore, we still use total probability formula, not Little's law, to derive the expected number of customers in system in the following Theorem 11.6.2. However, this computing method is different from that used in Sect. 11.6.4.1.

Theorem 11.6.2 *When the inventory is empty, a unique mixed strategy 'the customer chooses to enter the system with probability q_{ob}, balks with probability $1-q_{ob}$' exists and the probability is given by*

$$q_{ob}=\begin{cases}0, & \text{if } R\le\frac{c}{\alpha}+\frac{c}{\mu_h},\\ q_{ob}^*, & \text{if } \frac{c}{\alpha}+\frac{c}{\mu_h}<R<\frac{c}{\alpha}+\frac{c}{\mu_h-\lambda},\\ 1, & \text{if } R\ge\frac{c}{\alpha}+\frac{c}{\mu_h-\lambda},\end{cases}\tag{11.6.6}$$

where $q_{ob}^*=\frac{1}{\lambda}[\mu_h-\frac{c\alpha}{R\alpha-c}]$.

Proof If a tagged customer decides to enter the system when the inventory is empty upon arrival, the expected net benefit is

$$I_{ob}(q)=R-C\mathbb{E}[W_e]=R-C\frac{\mu_h+\alpha-\lambda q}{\alpha(\mu_h-\lambda q)}$$

and it is easy to show that $I_{ob}(q)$ is strictly decreasing in $q\in[0,1]$. If $I_{ob}(0)=R-c\frac{\mu_h+\alpha}{\alpha\mu_h}\le 0$ or $R\le\frac{c}{\alpha}+\frac{c}{\mu_h}$, the customer prefers to balk the system because the maximum benefit is nonpositive. The first branch of (11.6.6) is at hand. Furthermore if $I_{ob}(1) = R - C\frac{\mu_h+\alpha-\lambda}{\alpha(\mu_h-\lambda)} \geq 0$ or $R \geq \frac{C}{\alpha} + \frac{C}{\mu_h-\lambda}$, the customer prefers to join the system because the minimum benefit is nonnegative. The last branch of (11.6.6) is at hand. If $\frac{c}{\alpha} + \frac{c}{\mu_h} < R < \frac{c}{\alpha} + \frac{c}{\mu_h-\lambda}$, there exists a unique $q_{ob}^* \in (0, 1)$ such that $I_{ob}(q) = 0$. In this case, the tagged customer is indifferent between joining and balking the system.

Obviously, the joining probability q_{ob} is same as the result in Theorem 5 of Burnetas and Economou (2007) and wholly independent of S. This completes the proof. □

Followed similar analysis, when $\lambda \geq \mu_h$, the result in Theorem 11.6.2 also holds and a unique mixed strategy 'the customer chooses to enter the system with probability q_{ob}, balks with probability $1 - q_{ob}$' exists when the inventory is empty and the probability is given by

$$q_{ob} = \begin{cases} 0, & \text{if } R \leq \frac{c}{\alpha} + \frac{c}{\mu_h}, \\ q_{ob}^*, & \text{if } \frac{c}{\alpha} + \frac{c}{\mu_h} < R. \end{cases}$$

11.6.3.2 Cost Optimum

For a predetermined inventory threshold S, strategic customers make their optimal decisions with observed information level. And then, based on customers' behavior, the production manager has the motivation and right to determine the variable S to obtain the cost optimum. This forms a Stackelberg game among the manager and customers, where the manager is leader and customers are followers. In this part, we seek the optimal S^* to minimize the expected cost function of the system with strategic customers. The cost includes the waiting cost of customers in system and the holding cost of products in inventory. Therefore, we first present the expected number of customers in system and the expected number of products in inventory.

Theorem 11.6.3 *In the observable inventory case, the expected number of customers in system and the expected number of products in inventory are*

$$\mathbb{E}[N_{ob}] = \frac{\lambda q \mu_h(\lambda q + \alpha)(\mu_h + \alpha - \lambda q) p_{ob}(0, 2)}{\alpha^2(\mu_h - \lambda q)^2} \tag{11.6.7}$$

and

$$\mathbb{E}[I_{ob}] = \frac{q\varphi[1 - (q\varphi)^S - S(q\varphi)^S(1 - q\varphi)] p_{ob}(0, 2)}{(1 - q\varphi)^2} \tag{11.6.8}$$

$$= \frac{\mu_l[\lambda(\lambda^S - \mu_l^S) - S\mu_l^S(\lambda - \mu_l)] p_{ob}(0, 2)}{\lambda^S(\mu_l - \lambda)^2}. \tag{11.6.9}$$

Proof Note that $\sum_{i=1}^{\infty} ix^i = \frac{x}{(1-x)^2}$, then

$$
\begin{aligned}
\mathbb{E}[N_{ob}] &= \sum_{i=1}^{\infty} i p_{ob}(i,0) + \sum_{i=1}^{\infty} i p_{ob}(i,1) \\
&= p_{ob}(0,2)\left\{\sum_{i=1}^{\infty} i\left(\frac{\lambda q}{\lambda q+\alpha}\right)^i + \frac{\lambda q+\alpha}{\mu_h-\lambda q-\alpha}\sum_{i=1}^{\infty} i\left[\left(\frac{\lambda q}{\lambda q+\alpha}\right)^i - \left(\frac{\lambda q}{\mu_h}\right)^i\right]\right\} \\
&= p_{ob}(0,2)\left[\frac{\frac{\lambda q}{\lambda q+\alpha}}{\left(1-\frac{\lambda q}{\lambda q+\alpha}\right)^2}\left(1+\frac{\lambda q+\alpha}{\mu_h-\lambda q-\alpha}\right) - \frac{\lambda q+\alpha}{\mu_h-\lambda q-\alpha}\cdot\frac{\frac{\lambda q}{\mu_h}}{\left(1-\frac{\lambda q}{\mu_h}\right)^2}\right] \\
&= \frac{\lambda q\mu_h(\lambda q+\alpha)(\mu_h+\alpha-\lambda q)p_{ob}(0,2)}{\alpha^2(\mu_h-\lambda q)^2}.
\end{aligned}
$$

Since the inventory is non-empty only in state 2, applying $\sum_{i=1}^{S} ix^i = \frac{x[1-x^S-Sx^S(1-x)]}{(1-x)^2}$, we have

$$
\begin{aligned}
\mathbb{E}[I_{ob}] &= \sum_{i=-S}^{-1}(-i)p_{ob}(i,2) = \sum_{i=-S}^{-1}(-i)\left(\frac{\lambda}{\mu_l}\right)^i p_{ob}(0,2) = p_{ob}(0,2)\sum_{i=1}^{S} i\left(\frac{\mu_l}{\lambda}\right)^i \\
&= \frac{q\varphi[1-(q\varphi)^S - S(q\varphi)^S(1-q\varphi)]p_{ob}(0,2)}{(1-q\varphi)^2} \\
&= \frac{\mu_l[\lambda(\lambda^S-\mu_l^S) - S\mu_l^S(\lambda-\mu_l)]p_{ob}(0,2)}{\lambda^S(\mu_l-\lambda)^2}.
\end{aligned}
$$

This completes the proof. □

Denote f is the holding cost for one product per unit time, then the expected cost function of the system is

$$
F_{ob}(S) = C\mathbb{E}[N_{ob}] + f\mathbb{E}[I_{ob}]. \tag{11.6.10}
$$

Without loss of generality, let $C,\ f \geq 1$. To find the optimal threshold S^* to minimize the cost function, we need to analyze some properties of the performance measures $\mathbb{E}[N_{ob}]$ and $\mathbb{E}[I_{ob}]$. According to Eq. (11.6.7), the following theorem about $p_{ob}(0,2)$ and $\mathbb{E}[N_{ob}]$ is necessary.

Theorem 11.6.4 *In the observable inventory case, the following results hold true for* $S \geq 1$*:*

(1) $p_{ob}(0,2)$ *and* $\mathbb{E}[N_{ob}]$ *are all strictly decreasing in* S*;*

(2) If $\mu_l < \lambda$, $p_{ob}(0,2)$ *and* $\mathbb{E}[N_{ob}]$ *are all strictly convex in* S*;*

(3) If $\mu_l > \lambda$ *and* $\psi(\mu_l-\lambda) \leq \mu_l$, $p_{ob}(0,2)$ *and* $\mathbb{E}[N_{ob}]$ *are all strictly convex in* S*;*

(4) If $\mu_l > \lambda$ and $\psi(\mu_l - \lambda) > \mu_l$, $p_{ob}(0, 2)$ and $\mathbb{E}[N_{ob}]$ are all concave for $S \leq \max\{1, B\}$ and convex for $S \geq \max\{1, B\}$, where $B = \lfloor \log_{\frac{\mu_l}{\lambda}}(\psi(\frac{\mu_l}{\lambda} - 1) - \frac{\mu_l}{\lambda}) - 2 \rfloor$.

According to the definitions and propositions of quasi-convex and pseudo-convex functions in references Bazaraa et al. (2006), Mangasarian (1965), Mangasarian (1970), the following results are tractable through simple manipulation.

Theorem 11.6.5 *In the observable inventory case, $\mathbb{E}[I_{ob}]$ is strictly increasing in S.*

Proof From the Eqs. (11.6.1) and (11.6.8) we have

$$\mathbb{E}^S[I_{ob}] \triangleq \mathbb{E}[I_{ob}] = \frac{q\varphi[1 - (q\varphi)^S - S(q\varphi)^S(1 - q\varphi)]}{(q\varphi - 1)\{q\varphi[(q\varphi)^S - 1] + \psi(q\varphi - 1)\}}. \quad (11.6.11)$$

So

$$\begin{aligned}
&\mathbb{E}^{S+1}[I_{ob}] - \mathbb{E}^S[I_{ob}] \\
=&\frac{q\varphi}{(q\varphi - 1)}\left\{\frac{1 - (q\varphi)^{S+1} - (S+1)(q\varphi)^{S+1}(1 - q\varphi)}{q\varphi[(q\varphi)^{S+1} - 1] + \psi(q\varphi - 1)} - \frac{1 - (q\varphi)^S - S(q\varphi)^S(1 - q\varphi)}{q\varphi[(q\varphi)^S - 1] + \psi(q\varphi - 1)}\right\} \\
=&\frac{q\varphi}{(q\varphi - 1)\{q\varphi[(q\varphi)^{S+1} - 1] + \psi(q\varphi - 1)\}\{q\varphi[(q\varphi)^S - 1] + \psi(q\varphi - 1)\}} \times \\
&\Big\{[1 - (q\varphi)^{S+1} - (S+1)(q\varphi)^{S+1}(1 - q\varphi)]\{q\varphi[(q\varphi)^S - 1] + \psi(q\varphi - 1)\} \\
&- [1 - (q\varphi)^S - S(q\varphi)^S(1 - q\varphi)]\{q\varphi[(q\varphi)^{S+1} - 1] + \psi(q\varphi - 1)\}\Big\} \\
=&\frac{(q\varphi)^{S+1}\{\psi(S+1)(q\varphi - 1)^2 + q\varphi(q\varphi - 1)[q\varphi + (q\varphi)^2 + \cdots + (q\varphi)^S - S]\}}{\{q\varphi[(q\varphi)^{S+1} - 1] + \psi(q\varphi - 1)\}\{q\varphi[(q\varphi)^S - 1] + \psi(q\varphi - 1)\}}.
\end{aligned}$$

No matter $q\varphi > 1$ or $q\varphi < 1$, $(q\varphi - 1)[q\varphi + \cdots + (q\varphi)^S - S]$ and the denominator of the last equation are all positive, therefore $\mathbb{E}[I_{ob}]$ is strictly increasing in S. This completes the proof. □

Theorem 11.6.6 *In the observable inventory case, we have (1) $\mathbb{E}[I_{ob}]$ is quasi-convex in S if $(S+1)(\lambda - \mu_l) \leq \mu_l < \lambda$; (2) If S is a continuous variable, $\mathbb{E}[I_{ob}]$ is pseudo-convex in S if $(S+1)(\lambda - \mu_l) \leq \mu_l < \lambda$.*

Proof Obviously, the numerator and denominator of $\mathbb{E}[I_{ob}]$ defined in Eq. (11.6.11) are both positive for any $S > 0$ no matter $\varphi < 1$ or $\varphi > 1$. And the second order difference of numerator is

$$\begin{aligned}
&q\varphi\{[1 - (q\varphi)^{S+2} - (S+2)(q\varphi)^{S+2}(1 - q\varphi)] - 2[1 - (q\varphi)^{S+1} - (S+1)(q\varphi)^{S+1}(1 - q\varphi)] \\
&+ [1 - (q\varphi)^S - S(q\varphi)^S(1 - q\varphi)]\} = (q\varphi)^S(q\varphi - 1)^2[(S+1)(q\varphi - 1) + q\varphi].
\end{aligned}$$

Hence the numerator of $\mathbb{E}[I_{ob}]$ is convex in S if $(S+1)(q\varphi - 1) + q\varphi \geq 0$.

Moreover the second order difference of denominator is

$$(q\varphi - 1)\big\{\{q\varphi[(q\varphi)^{S+2} - 1] + \psi(q\varphi - 1)\} - 2\{q\varphi[(q\varphi)^{S+1} - 1] + \psi(q\varphi - 1)\}$$
$$+\{q\varphi[(q\varphi)^{S} - 1] + \psi(q\varphi - 1)\}\big\} = (q\varphi - 1)^3(q\varphi)^{S+1}.$$

Hence the denominator of $\mathbb{E}[I_{ob}]$ is concave in S if $q\varphi < 1$.

From the Ex. 3.61 of Bazaraa et al. (2006), we get the conclusion. This completes the proof. □

From Theorem 11.6.2, the equilibrium joining probability q_{ob} is independent of S. So, the properties of $\mathbb{E}[N_{ob}]$ and $\mathbb{E}[I_{ob}]$ on S would not be influenced after putting q_{ob} into the Eqs. (11.6.7) and (11.6.9). Therefore, from the Theorems 11.6.4 and 11.6.6, the following conclusion is reachable.

Theorem 11.6.7 *In the observable inventory case, (1) $F_{ob}(S)$ is quasi-convex in S if $(S+1)(\lambda - \mu_l) \le \mu_l < \lambda$; (2) If S is a continuous variable, $F_{ob}(S)$ is pseudo-convex in S if $(S+1)(\lambda - \mu_l) \le \mu_l < \lambda$.*

Theorem 11.6.7 illustrates the convexity of $F_{ob}(S)$ when $(S+1)(\lambda - \mu_l) \le \mu_l < \lambda$. Under this condition, the optimal threshold S^* must be a stationary point determined by $\nabla F_{ob}(S) = 0$ if S is a continuous variable; if S is a discrete variable, S^* can be found between two integral points which are close to the stationary point. Besides this, the following theorem establishes the boundedness of $F_{ob}(S)$ when $\mu_l < \lambda$.

Theorem 11.6.8 *If $\mu_l < \lambda$, $F_{ob}(S)$ is bounded in S.*

Proof From Theorems 11.6.4(1) and 11.6.5, we know $\mathbb{E}[N_{ob}]$ and $\mathbb{E}[I_{ob}]$ are monotone in S. Combining the following conclusions

$$\lim_{S\to+\infty} \mathbb{E}[N_{ob}] = \frac{\lambda q \mu_h(\lambda q + \alpha)(\lambda - \mu_l)(\mu_h + \alpha - \lambda q)}{\alpha^2(\mu_h - \lambda q)^2[\psi(\lambda - \mu_l) + \mu_l]},$$
$$\lim_{S\to+\infty} \mathbb{E}[I_{ob}] = \frac{\mu_l}{(\lambda - \mu_l)[\psi(\lambda - \mu_l) + \mu_l]},$$

$F_{ob}(S)$ is bounded obviously. This completes the proof. □

If $\mu_l > \lambda$, it is readily seen that

$$\lim_{S\to+\infty} \mathbb{E}[I_{ob}] = +\infty. \tag{11.6.12}$$

Hence $F_{ob}(S)$ is also unbounded in S. However, it is reasonable and available for production manager to set an upper bound $\overline{F}$ of the expected cost function $F_{ob}(S)$. Given this, we can find an interval containing the optimal S^* which can minimize the expected cost function $F_{ob}(S)$. Using the substitution method, we successively evaluate the function value $F_{ob}(S)$ with discretely variable S until the minimum point

S^* is achieved. This comes up with the following conclusion. For the convenience of description, we denote three coefficients as follows:

$$\begin{aligned} F_1 &= C\lambda q_{ob}\mu_h(\mu_l-\lambda)(\lambda q_{ob}+\alpha)(\mu_h+\alpha-\lambda q_{ob}) \geq 0, \\ F_2 &= \alpha^2\mu_l(\mu_h-\lambda q_{ob})^2 > 0, \\ F_3 &= \alpha\mu_h(\mu_h-\lambda q_{ob})(\mu_l-\lambda)(\lambda q_{ob}+\alpha) > 0. \end{aligned}$$

Theorem 11.6.9 *Denote $\overline{F}$ be an upper bound of the expected cost function $F_{ob}(S)$ when $\mu_l > \lambda$, then $\underline{S} \leq S^* \leq \overline{S}$, where $\overline{S} = \min\{S | f\mathbb{E}[I_{ob}] \geq \overline{F}\}$ and*

$$\underline{S} = \begin{cases} \lfloor \log_{\frac{\mu_l}{\lambda}} \frac{\overline{F}(F_2-F_3)+F_1}{\overline{F}F_2} \rfloor, & \text{if } \frac{\overline{F}(F_2-F_3)+F_1}{\overline{F}F_2} > \frac{\mu_l}{\lambda}, \\ 1, & \text{otherwise.} \end{cases}$$

Proof From Eqs. (11.6.10), (11.6.7) and (11.6.1), we have

$$\begin{aligned} F_{ob}(S) \geq\ & C\mathbb{E}[N_{ob}] \\ =\ & C\frac{\lambda q_{ob}\mu_h(\lambda q_{ob}+\alpha)(\mu_h+\alpha-\lambda q_{ob})}{\alpha^2(\mu_h-\lambda q_{ob})^2} \\ & \times\frac{\alpha(\mu_l-\lambda)(\mu_h-\lambda q_{ob})}{\alpha\mu_l(\mu_h-\lambda q_{ob})[(\frac{\mu_l}{\lambda})^S-1]+\mu_h(\mu_l-\lambda)(\lambda q_{ob}+\alpha)} \\ =\ & \frac{C\lambda q_{ob}\mu_h(\mu_l-\lambda)(\lambda q_{ob}+\alpha)(\mu_h+\alpha-\lambda q_{ob})}{\alpha(\mu_h-\lambda q_{ob})\{\alpha\mu_l(\mu_h-\lambda q_{ob})[(\frac{\mu_l}{\lambda})^S-1]+\mu_h(\mu_l-\lambda)(\lambda q_{ob}+\alpha)\}} \\ =\ & \frac{F_1}{F_2[(\frac{\mu_l}{\lambda})^S-1]+F_3} \geq \overline{F}. \end{aligned} \tag{11.6.13}$$

Then the last inequality of (11.6.13) above is equivalent to

$$S \leq \log_{\frac{\mu_l}{\lambda}} \frac{\overline{F}(F_2-F_3)+F_1}{\overline{F}F_2}.$$

If $\frac{\overline{F}(F_2-F_3)+F_1}{\overline{F}F_2} > \frac{\mu_l}{\lambda}$, we set $\underline{S} = \lfloor \log_{\frac{\mu_l}{\lambda}} \frac{\overline{F}(F_2-F_3)+F_1}{\overline{F}F_2} \rfloor$. Otherwise, $\underline{S} = 1$.

On the other hand, combining Theorem 11.6.5 and Eq. (11.6.12), there exists a smallest positive integer $\overline{S}$ such that $f\mathbb{E}[I_{ob}] \geq \overline{F}$. Moreover, the following inequality

$$F_{ob}(S) \geq f\mathbb{E}[I_{ob}] \geq \overline{F}$$

also holds for $S \geq \overline{S}$.

Therefore, $F_{ob}(S) \geq \overline{F}$ when $S \leq \underline{S}$ or $S \geq \overline{S}$. That is to say, $\underline{S} \leq S^* \leq \overline{S}$. □

If $\underline{S} > \overline{S}$, we have $F_{ob}(S) > \overline{F}$ for any $S \geq 1$, i.e., the expected cost function is greater than $\overline{F}$ for any S. So, from the results above, we can always set an upper bound $\overline{F}$ of the expected cost function $F_{ob}(S)$ no matter $\mu_l < \lambda$ or $\mu_l > \lambda$. Therefore, an interval $[\underline{S}, \overline{S}]$ which contain the optimal threshold S^* is available. Using the substitution method, we successively compare the function value $F_{ob}(S)$ and summarize the following algorithm.

Algorithm of seeking for S^*:

Step 1: Set $\overline{F}$ be an upper bound of the expected cost function $F_{ob}(S)$.

Step 2: Determine $\underline{S}$ and $\overline{S}$: if $\mu_l > \lambda$, $\underline{S}$ and $\overline{S}$ are determined from Theorem 11.6.9; if $\mu_l < \lambda$, define $\underline{S} = 1$ and $\overline{S}$ arbitrarily large.

Step 3: Denote initial $S^* = \underline{S}$, $S = \underline{S} + 1$.

Step 4: If $F_{ob}(S^*) \leq F_{ob}(S)$ and $S = \overline{S}$, yield S^* and stop.

If $F_{ob}(S^*) \leq F_{ob}(S)$ and $S < \overline{S}$, set $S = S + 1$ and carry out Step 4 again.

If $F_{ob}(S^*) > F_{ob}(S)$ and $S = \overline{S}$, set $S^* = \overline{S}$ and stop.

If $F_{ob}(S^*) > F_{ob}(S)$ and $S < \overline{S}$, set $S^* = S$, $S = S + 1$ and carry out Step 4 again.

11.6.4 *Unobservable Inventory Case*

In this section, customers can not observe the number of products which implies that neither the server's state nor the number of customers in system is observable to them. Similar to previous section, homogeneous customers follow a mixed strategy "enter with probability q and balk with probability $1 - q$" and the corresponding transition diagram can be seen in Fig. 11.12.

We first explore the individual behavior of customer who only considers his own benefit, and then investigate the Stackelberg game between the production manager and selfish customers, seek the optimal threshold to minimize the expected cost function. To get the results above, the steady state probabilities $\{p_{un}(i, k) : (i, k) \in \Omega\}$ of the system are necessary. Similar to Sect. 11.6.3, these probabilities exist under the condition $\lambda q < \mu_h$. For brevity we consider again only the typical case where $\lambda q \neq \mu_l$ and $\lambda q + \alpha \neq \mu_h$.

Lemma 11.6.2 *In the unobservable inventory case, the steady state probabilities* $\{p_{un}(i, k) : (i, k) \in \Omega\}$ *are given as follows:*

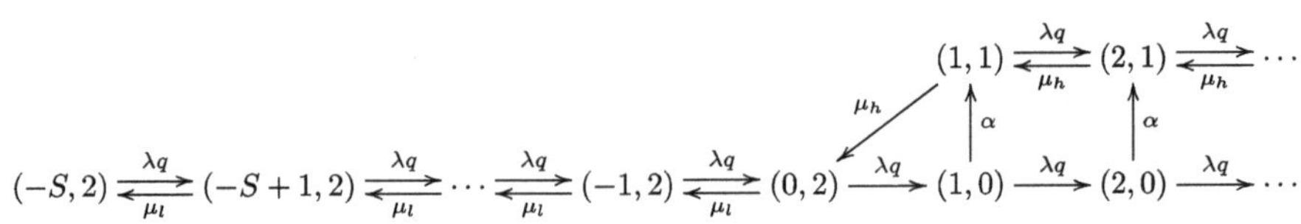

Fig. 11.12 Transition rate diagram of the system for unobservable inventory case

$$p_{un}(0,2) = \left\{ \frac{\mu_l \left[\left(\frac{\lambda q}{\mu_l} \right)^{-s} - 1 \right]}{\mu_l - \lambda q} + \frac{\mu_h(\lambda q + \alpha)}{\alpha(\mu_h - \lambda q)} \right\}^{-1}, \tag{11.6.14}$$

$$p_{un}(i,2) = \left(\frac{\lambda q}{\mu_l} \right)^i p_{un}(0,2), \quad i = -S, \ldots, -1, \tag{11.6.15}$$

$$p_{un}(i,0) = \left(\frac{\lambda q}{\lambda q + \alpha} \right)^i p_{un}(0,2), \quad i = 1, 2, \ldots, \tag{11.6.16}$$

$$p_{un}(i,1) = \frac{\lambda q + \alpha}{\mu_h - \lambda q - \alpha} \left[\left(\frac{\lambda q}{\lambda q + \alpha} \right)^i - \left(\frac{\lambda q}{\mu_h} \right)^i \right] p_{un}(0,2), \quad i = 1, 2, \ldots \tag{11.6.17}$$

Proof The proof is similar to Lemma 11.6.1 and therefore it is omitted here. □

11.6.4.1 Individual Optimum

In this part, we establish the individual equilibrium joining probability q_{un} and so the expected sojourn time is crucial to an arriving customer with no information. Similar to Theorem 11.6.1, we can derive the following result.

Theorem 11.6.10 *In the unobservable inventory case, the expected sojourn time of an arriving customer who decides to join the system is*

$$\mathbb{E}[W_{un}] = \frac{\mu_h(\lambda q + \alpha)(\mu_h + \alpha - \lambda q) p_{un}(0,2)}{\alpha^2(\mu_h - \lambda q)^2}. \tag{11.6.18}$$

Based on Little's law, the expected number of customers in system is

$$\mathbb{E}[N_{un}] = \lambda q \mathbb{E}[W_{un}] = \frac{\lambda q \mu_h(\lambda q + \alpha)(\mu_h + \alpha - \lambda q) p_{un}(0,2)}{\alpha^2(\mu_h - \lambda q)^2}. \tag{11.6.19}$$

To illustrate the unique individual equilibrium joining probability, the following theorem shows that $\mathbb{E}[W_{un}]$ is strictly increasing in q.

Theorem 11.6.11 *In the unobservable inventory case, the expected sojourn time of an arriving customer $\mathbb{E}[W_{un}]$ is strictly increasing in $q \in [0, 1]$.*

Proof From the Eqs. (11.6.14) and (11.6.18), we have

$$\begin{aligned} \mathbb{E}[W_{un}] &= \frac{\mu_h}{\alpha} \cdot \frac{(\lambda q)^S(\lambda q + \alpha)(\mu_l - \lambda q)(\mu_h + \alpha - \lambda q)}{\alpha \mu_l [\mu_l^S - (\lambda q)^S](\mu_h - \lambda q)^2 + \mu_h (\lambda q)^S (\lambda q + \alpha)(\mu_l - \lambda q)(\mu_h - \lambda q)} \\ &\triangleq \frac{\mu_h}{\alpha} \cdot \frac{nu}{de}. \end{aligned} \tag{11.6.20}$$

And furthermore,

$$
\begin{aligned}
(nu)' = \frac{d(nu)}{dq} =& S\lambda(\lambda q)^{S-1}(\lambda q+\alpha)(\mu_l-\lambda q)(\mu_h+\alpha-\lambda q)\\
&+\lambda(\lambda q)^S[\mu_l\mu_h-2\lambda q\mu_l-2\lambda q\mu_h+3(\lambda q)^2-\alpha\mu_h-\alpha^2],\\
(de)' = \frac{d(de)}{dq} =& S\lambda(\lambda q)^S(\mu_h-\lambda q)(\mu_l\mu_h-\lambda q\mu_h+\alpha\mu_l-\alpha\mu_h)\\
&-2\lambda\alpha\mu_l(\mu_h-\lambda q)[\mu_l^S-(\lambda q)^S]\\
&+\lambda\mu_h(\lambda q)^S[\mu_l\mu_h-2\lambda q\mu_l-2\lambda q\mu_h+3(\lambda q)^2-\alpha\mu_h+2\lambda q\alpha-\alpha\mu_l],\\
(nu)'\cdot(de) =& S\lambda\alpha\mu_l(\lambda q)^{S-1}(\lambda q+\alpha)(\mu_l-\lambda q)(\mu_h+\alpha-\lambda q)(\mu_h-\lambda q)^2[\mu_l^S-(\lambda q)^S]\\
&+S\lambda\mu_h(\lambda q)^{2S-1}(\lambda q+\alpha)^2(\mu_l-\lambda q)^2(\mu_h+\alpha-\lambda q)(\mu_h-\lambda q)\\
&+\lambda\alpha\mu_l(\lambda q)^S[\mu_l^S-(\lambda q)^S](\mu_h-\lambda q)^2[\mu_l\mu_h-2\lambda q\mu_l-2\lambda q\mu_h\\
&+3(\lambda q)^2-\alpha\mu_h-\alpha^2]+\lambda\mu_h(\lambda q)^{2S}(\lambda q+\alpha)(\mu_l-\lambda q)(\mu_h-\lambda q)\\
&\times[\mu_l\mu_h-2\lambda q\mu_l-2\lambda q\mu_h+3(\lambda q)^2-\alpha\mu_h-\alpha^2],\\
(nu)\cdot(de)' =& S\lambda(\lambda q)^{2S}(\lambda q+\alpha)(\mu_l-\lambda q)(\mu_h+\alpha-\lambda q)(\mu_l\mu_h-\lambda q\mu_h+\alpha\mu_l-\alpha\mu_h)\\
&-2\lambda\alpha\mu_l(\lambda q)^S(\lambda q+\alpha)(\mu_l-\lambda q)(\mu_h+\alpha-\lambda q)(\mu_h-\lambda q)[\mu_l^S-(\lambda q)^S]\\
&+\lambda\mu_h(\lambda q)^{2S}(\lambda q+\alpha)(\mu_l-\lambda q)(\mu_h+\alpha-\lambda q)[\mu_l\mu_h-2\lambda q\mu_l-2\lambda q\mu_h\\
&+3(\lambda q)^2-\alpha\mu_h+2\lambda q\alpha-\alpha\mu_l].
\end{aligned}
$$

Then we have

$$
\begin{aligned}
&(nu)'\cdot(de)-(nu)\cdot(de)'\\
=\;& \lambda\alpha\mu_h(\lambda q)^{2S}(\lambda q+\alpha)^2(\mu_l-\lambda q)^2\\
&+S\lambda\alpha\mu_l(\lambda q)^{2S-1}(\lambda q+\alpha)(\mu_l-\lambda q)(\mu_h+\alpha-\lambda q)(\mu_h-\lambda q)^2\\
&+S\lambda\alpha\mu_l(\lambda q)^{S-1}(\lambda q+\alpha)(\mu_l-\lambda q)(\mu_h+\alpha-\lambda q)(\mu_h-\lambda q)^2[\mu_l^S-(\lambda q)^S]\\
&+\lambda\alpha\mu_l(\lambda q)^S(\mu_h-\lambda q)[\mu_l^S-(\lambda q)^S][\mu_h(\mu_h-\lambda q)(\mu_l-\lambda q)+\alpha(\mu_h+\alpha)(\mu_l-\lambda q)]\\
&-\lambda\alpha\mu_l(\lambda q)^S(\mu_h-\lambda q)[\mu_l^S-(\lambda q)^S][\lambda q(\mu_h-\lambda q)^2+\alpha(\mu_h+\alpha)(\mu_h-\mu_l)]\\
=\;& \lambda\alpha\mu_h(\lambda q)^{2S}(\lambda q+\alpha)^2(\mu_l-\lambda q)^2\\
&+\lambda\alpha\mu_l(\lambda q)^{2S}\Big\{S(\lambda q+\alpha)(\varphi-1)(\mu_h+\alpha-\lambda q)(\mu_h-\lambda q)^2\\
&+S(\lambda q+\alpha)(\varphi-1)(\mu_h+\alpha-\lambda q)(\mu_h-\lambda q)^2(\varphi^S-1)\\
&+(\mu_h-\lambda q)(\varphi^S-1)(\mu_l-\lambda q)[\mu_h(\mu_h-\lambda q)+\alpha(\mu_h+\alpha)]\\
&-(\mu_h-\lambda q)(\varphi^S-1)[\lambda q(\mu_h-\lambda q)^2+\alpha(\mu_h+\alpha)(\mu_h-\mu_l)]\Big\}\\
=\;& \lambda\alpha\mu_h(\lambda q)^{2S}(\lambda q+\alpha)^2(\mu_l-\lambda q)^2\\
&+\lambda\alpha\mu_l(\lambda q)^{2S}(\mu_h-\lambda q)\Big\{S\varphi^S(\lambda q+\alpha)(\varphi-1)(\mu_h+\alpha-\lambda q)(\mu_h-\lambda q)\\
&+(\varphi^S-1)(\mu_l-\lambda q)[\mu_h(\mu_h-\lambda q)+\alpha(\mu_h+\alpha)]\\
&-(\varphi^S-1)[\lambda q(\mu_h-\lambda q)^2+\alpha(\mu_h+\alpha)(\mu_h-\mu_l)]\Big\}.
\end{aligned}
$$

Considering Proposition A stated in Appendix, we have $\frac{d\mathbb{E}[W_{un}]}{dq} = \frac{\mu_h}{\alpha} \cdot \frac{(nu)' \cdot (de) - (nu) \cdot (de)'}{(de)^2} > 0$, $\mathbb{E}[W_{un}]$ is strictly increasing in $q \in [0, 1]$. This completes the proof. □

If the other customers refuse to join the system, a tagged customer must enter the system. This is because of $\lim_{q \to 0} p(-S, 2) = 1$ which coincides with the fact, from Eqs. (11.6.18) or (11.6.20), the expected sojourn time of the tagged customer $\mathbb{E}[W_{un}]$ tends to zero as $q \to 0$. Therefore $q = 0$ is not an equilibrium joining probability and this fact is quite different from the other existing papers.

Next, we seek the nonzero equilibrium joining probability. Note that the expected sojourn time of an arriving customer reaches its maximum $\overline{W} = \frac{\mu_h \lambda^S (\lambda+\alpha)(\mu_l - \lambda)(\mu_h + \alpha - \lambda)}{\alpha^2 \mu_l (\mu_l^S - \lambda^S)(\mu_h - \lambda)^2 + \alpha \mu_h \lambda^S (\lambda + \alpha)(\mu_l - \lambda)(\mu_h - \lambda)}$ when $q = 1$, then we have the following theorem.

Theorem 11.6.12 *In the unobservable inventory case, a unique mixed strategy "the customer chooses to enter the system with probability q_{un}, balks with probability $1 - q_{un}$" exists and the probability is given by*

$$q_{un} = \begin{cases} q_{un}^*, & \text{if} \quad R < C\overline{W}, \\ 1, & \text{if} \quad R \geq C\overline{W}, \end{cases} \tag{11.6.21}$$

where q_{un}^* is the unique solution of $I_{un}(q) = R - C\mathbb{E}[W_{un}] = 0$.

Proof From Eq. (11.6.20), the expected net benefit of a tagged customer who decides to enter the system at his arrival instant is

$$\begin{aligned} I_{un}(q) = & \; R - C\mathbb{E}[W_{un}] \\ = & \; R - C\frac{\mu_h}{\alpha} \cdot \frac{(\lambda q)^S (\lambda q + \alpha)(\mu_l - \lambda q)(\mu_h + \alpha - \lambda q)}{\alpha \mu_l [\mu_l^S - (\lambda q)^S](\mu_h - \lambda q)^2 + \mu_h (\lambda q)^S (\lambda q + \alpha)(\mu_l - \lambda q)(\mu_h - \lambda q)}, \end{aligned}$$

and it is easy to show that $I_{un}(q)$ is strictly decreasing in $q \in [0, 1]$. If $I_{un}(1) = R - C\overline{W} \geq 0$ or $R \geq C\overline{W}$, the customer prefers to join the system because the minimum benefit is nonnegative. The second branch of (11.6.21) is at hand. If $R < C\overline{W}$, then there exists a unique $q_{un}^* \in (0, 1)$ such that $I_{un}(q) = 0$. In this case, the tagged customer is indifferent between joining and balking the system.

This completes the proof. □

Follow a similar analysis, when $\lambda \geq \mu_h$, it is readily shown that the result in Theorem 11.6.12 also holds and the unique mixed strategy "the customer chooses to enter the system with probability q_{un}, balks with probability $1 - q_{un}$" exists and the probability is given by $q_{un} = q_{un}^*$.

11.6.4.2 Cost Optimum

Similar to the Sect. 11.6.3.2, we seek for the optimal threshold to minimize the expected cost function

$$F_{un}(S) = C\mathbb{E}[N_{un}] + f\mathbb{E}[I_{un}], \tag{11.6.22}$$

where $\mathbb{E}[N_{un}]$ is given in Eq. (11.6.19) and $\mathbb{E}[I_{un}]$ is the expected number of products in inventory. Similar to Theorems 11.6.3–11.6.6, we also have the similar conclusions and omit their proofs.

Theorem 11.6.13 *In the unobservable inventory case, the expected number of products in inventory is*

$$\mathbb{E}[I_{un}] = \frac{\varphi[(1-\varphi^S) - S\varphi^S(1-\varphi)]p_{un}(0,2)}{(1-\varphi)^2} \tag{11.6.23}$$

$$= \frac{\mu_l\{\lambda q[(\lambda q)^S - \mu_l^S] - S\mu_l^S(\lambda q - \mu_l)\}p_{un}(0,2)}{(\lambda q)^S(\mu_l - \lambda q)^2}. \tag{11.6.24}$$

Theorem 11.6.14 *In the unobservable inventory case, the following results hold true for $S \geq 1$:*

(1) $p_{un}(0,2)$, $\mathbb{E}[W_{un}]$ and $\mathbb{E}[N_{un}]$ are all strictly decreasing in S;

(2) If $\mu_l < \lambda q$, $p_{un}(0,2)$, $\mathbb{E}[W_{un}]$ and $\mathbb{E}[N_{un}]$ are all strictly convex in S;

(3) If $\mu_l > \lambda q$ and $\psi(\mu_l - \lambda q) \leq \mu_l$, $p_{un}(0,2)$, $\mathbb{E}[W_{un}]$ and $\mathbb{E}[N_{un}]$ are all strictly convex in S;

(4) If $\mu_l > \lambda q$ and $\psi(\mu_l - \lambda q) > \mu_l$, $p_{un}(0,2)$, $E[W_{un}]$ and $\mathbb{E}[N_{un}]$ are all concave for $S \leq \max\{1, C\}$ and convex for $S \geq \max\{1, C\}$, where $C = \lfloor \log_{\frac{\mu_l}{\lambda q}}(\psi(\frac{\mu_l}{\lambda q} - 1) - \frac{\mu_l}{\lambda q}) - 2 \rfloor$.

Theorem 11.6.15 *In the unobservable inventory case, $\mathbb{E}[I_{un}]$ is strictly increasing in S.*

Theorem 11.6.16 *In the unobservable inventory case, (1) $\mathbb{E}[I_{un}]$ is quasi-convex in S if $(S+1)(\lambda q - \mu_l) \leq \mu_l < \lambda q$; (2) If S is a continuous variable, $\mathbb{E}[I_{un}]$ is pseudo-convex in S if $(S+1)(\lambda q - \mu_l) \leq \mu_l < \lambda q$.*

Proof From the Eqs. (11.6.14) and (11.6.23) we have

$$\mathbb{E}[I_{un}] = \frac{\varphi[(1-\varphi^S) - S\varphi^S(1-\varphi)]}{(\varphi-1)[\varphi(\varphi^S-1) + \psi(\varphi-1)]}.$$

Obviously, the numerator and denominator of $\mathbb{E}[I_{un}]$ is positive for any $S > 0$ no matter $\varphi < 1$ or $\varphi > 1$. And the second order difference of numerator is

$$\varphi\{[(1-\varphi^{S+2}) - (S+2)\varphi^{S+2}(1-\varphi)] - 2[(1-\varphi^{S+1}) - (S+1)\varphi^{S+1}(1-\varphi)] \\ +[(1-\varphi^{S}) - S\varphi^{S}(1-\varphi)]\} = \varphi^{S}(1-\varphi)^2[(S+1)(\varphi-1)+\varphi].$$

Hence the numerator of $\mathbb{E}[I_{un}]$ is convex if $(S+1)(\varphi-1)+\varphi \geq 0$.
Moreover the second order difference of denominator is

$$(\varphi-1)\{[\varphi(\varphi^{S+2}-1) + \psi(\varphi-1)] - 2[\varphi(\varphi^{S+1}-1) \\ +\psi(\varphi-1)] + [\varphi(\varphi^{S}-1) + \psi(\varphi-1)]\} = (\varphi-1)^3\varphi^{S+1}.$$

Hence the denominator of $\mathbb{E}[I_{un}]$ is concave on S if $\varphi < 1$.
In virtue of the property of pseudo-convex functions (Appendix provides this property named "Proposition A" and also its proof), we get the conclusion. □

However, q_{un} is an implicit function in S based on Theorem 11.6.12, and hence it is intractable to establish the properties of F_{un} in S after putting q_{un} into the Eq. (11.6.22). So we have some numerical experiments and make some comparison with the results of observable case in the next section.

11.7 Numerical Experiments

In this section we show the tendency of the cost functions $F_{ob}(S)$, $F_{un}(S)$ with respect to inventory threshold S and then seek the optimal S^* that minimizes the cost functions $F_{ob}(S)$ and $F_{un}(S)$, respectively. To this end, we first present the figures of some performance measures, including equilibrium joining probabilities q_{ob}, q_{un}, the expected number of customers in system $\mathbb{E}[N_{ob}]$, $\mathbb{E}[N_{un}]$, the expected number of products in inventory $\mathbb{E}[I_{ob}]$, $\mathbb{E}[I_{un}]$. Since the results are similar regardless of whether $\mu_l > \lambda$ or $\mu_l < \lambda$, we focus on the case $\mu_l > \lambda$ and the related parameters setting are $\lambda = 5.5$, $\mu_l = 6$, $\mu_h = 6.5$, $\alpha = 2.5$, $R = 7$.

Based on the parameters above, from the Eq. (11.6.6), it is easy to show that $q_{ob} = 1$ holds true when $C \leq 5$. Similarly, $q_{ob} = 0$ if $c \geq 12.64$. Thus we pay more attention to the strategic behavior of customers for four different values when $C = 5, 6.5, 8, 9.5$ within the interval $[5, 12]$. Combining the monotonicity of $\mathbb{E}[W_{un}]$ proved in Theorem 11.6.14 with the definition of $\overline{W}$, we know $\overline{W}$ is also decreasing in S. Therefore, from (11.6.21), we also have $q_{un} = 1$ if $C < 5.26$ which implies that the observable case and the unobservable case are entirely identical when $C \leq 5$.

Since the parameter f has no influence on the strategic behavior of customer, considering Eqs. (11.6.10) or (11.6.22), we fix $f = 8$ and adopt variational cost coefficient C to explore the optimal threshold S^*.

Firstly, it is understandable that: (1) The joining probability q_{ob} in observable case is independent of S from Theorem 11.6.2 and thus q_{ob} is constant for any S; (2) Although the joining probability q_{un} in unobservable case has no obvious

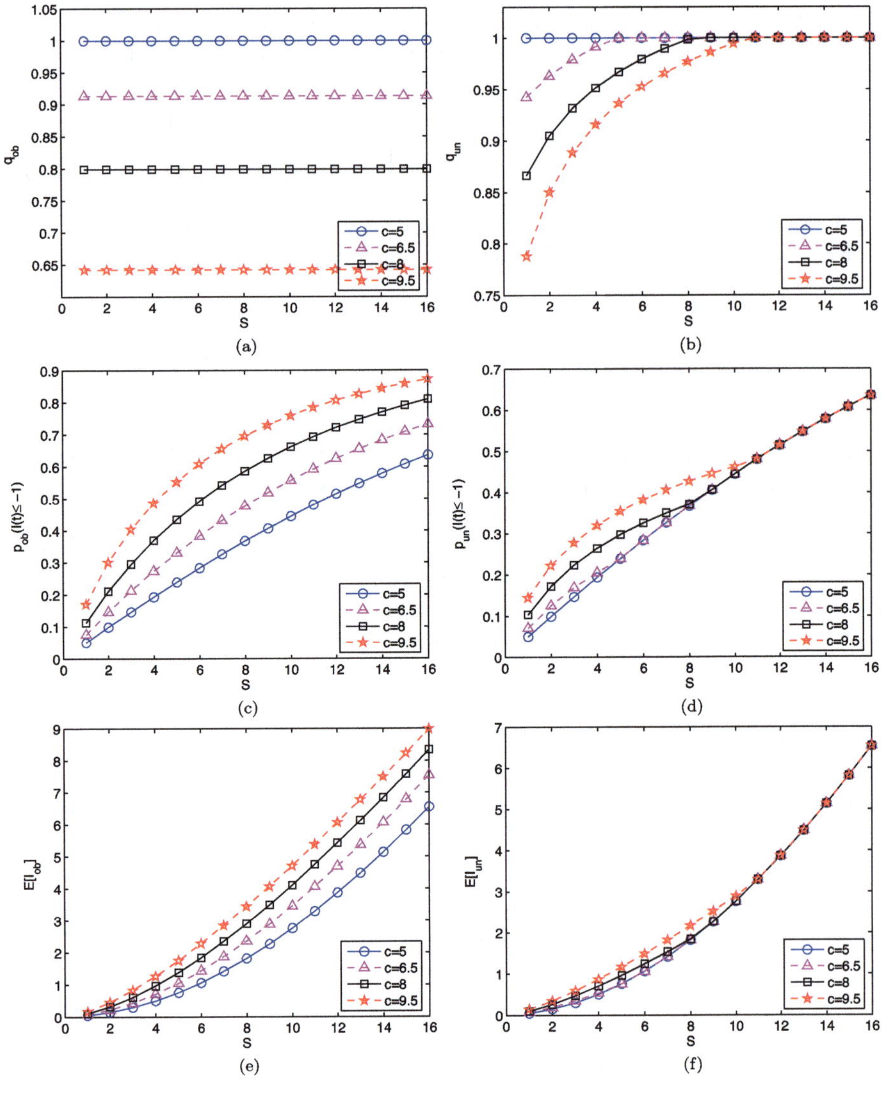

Fig. 11.13 The joining probability q_{ob}, q_{un}, the probability of nonempty inventory $p_{ob}(I(t) \leq -1)$, $p_{un}(I(t) \leq -1)$, and the expected number of products in inventory $\mathbb{E}[I_{ob}]$, $\mathbb{E}[I_{un}]$ versus S

difference for greater S, q_{un} is increasing in S because a customer has a higher probability of taking one product away without delay; (3) Similar to (2), with greater S, the probabilities of nonempty inventory $p_{ob}(I(t) \leq -1)$ and $p_{un}(I(t) \leq -1)$ are both increasing in S; (4) q_{ob} is the joining probability when the inventory is empty while q_{un} is the joining probability when the inventory is unobservable, thus q_{ob} is not greater than q_{un}. For different c, all these conclusions can be verified from the Fig. 11.13a–d.

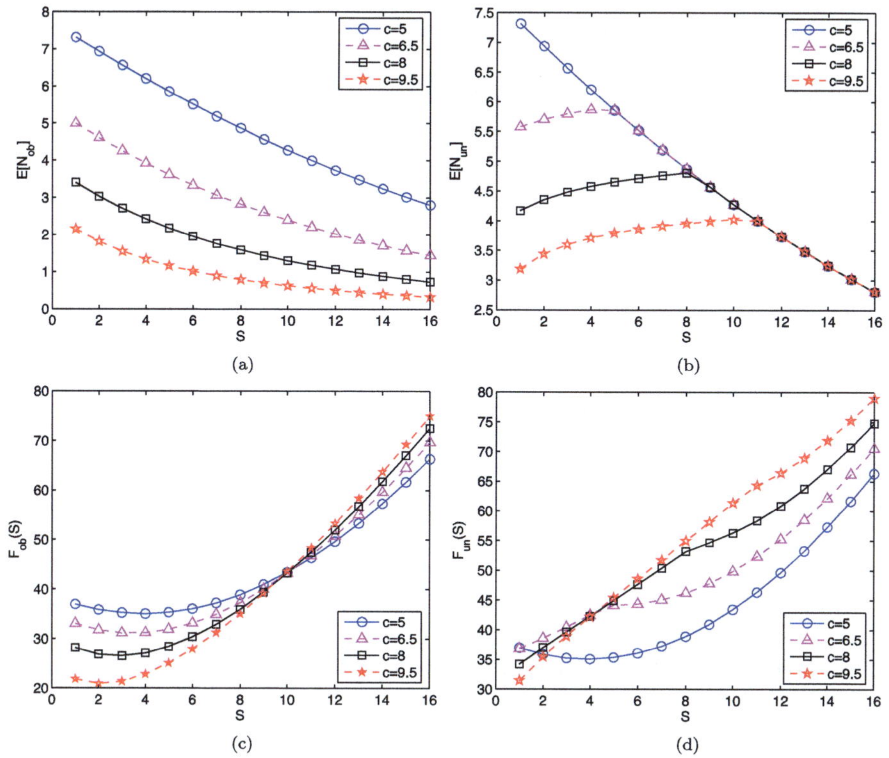

Fig. 11.14 The expected number of customers in system $\mathbb{E}[N_{ob}]$, $\mathbb{E}[N_{un}]$, and the expected cost function F_{ob}, F_{un} versus S

Secondly, with S increases, the probability $p_{ob}(I(t) \leq -1)$ or $p_{un}(I(t) \leq -1)$ is increasing in S and therefore the expected number of products in inventory also increases in S. This tendency can be seen in Fig. 11.13e, f and is in accordance with the results in Theorems 11.6.5 and 11.6.15. For the same reason, the expected number of customers in system $\mathbb{E}[N_{ob}]$ is decreasing in S and the tendency of $\mathbb{E}[N_{ob}]$ in Fig. 11.14a also coincides with Theorem 11.6.4(1).

Remarkably, this is not the case for $\mathbb{E}[N_{un}]$. Although $\mathbb{E}[N_{un}]$ is proved to be decreasing in S from Theorem 11.6.14(1), we find that $\mathbb{E}[N_{un}]$ is not decreasing in S from Fig. 11.14b. This is not self-contradictory: In Theorem 11.6.14(1), q is a constant, independent of the S and then $p_{un}(0, 2)$ and $\mathbb{E}[N_{un}]$ have the same monotonicity. However, from Eq. (11.6.21), $q = q_{un}$ is dependent of S in Sect. 11.6.4.1 and therefore the monotonicity of $\mathbb{E}[N_{un}]$ would be changed after q_{un} is drawn into the Eq. (11.6.19). Comparing Fig. 11.13b with Fig. 11.14b, if $q_{un} = 1$ (independent of S), $\mathbb{E}[N_{un}]$ is decreasing in S because of Theorem 11.6.14(1). On the contrary, if $q_{un} \neq 1$ (dependent of S), greater proportion of customers prefer to join the system with S increases. This tendency can be seen in Fig. 11.13b. Thus $\mathbb{E}[N_{un}]$ is increasing in S until all the customers choose to join the system, i.e., $\mathbb{E}[N_{un}]$ is decreasing in S once joining probability q_{un} equals to 1.

Table 11.4 Performance measures of the system when $S = 0$

c	5	6.5	8	9.5
$q_{ob} = q_{un}$	1	0.9132	0.7990	0.6420
$\mathbb{E}[N_{ob}] = \mathbb{E}[N_{un}]$	7.7000	5.4091	3.8454	2.6020
$F_{ob}(S) = F_{un}(S)$	38.5000	35.1591	30.7632	24.7188

Thirdly, we compare the system cost between the cases $S = 0$ and $S \neq 0$. As stated before, the model degenerates into Make-to-Order case when $S = 0$.

On one hand, the cost functions $F_{ob}(S)$ and $F_{un}(S)$ are exactly the same with each other when $c \leq 5$. On the other hand, $F_{un}(S)$ in Fig. 11.14d is far beyond the cost in Table 11.4 for greater C. Therefore, to minimize the system cost, we only need to make comparison of the cost function $F_{ob}(S)$ in Table 11.4 with that in Fig. 11.14c. It's obvious that Make-to-Stock policy does not always dominate Make-to-Order policy. In Fig. 11.15, we depict a demarcation line to separate Make-to-Stock policy from Make-to-Order policy aiming at different delay parameter c. For a certain C, preparing some products in advance is better for the whole system, and on the contrary more products would increase the burden of the system after passing through this line. Therefore, it is more economic to reserve none or less products rather than to prepare more products.

Finally, as C increases, customers have less motivation to join the system and so the joining probability is monotone decreasing. And then the expected number of customers in system $\mathbb{E}[N_{ob}]$ and the expected number of products in inventory $\mathbb{E}[I_{ob}]$ have the distinct monotonicity which implies the cost function $F_{ob}(S)$ does not possess monotonicity in C. This tendency can be seen from Figs. 11.13 and 11.14. Furthermore, this result also illustrates the demarcation point in Fig. 11.15 is not monotone in C.

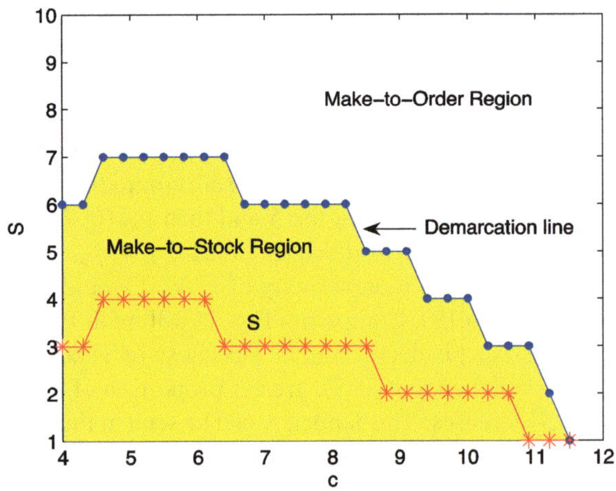

Fig. 11.15 Make-to-Stock region, Make-to-Order region and the optimal threshold S^*

Beyond this, based on the Theorem 11.6.7, $F_{ob}(S)$ possesses quasi-convexity for smaller S and this tendency can be seen in Fig. 11.14c. Combining its boundedness proved in Theorem 11.6.8, the optimal threshold S^* can be found in the Make-to-Stock region of Fig. 11.15.

11.8 Conclusions

We consider a Make-to-Stock production system with one product type and dynamic service policy. According to whether the number of customers is zero or not, the system operates with low or high production rate. Under the observable inventory and unobservable inventory cases, we explore the strategic behavior of customers and then establish a Stackelberg game between production manager and customers. The numerical results demonstrate that if the customers are patient enough, the expected cost functions in these two cases are entirely identical. And with increasing of customers' delay sensitivity, the expected cost function in observable case is superior to the unobservable case. Finally, a demarcation line is depicted to differentiate Make-to-Stock policy from Make-to-Order policy and the optimal inventory threshold is found within the Make-to-Stock region.

11.9 Appendix

Proof of Lemma 11.6.1 Considering the Fig. 11.11, we have the following balance equations

$$\lambda p_{ob}(i,2) = \mu_l p_{ob}(i+1,2), \ i=-S,\ldots,-2, \tag{11.9.1}$$
$$\lambda p_{ob}(-1,2) = \mu_l p_{ob}(0,2), \tag{11.9.2}$$
$$(\lambda q+\alpha) p_{ob}(i,0) = \lambda q p_{ob}(i-1,0), \ i\geq 1, \tag{11.9.3}$$
$$\lambda q p_{ob}(0,2) = \mu_h p_{ob}(1,1), \tag{11.9.4}$$
$$(\lambda q+\mu_h) p_{ob}(1,1) = \mu_h p_{ob}(2,1)+\alpha p_{ob}(1,0), \tag{11.9.5}$$
$$(\lambda q+\mu_h) p_{ob}(i,1) = \lambda q p_{ob}(i-1,1)+\mu_h p_{ob}(i+1,1)+\alpha p_{ob}(i,0), \ i\geq 2. \tag{11.9.6}$$

From Eqs. (11.9.1)–(11.9.3), we have the results of Eqs. (11.6.2)–(11.6.3). According to Eqs. (11.9.6) and (11.6.3), the steady probabilities $p_{ob}(i,1), \ i = 2,3,\ldots$ are the solution of the nonhomogeneous linear difference equation with constant coefficients as follows

$$\begin{aligned} &\mu_h p_{ob}(i+1,1) - (\lambda q+\mu_h) p_{ob}(i,1) + \lambda q p_{ob}(i-1,1) \\ &= -\alpha p_{ob}(i,0) = -\alpha\left(\frac{\lambda q}{\lambda q+\alpha}\right)^i p_{ob}(0,2), \ i=2,3,\ldots, \end{aligned} \tag{11.9.7}$$

where the corresponding characteristic equation is

$$\mu_h r^2 - (\lambda q + \mu_h) r + \lambda q = 0$$

with roots $r_1 = \frac{\lambda q}{\mu_h}$ and $r_2 = 1$. Hence the general solution of the corresponding homogeneous linear difference equation of (11.9.7) is

$$p_{ob}^{hom}(i, 1) = \varepsilon_1 (\frac{\lambda q}{\mu_h})^i + \varepsilon_2 1^i, \quad i = 1, 2, \ldots,$$

where ε_1 and ε_2 are coefficients to be determined. Because the nonhomogeneous part of Eq. (11.9.7) is $-\alpha(\frac{\lambda q}{\lambda q+\alpha})^i p_{ob}(0, 2)$ that is geometric with $\frac{\lambda q}{\lambda q+\alpha}$, the specific solution of Eq. (11.9.7) is the form of $\varepsilon_3(\frac{\lambda q}{\lambda q+\alpha})^i$. Hence the general solution of Eq. (11.9.7) is

$$p_{ob}(i, 1) = \varepsilon_1 \left(\frac{\lambda q}{\mu_h}\right)^i + \varepsilon_2 1^i + \varepsilon_3 \left(\frac{\lambda q}{\lambda q + \alpha}\right)^i, \quad i = 1, 2, \ldots,$$

where ε_3 is coefficient to be determined. We note that $r_2 = 1$, then ε_2 should be necessarily 0 because $p_{ob}(i, 1)$, $i = 1, 2, \ldots$, are probabilities. Then $p_{ob}(i, 1)$ can be simplified as

$$p_{ob}(i, 1) = \varepsilon_1 \left(\frac{\lambda q}{\mu_h}\right)^i + \varepsilon_3 \left(\frac{\lambda q}{\lambda q + \alpha}\right)^i, \quad i = 1, 2, \ldots. \tag{11.9.8}$$

Putting it into the Eq. (11.9.7) we get $\varepsilon_3 = \frac{(\lambda q+\alpha) p_{ob}(0,2)}{\mu_h - \lambda q - \alpha}$. Furthermore, we have $\varepsilon_1 = -\varepsilon_3$ from Eq. (11.9.4). So the result of Eq. (11.6.4) is at hand. Finally, the remaining probability, $p_{ob}(0, 2)$, can be found from the normalization equation

$$1 = \sum_{i=-S}^{-1} p_{ob}(i, 2) + \sum_{i=0}^{\infty} p_{ob}(i, 0) + \sum_{i=1}^{\infty} p_{ob}(i, 1).$$

This completes the proof. □

Proof of Theorem 11.6.4 From the definition (11.6.7), we just prove the conclusion for $p_{ob}(0, 2)$. By definition (11.6.1),

$$p_{ob}^S(0, 2) \triangleq p_{ob}(0, 2) = \frac{q\varphi - 1}{q\varphi[(q\varphi)^S - 1] + \psi(q\varphi - 1)}. \tag{11.9.9}$$

Note that the numerator and denominator of $p_{ob}^S(0, 2)$ have the same sign no matter $q\varphi < 1$ or $q\varphi > 1$. Then the first order difference Λ^1 and the second order difference Λ^2 of $p_{ob}^S(0, 2)$ in S are

$$
\begin{aligned}
\Lambda^1 =& \; p_{ob}^{S+1}(0,2) - p_{ob}^{S}(0,2)\\
=& \; \frac{q\varphi-1}{q\varphi[(q\varphi)^{S+1}-1]+\psi(q\varphi-1)} - \frac{q\varphi-1}{q\varphi[(q\varphi)^{S}-1]+\psi(q\varphi-1)}\\
=& \; \frac{(q\varphi-1)\{q\varphi[(q\varphi)^S-1]+\psi(q\varphi-1)-q\varphi[(q\varphi)^{S+1}-1]-\psi(q\varphi-1)\}}{\{q\varphi[(q\varphi)^{S+1}-1]+\psi(q\varphi-1)\}\{q\varphi[(q\varphi)^{S}-1]+\psi(q\varphi-1)\}}\\
=& \; -\frac{(q\varphi)^{S+1}(q\varphi-1)^2}{\{q\varphi[(q\varphi)^{S+1}-1]+\psi(q\varphi-1)\}\{q\varphi[(q\varphi)^{S}-1]+\psi(q\varphi-1)\}}
\end{aligned}
$$

and

$$
\begin{aligned}
\Lambda^2 =& \; p_{ob}^{S+2}(0,2) - 2p_{ob}^{S+1}(0,2) + p_{ob}^{S}(0,2)\\
=& \; [p_{ob}^{S+2}(0,2) - p_{ob}^{S+1}(0,2)] - [p_{ob}^{S+1}(0,2) - p_{ob}^{S}(0,2)] = \frac{\Lambda_n^2}{\Lambda_d^2}.
\end{aligned}
$$

where $\Lambda_n^2 = (q\varphi)^{S+1}(q\varphi-1)^3[(q\varphi)^{S+2}+q\varphi-\psi(q\varphi-1)]$, $\Lambda_d^2 = \{q\varphi[(q\varphi)^{S+2}-1]+\psi(q\varphi-1)\}\{q\varphi[(q\varphi)^{S+1}-1]+\psi(q\varphi-1)\}\{q\varphi[(q\varphi)^{S}-1]+\psi(q\varphi-1)\}$.

It is obvious that the conclusion (1) holds true from the definition of $p_{ob}^{S}(0,2)$ and Λ^1.

If $q\varphi < 1$, then $\Lambda^2 > 0$ and the conclusion (2) holds true.

If $q\varphi > 1$ and $\psi(q\varphi-1) \le q\varphi$, then $(q\varphi)^{S+2}+q\varphi-\psi(q\varphi-1) > 0$ and $\Lambda^2 > 0$. Therefore the conclusion (3) holds true.

If $q\varphi > 1$ and $\psi(q\varphi-1) > q\varphi$, then $p_{ob}^{S}(0,2)$ is convex (concave) $\Leftrightarrow$ $\Lambda^2 \ge 0 (\le 0)$ $\Leftrightarrow$ $(q\varphi)^{S+2}+q\varphi-\psi(q\varphi-1) \ge 0 (\le 0)$ $\Leftrightarrow$ $S \ge (\le) \log_{q\varphi}(\psi(q\varphi-1)-q\varphi)-2$. Denote $B = \lfloor \log_{q\varphi}(\psi(q\varphi-1)-q\varphi)-2 \rfloor$, then the conclusion (4) holds true. This completes the proof. □

Proof of Theorem 11.6.18 If the system is in state $(i,2)$, $i \le -1$ upon arrival, the sojourn time of an arriving customer is zero. If the system is in state $(0,2)$ upon arrival, the sojourn time of an arriving customer is $\frac{1}{\alpha}+\frac{1}{\mu_h}$. If the system is in state $(i,0)$, $i \ge 1$ upon arrival, the sojourn time of an arriving customer is $\frac{1}{\alpha}+\frac{i+1}{\mu_h}$. If the system is in state $(i,1)$, $i \ge 1$ upon arrival, the sojourn time of an arriving customer is $\frac{i+1}{\mu_h}$. Thus, similar to Theorem 11.6.1, the expected sojourn time of an arriving customer is

$$
\begin{aligned}
\mathbb{E}[W_{un}] =& p_{un}(0,2)\left(\frac{1}{\alpha}+\frac{1}{\mu_h}\right) + \sum_{i=1}^{\infty} p_{un}(i,0)\left(\frac{1}{\alpha}+\frac{i+1}{\mu_h}\right) + \sum_{i=1}^{\infty} p_{un}(i,1)\frac{i+1}{\mu_h}\\
=& p_{un}(0,2)\left\{\sum_{i=0}^{\infty}\left(\frac{\lambda q}{\lambda q+\alpha}\right)^i\left(\frac{1}{\alpha}+\frac{i+1}{\mu_h}\right) + \sum_{i=1}^{\infty}\frac{\lambda q+\alpha}{\mu_h-\lambda q-\alpha}\left[\left(\frac{\lambda q}{\lambda q+\alpha}\right)^i - \left(\frac{\lambda q}{\mu_h}\right)^i\right]\frac{i+1}{\mu_h}\right\}\\
=& p_{un}(0,2)\left\{\frac{1}{\alpha}\sum_{i=0}^{\infty}\left(\frac{\lambda q}{\lambda q+\alpha}\right)^i + \frac{1}{\mu_h}\sum_{i=0}^{\infty}(i+1)\left(\frac{\lambda q}{\lambda q+\alpha}\right)^i\right.\\
&\left. + \frac{\lambda q+\alpha}{\mu_h(\mu_h-\lambda q-\alpha)}\sum_{i=1}^{\infty}(i+1)\left[\left(\frac{\lambda q}{\lambda q+\alpha}\right)^i - \left(\frac{\lambda q}{\mu_h}\right)^i\right]\right\}.
\end{aligned}
$$

Note that $\sum_{i=0}^{\infty}(i+1)x^i = \frac{1}{(1-x)^2}$ and $\sum_{i=1}^{\infty}(i+1)x^i = \frac{1}{(1-x)^2} - 1$, then

$$\begin{aligned}\mathbb{E}[W_{un}] =& p_{un}(0,2)\Big\{\frac{\lambda q+\alpha}{\alpha^2} + \frac{(\lambda q+\alpha)^2}{\mu_h\alpha^2} + \frac{\lambda q+\alpha}{\mu_h(\mu_h-\lambda q-\alpha)}[\frac{(\lambda q+\alpha)^2}{\alpha^2} - \frac{\mu_h^2}{(\mu_h-\lambda q)^2}]\Big\}\\ =& p_{un}(0,2)\Big\{\frac{(\lambda q+\alpha)(\mu_h+\lambda q+\alpha)}{\mu_h\alpha^2} + \frac{\lambda q(\lambda q+\alpha)[(\lambda q+\alpha)(\mu_h-\lambda q)+\alpha\mu_h]}{\mu_h\alpha^2(\mu_h-\lambda q)^2}\Big\}\\ =& \frac{\mu_h(\lambda q+\alpha)(\mu_h+\alpha-\lambda q)p_{un}(0,2)}{\alpha^2(\mu_h-\lambda q)^2}.\end{aligned}$$

This completes the proof. □

Proposition A (property of pseudo-convex functions) For any φ,

$$\begin{aligned} g =\ & S\varphi^S(\lambda q+\alpha)(\varphi-1)(\mu_h+\alpha-\lambda q)(\mu_h-\lambda q) + (\varphi^S-1)(\mu_l-\lambda q)[\mu_h(\mu_h-\lambda q)+\alpha(\mu_h+\alpha)]\\ & -(\varphi^S-1)[\lambda q(\mu_h-\lambda q)^2+\alpha(\mu_h+\alpha)(\mu_h-\mu_l)]\\ >\ & 0.\end{aligned}$$

Proof of Proposition A: For $\varphi > 1$, $S\varphi^S > 1+\rho+\ldots+\varphi^{S-1}$ and then $S\varphi^S(\varphi-1) > (\varphi-1)(1+\rho+\cdots+\varphi^{S-1}) = \varphi^S - 1 > 0$. Therefore we have

$$\begin{aligned} g >\ & (\varphi^S-1)\Big\{(\lambda q+\alpha)(\mu_h+\alpha-\lambda q)(\mu_h-\lambda q) + (\mu_l-\lambda q)[\mu_h(\mu_h-\lambda q)+\alpha(\mu_h+\alpha)]\\ & -[\lambda q(\mu_h-\lambda q)^2+\alpha(\mu_h+\alpha)(\mu_h-\mu_l)]\Big\}\\ =\ & (\varphi^S-1)(\mu_l-\lambda q)[\mu_h(\mu_h-\lambda q)+2\alpha(\mu_h+\alpha)] > 0.\end{aligned}$$

For $\varphi < 1$, $S\varphi^S < 1+\rho+\cdots+\varphi^{S-1}$ and then $S\varphi^S(\varphi-1) > (\varphi-1)(1+\rho+\cdots+\varphi^{S-1}) = \varphi^S - 1$. Therefore we have

$$\begin{aligned} g >\ & (\varphi^S-1)\Big\{(\lambda q+\alpha)(\mu_h+\alpha-\lambda q)(\mu_h-\lambda q) + (\mu_l-\lambda q)[\mu_h(\mu_h-\lambda q)+\alpha(\mu_h+\alpha)]\\ & -[\lambda q(\mu_h-\lambda q)^2+\alpha(\mu_h+\alpha)(\mu_h-\mu_l)]\Big\}\\ =\ & (\varphi^S-1)(\mu_l-\lambda q)[\mu_h(\mu_h-\lambda q)+2\alpha(\mu_h+\alpha)] > 0.\end{aligned}$$

This completes the proof. □

References

Baek, J. W., and S. K. Moon. 2014. The M/M/1 queue with a production-inventory system and lost sales. *Applied Mathematics and Computation* 233: 534–544.

Bazaraa, Mokhtar S., Hanif D. Sherali, and Chitharanjan M. Shetty. 2006. *Nonlinear Programming: Theory and Algorithms*. Wiley.

Berman, Oded, and Eungab Kim. 1999. Stochastic models for inventory management at service facilities. *Stochastic Models* 15 (4): 695–718.

Berman, Oded, and K. P. Sapna. 2000. Inventory management at service facilities for systems with arbitrarily distributed service times. *Stochastic Models* 16 (3&4): 343–360.

Bertsimas, Dimitri, and I. C. Paschalidis. 2001. Probabilistic service level guarantees in make-to-stock manufacturing systems. *Operations Research* 49 (1): 119–133.

Bijvank, M., and I. F. A. Vis. 2011. Lost-sales inventory theory: A review. *European Journal of Operational Research* 215 (1): 1–13.

Boudali, O., and Antonis Economou. 2012. Optimal and equilibrium balking strategies in the single server Markovian queue with catastrophes. *European Journal of Operational Research* 218: 708–715.

Burnetas, A., and Antonis Economou. 2007. Equilibrium customer strategies in a single server Markovian queue with setup times. *Queueing Systems* 56: 213–228.

Carlisle, A., and G. Dozier. 2001. An off-the-Shelf PSO. In *Proceeding of the Workshop on Particle Swarm Optimization*. Indianapolis: Purdue School of Engineering and Technology.

Chen, Hong, and Murray Frank. 2001. State dependent pricing with a queue. *IIE Transactions* 33: 847–860.

Chen, Hong, and Murray Frank. 2004. Monopoly pricing when customers queue. *IIE Transactions* 36: 569–581.

Chen, X., and D. S. Levi. 2004. Coordinating inventory control and pricing strategies with random demand and fixed ordering cost: The finite horizon case. *Operations Research* 52 (6): 887–896.

Chen, X., and D. S. Levi. 2006. Coordinating inventory control and pricing strategies: The continuous review model. *Operations Research Letters* 34: 323–332.

Chen, Hong, and Yat-wah Wan. 2003. Price competition of Make-to-Order firms. *IIE Transactions* 35: 817–832.

Chen, Y. H., S. Ray, and Y. Y. Song. 2006. Optimal pricing and inventory control policy in periodic-review systems with fixed ordering cost and lost sales. *Naval Research Logistics* 53 (2): 117–136.

Chen, Jian, Shuo Huang, Refael Hassin, and Nan Zhang. 2015. Two back-order compensation mechanisms in inventory systems with impatient customers. *Production and Operations Management* 24 (10): 1640–1656.

Economopoulos, Angelos A., Vassilis S. Kouikoglou, and Evangelos Grigoroudis. 2011. The base stock/base backlog control policy for a make-to-stock system with impatient customers. *IEEE Transactions on Automation Science and Engineering* 8 (1): 243–249.

Economou, Antonis, and Spyridoula Kanta. 2008. Equilibrium balking strategies in the observable single-server queue with breakdowns and repairs. *Operations Research Letters* 36: 696–699.

Economou, Antonis, and Spyridoula Kanta. 2011. Equilibrium customer strategies and social-profit maximization in the single-server constant retrial queue. *Naval Research Logistics* 58: 107–122.

Economou, Antonis, and Athanasia Manou. 2013. Equilibrium balking strategies for a clearing queueing system in alternating environment. *Annals of Operations Research* 208: 489–514.

Economou, Antonis, Antonio Gómez-Corral, and Spyridoula Kanta. 2011. Optimal balking strategies in single-server queues with general service and vacation times. *Performance Evaluation* 68: 967–982.

Edelson, Noel M., and David K. Hildebrand. 1975. Congestion tolls for Poisson queueing processes. *Econometrica* 43: 81–92.

Elcan, A. 1994. Optimal customer return rate for an M/M/1 queueing system with retrials. *Probability in the Engineering and Informational Sciences* 8: 521–539.

Gavish, B., and S. C. Graves. 1980. A one-product production/inventory problem under continuous review policy. *Operations Research* 28 (5): 1228–1236.

Gavish, B., and S. C. Graves. 1981. Production/inventory systems with a stochastic production rate under a continuous review policy. *Computers & Operations Research* 8 (3): 169–183.

Graves, S. C., and J. Keilson. 1981. The compensation method applied to a one-product production/inventory problem. *Mathematics of Operations Research* 6 (2): 246–262.

Guo, Pengfei, and Refael Hassin. 2011. Strategic behavior and social optimization in Markovian vacation queues. *Operations Research* 59: 986–997.
Guo, Pengfei, and Refael Hassin. 2012. Strategic behavior and social optimization in Markovian vacation queues: The case of heterogeneous customers. *European Journal of Operational Research* 222: 278–286.
Guo, Pengfei, and Qingying Li. 2013. Strategic behavior and social optimization in partially-observable Markovian vacation queue. *Operations Research Letters* 41: 277–284.
Ha, Albert Y. 1997. Optimal dynamic scheduling policy for a make-to-stock production system. *Operations Research* 45 (1): 42–53.
Hassin, Refael. 2016. *Rational Queueing*. Chapman and Hall/CRC Press.
Hassin, Refael, and Moshe Haviv. 1996. On optimal and equilibrium retrial rates in a queueing system. *Probability in the Engineering and Informational Sciences* 10: 223–227.
Hassin, Refael, and Moshe Haviv. 2003. *To Queue or Not to Queue: Equilibrium Behavior in Queueing Systems*. Boston: Kluwer Academic Publishers.
Huh, W. T., and G. Janakiraman. 2008. (s, S) optimality in joint inventory-pricing control: An alternate approach. *Operations Research* 56 (3): 783–790.
Iravani, S. M. R., T. Liu, and D. Simchi-Levi. 2012. Optimal production and admission policies in make-to-stock/make-to-order manufacturing systems. *Production and Operations Management* 21 (2): 224–235.
Kennedy, J. and R. Eberhart. 1995. Particle swarm optimization. In *Proceedings of ICNN'95-International Conference on Neural Networks*, 4, 1942–1948. IEEE.
Knudsen, N. C. 1972. Individual and social optimization in a multiserver queue with a general cost-benefit structure. *Econometrica* 40: 515–528.
Krishnamoorthy, A., S. S. Nair, and V. C. Narayanan. 2012. An inventory model with server interruptions and retrials. *Operational Research* 12 (2): 151–171.
Li, L. 1992. The role of inventory in delivery-time competition. *Management Science* 38 (2): 182–197.
Li, Na., and Zhibin Jiang. 2013. Modeling and optimization of a product-service system with additional service capacity and impatient customers. *Computers & Operations Research* 40 (8): 1923–1937.
Li, Xiangyu, Jinting Wang, and F. Zhang. 2014. New results on equilibrium balking strategies in the single-server queue with breakdowns and repairs. *Applied Mathematics and Computation* 241: 380–388.
Li, Qingying, P. Guo, C. L. Li, and J. S. Song. 2016. Equilibrium joining strategies and optimal control of a make-to-stock queue. *Production and Operations Management* 25 (9): 1513–1527.
Liu, Weiqi, Yan Ma, and Jihong Li. 2012. Equilibrium threshold strategies in observable queueing systems under single vacation policy. *Applied Mathematical Modelling* 36: 6186–6202.
Ma, Yan, Weiqi Liu, and Jihong Li. 2012. Equilibrium balking behavior in the Geo/Geo/1 queueing system with multiple vacations. *Applied Mathematical Modelling* 37: 3861–3878.
Mangasarian, O. L. 1965. Pseudo-convexity functions. *Journal of SIAM Control* A3: 281–290.
Mangasarian, O. L. 1970. Convexity, pseudo-convexity and quasi-convexity of composite functioins. *Cahiers du Centre d'Études de Recherche Opérationelle* 12: 114–122.
Mendelson, H., and S. Whang. 1990. Optimal incentive-compatible priority pricing for the M/M/1 queue. *Operations Research* 38: 870–883.
Naor, Pinhas. 1969. The regulation of queue size by levying tolls. *Econometrica* 37: 15–24.
Schwarz, M., C. Sauer, H. Daduna, R. Kulik, and R. Szekli. 2006. M/M/1 queueing systems with inventory. *Queueing Systems* 54: 55–78.
Shortle, J.F., J.M. Thompson, D. Gross, and C.M. Harris. 2018. *Fundamentals of Queueing Theory*, 5th ed. Wiley.
Srinivasan, M. M., and H.-S. Lee. 1991. Random review production/inventory systems with compound Poisson demands and arbitrary processing times. *Management Science* 37 (7): 813–833.

Stidham, S., Jr. 2009. *Optimal Design of Queueing Systems*. Taylor and Francis, Boca Raton: CRC Press.
Veatch, M. H., and L. M. Wein. 1996. Scheduling a make-to-stock queue: Index policies and hedging points. *Operations Research* 44 (4): 643–647.
Wang, Jinting, Zhongbin Wang, and Y. Chen. 2021a. Equilibrium strategies and optimal pricing in an online retailing queueing system. *Naval Research Logistics* 68 (5): 556–576.
Wang, Zhongbin, Luyi Yang, Shiliang Cui, and Jinting Wang. 2021b. In-queue priority purchase: A dynamic game approach. *Queueing Systems* 97 (3–4): 343–381.
Wang, Jinting, and K. Sun. 2022. Optimal pricing and capacity sizing for online service systems with free trials. *OR Spectrum* 44 (1): 57–86.
Wang, Jinting, and F. Zhang. 2011. Equilibrium analysis of the observable queues with balking and delayed repairs. *Applied Mathematics and Computation* 218: 2716–2729.
Wang, Jinting, and F. Zhang. 2013. Strategic joining in M/M/1 retrial queues. *European Journal of Operational Research* 230: 76–87.
Wang, Jinting, and Feng Zhang. 2015. Monopoly pricing in a retrial queue with delayed vacations for local area network applications. *IMA Journal of Management Mathematics* 27 (2): 315–334.
Wang, Jinting, and Xuelu Zhang. 2017. Optimal pricing in a service-inventory system with delay-sensitive customers and lost sales. *International Journal of Production Research* 55 (22): 6883–6902.
Wang, Zhongbin, Jinting Wang, Yang Zhang, and Dusit Niyato. 2019. Strategic access and pricing in Internet of Things (IoT) service with energy harvesting. *IEEE Access* 7: 34655–34674.
Wang, Jinting, Pengfei Guo, Yilin Wang, and Lingjiao Zhang. 2024. How should restaurants operate in the omnichannel era? A queueing game approach. *International Journal of Production Economics* 274: 109306.
Wein, L. M. 1992. Dynamic scheduling of a make-to-stock queue. *Operations Research* 40 (4): 724–735.
Whitin, T. M. 1955. Inventory control and price theory. *Management Science* 2: 61–80.
Yu, Miaomiao, Yinghui Tang, and Wu. Wenqing. 2014. Individually and socially optimal joining rules for an Egalitarian Processor-Sharing queue under different information scenarios. *Computers & Industrial Engineering* 78: 26–32.
Zhang, Xuelu, and Jinting Wang. 2019. Optimal inventory threshold for a dynamic service Make-to-Stock system with strategic customers. *Applied Stochastic Models in Business and Industry* 35 (4): 1103–1123.
Zhang, F., Jinting Wang, and Bin Liu. 2012. On the optimal and equilibrium retrial rates in an unreliable retrial queue with vacations. *Journal of Industrial and Management Optimization* 8: 861–875.
Zhang, F., Jinting Wang, and Bin Liu. 2013. Equilibrium balking strategies in Markovian queues with working vacations. *Applied Mathematical Modelling* 37: 8264–8282.
Zheng, Y. S., and Paul Zipkin. 1990. A queueing model to analyze the value of centralized inventory information. *Operations Research* 38: 296–307.
Zipkin, P.H. 2000. *Foundations of Inventory Management*. McGraw-Hill.

Chapter 12
Healthcare Systems

Recently, due to the growing demand in public healthcare system subject to congestions, the service speed as well as the service quality have arose an extensive attention of social manager. It is well believed that longer service times generally result in longer waits for patients, which increases the congestion of overall system. For instance, it is reported that the comprehensive hospitals of China were providing services to almost 50% of the total number of patients, even though they only account for 7.7% of all hospitals in China.[1] A typical interaction between service speed and service quality can be found in the healthcare system with comprehensive hospital and primary hospital, where the former provides better service quality (with lower service speed) and the latter provides higher efficiency (with inferior service quality), respectively. The two medical channels mentioned above play different roles in the healthcare system towards different types of patients. The primary hospital is dedicated to provide the preliminary diagnosis and treatment. If the health problem of patients is found to be severe, they will be suggested to transfer to the comprehensive hospital. Therefore, considering the limited budget provided by government, it is imperative to combine comprehensive and primary hospitals to balance the quality and efficiency in healthcare system, i.e., the efficiency-quality trade-off, so as to improve the social welfare.

The effects of service quality in service systems have been studied broadly in existing literatures. In most works, the service quality was naturally assumed to be positively correlated to the length of service time, which determines the service value for patients. Hassin (2007) stresses that the quality of service always varied due to the random price. Milgrom and Roberts (1986) show how service price serves as a signal for service quality. Sekhri et al. (2011) investigate a comprehensive-primary integrated partnership and concluded that the partnership might improve access, quality, and efficiency in health care. From the viewpoint of customers, Anand et

[1] http://www.moh.gov.cn/zwgkzt/pwstj/list.shtml.

J. Wang, *Fundamentals of Queueing-Game Models*,
https://doi.org/10.1007/978-981-95-0261-5_12

al. (2011) study a customer-intensive service system where customers had to make the optimal quality-speed trade-off.

On the other hand, in a competitive situation, Li and Lee (1994) derive the Nash equilibrium under the competition between two service providers in terms of price quality, and service rate. In their model, the time-sensitive customers can observe the queue lengths and select the firm to maximize their utilities. Under the optimal pricing strategies of two firms, they found that the firm with a higher processing rate always enjoys a price premium. Chen et al. (2016) discuss the price competition issue in a system with two private service providers. They assume that the service providers in the system have different quality levels which are measured by the scores over [0, 1]. In practice, these scores could be judged by making a customer survey, and they can be used as a quantitative measure of the impact of service quality on customer. Sadat et al. (2015) consider a duopoly competition between two hospitals with different perceived quality of care. Their results disclosed that hospitals' actions are dependent on their capacity, market demand and the perceived quality of care. However, most works mentioned above ignored the constrained budget provided by government. In reality, government budget is an important factor in the management of comprehensive service system because it determines the ceilings of both service capacity (efficiency) and service quality. Thus, how to allocate the comprehensive resource effectively becomes a critical issue in the trade-off problem between efficiency and quality. Exceptionally, when a reimbursement is provided by government, Guo et al. (2019) compare the equilibrium outcomes (the patient welfare, readmission rate and waiting time) associated with two payment schemes with endogenous patient arrivals and readmissions in an M/M/1 queueing system. Wang and Wang (2021) investigate a healthcare system under two priority charging schemes with reimbursement.

To capture the interactions between the two medical channels (i.e., comprehensive hospital and primary hospital), it is usually adopted in the literature to establish a two-level (or two-tier) queueing network with two types service providers. The two-tier system has been investigated extensively in the past years because of its prevalent applications in practice. In particular, Chen et al. (2015) analyze the equilibrium outcome in a two-tier public service system, where a government subsidy policy is provided to encourage customers to pay for service. Wan and Wang (2017) extend that work to the case of general service times (i.e., M/G/1 queueing system) and analyze the optimization problem from both the perspective of minimizing all patients' total waiting costs and the perspective of maximizing the public service provider's social gain or the private service provider's revenue.

Another research stream of research focuses on the competition and coordination issues for two-tier service systems with customers who are sensitive to both service quality and delay, see Chen et al. (2020), Hua et al. (2016), Zhang et al. (2020), among others. Qian and Zhuang (2017) derive the tax/subsidy and capacity decisions of the public sector in a two-tier healthcare system so as to achieve welfare redistribution. Nevertheless, the service quality has never been considered in these works. Recently, Wang et al. (2020) explore the impact of referral payment scheme on implementing the healthcare referral program in the two-tier service systems, where the quality level of healthcare services are examined. But the efficiency-quality trade-off issue in the referral process is ignored in that work.

This chapter is adapted from Wang et al. (2022), where the authors focus on the resource (or budget) allocating problem of government for two-tier healthcare system with two types of patients. One channel of hospitals (e.g., comprehensive hospital) is with high service quality, but is with low service capacity. The number of another channel of hospitals (e.g., primary hospital) is much larger, then it is with a higher service capacity, but the service quality cannot be guaranteed due to the shortage of comprehensive equipments. Thus, the trade off between the two channels should be considered, i.e., quality or speed. In the literature of queueing game applications, this work is the first to investigate the interactions between comprehensive hospital and primary hospital with public reimbursement. In summary, this chapter makes the following contributions.

- **Socially optimal policy**. From the view of social planner, the socially optimal budget policy as well as joining strategy of patients are obtained. The sensitivity analysis is conducted, and our results indicate that the optimal amount of budget to primary hospital is monotonically increasing and decreasing in the potential arrival rate of patients with severe and uncertain health problem, respectively. However, the optimal budget and the corresponding optimal social welfare are non-monotone in the effectiveness parameter that measures the sensitivity of medical level to the budget.
- **Comparison of two scenarios**. The individually equilibrium strategy of patients is fully characterized and it is compared to the socially optimal strategy. The comparison result discloses an interesting insight that individual behavior of patients does not necessarily lead to systems more congested than what is socially desirable, which is a sharp contrast to the classical literature result. Moreover, the feasible pricing polices are conducted to regulate the patients' behavior to achieve the socially optimal strategy.
- **Extensions to multi-channels**. The base model is extended in two directions: a three-level system and a two-level system with multiple primary hospitals. In the two extensions, the socially optimal policy is derived and the equilibrium strategy of patients is characterized. And some insights in terms of the optimal budget policy and equilibrium strategy are provided to demonstrate the robustness of our base model.

12.1 Model Preliminaries

Consider a healthcare system consisting of two types of service providers where patients arrive according to a Poisson process with total rate Λ. This system has two types hospitals: the $high$ quality (comprehensive) and the low quality (primary) hospital, denoted by H-hospital and L-hospital, respectively. We classify the potential patients into two types, i.e., $\Lambda = \Lambda_s + \Lambda_m$:

(i) The s-type patient arrives with rate Λ_s, who is with severe illness, and will certainly join the H-hospital to be cured (i.e., the illness will be healed up) definitely.
(ii) The m-type patient arrives with rate Λ_m, who is with uncertain (or minor) health problem. These patients can choose to join either the H-hospital or L-hospital. If they join the H-hospital, they will be cured definitely since a more comprehensive service is received. On the other hand, if they join the L-hospital to have a preliminary diagnosis, then two cases may happen. (i) If they are diagnosed with a minor or no health problem, they can receive a treatment in the L-hospital successfully and then leave this system. (ii) Otherwise, if they diagnosed with a severe illness or they are undiagnosed and need further comprehensive diagnosis, they would be suggested to refer to the H-hospital, and the corresponding probability is defined as q_{lh}, which measures the medical level of L-hospital.

We assume that x-type patient can get a reward R_x $(x = s, m)$ at the completion of service (also called a service value) and all patients incur a delay cost at rate h during their total sojourn time. Assume that the proportion of arriving m-type patients choosing the H-hospital is $q \in [0, 1]$ (accordingly $1 - q$ of patients choosing L-hospital).[2] The service rates for H-hospital and L-hospital are μ_h and μ_l respectively, where μ_X can be regarded as the workforce level of Y-hospital $(Y = H, L)$. Balking is not allowed in our heathcare system since we need to ensure that all patients can be cured. In practice, the H-hospital is more reliable but is more congested, the L-hospital is less congested but the medical level cannot be guaranteed. We assume that the m-type patients are strategic and delay sensitive, thus, they need to balance their waiting time cost and service quality to maximize their expected utilities. As a result, the joining rate of s-type patients is exogenous, while the joining rate of m-type patients is endogenous. Figure 12.1 presents the patient flows of the healthcare system through a queueing network.

Our study focuses on the allocation policy for a fixed budget from the government, which aims to improve the social welfare of this healthcare system. Recall that in practice, the workforce level of L-hospital is high since there are plenty of L-hospitals. Thus, it is not necessary to expand its capacity further from the perspective of government. And the budget to L-hospital should be used to improve its medical level, i.e., the cure rate $1 - q_{lh}$. By contrast, the H-hospitals are big and fewer, their medical level is high enough, thus the received budget can be used to improve their workforce level, i.e., μ_h. Assume that the amount of budget is B. We face a problem that how to make use of the budget B to maximize the overall social welfare of the system. Denote by Δ the budget allocated to L-hospital (improving the medical level), then $B - \Delta$ is allocated to the H-hospital (improving the workforce level). We assume that the workforce level of H-hospital is linear to the budget,[3] i.e., $\mu_h = (B - \Delta)/c_h$, where c_h is unit cost of service rate. The similar assumption can be found in many queueing papers, e.g., Chen and Frank (2004), Gilbert and Wang (1998). On the

[2] Note that q can also be regarded as the joining probability of patients to L-hospital.

[3] The linear form of workforce level is intuitive since we can approximately achieve a k-times workforce level by hiring about k-times labors in hospitals.

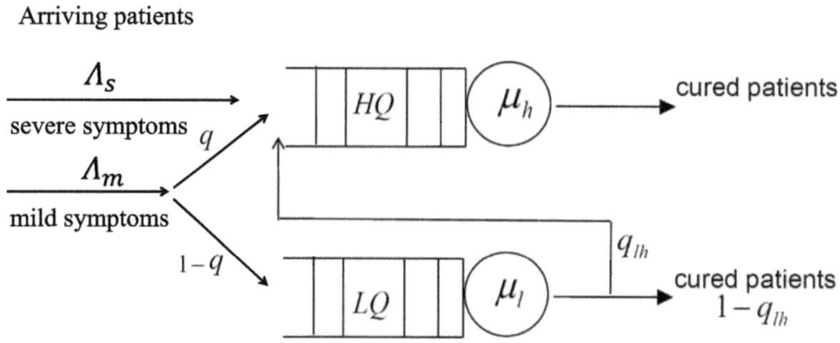

Fig. 12.1 A schematic of the model for two-level healthcare system

other hand, we assume that the medical level $1 - q_{lh}(\Delta)$ is increasing and concave in Δ (i.e., $q_{lh}(\Delta)$ is decreasing and convex in Δ). That is to say, $q'_{lh}(\Delta) < 0$ and $q''_{lh}(\Delta) > 0$. It is consistent with the empirical evidence from the outpatient practice that the medical level can be increased with the amount of reimbursement, and the medical level is less sensitive to the change in Δ when Δ is large. When no budget is provided, the basic medical level is $q_{lh}(0) = \alpha \in (0, 1)$. We assume that the medical level function is with an exponential form, i.e., $q_{lh}(\Delta) = \alpha e^{-\beta\Delta}$, where β is the effectiveness parameter which measures the sensitivity of medical level to the budget Δ. The exponential form medical level allows us to investigate the sensitivity of budget by focusing on a single sensitivity parameter β, which makes the model more elegant and the results more insightful. The exponential form function is widely used to identify those utility function with properties $q'_{lh}(\Delta) < 0$ and $q''_{lh}(\Delta) > 0$, see Gerber (1998), and particularly applied in healthcare system with reimbursement policy, see Guo et al. (2019), Wang et al. (2021).

12.1.1 System Performance

In this subsection, we tend to derive the system performance measures (e.g., expected waiting time, mean queue length and social welfare) under any fixed strategy $q \in [0, 1]$, and then identify the optimal reimbursement policy to maximize the social welfare as well as the socially optimal joining strategy of patients. In the next section, we will characterize the equilibrium joining probability q by considering the utility function of patients.

Denote by λ_h and λ_l the effective arrival rates to the H-hospital and L-hospital respectively. For a given proportion q, the traffic equations are given by

$$\lambda_h = \Lambda_s + \Lambda_m q + \lambda_l q_{lh}, \tag{12.1.1}$$
$$\lambda_l = \Lambda_m(1 - q), \tag{12.1.2}$$

or equivalently,

$$\lambda_h = \Lambda_s + \Lambda_m[q + q_{lh}(1-q)], \tag{12.1.3}$$

$$\lambda_l = \lambda(1-q). \tag{12.1.4}$$

Under steady state, the expected total number of patients in the system can be regarded as the summation of queue length for two independent $M/M/1$ queues with arrival rate λ_l, λ_h:

$$\begin{aligned} N = N_h(q) + N_l(q) &= \frac{\frac{\lambda_h}{\mu_h}}{1-\frac{\lambda_h}{\mu_h}} + \frac{\frac{\lambda_l}{\mu_l}}{1-\frac{\lambda_l}{\mu_l}} \\ &= \frac{\Lambda_s + \Lambda_m[q + q_{lh}(1-q)]}{\mu_h - \Lambda_s - \Lambda_m[q + q_{lh}(1-q)]} + \frac{(1-q)\Lambda_m}{\mu_l - (1-q)\Lambda_m}. \end{aligned} \tag{12.1.5}$$

Then the social welfare can be expressed as the total net benefits of all patients seeking service per time unit: $S(\Delta) = \Lambda_s R_s + \Lambda_m R_m - hN$, which is a function of the budget allocation to L-hospital (i.e., Δ). Thus we have the following optimization problem:

$$\max_{0\le\Delta<B} S(\Delta) = \Lambda_s R_s + \Lambda_m R_m - h\left[\frac{\Lambda_s + \Lambda_m[q + q_{lh}(\Delta)(1-q)]}{\mu_h(\Delta) - \Lambda_s - \Lambda_m[q + q_{lh}(\Delta)(1-q)]} + \frac{(1-q)\Lambda_m}{\mu_l - (1-q)\Lambda_m}\right] \tag{12.1.6}$$

$$\text{s.t.}\ \mu_h(\Delta) > \Lambda_s + \Lambda_m[q + q_{lh}(\Delta)(1-q)],$$

$$\mu_l > (1-q)\Lambda_m.$$

The constraint conditions are needed to ensure the stability of system. It should be noted that the social welfare consists of three parts: the service value and the waiting costs occur in two channels. Intuitively, the waiting cost in L-hospital is not affected by the amount Δ. Then the reimbursement to L-hospital Δ is actually used to reduce the transfer rate of patients to H-hospital, which helps to reduce the congestion of H-hospital. It should be noted that the system is stable if and only if $\frac{\Lambda_s + \Lambda_m[q + q_{lh}(\Delta)(1-q)]}{\mu_h(\Delta)} < 1$ and $\mu_l > (1-q)\Lambda_m$. Thus, we just consider the case that $\mu_l > (1-q)\Lambda_m$ and $\min\limits_{\Delta\in[0,B)} \frac{\Lambda_s + \Lambda_m[q + q_{lh}(\Delta)(1-q)]}{\mu_h(\Delta)} < 1$, which guarantees that there exist some non-empty feasible region such that the system can be stable. Otherwise, if $\min_{\Delta\in[0,B)} \frac{\Lambda_s + \Lambda_m[q + q_{lh}(\Delta)(1-q)]}{\mu_h(\Delta)} \ge 1$, the system cannot be stable for any $\Delta \in [0, B)$. Rewriting (12.1.6) after some manipulations gives

$$\max_{0\le\Delta<B} S(\Delta) = \Lambda_s R_s + \Lambda_m R_m - h\left(\frac{f_1(\Delta)}{1 - f_1(\Delta)} + \frac{(1-q)\Lambda_m}{\mu_l - (1-q)\Lambda_m}\right),$$

where $f_1(\Delta) = \frac{([q + \alpha e^{-\beta\Delta}(1-q)]\Lambda_m + \Lambda_s)c_h}{B-\Delta} < 1$ is the system workload in H-hospital. Therefore, (12.1.6) is equivalent to

$$\min_{0\le\Delta<B} f_1(\Delta) = \frac{([q+\alpha e^{-\beta\Delta}(1-q)]\Lambda_m + \Lambda_s)c_h}{B-\Delta}. \tag{12.1.7}$$

The following proposition gives the property of social welfare in Δ and the optimal allocation amount Δ^*.

Proposition 12.1.1 *For any fixed q, the social welfare is concave in Δ, and there exists a unique Δ^* that maximizes the social welfare, which is given as follow:*

$$\Delta^* = \begin{cases} 0, \ \textit{if}\ \beta \le \frac{c_h\Lambda_m[q+\alpha(1-q)]+\Lambda_s}{\alpha(1-q)c_h\Lambda_m B}; \\ \widehat{\Delta}, \ \textit{if}\ \beta > \frac{c_h\Lambda_m[q+\alpha(1-q)]+\Lambda_s}{\alpha(1-q)c_h\Lambda_m B}. \end{cases} \tag{12.1.8}$$

where $\widehat{\Delta}$ uniquely solves $c_h\Lambda_m(e^{\beta\Delta}q + (1-q)\alpha[1-(B-\Delta)\beta]) + \Lambda_s = 0$.

Remark 12.1.1 Proposition 12.1.1 indicates that when it is not effective to improve the service quality of L-hospital, i.e., β is small, the social manager will not assign any budge to L-hospital, and all budget will be allocated to H-hospital. It is because that by improving the service quality, the decreased readmission rate to H-hospital cannot compensate for the reduced capacity expansion. Thus it is optimal to use all the budget to expand the service capacity. On the other hand, when β is large, it is efficient to reduce the congestion by improving the service quality of L-hospital, then it gives $\Delta^* > 0$. At first glance, Δ^* should increase with demand size of m-type patients Λ_m, and it seems that Δ^* will increase with β. That is, more budget should be allocated to L-hospital when the effectiveness parameter β increases. Are these intuitions true? To answer these questions, we conduct the sensitivity analysis for Δ^* in the following subsection.

12.1.2 Sensitivity Analysis

Proposition 12.1.2 *For any fixed q, the optimal allocation Δ^* has the following properties:*

(i) Δ^* is weakly increasing in α, but is non-monotone in β.
(ii) Δ^* is weakly increasing and decreasing in Λ_s and Λ_m, respectively.

Proposition 12.1.2 claims that more budget should be allocated to L-hospital when the scale parameter α increases to reduce the excessive readmissions, which is quite intuitive. However, counter intuitively, Δ^* is found to be not necessarily increasing in β. We provide the explanation below. When β is relatively large, the increment of service quality becomes less sensitive, then less service quality can be improved when the same amount of incremental budget is assigned. Thus the increased medical level of L-hospital cannot compensate for the reduced service capacity of H-hospital. As a result, it is better to transfer more budget to H-hospital to increase the efficiency of service when β is large.

Proposition 12.1.3 *The optimal allocation Δ^* is weakly decreasing in $q \in [0, 1]$.*

Proposition 12.1.3 indicates an intuitive result that more budgets should be allocated to H-hospital when the proportion of patients to H-hospital increases. In other words, when the level of congestion in H-hospital increases, it is more efficient to improve its service speed rather than to control its admissions.

Proposition 12.1.4 *For any fixed $\Delta \in [0, B]$, the socially optimal joining probability $q_s(\Delta)$ is given by*

$$q_s(\Delta) = \begin{cases} 0, & \text{if } B \leq \underline{B}(\Delta); \\ \hat{q}(\Delta), & \text{if } \underline{B}(\Delta) < B < \bar{B}(\Delta); \\ 1, & \text{if } B \geq \bar{B}(\Delta), \end{cases}$$

where $\underline{B}(\Delta) = \Delta + c_h(\Lambda_s + \Lambda_m \alpha e^{-\beta\Delta}) + \frac{1+\sqrt{1+4(\Lambda_s+\Lambda_m\alpha e^{-\beta\Delta})\mu_l/[(1-q_{lh})(\mu-\Lambda_m)^2]}}{2\mu_l/[(1-q_{lh})(\mu-\Lambda_m)^2]}$, $\bar{B}(\Delta) = \Delta + c_h[2(\Lambda_s + \Lambda_m) + \mu_l(1 - \alpha e^{-\beta\Delta}) + \sqrt{\mu_l(1-\alpha)[4(\Lambda_s + \Lambda_m) + \mu_l(1-\alpha)]}]/2$, *and*

$$\hat{q}(\Delta) = \frac{c_h(\mu_l[c_h(\Lambda_s + \Lambda_m\alpha e^{-\beta\Delta}) - 2(B-\Delta)] + \Lambda_m(B-\Delta)) + \sqrt{\frac{c_h\mu_l(B-\Delta)(c_h(\Lambda_m+\Lambda_s+\mu_l(\alpha e^{-\beta\Delta}-1))-B+\Delta)^2}{1-\alpha e^{-\beta\Delta}}}}{c_h\Lambda_m(B + c_h\mu_l(\alpha e^{-\beta\Delta}-1) - \Delta)}.$$

Proposition 12.1.4 identifies the socially optimal joining strategy for given any reimbursement policy. Intuitively, when B increases, the proportion of budget assigned to H-hospital improves, which expands its service capacity efficiently, and attract more m-type patients to join. Thus, not surprisingly, we can find that the socially optimal joining probability to H-hospital increases in the total budget B. Furthermore, denote by q_s the socially optimal joining probability under optimal allocation policy Δ^S, we can have the following result.

Corollary 12.1.1 *There exists a certain $\tilde{B}$ such that the socially optimal proportion of m-type patients to visit H-hospital $q_s < 1$ if $B < \bar{B}$.*

Corollary 12.1.1 discloses the fact that when the total amount of budget is not so large, both H-hospital and L-hospital can exist. That is, there still exist some patients choosing to join the L-hospital to have a preliminary diagnosis. It is quite intuitive since when the budget provided to H-hospital is not abundant to expand the service capacity or workforce level, splitting some m-type patients to L-hospital is an efficient way to relegate the workload of H-hospital, and thus to optimize the overall social welfare.

12.1.3 Numerical Examples

We illustrate social welfare function in Δ in Fig. 12.2 for a given set of parameters. It can be observed that $S(\Delta)$ is unimodal in Δ and it can be maximized at a curtain

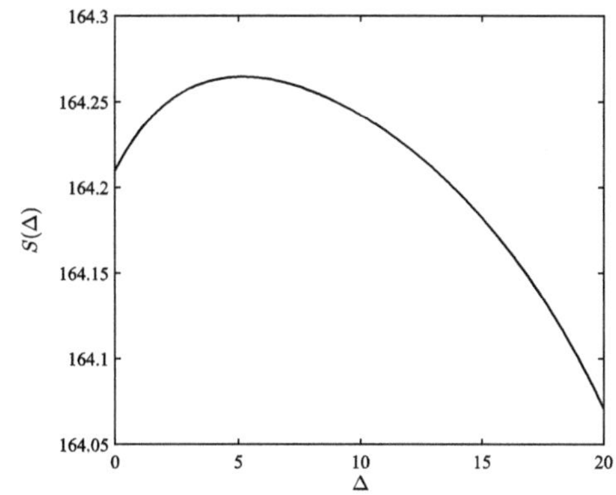

Fig. 12.2 Social welfare with respect to: Δ, for $q = 0.5, \alpha = 0.6, \beta = 0.3, R_s = R_m = 15, \Lambda_m = 10, \Lambda_s = 1, \mu_l = 15, c_h = 1, h = 1, B = 40$

$\Delta^* \in [0, B)$. It is consistent with our results in Proposition 12.1.1. Furthermore, through numerical examples in Figs. 12.4 and 12.5, we investigate the impacts of some important parameters (e.g., the scale parameter α, effectiveness parameter β and the proportion of patients to H-hospital) on the optimal budget allocation Δ^*. Figure 12.3a shows that the optimal allocation amount to L-hospital is increasing in α, which verifies the result in Proposition 12.1.2. Also, from Fig. 12.3b, we can see that the optimal social welfare is decreasing in α because of the reduction of cure rate in L-hospital.

In Fig. 12.4a, we find that Δ^* is unimodal in β. It firstly increases with β and then decreases with it, which is consistent with our Proposition 12.1.2. The main reason for such a phenomenon is that there exist two types of effects on the system when Δ

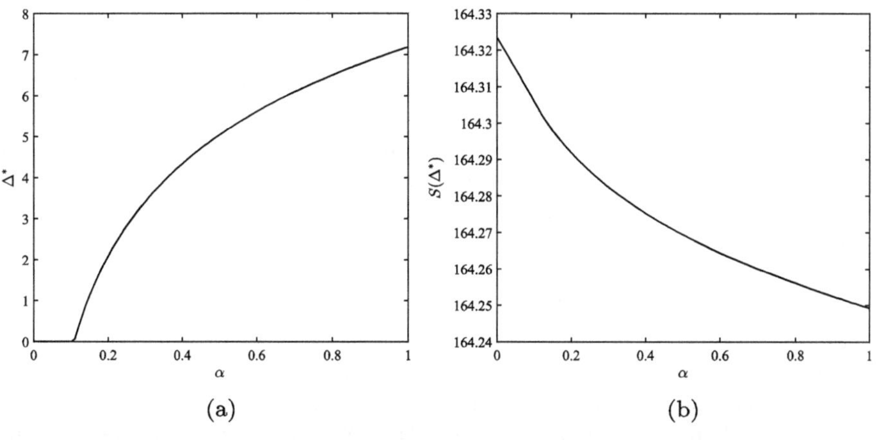

Fig. 12.3 Optimal allocation policy and the corresponding maximal social welfare with respect to: α, for $q = 0.5, \beta = 0.3, R_s = R_m = 15, \Lambda_m = 10, \Lambda_s = 1, \mu_l = 15, c_h = 1, h = 1, B = 40$

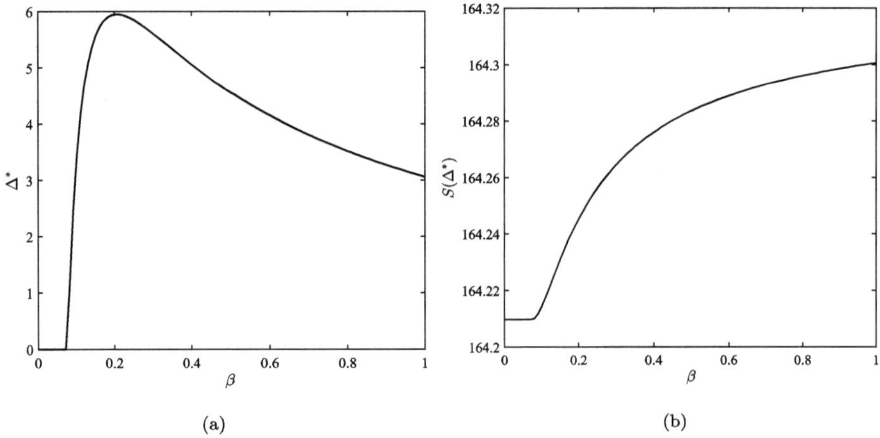

Fig. 12.4 Optimal allocation policy and the corresponding maximal social welfare with respect to: β, for $q = 0.5$, $\alpha = 0.6$, $R_s = R_m = 15$, $\Lambda_m = 10$, $\Lambda_s = 1$, $\mu_l = 15$, $c_h = 1$, $h = 1$, $B = 40$

increases. The first effect is the reduction in q_{lh} (i.e., the increased service quality for the L-hospital); and the second effect is the reduction in capacity expansion for the H-hospital. Clearly, the first effect is positive on social welfare and the second is negative on social welfare. The shape of Δ^* curve (Fig. 12.4a) shows that when β is small, the first effect dominates the second one so that Δ^* increases with β, indicating that allocating to the L-hospital is more beneficial. When β exceeds a certain value, the reduced readmission cannot compensate for the reduction in capacity of H-hospital, so Δ^* becomes to be decreasing in β. Note that the non-monotone property in β is robust, which can also be observed in the presence of multiple L-hospitals, see Sect. 12.3.2. Intuitively, the optimal social welfare $S(\Delta^*)$ can be increased when β increases because less admissions to H-hospital can be resulted ($q_{lh} \downarrow$), which can be verified in Fig. 12.4b.

Figure 12.5 illustrates how Δ^* or $S(\Delta^*)$ changes with q. It is not difficult to see that Δ^* decreases with q because more budget should be allocated to the H-hospital when more patients choose the H-hospital so as to reduce the congestion, which is consistent with Proposition 12.1.3. On the other hand, the corresponding optimal social welfare $S(\Delta^*)$ is non-monotone in q, and there exists an optimal joining probability q_s maximizes the social welfare. This is resulted from the double effects of two-tier service model. When q is small, the optimal resource allocation to H-hospital will increase rapidly with q to expand the service capacity so as to reduce the overall congestion, which leads to a higher social welfare. However, when q is large, the delay cost in L-hospital is too small, and the increased reimbursement to H-hospital is less effective. Finally, when q is between 0.95 and 1, Δ^* is constantly 0 to guarantee a sufficient service capacity in H-hospital channel, in which the decreased waiting cost in L-hospital dominates that in H-hospital. Then $S(\Delta^*)$ will increase with q again.

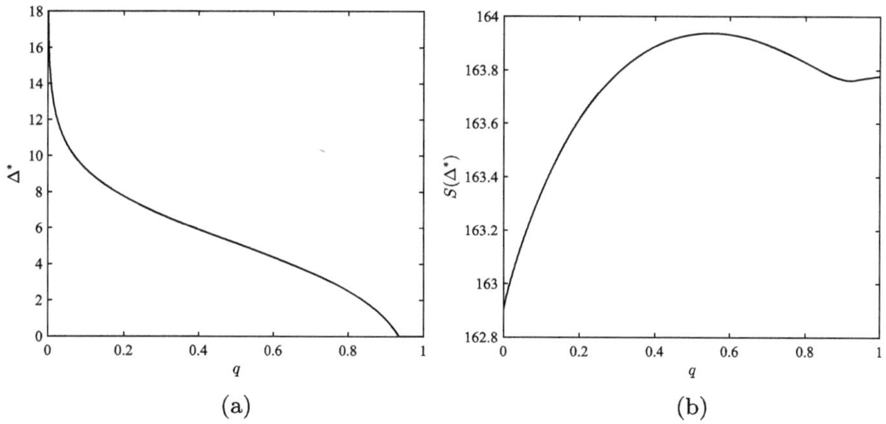

Fig. 12.5 Optimal allocation policy and the corresponding maximal social welfare with respect to: q, for $\alpha = 0.8$, $\beta = 0.3$, $R_s = R_m = 15$, $\Lambda_m = 10$, $\Lambda_s = 1$, $\mu_l = 15$, $c_h = 2$, $h = 1$, $B = 40$

Remark 12.1.2 This section establishes the benchmark and preliminaries of our base model. It should be pointed out that in this section, the behavior of patients is exogenously determined by social planner rather than themselves. It can be regarded as the first step to examine the functional relation between the social welfare and the budget allocation policy, and the sensitivity of optimal social welfare. In the next subsection, we proceed to study the impact of individually equilibrium strategy of m-type patients on the system, where the joining probability q determined by the m-type individuals that characterizes the equilibrium joining probability q_e.

12.2 The Impact of Equilibrium Behavior

In this section, the equilibrium joining strategy of m-type patients is studied and a two-stage Stackelberg game is established to determine the optimal budget allocation policy. In such dynamic game, government is the leader who selects the amount Δ and the arriving m-type patients are the followers who determine the joining strategy. When the joining probability of m-type patients to H-hospital is q (with probability to L-hospital $1 - q$), combining the effective arrival rates to H-hospital and L-hospital that have been obtained in previous Sect. 12.1 [see (12.1.1)–(12.1.2)], the expected sojourn times of patients in H-hospital and L-hospital (as $M/M/1$ queues) can be derived by using Little's formula: $W_h(q) = N_h(q)/\lambda_h$ and $W_l(q) = N_l(q)/\lambda_l + q_{lh} W_h(q)$. Then it gives

$$W_h(q) = \frac{1}{\mu_h - \Lambda_s - (q + q_{lh}(1-q))\Lambda_m}, \tag{12.2.1}$$

$$W_l(q) = \frac{1}{\mu_l - (1-q)\Lambda_m} + \frac{q_{lh}}{\mu_h - \Lambda_s - (q + q_{lh}(1-q))\Lambda_m}. \tag{12.2.2}$$

Assume that the service fees for L-hospital and H-hospital as P_l and P_h, respectively, then the net utility of m-type patients choose to join H-hospital or L-hospital is given by $R_m - hW_h(q) - P_h$ or $R_m - hW_l(q) - P_l - q_{lh}P_h$, respectively. A patient joins H-hospital if and only if $R_m - hW_h(q) - P_h > R_m - hW_l(q) - P_l - q_{lh}P_h$. For any $\Delta \in [0, B)$ selected by the government, the value of q_{lh} and μ_h can be determined correspondingly. Obviously, the equilibrium joining strategy is dependent on the budget allocation policy Δ, therefore, we denote by $q_e(\Delta)$ the equilibrium joining probability under Δ. Define $D(q) \triangleq W_h(q) - W_l(q)$ as the reduced amount of delay by joining L-hospitals. Before characterizing the equilibrium of m-type patients, we first give a lemma as follow.

Lemma 12.2.1 *$D(q)$ is increasing in $q \in [0, 1]$.*

Lemma 12.2.1 claims that the gap of delay between two hospitals increases with q, which is intuitive because the H-hospital is more congested when q increases. For a given allocation policy Δ, the patient's decentralized joining behavior have the following three cases:
Case (i): $hD(0) > P_l - (1 - q_{lh})P_h$
Case (ii): $hD(1) < P_l - (1 - q_{lh})P_h$.
Case (iii): $hD(0) \leq P_l - (1 - q_{lh})P_h \leq hD(1)$,

In case (i), all patients will join L-hospital, and no one has incentive to deviate from it, i.e., $q_e(\Delta) = 0$ is an equilibrium. Similarly, in case (ii), $q_e(\Delta) = 1$ is an equilibrium. Otherwise, in case (iii), there exists unique $q_e(\Delta)$ such that patients are indifferent in choosing either L-hospital or H-hospital, i.e., $hD(q_e(\Delta)) = P_l - (1 - q_{lh})P_h$. The uniqueness can be guaranteed by Lemma 12.2.1. The explicit equilibrium of patients when $hD(0) \leq P_l - (1 - q_{lh})P_h \leq hD(1)$ is given in the following proposition.

Proposition 12.2.1 *(i) When $P_l - (1 - q_{lh})P_h < hD(0)$, $q_e(\Delta) = 0$; (ii) When $P_l - (1 - q_{lh})P_h > hD(1)$, $q_e(\Delta) = 1$; (iii) When $P_l - (1 - q_{lh})P_h \in [hD(0), hD(1)]$, we have*

$$q_e(\Delta) = \begin{cases} \min\{\max\{0, \frac{(\mu_h - \Lambda_s - q_{lh}\Lambda_m) - (1 - q_{lh})(\mu_l - \Lambda_m)}{2(1 - q_{lh})\Lambda_m}\}, 1\}, & \textit{if } (1 - q_{lh})P_h - P_l = 0; \\ \min\{\frac{-b - \sqrt{b^2 - 4ac}}{2a}, 1\}, & \textit{if } (1 - q_{lh})P_h - P_l \neq 0. \end{cases}$$

where $a = ((1 - q_{lh})P_h - P_l)(1 - q_{lh})\Lambda_m^2$, $b = \Lambda_m[((1 - q_{lh})P_h - P_l)((1 - q_{lh})(\mu_l - \Lambda_m) - (\mu_h - \Lambda_s - q_{lh}\Lambda_m)) - 2h(1 - q_{lh})]$, *and* $c = h[(\mu_h - \Lambda_s - q_{lh}\Lambda_m) - (1 - q_{lh})(\mu_l - \Lambda_m)] - ((1 - q_{lh})P_h - P_l)(\mu_h - \Lambda_s - q_{lh}\Lambda_m)(\mu_l - \Lambda_m)$.

So far, for any given budget Δ, the unique equilibrium joining probability is characterized. The next proposition discloses how q_e changes with budget Δ.

Proposition 12.2.2 *In equilibrium, q_e is weakly decreasing in Δ. Furthermore, q_e is weakly decreasing and increasing in P_h and P_l, respectively.*

Proposition 12.2.2 indicates that a smaller equilibrium joining probability to H-hospital q_e should be adopted by patients when Δ improves. It is intuitive because when Δ increases, the service quality (capacity) for L-hospital (H-hospital) will increase (decrease), then the patients are more inclined to join the L-hospital, which results in a lower q_e. Also, since the patients are price sensitive, more m-type patients tend to join X-hospital if P_X is reduced (or, P_{-X} is increased), where $X = H, L$ and $-X = L, H$. Under equilibrium q_e, the object function of the government who aims to maximizes the social welfare becomes to

$$\begin{aligned}
\max_{0\le\Delta<B}\ S(\Delta) =& \Lambda_m R_m + \Lambda_s R_s - h\left[\frac{(q_e(\Delta)+\alpha e^{-\beta\Delta}[1-q_e(\Delta)])\Lambda_m + \Lambda_s}{\frac{B-\Delta}{c_h} - \Lambda_s - (q_e(\Delta)+\alpha e^{-\beta\Delta}[1-q_e(\Delta)])\Lambda_m} + \frac{(1-q_e(\Delta))\lambda}{\mu_l - (1-q_e(\Delta))\Lambda_m}\right]\\
\text{s.t. } \frac{B-\Delta}{c_h} >& (q_e(\Delta)+\alpha e^{-\beta\Delta}[1-q_e(\Delta)])\Lambda_m + \Lambda_s,\\
\mu_l >& [1-q_e(\Delta)]\Lambda_m,\\
q_e(\Delta) =& \begin{cases} \min\{\max\{0, \frac{(\mu_h-\Lambda_s-q_{lh}\Lambda_m)-(1-q_{lh})(\mu_l-\Lambda_m)}{2(1-q_{lh})\Lambda_m}\}, 1\}, & \text{if } (1-q_{lh})P_h - P_l = 0;\\ \min\{\frac{-b-\sqrt{b^2-4ac}}{2a}, 1\}, & \text{if } (1-q_{lh})P_h - P_l \neq 0. \end{cases}
\end{aligned}$$

where $q_e(\Delta)$ is given in Proposition 12.2.1. Unfortunately, it is untractable to solve this optimization problem analytically because of the complexity of expression $S(\Delta; q_e)$. However, we can derive the optimal allocation policy numerically by using Particle Swarm Optimization (PSO) algorithm, where the solution space is a finite 1-dimensional space and every position of particle is a potential optimal solution to the considered problem. Denote $x_i \in [0, B)$ as the position of particle i, the corresponding improved direction and step-size, i.e., velocity is v_i. Then the formulas that update the position x_i and velocity v_i is

$$\begin{aligned}
v_i &= v_i + 2 * rand() * (p_i - x_i) + 2 * rand() * (p_g - x_i),\\
x_i &= x_i + v_i,
\end{aligned}$$

where $rand()$ is a unform random variable in $[0, 1]$, p_i is the best position of particle i found so fare and p_g is the globally best position of all the particles, more details of the PSO algorithm can be referred to Kennedy and Eberhart (1995) and Carlisle and Dozier (2001).

When the optimal policy Δ^* is found, the corresponding equilibrium outcome for the arriving patient can be derived immediately by plugging Δ^* into Proposition 12.2.1. On the other hand, recall that for any fixed Δ, we can always obtain the socially optimal joining probability of patients $q_s(\Delta)$, see Proposition 12.1.4. The following result indicates that in equilibrium, the socially optimal strategy can always be induced by setting suitable prices.

Proposition 12.2.3 *For any fixed Δ, in equilibrium, the socially optimal joining probability can be attained, i.e., $q_e = q_s$, by adjusting the prices $(\hat{P}_h(\Delta), \hat{P}_l(\Delta)) \in \Omega(\Delta)$, where*

$$\Omega(\Delta) = \begin{cases} \{(P_h, P_l) | P_l - (1 - q_{lh}) P_h < hD(0)\}, & \text{if } B \leq \underline{B}(\Delta); \\ \{(P_h, P_l) | P_l - (1 - q_{lh}) P_h = hD(\hat{q}_S(\Delta)), & \text{if } \underline{B}(\Delta) < B < \bar{B}(\Delta); \\ \{(P_h, P_l) | P_l - (1 - q_{lh}) P_h > hD(1)\}, & \text{if } B \geq \bar{B}(\Delta), \end{cases}$$

where $\hat{q}_S(\Delta)$ is given in Proposition 12.1.4.

We explain this proposition as follows. When the total budget B is small, the socially optimal joining probability to H-hospital is 0, then the P_h should be set as large as possible to screen out all the m-type patients deciding to join H-hospital. In contrast, when the budget B is relatively large, the socially optimal joining probability to H-hospital is 1, then setting a large price P_l is an optimal policy to attract all m-type patients to join L-hospital can maximize the social welfare. Interestingly, when B is intermediate, the prices should be selected carefully, and there exists unique gap between P_l and $(1 - q_{lh})P_h$ such that the coordination between socially optimal strategy and equilibrium strategy of m-type patients can be attained. Nevertheless, it should be noted that the socially optimal pricing policy is not necessarily to be unique. That is to say, the healthcare system can achieve the optimal social welfare if the prices charged by both hospitals are improved (or reduced) at the same time without changing the gap between P_l and $(1 - q_{lh})P_h$. More discussions about the coordination between socially optimal strategy and individually equilibrium strategy of m-type patients are provided in Fig. 12.8.

12.2.1 Numerical Examples

In this subsection, we conduct numerical experiments to study the impact of equilibrium strategy. In Figs. 12.6 and 12.7, we examine how the equilibrium joining probability and the corresponding optimal social welfare vary with effectiveness parameter β and budget size Δ, respectively. In particular, Figure 12.6 indicates that q_e is decreasing in β, which is intuitive because when β increases, the service quality of L-hospital can be improved more efficiently, which can attract more m-type patients to join, i.e., $(1 - q_e) \uparrow$. In Fig. 12.7, we can observe that $S(\Delta; q_e)$ is unimodal in Δ, and there still exists a unique optimal budget allocation policy to maximize the social welfare.

We compare the socially optimal joining probability q_s and equilibrium (i.e., individually optimal) joining probability q_e for different budget size Δ in Fig. 12.8. Even though both q_s and q_e are monotonically decreasing in Δ, their relationship is uncertain. In particular, when Δ is small (i.e., $B - \Delta$ is large), we have $q_s > q_e$, i.e., a larger proportion to H-hospital is more desirable for social planner. It is because that when the budget provided to H-hospital is large, the capacity of H-hospital is expanded to some extend. Considering that the uncertain medical rate of L-hospital will incur more transfers to H-hospital and result in potential negative externalities to the system, the social planner should favor more patients to visit H-hospital directly. By contrast, from the view of individuals, they can always try their

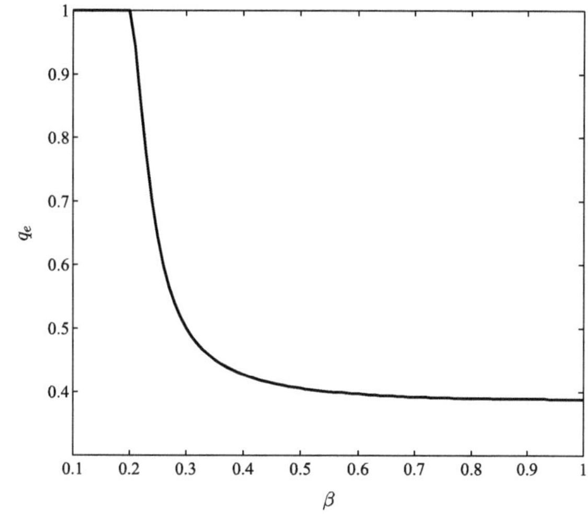

Fig. 12.6 Equilibrium strategy with respect to: β, for $P_l = 3$, $P_h = 6$, $\Delta = 6$, $\alpha = 2$, $R = 15$, $\Lambda_m = 10$, $\mu_l = 7$, $c_h = 3$, $h = 3$, $B = 40$

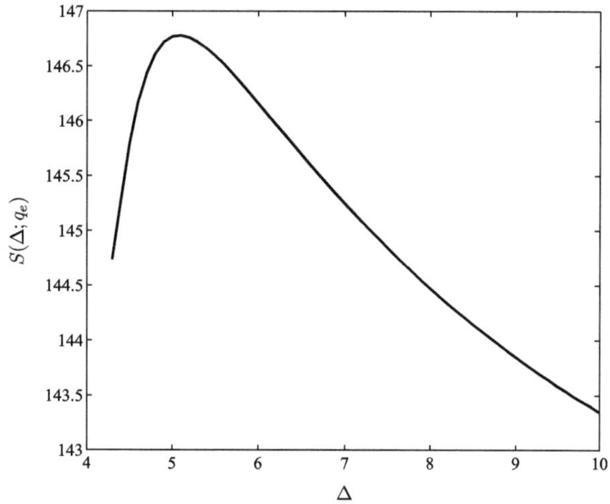

Fig. 12.7 Social welfare with respect to: Δ, for $P_l = 3$, $P_h = 6$, $\beta = 0.3$, $\alpha = 2$, $R = 15$, $\Lambda_m = 10$, $\mu_l = 7$, $c_h = 3$, $h = 3$, $B = 40$

luck by visiting the L-hospital since the sojourn time during which is quite shorter than visiting H-hospital. On the other hand, when Δ is large, more m-type patients will choose to visit L-hospital since the significant improvement of its medical level, and we have $q_s < q_e$, that is, patients are more inclined to join H-hospital from the perspective of individual comparing to that of social planner in the presence of large Δ. This is because that when more patients join the L-hospital, both H-hospital and L-hospital are congested, then the advantage of fast service in L-hospital disappears, which motivates some of them to visit H-hospital directly. As a result, a "pull-to-center" effect can found by investigating the individual behavior of patients, where the patients would decrease (increase) the socially optimal joining probability to H-hospital if the budget is low (high). Furthermore, comparing the socially optimal

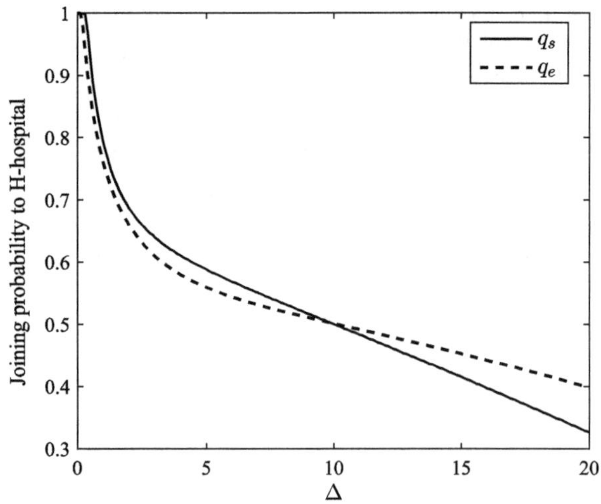

Fig. 12.8 Comparison of q_s and q_e for different Δ, for $P_l = 6$, $P_h = 3$, $\beta = 0.5$, $\alpha = 0.8$, $R = 15$, $\Lambda_m = 15$, $\mu_l = 15$, $c_h = 2$, $h = 1$, $B = 40$

and individually equilibrium strategy discloses that the government should urge the H-hospital to reduce (resp. increase) its price (i.e., P_h) when the budget to L-hospital is small (resp. large), so as to achieve the optimal social welfare.

12.3 Extensions

In this section, we consider two extensions to demonstrate the robustness of our model. We first extend our analysis to a three-level system with H-hospital, M-hospital, and L-hospital. And then investigate a system with a single H-hospital with n parallel L-hospitals.

12.3.1 Model with Three-Level System

Similar to our preassumptions that there are two types of patients, where the s-type patient arrives with rate Λ_s and they will certainly join the H-hospital. The m-type patients are flexible, and the proportions of m-type patients who join L-hospital, M-hospital and H-hospital are denoted by q_L, q_M and q_H, respectively, which satisfy $q_L + q_M + q_H = 1$. H-hospital offers the guaranteed service (patient can be completely treated). M-hospital provides a service with a medium level (between H-hospital and L-hospital), whereas the patients have a higher probability to be cured than the L-hospital, and the patient receives service in M-hospital may be transferred to H-hospital with probability q_{MH}. Likewise, a patient joins L-hospital will be transferred to M-hospital with probability q_{LM} after service. The service rates for the three hospitals are μ_L, μ_M and μ_H respectively. Such a three-level system is illustrated in Fig. 12.9. In particular, when $q_L = 0$, it degenerates to our base model with two-level system. Denote by Δ_L, Δ_M and $\Delta_H = B - \Delta_L - \Delta_M$ the budgets allocated to L-, M- and H-hospitals, respectively. Under the budget imposed on three different hospitals, the cured probabilities for L-hospital and M-hospital are described as $1 - q_{LM} = 1 - \alpha_L e^{-\beta_L \Delta_L}$ and $1 - q_{MH} = 1 - \alpha_M e^{-\beta_M \Delta_M}$, which are increasing in budgets Δ_L and Δ_M, respectively. They can be regarded as the service qualities for L-hospital and M-hospital. On the other hand, $\mu_H = (B - \Delta_L - \Delta_M)/c_h$ is the service capacity of H-hospital, which decreases with the amount of Δ_L and Δ_M. Our aim is to determine the optimal Δ_L and Δ_M so as to maximize the social welfare of overall heathcare system.

Similar to previous section, we firstly study the social welfare maximization when the proportions of patients are fixed (i.e., q_L, q_M and q_H are exogenously determined). Denote by λ_H, λ_M and λ_L the effective arrival rates to the H-, M-, and L-hospitals, respectively. Through the description of queueing network in Fig. 12.9, they are given by

$$\lambda_H = \Lambda_s + \Lambda_m q_H + \lambda_M q_{MH}, \tag{12.3.1}$$
$$\lambda_M = \Lambda_m q_M + \lambda_L q_{LM}, \tag{12.3.2}$$
$$\lambda_L = \Lambda_m q_L. \tag{12.3.3}$$

or equivalently,

$$\lambda_H = \Lambda_s + \Lambda_m[q_H + (q_M + q_L q_{LM}) q_{MH}], \tag{12.3.4}$$
$$\lambda_M = \Lambda_m (q_M + q_L q_{LM}), \tag{12.3.5}$$
$$\lambda_L = \Lambda_m q_L. \tag{12.3.6}$$

Similar to our base model, the expected total number of patients in the system can be regarded as the summation of queue length for three independent M/M/1 queues:

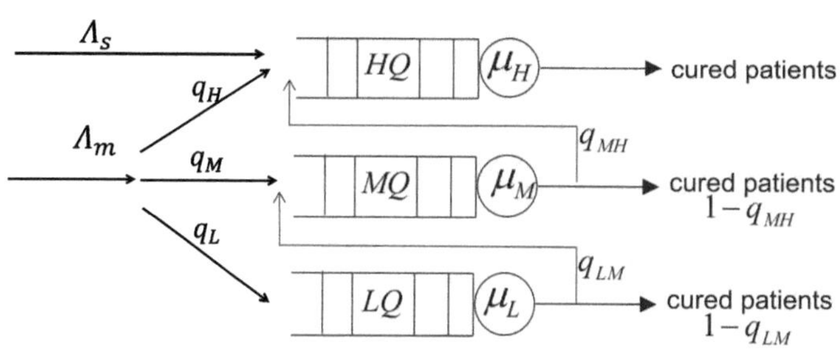

Fig. 12.9 A schematic of the model for three-level healthcare system

$$
\begin{aligned}
N_0 &= \frac{\frac{\lambda_H}{\mu_H}}{1-\frac{\lambda_H}{\mu_H}} + \frac{\frac{\lambda_M}{\mu_M}}{1-\frac{\lambda_M}{\mu_M}} + \frac{\frac{\lambda_L}{\mu_L}}{1-\frac{\lambda_L}{\mu_L}} \\
&= \frac{\Lambda_s + \Lambda_m[q_H + (q_M + q_L q_{LM})q_{MH}]}{\mu_H - \Lambda_s - \Lambda_m[q_H + (q_M + q_L q_{LM})q_{MH}]} \\
&\quad + \frac{(q_M + q_L q_{LM})\Lambda_m}{\mu_m - (q_M + q_L q_{LM})\Lambda_m} + \frac{q_L \Lambda_m}{\mu_L - q_L \Lambda_m}.
\end{aligned} \tag{12.3.7}
$$

The optimization problem for social welfare can be written as

$$
\begin{aligned}
\max_{0 \le \Delta_L, \Delta_M < B} S(\Delta_L, \Delta_M) &= \Lambda_s R_s + \Lambda_m R_m - hN_0 \\
\text{s.t.} \Delta_L + \Delta_M &< B, \\
\frac{B - \Delta_L - \Delta_M}{c_h} &> [q_H + (q_M + q_L \alpha_L e^{-\beta_L \Delta_L})\alpha_M e^{-\beta_M \Delta_M}]\Lambda_m + \Lambda_s, \\
\mu_M &> (q_M + q_L \alpha_L e^{-\beta_L \Delta_L})\Lambda_m, \\
\mu_L &> q_L \Lambda_m,
\end{aligned} \tag{12.3.8}
$$

where N_0 is given by (12.3.7). To solve the maximization problem in (12.3.8), we provide the following lemma to help us analyze the holding cost term (i.e., N_0) in (12.3.8).

Lemma 12.3.1 *Consider continuous functions $f(x, y)$ and $g(x, y) = \frac{f(x,y)}{b - f(x,y)}$, where $b > f(x, y)$ for $x, y \in \mathbb{R}$. If $\min f(x, y) = f(x_0, y_0)$, then we have that $\min g(x, y) = g(x_0, y_0)$. If $f(x, y)$ is jointly convex in (x, y), then $g(x, y)$ is jointly convex, too.*

The social welfare function of $S(\Delta_L, \Delta_M)$ in (12.3.8) can be rewritten in below

$$
S(\Delta_L, \Delta_M) = \Lambda_s R_s + \Lambda_m R_m - h\left[g_1(\Delta_L) + g_2(\Delta_L, \Delta_M) + \frac{q_L \Lambda_m}{\mu_L - q_L \Lambda_m}\right], \tag{12.3.9}
$$

where

$$
\begin{aligned}
g_1(\Delta_L) &= \frac{f_1(\Delta_L)}{\mu_M/\Lambda_m - f_1(\Delta_L)}, \\
f_1(\Delta_L) &= q_M + q_L \alpha_L e^{-\beta_L \Delta_L}, \\
g_2(\Delta_L, \Delta_M) &= \frac{f_2(\Delta_L, \Delta_M)}{1/(\Lambda_m c_h) - f_2(\Delta_L, \Delta_M)}, \\
f_2(\Delta_L, \Delta_M) &= \frac{q_H + (q_M + q_L \alpha_L e^{-\beta_L \Delta_L})\alpha_M e^{-\beta_M \Delta_M} + \Lambda_s/\Lambda_m}{B - \Delta_L - \Delta_M}.
\end{aligned}
$$

Based on Lemma 12.3.1, the following proposition is provided to show the existence and uniqueness of optimal value (Δ_L^*, Δ_M^*) which maximizes the social welfare.

Proposition 12.3.1 *There exists a unique (Δ_L^*, Δ_M^*) that maximizes $S(\Delta_L, \Delta_M)$. In particular, if (1) $0 \le \hat{\Delta}_L + \hat{\Delta}_M < B$, (2) $\frac{B - \hat{\Delta}_L - \hat{\Delta}_M}{c_h} > [q_H + (q_M +$*

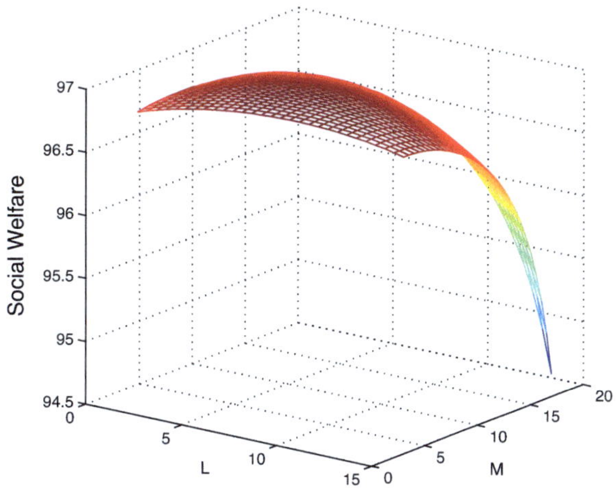

Fig. 12.10 Social welfare with respect to: Δ_L and Δ_M, for $q_L = 0.3$, $q_M = 0.5$, $\mu_L = 12$, $\mu_M = 20$, $\alpha_L = 0.8$, $\beta_L = 0.08$, $\alpha_M = 0.2$, $\beta_M = 0.2$, $R_s = R_m = 10$, $\Lambda_m = 9$, $\Lambda_s = 1$, $c_h = 2$, $h = 3$, $B = 40$

$q_L\alpha_L e^{-\beta_L\hat{\Delta}_L})\alpha_M e^{-\beta_M\hat{\Delta}_M}]\Lambda_m + \Lambda_s$, *(3)* $\mu_M > (q_M + q_L\alpha_L e^{-\beta_L\hat{\Delta}_L})\Lambda_m$, *then* $(\Delta_L^*, \Delta_M^*) = (\hat{\Delta}_L, \hat{\Delta}_M)$, *where* $(\hat{\Delta}_L, \hat{\Delta}_M)$ *uniquely solves*

$$\begin{cases} \frac{q_H}{q_M+q_L\alpha_L e^{-\beta_L\Delta_L}} + \alpha_M e^{-\beta_M\Delta_M} = \alpha_M\beta_M e^{-\beta_M\Delta_M}(B - \Delta_L - \Delta_M), \\ c_h\mu_M q_L\alpha_L\beta_L e^{-\beta_L\Delta_L}(\frac{B-\Delta_L-\Delta_M}{c_h} - (q_H + (q_M + q_L\alpha_L e^{-\beta_L\Delta_L})\alpha_M e^{-\beta_M\Delta_M})\Lambda_m + \Lambda_s)^2 = \\ (q_M\beta_M + (\beta_m - \beta_L)q_L\alpha_L e^{-\beta_L\Delta_L})\alpha_M e^{-\beta_M\Delta_M}(B - \Delta_L - \Delta_M)(\mu_m - (q_M + q_L\alpha_L e^{-\beta_L\Delta_L})\Lambda_m)^2, \end{cases}$$

Proposition 12.3.1 characterizes the optimal allocation budget explicitly. In particular, the optimal policy (Δ_L^*, Δ_M^*) can be attained at the extreme point $(\hat{\Delta}_L, \hat{\Delta}_M)$ if it drops into the feasible region. Otherwise, the optimal solution will be reached at the boundary of such feasible region, that is, $\Delta_L = 0$ or $\Delta_M = 0$, thus it degenerates to our base model.

12.3.1.1 Numerical Examples

The social welfare is shown in Fig. 12.10. It shows that $S(\Delta_L, \Delta_M)$ is concave in (Δ_L, Δ_M), which is consistent with Proposition 12.3.1. And we can observe that the maximum social welfare can be achieved in $(\Delta_L^* = 9, \Delta_M^* = 5)$,

Figures 12.11 and 12.12 present the effects of β_L and β_M on the optimal allocation policy. From Fig. 12.11, we observe that when β_L increases, the Δ_L^* will decrease because a lower reimbursement is needed to improve the service quality for L-hospital. On the other hand, Δ_M^* shows a different trend, which is increasing in β_L. It is because that the increased amount to M-hospital can be used to compensate for the reduced reimbursement to L-hospital so as to improve the service quality overally. However, the total amount of Δ_L^* and Δ_M^* (i.e., $\Delta_L^* + \Delta_M^*$) is decreasing in β_L, and more reimbursement should be allocated to H-hospital to reduce the congestion, i.e., $\Delta_H^* \uparrow$. It is an interesting observation that when the effectiveness parameter (i.e.,

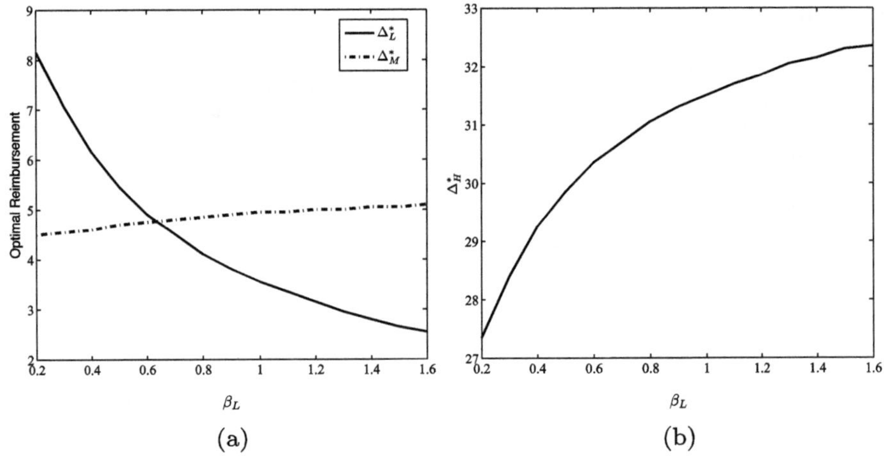

Fig. 12.11 Optimal allocation policy with respect to: β_L, for $\beta_M = 0.2, q_L = 0.3, q_M = 0.5, \mu_L = 12, \mu_M = 20, \alpha_L = 0.8, \alpha_M = 0.2, R_s = R_m = 10, \Lambda_m = 9, \Lambda_s = 1, c_h = 2, h = 3, B = 40$

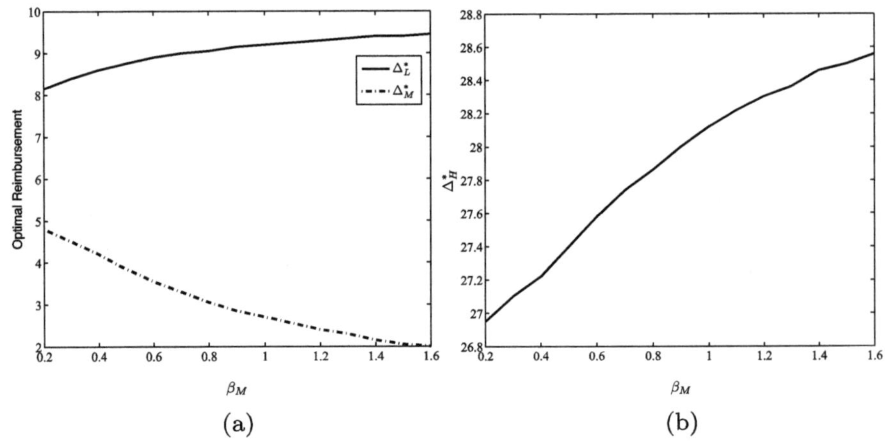

Fig. 12.12 Optimal allocation policy with respect to: β_M, for $\beta_L = 0.2, q_L = 0.3, q_M = 0.5, \mu_L = 12, \mu_M = 20, \alpha_L = 0.8, \alpha_M = 0.2, R_s = R_m = 10, \Lambda_m = 9, \Lambda_s = 1, c_h = 2, h = 3, B = 40$

β_L) for L-hospital increases, more budgets should be allocated to H-hospital rather than to the L-hospital. It can be explained by that the increased service quality has a second-order effect comparing to the reduced delay costs, then it is more effective to reduce the congestion by improving Δ_H. By contrast, in Fig. 12.12, for a fixed β_M, the similar phenomenon ($\Delta_H^* \uparrow$) can be observed when β_M changes.

12.3.1.2 Equilibrium Joining Strategy

Now we consider the case where patients can make their decisions based on the delay information (their expected waiting times). We still consider a two-stage Stackelberg game to characterize the equilibrium of both patients and government, where the government is the leader and patients are the followers.

For joining probabilities q_L, q_M and q_H (where $q_L + q_M + q_H = 1$), the effective arrival rates are given by (12.3.1)–(12.3.3). Based on the similar argument in previous Sect. 12.2, the expected waiting times for patients choose to join H-, M-, and L-hospitals can be given by

$$W_h(q_L, q_M) = \frac{1}{\mu_H - \Lambda_s - \Lambda_m[q_H + (q_M + q_L q_{LM})q_{MH}]},$$
$$W_m(q_L, q_M) = \frac{1}{\mu_m - (q_M + q_L q_{LM})\Lambda_m} + \frac{q_{MH}}{\mu_H - \Lambda_s - \Lambda_m[q_H + (q_M + q_L q_{LM})q_{MH}]},$$
$$W_l(q_L, q_M) = \frac{1}{\mu_L - q_L\Lambda_m} + \frac{q_{LM}}{\mu_m - (q_M + q_L q_{LM})\Lambda_m} + \frac{q_{LM}q_{MH}}{\mu_H - \Lambda_s - \Lambda_m[q_H + (q_M + q_L q_{LM})q_{MH}]},$$

respectively. Denote by $U_H(q_L, q_M) = R_m - P_H - hW_h(q_L, q_M)$, $U_M(q_L, q_M) = R_m - P_m - q_{MH}P_H - hW_m(q_L, q_M)$ and $U_L(q_L, q_M) = R_m - P_L - q_{LM}P_m - q_{LM}q_{MH}P_H - hW_l(q_L, q_M)$ the expected utilities of patients choose to join H-, M-, and L-hospitals, respectively, then the equilibrium (q_L^e, q_M^e) can be characterized in the following proposition.

Proposition 12.3.2 *The equilibrium strategy (q_L^e, q_M^e) is determined as follow:*

$$(q_L^e, q_M^e) = \begin{cases} (0,0), & \text{if } U_L(0,0) < U_H(0,0) \text{ and } U_M(0,0) < U_H(0,0); \\ (0,1), & \text{if } U_L(0,1) < U_M(0,1) \text{ and } U_H(0,1) < U_M(0,1); \\ (1,0), & \text{if } U_M(1,0) < U_L(1,0) \text{ and } U_H(1,0) < U_L(1,0); \\ (q_L^{(1)}, 0), & \text{if } U_M(q_L^{(1)},0) < U_L(q_L^{(1)},0) = U_H(q_L^{(1)},0), \text{ where } q_L^{(1)} \in (0,1); \\ (0, q_M^{(1)}), & \text{if } U_L(0,q_M^{(1)}) < U_M(0,q_M^{(1)}) = U_H(0,q_M^{(1)}), \text{ where } q_M^{(1)} \in (0,1); \\ (q_L^{(2)}, 1-q_L^{(2)}), & \text{if } U_L(q_L^{(2)},1-q_L^{(2)}) = U_M(q_L^{(2)},1-q_L^{(2)}) > U_H(q_L^{(2)},1-q_L^{(2)}), \text{ where } q_L^{(2)} \in (0,1). \\ (q_L^{(3)}, q_M^{(3)}), & \text{otherwise.} \end{cases}$$

where $q_L^{(1)}$ uniquely solves $U_L(q_L^{(1)}, 0) = U_H(q_L^{(1)}, 0)$, $q_M^{(1)}$ uniquely solves $U_M(0, q_M^{(1)}) = U_H(0, q_M^{(1)})$ and $(q_L^{(2)}, 1 - q_L^{(2)})$ uniquely solves $U_L(q_L^{(2)}, 1 - q_L^{(2)}) = U_M(q_L^{(2)}, 1 - q_L^{(2)})$; $(q_L^{(3)}, q_M^{(3)})$ is determined by $U_L(q_L^e, q_M^e) = U_M(q_L^e, q_M^e) = U_H(q_L^e, q_M^e)$.

It is noted that when $q_L^e \cdot q_M^e \cdot (1 - q_L^e - q_M^e) = 0$ (cases (1)-(6)), there always exist at least one hospital such that there is no arriving patients choose to join. For a curtain reimbursement policy: (Δ_L, Δ_M), we have $q_{LM} = \alpha_L e^{-\beta_L \Delta_L}$, $q_{MH} = \alpha_M e^{-\beta_M \Delta_M}$, and $\mu_H = (B - \Delta_L - \Delta_M)/c_h$, then the equilibrium probabilities (q_L^e, q_M^e) can be derived by examining the cases in Proposition 12.3.2.

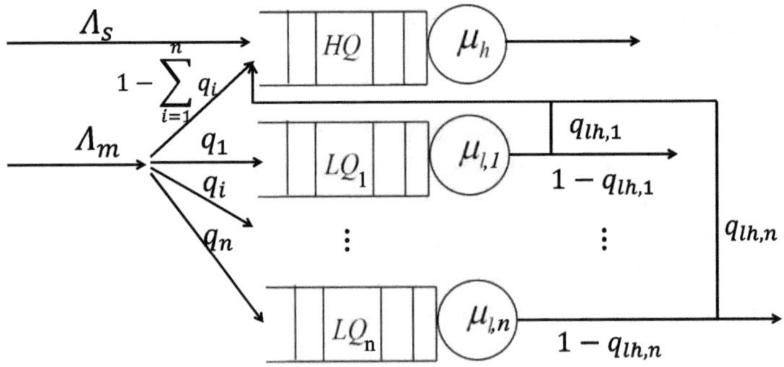

Fig. 12.13 A schematic of the model with n L-hospitals

12.3.2 Model with Multiple L-Hospitals

In this section, we extend our benchmark to a model with multiple L-hospitals. In particular, we still assume that there is only one H-hospital, and n ($n \geq 1$) homogenous L-hospitals can exist, that share the same initial service medical level α and sensitivity parameter β. The budget provided to i-th L-hospital is denoted by Δ_i, which satisfies $\sum_i^n \Delta_i \leq B$. Then the transfer rate of i-th L-hospital satisfies $q_{lh,i} = \alpha e^{-\beta \Delta_i}$ for $i = 1, 2, \ldots, n$ and the service rate of H-hospital is $(B - \sum_i^n \Delta_i)/c_h$. The queueing network with n L-hospitals is illustrated in Fig. 12.13.

Denote by λ_h and $\lambda_{l,i}$ the effective arrival rates to the H-hospital and i-th L-hospital, respectively. For given joining strategy vector $\mathbf{q} = (q_1, \ldots, q_n)$, the traffic equations are given by

$$\begin{aligned}
\lambda_h &= \Lambda_s + \Lambda_m \left(1 - \sum_{i=1}^{n} q_i\right) + \sum_{i=1}^{n} \lambda_{l,i} q_{lh,i}, \\
\lambda_{l,i} &= \Lambda_m q_i,
\end{aligned}$$

for $i = 1, 2, \ldots, n$, which gives

$$\begin{aligned}
\lambda_h &= \Lambda_s + \Lambda_m \left(q + \sum_{i=1}^{n} q_i q_{lh,i}\right), \\
\lambda_{l,i} &= \Lambda_m q_i,
\end{aligned}$$

where $q = 1 - \sum_{i=1}^{n} q_i$. Under steady state, the expected total number of patients in the system is given by

$$N_n = N_h(\mathbf{q}) + \sum_{i=1}^{n} N_{l,i}(q_i) = \frac{\frac{\lambda_h}{\mu_h}}{1-\frac{\lambda_h}{\mu_h}} + \sum_{i=1}^{n} \frac{\frac{\lambda_{l,i}}{\mu_l}}{1-\frac{\lambda_{l,i}}{\mu_l}}$$
$$= \frac{\Lambda_s + \Lambda_m \left(q + \sum_{i=1}^{n} q_i q_{lh,i}\right)}{\mu_h - \Lambda_s - \Lambda_m \left(q + \sum_{i=1}^{n} q_i q_{lh,i}\right)} + \sum_{i=1}^{n} \frac{q_i \Lambda_m}{\mu_l - q_i \Lambda_m}.$$

Then the social welfare can be expressed as the total net benefits of all patients seeking service per time unit: $S(\Delta) = \Lambda_s R_s + \Lambda_m R_m - hN_n$, which is a function of the budget allocation to L-hospitals, i.e., $\Delta = (\Delta_1, \ldots, \Delta_n)$. Thus we have the following optimization problem:

$$\max_{0 \le \sum_i^n \Delta_i < B} \quad S(\Delta) = \Lambda_s R_s + \Lambda_m R_m - h \left[\frac{\Lambda_s + \Lambda_m \left(q + \sum_{i=1}^{n} q_i q_{lh,i}\right)}{\mu_h - \Lambda_s - \Lambda_m \left(q + \sum_{i=1}^{n} q_i q_{lh,i}\right)} + \sum_{i=1}^{n} \frac{q_i \Lambda_m}{\mu_l - q_i \Lambda_m} \right] \quad (12.3.10)$$
$$\text{s.t. } \mu_h(\Delta) > \Lambda_s + \Lambda_m \left(q + \sum_{i=1}^{n} q_i q_{lh,i}\right),$$
$$\mu_l > q_i \Lambda_m, \quad i = 1, \ldots, n.$$

The constraint conditions are needed to ensure the stability of system. It should be noted that the social welfare consists of three parts: the service value and the waiting costs occur in two hospitals. Intuitively, the waiting cost in L-hospital is not affected by the budget vector Δ. Thus, rewriting (12.3.10) after some manipulations gives

$$\max_{0 \le \sum_i^n \Delta_i < B} \quad S(\Delta) = \Lambda_s R_s + \Lambda_m R_m - h \left(\frac{f_n(\Delta)}{1 - f_n(\Delta)} + \sum_{i=1}^{n} \frac{q_i \Lambda_m}{\mu_l - q_i \Lambda_m} \right),$$

where $f_n(\Delta) = \frac{([q + \sum_{i=1}^{n} q_i \alpha e^{-\beta \Delta_i}]\Lambda_m + \Lambda_s) c_h}{B - \sum_i^n \Delta_i} < 1$ is the system workload in H-hospital. Therefore, (12.3.10) is equivalent to

$$\min_{0 \le \sum_i^n \Delta_i < B} \quad f_n(\Delta) = \frac{[q + \sum_{i=1}^{n} q_i \alpha e^{-\beta \Delta_i}]\Lambda_m + \Lambda_s}{B - \sum_i^n \Delta_i}. \quad (12.3.11)$$

We conduct the numerical experiment to investigate the sensitivity of optimal budget with respect to the joining probability and sensitivity parameter in Table 12.1 when $\mathbf{q} = (0.1, 0.2, 0.3)$.

Remark 12.3.1 Table 12.1 discloses some insights. First, we can observe that with the increase of sensitivity parameter β, $f_n(\Delta^*)$ decreases. It indicates that the optimal social welfare is increasing in β, which is intuitive and is consistent with our result in base model. On the other hand, when the proportions to each L-hospital satisfies $q_1 < q_2 < q_3$, the optimal budget should satisfy $\Delta_1^* < \Delta_2^* < \Delta_3^*$, which is intuitive since the government should always reimburse those L-hospitals with large demand. However, we find that the optimal budget to a single L-hospital is not always

Table 12.1 Optimal budget $\Delta^* = (\Delta_1^*, \Delta_2^*, \Delta_3^*)$ for different β

β	1.0	1.1	1.2	1.3	1.4	1.5	1.6	1.7	1.8	1.9	2.0
Δ_1^*	0	0	0	0	0.049	0.098	0.148	0.185	0.210	0.234	0.246
Δ_2^*	0.381	0.443	0.492	0.529	0.554	0.566	0.578	0.590	0.590	0.590	0.590
Δ_3^*	0.787	0.812	0.824	0.836	0.836	0.836	0.836	0.824	0.812	0.812	0.800
$f_n(\Delta^*)$	0.820	0.809	0.799	0.789	0.779	0.770	0.761	0.753	0.744	0.737	0.729

Note $\mathbf{q} = (0.1, 0.2, 0.3)$, $\alpha = 0.6$, $\Lambda_m = 10$, $\Lambda_s = 1$, $B = 10$

increasing in the sensitivity parameter β. For example, when β is small, Δ_3^* increases with β, indicating that allocating more to the 3-th L-hospital is more beneficial. But when β exceeds a certain value, the cure rate of this L-hospital becomes to be less sensitive to β, then less budget should be allocated to it with the increase of β. This observation is also consistent with our base model (see Proposition 12.1.2).

12.3.2.1 Equilibrium Joining Strategy

Denote by $q = 1 - \sum_{i=1}^{n} q_i$ the proportion of patients join H-hospital. Without loss of generality, we assume that the budget amount to L-hospitals satisfies $\Delta_1 \geq \Delta_2 \cdots \geq \Delta_n$. Based on the similar argument in previous section, the expected waiting times for patients choose to join H- and i-th L-hospital can be given by

$$W_h(\mathbf{q}) = \frac{1}{\mu_h - \Lambda_s - [q + \sum_i^n q_{lh,i} q_i]\Lambda_m},$$

$$W_{l,i}(\mathbf{q}) = \frac{1}{\mu_l - q_i \Lambda_m} + \frac{q_{lh,i}}{\mu_h - \Lambda_s - [q + \sum_i^n q_{lh,i} q_i]\Lambda_m},$$

for $i = 1, \cdots, n$. Denote by $U_H(\mathbf{q}) = R_m - P_H - hW_h(\mathbf{q})$ and $U_{L,i}(\mathbf{q}) = R_m - P_L - q_{lh,i} P_H - hW_{l,i}(\mathbf{q})$ the expected utilities of patients choose to join H- and i-th L-hospitals, respectively. The following proposition establishes the structure of equilibrium.

Proposition 12.3.3 *The equilibrium joining strategy satisfies* $q_1^e \geq q_2^e \cdots \geq q_n^e$.

Proposition 12.3.3 indicates that the equilibrium proportion of patients to L-hospital decreases with the budget amount. It is quite intuitive since the L-hospital will be more attractive if its medical level improves.

12.4 Concluding Remarks

This section studies how a reimbursement policy affects the healthcare system with multi-level hospitals. Our base model is a two-level system characterized by two hospitals with different service qualities and service capacities. Two types of patients exist, where the patients with severe illness will certainly join the hospital with guaranteed service quality and the ones with mild health problem are delay sensitive and strategically decide which hospital to join so as to maximize their expected utilities. To our best knowledge, it is the first time to investigate the reimbursement policy by considering the trade-off between efficiency and service quality issues.

The optimal allocation policy as well as the equilibrium strategy of patients are derived and compared. And some interesting results can be observed, e.g, the optimal budget to L-hospital is non-monotone in the effectiveness parameter while the optimal social welfare is non-monotone in the joining strategy of patients, and we provide the explanations for these observations. By comparing the socially optimal strategy with the individually equilibrium strategy, we find that, the individual behavior of patients does not necessarily lead to systems more congested than what is socially desirable, which is a sharp contrast to the classical literature result (e.g., Naor 1969). The detailed explanations in the reason in the end of Section 3. We also demonstrate the robustness of our base model by extending it to a three-level system and a system with multiple homogenous L-hospitals. Our work provides some managerial insights for the social planner (government):

(1) First, from the view of social planner, more (resp. less) budget should be assigned to L-hospital when the potential arrival rate of s-type (resp. m-type) patients increases, so as to meet the demand of patients adequately.
(2) Second, the optimal budget to L-hospital is not necessarily increasing in its effectiveness parameter (which measures the sensitivity of medical level to the budget). In particular, when the effectiveness parameter is large, with the increase of effectiveness parameter, less budget should be assigned to L-hospital. That is to say, the government should carefully determine the budget to L-hospital.
(3) Third, comparing the socially optimal and individually equilibrium strategy discloses that the government should urge the H-hospital to reduce (resp. increase) its price (i.e., P_h) when the budget to L-hospital is small (resp. large), so as to achieve the optimal social welfare.

A possible future research direction is to consider a duopoly model, in which H-hospital and L-hospital are competitive to maximize their own profits rather than the overall social welfare. And some new insights may be found by comparing to our base model.

12.5 Appendix: Technical Proofs

Proof of Proposition 12.1.1 The first and second derivatives of $f_1(\Delta)$ are

$$\frac{df_1(\Delta)}{d\Delta} = \frac{c_h \Lambda_m [q + \alpha e^{-\beta\Delta}(1-q) - (B-\Delta)\alpha\beta e^{-\beta\Delta}(1-q)]}{(B-\Delta)^2} + \frac{\Lambda_s}{(B-\Delta)^2} \tag{12.5.1}$$

and

$$\frac{d^2 f_1(\Delta)}{d\Delta^2} = \frac{c_h \Lambda_m f(\Delta) + 2\Lambda_s}{(B-\Delta)^3}, \tag{12.5.2}$$

respectively, where

$$f(\Delta) = (B-\Delta)^2 \alpha\beta^2 e^{-\beta\Delta}(1-q) + 2[q + \alpha e^{-\beta\Delta}(1-q) - (B-\Delta)\alpha\beta e^{-\beta\Delta}(1-q)]. \tag{12.5.3}$$

We can interpret the function of $f(\Delta)$ as

$$\begin{aligned} f(\Delta) = \; & [(B-\Delta)\sqrt{\alpha\beta^2 e^{-\beta\Delta}(1-q)} - \sqrt{2(q+\alpha e^{-\beta\Delta}(1-q))}]^2 + \\ & 2(B-\Delta)\sqrt{\alpha\beta^2 e^{-\beta\Delta}(1-q)}[\sqrt{2(q+\alpha e^{-\beta\Delta}(1-q))} - \sqrt{\alpha e^{-\beta\Delta}(1-q)}] \geq 0. \end{aligned} \tag{12.5.4}$$

Thus $\frac{df_1(\Delta)}{d\Delta}$ is increasing in $\Delta \in [0, B]$, i.e., $f_1(\Delta)$ is convex in $\Delta \in [0, B]$. Notice that $\frac{df_1(\Delta)}{d\Delta}|_{\Delta=0} = \frac{c_h\Lambda_m[q+\alpha(1-q)(1-B\beta)]+\Lambda_s}{B^2}$ and $\frac{df_1(\Delta)}{d\Delta}|_{\Delta=B^-} = +\infty > 0$, we consider two cases below:

Case (1): If $c_h\Lambda_m[q+\alpha(1-q)(1-B\beta)]+\Lambda_s \geq 0$, then $f_1(\Delta)$ is increasing in $\Delta \in [0, B)$, which implies that $\frac{dS(\Delta)}{d\Delta} \leq 0$ in $\Delta \in [0, B)$. Based on our preassumption that $\frac{\Lambda_s+\Lambda_m[q+q_{lh}(\Delta)(1-q)]}{\mu_h(\Delta)} < 1$, we have $f_1(0) < 1$, then the system can be stable when $\Delta = 0$. Thus the maximal $S(\Delta)$ can be attained at $\Delta^* = 0$.

Case (2): If $c_h\Lambda_m[q+\alpha(1-q)(1-B\beta)]+\Lambda_s < 0$, then $f_1(\Delta)$ is decreasing firstly and increasing in $\Delta \in [0, B)$, there exists unique $\widehat{\Delta} \in [0, B)$ satisfying that $\frac{dS(\Delta)}{d\Delta}|_{\Delta=\widehat{\Delta}} = \frac{df_1(\Delta)}{d\Delta}|_{\Delta=\widehat{\Delta}} = 0$. And $S(\Delta)$ is increasing in $[0, \widehat{\Delta})$ and decreasing in $(\widehat{\Delta}, B)$. Also, it implies that $\min\limits_{\Delta\in[0,B)} \frac{\Lambda_s+\Lambda_m[q+q_{lh}(\Delta)(1-q)]}{\mu_h(\Delta)} = f_1(\widehat{\Delta})$. Based on our preassumption that $\frac{\Lambda_s+\Lambda_m[q+q_{lh}(\Delta)(1-q)]}{\mu_h(\Delta)} < 1$, the system is stable when $\Delta = \widehat{\Delta}$, thus the maximal $S(\Delta)$ can be attained at $\Delta^* = \widehat{\Delta}$. □

Proof of Proposition 12.1.2 Firstly, from Proposition 12.1.1, when $\beta \leq \frac{c_h\Lambda_m[q+\alpha(1-q)]+\Lambda_s}{\alpha(1-q)c_h\Lambda_m B}$, we have $\Delta^* = 0$, which is constant in β. Thus we just consider the case that $\beta > \frac{c_h\Lambda_m[q+\alpha(1-q)]+\Lambda_s}{\alpha(1-q)c_h\Lambda_m B}$. In this case, the optimal Δ^* is uniquely determined by $c_h\Lambda_m(e^{\beta\Delta}q + (1-q)\alpha[1-(B-\Delta)\beta]) + \Lambda_s = 0$. Define $\Psi(\alpha, \beta, q, \Delta) = c_h\Lambda_m(e^{\beta\Delta}q + (1-q)\alpha[1-(B-\Delta)\beta]) + \Lambda_s$, we have

$$\frac{\partial \Psi(\alpha, \beta, q, \Delta)}{\partial \Delta} = c_h \Lambda_m [e^{\beta \Delta} \beta q + (1-q)\alpha\beta],$$
$$\frac{\partial \Psi(\alpha, \beta, q, \Delta)}{\partial \alpha} = c_h \Lambda_m (1-q)[1-(B-\Delta)\beta] < 0,$$
$$\frac{\partial \Psi(\alpha, \beta, q, \Delta)}{\partial \beta} = c_h \Lambda_m [e^{\beta \Delta} \Delta q - (1-q)\alpha(B-\Delta)].$$

Then it gives

$$\frac{d\Delta^*(\alpha)}{d\alpha} = -\frac{\partial \Psi(\alpha, \beta, q, \Delta)/\partial \alpha}{\partial \Psi(\alpha, \beta, q, \Delta)/\partial \Delta} = \frac{(1-q)[(B-\Delta)\beta - 1]}{e^{\beta\Delta}\beta q + (1-q)\alpha\beta} > 0.$$

Thus $\Delta^*(\alpha)$ is increasing in α. Similarly, we can get

$$\frac{d\Delta^*(\beta)}{d\beta} = -\frac{\partial \Psi(\alpha, \beta, q, \Delta)/\partial \beta}{\partial \Psi(\alpha, \beta, q, \Delta)/\partial \Delta} = \frac{(1-q)\alpha(B-\Delta) - e^{\beta\Delta}\Delta q}{e^{\beta\Delta}\beta q + (1-q)\alpha\beta}.$$

Notice that $\frac{\partial \Psi(\alpha,\beta,q,\Delta)}{\partial \Delta} > 0$. Therefore, $\frac{d\Delta^*(\beta)}{d\beta} > 0$ if and only if $(1-q)\alpha(B-\Delta) - e^{\beta\Delta}\Delta q > 0$. Because $\frac{d((1-q)\alpha(B-\Delta)-e^{\beta\Delta}\Delta q)}{d\beta} = -e^{\beta\Delta}\Delta^2 q < 0$, then $(1-q)\alpha(B-\Delta) - e^{\beta\Delta}\Delta q$ is decreasing in β. When $\beta = \frac{q+\alpha(1-q)}{\alpha(1-q)B}$, we have $\Delta^* = 0$, it gives $\frac{d((1-q)\alpha(B-\Delta)-e^{\beta\Delta}\Delta q)}{d\beta}|_{\beta=\frac{q+\alpha(1-q)}{\alpha(1-q)B}} = (1-q)\alpha B > 0$. Notice that $\frac{d((1-q)\alpha(B-\Delta)-e^{\beta\Delta}\Delta q)}{d\beta}|_{\beta=\infty} = -\infty$, then there exists unique $\overline{\beta}$ such that $\Delta^*(\beta)$ is increasing in $\beta \in (\frac{q+\alpha(1-q)}{\alpha(1-q)B}, \overline{\beta})$, and is decreasing in $\beta \in [\overline{\beta}, \infty)$.

Similarly, since

$$\frac{\partial \Psi(\alpha, \beta, q, \Delta)}{\partial \Lambda_m} = c_h(e^{\beta\Delta} q + (1-q)\alpha[1-(B-\Delta)\beta]),$$
$$\frac{\partial \Psi(\alpha, \beta, q, \Delta)}{\partial \Lambda_s} = 1,$$

and $c_h \Lambda_m (e^{\beta\Delta} q + (1-q)\alpha[1-(B-\Delta)\beta])|_{\Delta=\Delta^*} = -\Lambda_s < 0$, it gives

$$\frac{d\Delta^*(q)}{d\Lambda_m} = -\frac{\partial \Psi(\alpha, \beta, q, \Delta)/\partial \Lambda_m}{\partial \Psi(\alpha, \beta, q, \Delta)/\partial \Delta} = \frac{e^{\beta\Delta} q + (1-q)\alpha[1-(B-\Delta)\beta]}{\Lambda_m[e^{\beta\Delta}\beta q + (1-q)\alpha\beta]} < 0.$$

$$\frac{d\Delta^*(q)}{d\Lambda_s} = -\frac{\partial \Psi(\alpha, \beta, q, \Delta)/\partial \Lambda_s}{\partial \Psi(\alpha, \beta, q, \Delta)/\partial \Delta} = \frac{1}{c_h\Lambda_m[e^{\beta\Delta}\beta q + (1-q)\alpha\beta]} > 0.$$

□

Proof of Proposition 12.1.3 Firstly, from Proposition 12.1.1, when $q \geq \frac{\alpha\beta B c_h \Lambda_m - \alpha c_h \Lambda_m - \Lambda_s}{\alpha\beta B c_h \Lambda_m - \alpha c_h \Lambda_m + c_h \Lambda_m}$, we have $\Delta^* = 0$, which is constant in q. Thus we just consider the case that $q < \frac{\alpha\beta B c_h \Lambda_m - \alpha c_h \Lambda_m - \Lambda_s}{\alpha\beta B c_h \Lambda_m - \alpha c_h \Lambda_m + c_h \Lambda_m}$. In this case, the optimal Δ^* is uniquely determined by $c_h\Lambda_m(e^{\beta\Delta}q + (1-q)\alpha[1-(B-\Delta)\beta]) + \Lambda_s = 0$. Define $\Psi(\alpha, \beta, q, \Delta) = c_h\Lambda_m(e^{\beta\Delta}q + (1-q)\alpha[1-(B-\Delta)\beta]) + \Lambda_s$, we have

$$\frac{\partial\Psi(\alpha, \beta, q, \Delta)}{\partial\Delta} = c_h\Lambda_m[e^{\beta\Delta}\beta q + (1-q)\alpha\beta],$$
$$\frac{\partial\Psi(\alpha, \beta, q, \Delta)}{\partial q} = c_h\Lambda_m(e^{\beta\Delta} - \alpha[1-(B-\Delta)\beta]).$$

Then it gives

$$\frac{d\Delta^*(q)}{dq} = -\frac{\partial\Psi(\alpha, \beta, q, \Delta)/\partial q}{\partial\Psi(\alpha, \beta, q, \Delta)/\partial\Delta} = \frac{\alpha[1-(B-\Delta)\beta] - e^{\beta\Delta}}{e^{\beta\Delta}\beta q + (1-q)\alpha\beta}.$$

Recall that Δ^* is uniquely determined by $c_h\Lambda_m(e^{\beta\Delta}q + (1-q)\alpha[1-(B-\Delta)\beta]) + \Lambda_s = 0$, then we must have $\alpha[1-(B-\Delta^*)\beta] < 0$ for any given q. Then it gives

$$\frac{d\Delta^*(q)}{dq} = \frac{\alpha[1-(B-\Delta)\beta] - e^{\beta\Delta}}{e^{\beta\Delta}\beta q + (1-q)\alpha\beta} < 0.$$

Therefore, Δ^* is decreasing in

$$q \in [0, \frac{\alpha\beta Bc_h\Lambda_m - \alpha c_h\Lambda_m - \Lambda_s}{\alpha\beta Bc_h\Lambda_m - \alpha c_h\Lambda_m + c_h\Lambda_m})$$

and constant in

$$q \in \left[\frac{\alpha\beta Bc_h\Lambda_m - \alpha c_h\Lambda_m - \Lambda_s}{\alpha\beta Bc_h\Lambda_m - \alpha c_h\Lambda_m + c_h\Lambda_m}, 1\right].$$

□

Proof of Proposition 12.1.4 Notice that for any fixed Δ, the social welfare is

$$S(\Delta, q) = \Lambda_s R_s + \Lambda_m R_m - h\left[\frac{\Lambda_s + \Lambda_m[q + q_{lh}(\Delta)(1-q)]}{\mu_h(\Delta) - \Lambda_s - \Lambda_m[q + q_{lh}(\Delta)(1-q)]} + \frac{(1-q)\Lambda_m}{\mu_l - (1-q)\Lambda_m}\right],$$

and

$$\frac{dS(\Delta, q)}{dq} = h\Lambda_m\left(\frac{\mu_l}{(\mu_l - (1-q)\Lambda_m)^2} + \frac{\mu_h(\Delta)(q_{lh}(\Delta) - 1)}{(\mu_h(\Delta) - \Lambda_s - \Lambda_m[q + q_{lh}(\Delta)(1-q)])^2}\right),$$
$$\frac{d^2S(\Delta, q)}{dq^2} = -2h\Lambda_m^2\left(\frac{\mu_h(\Delta)(q_{lh}(\Delta) - 1)^2}{(\mu_h(\Delta) - \Lambda_s - \Lambda_m[q + q_{lh}(\Delta)(1-q)])^3} + \frac{\mu_l}{(\mu_l - (1-q)\Lambda_m)^3}\right).$$

That is, $S(\Delta, q)$ is strictly concave in q.

consider three cases below.

Case (i): $q_s(\Delta) = 1$ if and only if $\frac{dS(\Delta,q)}{dq}|_{q=1} \geq 0$, i.e., $\frac{1}{\mu_l} \geq \frac{\mu_h(\Delta)(1-q_{lh}(\Delta))}{(\mu_h(\Delta)-\Lambda_s-\Lambda_m)^2} \Leftrightarrow B \geq \bar{B}(\Delta) = \Delta + c_h[2(\Lambda_s + \Lambda_m) + \mu_l(1 - \alpha e^{-\beta\Delta}) + \sqrt{\mu_l(1-\alpha)[4(\Lambda_s + \Lambda_m) + \mu_l(1-\alpha)]}]/2$, which increases in Δ.

Case (ii): $q_s(\Delta) = 0$ if and only if $\frac{dS(\Delta,q)}{dq}|_{q=0} \leq 0$, i.e.,

$$B \leq \underline{B}(\Delta) = \Delta + c_h(\Lambda_s + \Lambda_m\alpha e^{-\beta\Delta}) + \frac{1 + \sqrt{1 + 4(\Lambda_s + \Lambda_m\alpha e^{-\beta\Delta})\mu_l/[(1-q_{lh})(\mu - \Lambda_m)^2]}}{2\mu_l/[(1-q_{lh})(\mu - \Lambda_m)^2]}.$$

Case (iii): $q_s(\Delta) < (0, 1)$ if and only if $B \in (\underline{B}(\Delta), \bar{B}(\Delta))$, and $q_s(\Delta)$ uniquely solves $\frac{dS(\Delta,q)}{dq} = 0$, which gives

$$q_s(\Delta) = \frac{c_h(\mu_l[c_h(\Lambda_s + \Lambda_m\alpha e^{-\beta\Delta}) - 2(B-\Delta)] + \Lambda_m(B-\Delta)) + \sqrt{\frac{c_h\mu_l(B-\Delta)(c_h(\Lambda_m+\Lambda_s+\mu_l(\alpha e^{-\beta\Delta}-1)) - B+\Delta)^2}{1-\alpha e^{-\beta\Delta}}}}{c_h\Lambda_m(B + c_h\mu_l(\alpha e^{-\beta\Delta} - 1) - \Delta)}.$$

□

Proof of Corollary 12.1.1 By the analysis in the proof of Proposition 12.1.4, if $B < \tilde{B} = c_h[2(\Lambda_s + \Lambda_m) + \mu_l(1-\alpha) + \sqrt{\mu_l(1-\alpha)[4(\Lambda_s + \Lambda_m) + \mu_l(1-\alpha)]}]/2$, we have that $q_s(\Delta) < 1$ for any fixed $\Delta \in [0, B]$, which implies that $q_s < 1$. Thus, we can conclude that $q_s < 1$ if $B < \tilde{B}$. □

Proof of Lemma 12.2.1 It is noted that $D(q) = \frac{1-q_{lh}}{\mu_h-\Lambda_s-(q+q_{lh}(1-q))\Lambda_m} - \frac{1}{\mu_l-(1-q)\Lambda_m}$. By taking a deviation to $D(q)$, we have

$$\frac{dD(q)}{dq} = \frac{\lambda(1-q_{lh})^2}{(\mu_h - \Lambda_s - (q + q_{lh}(1-q))\Lambda_m)^2} + \frac{\Lambda_m}{(\mu_l - (1-q)\Lambda_m)^2} > 0,$$

which completes this proof. □

Proof of Proposition 12.2.1 According to equation $R_m h W_h(q) - P_h = R_m h W_l(q) - P_l - q_{lh} P_h$, we can solve for the equilibrium q by rewriting it as

$$aq^2 + bq + c = 0, \quad (12.5.5)$$

where $a = ((1-q_{lh})P_h - P_l)(1-q_{lh})\Lambda_m^2$, $b = \lambda[((1-q_{lh})P_h - P_l)((1-q_{lh})(\mu_l - \Lambda_m) - (\mu_h - \Lambda_s - q_{lh}\Lambda_m)) - 2h(1-q_{lh})]$, and $c = h[(\mu_h - \Lambda_s - q_{lh}\Lambda_m) - (1-q_{lh})(\mu_l - \Lambda_m)] - ((1-q_{lh})P_h - P_l)(\mu_h - \Lambda_s - q_{lh}\Lambda_m)(\mu_l - \Lambda_m)$.

When $(1-q_{lh})P_h - P_l = 0$, Eq. (12.5.5) has at most one root and the equilibrium probability of patients entering the H-hospital can be solved as $q_e = \min\{\max\{0, \frac{(\mu_h-\Lambda_s-q_{lh}\Lambda_m)-(1-q_{lh})(\mu_l-\Lambda_m)}{2(1-q_{lh})\Lambda_m}\}, 1\}$.

When $(1-q_{lh})P_h - P_l \neq 0$, it is clear that

$$b^2 - 4ac = [(1-q_{lh})P_h - P_l]^2[(1-q_{lh})(\mu_l - \Lambda_m) + (\mu_h - \Lambda_s - q_{lh}\Lambda_m)]^2\Lambda_m^2 + 4h^2(1-q_{lh})^2\Lambda_m^2 > 0.$$

So the equation (12.5.5) has two roots where $q_{1,2} = \frac{-b\pm\sqrt{b^2-4ac}}{2a}$. More concretely, when $(1-q_{lh})P_h - P_l > 0$, only one root satisfies $q_e = q_2$ for both two conditions where

$$\begin{cases} ((1-q_{lh})q_e + q_{lh})\Lambda_m + \Lambda_s < \mu_h, \\ (1-q_e)\Lambda_m < \mu_l. \end{cases}$$

When $(1-q_{lh})P_h - P_l < 0$, Eq. (12.5.5) also has only one root $q_e = q_2$ which satisfies the above conditions. □

Proof of Proposition 12.2.2 We just consider the cases that $q_e(\Delta)$ is not a constant, because when $q_e(\Delta)$ is constant in Δ, we must have that $\frac{dq_e(\Delta)}{d\Delta} = 0$.

When $(1-q_{lh})P_h - P_l = 0$, $q_e(\Delta) = \min\{\max\{0, \frac{(\mu_h-\Lambda_s-q_{lh}\Lambda_m)-(1-q_{lh})(\mu_l-\Lambda_m)}{2(1-q_{lh})\Lambda_m}\}, 1\}$. For $q_e(\Delta) \in (0,1)$, we have $\frac{dq_e(\Delta)}{d\Delta} = -\frac{(1-\alpha e^{-\beta\Delta})+\alpha\beta e^{-\beta\Delta}[B-\Delta-c_h(\Lambda_m+\Lambda_s)]}{2c_h\Lambda_m(1-\alpha e^{-\beta\Delta})^2} < 0$ because $1-\alpha e^{-\beta\Delta} > 0$ and $B - \Delta - c_h(\Lambda_m + \Lambda_s) > 0$ in steady state. Thus we can obtain that $q_e(\Delta)$ is weakly decreasing in Δ (constant in Δ when $q_e = 0$ or 1).

When $(1-q_{lh})P_h - P_l \neq 0$, $q_e(\Delta)$ is the solution of $hD(q) - P_l + (1 - q_{lh}P_h) = 0$. Define the implication function $F(q,\Delta) = hD(q) - P_l + (1-q_{lh}P_h)$, we have

$$\frac{dq_e}{d\Delta} = -\frac{\partial F(q,\Delta)/\partial\Delta}{\partial F(q,\Delta)/\partial q}.$$

By Lemma 12.2.1, we have $\frac{\partial F(q,\Delta)}{\partial q} = \frac{\lambda(1-q_{lh})^2}{(\mu_h-\Lambda_s-[q_e+\alpha e^{-\beta\Delta}(1-q_e)]\Lambda_l)^2} + \frac{\Lambda_m}{(\mu_l-(1-q_e)\Lambda_m)^2} > 0$. Notice that $\frac{\partial F(q,\Delta)}{\partial\Delta} = \frac{2c_h^2\left(1-\alpha e^{-\beta\Delta}\right)\Lambda_m\left(1-\alpha e^{-\beta\Delta}+\alpha e^{-\beta\Delta}\beta[B-\Delta-c_h(\Lambda_m+\Lambda_s)]\right)}{\left([B-\Delta-c_h(q_e\Lambda_m+\Lambda_s)]-c_h(1-q_e)\alpha e^{-\beta\Delta}\Lambda_m\right)^3} + \alpha\beta e^{-\beta\Delta}P_h > 0$ because $1-\alpha e^{-\beta\Delta} > 0$ and $B - \Delta - c_h(\Lambda_s+\Lambda_m) > 0$ in equilibrium, then it gives that $\frac{dq_e}{d\Delta} = -\frac{F_\Delta}{F_{q_e}} < 0$. That is, q_e is also decreasing in Δ in this case.

When $(1-q_{lh})P_h - P_l = 0$, q_e is constant in P_h and P_l. Then we just consider the case that $(1-q_{lh})P_h - P_l \neq 0$, under which

$$\frac{\partial q_e}{\partial P_h} = \frac{-[(1-q_{lh})(\mu_l-\Lambda_m)+(\mu_h-\Lambda_s-q_{lh}\Lambda_m)]^2 h(1-q_{lh})}{\Lambda_m[2h(1-q_{lh})+A]} < 0, \tag{12.5.6}$$

$$\frac{\partial q_e}{\partial P_l} = \frac{[(1-q_{lh})(\mu_l-\Lambda_m)+(\mu_h-\Lambda_s-q_{lh}\Lambda_m)]^2 h}{\Lambda_m[2h(1-q_{lh})+A]} > 0, \tag{12.5.7}$$

where $A = \sqrt{[(1-q_{lh})P_h - P_l]^2[(1-q_{lh})(\mu_l-\Lambda_m)+(\mu_h-\Lambda_s-q_{lh}\Lambda_m)]^2 + 4h^2(1-q_{lh})^2}$. □

Proof of Proposition 12.2.3 When $B \leq \underline{\Delta}$, we have $q_s(\Delta) = 0$ by Proposition 12.1.4. Then it gives $q_e(\Delta) = q_s(\Delta) = 0$ if and only if $P_l - (1-q_{lh})P_h < hD(0)$ by Proposition 12.2.1. When $B \geq \bar{\Delta}$, we have $q_s(\Delta) = 1$ by Proposition 12.1.4. Then it gives $q_e(\Delta) = q_s(\Delta) = 1$ if and only if $P_l - (1-q_{lh})P_h > hD(1)$ by Proposition 12.2.1. Otherwise, when $B \in (\underline{\Delta}, \bar{\Delta})$, we have that $q_s(\Delta) = \hat{q}_S(\Delta) \in (0,1)$ by Proposition 12.1.4. Recall that the equilibrium uniquely solves $P_l - (1-q_{lh})P_h =$

$hD(q_e)$, where $D(q)$ increases in q. Therefore, for any prices pair (P_h, P_l) satisfies $P_l - (1 - q_{lh})P_h = hD(\hat{q}_S(\Delta))$, we have $q_e = \hat{q}_S(\Delta)$.

□

Proof of Lemma 12.3.1 Let $\underline{t}$ and $\overline{t}$ as the lower bound and upper bound of $f(x, y)$, obviously, we have $\overline{t} < b$ based on our preassumption. Denote by $F(t) = t/(b - t)$ for $t \in (\underline{t}, \overline{t})$, it is not difficult to verify that $F(t)$ is increasing in $t \in (\underline{t}, \overline{t})$, then it can be minimized at $\underline{t} = f(x_0, y_0)$. Therefore, $g(x, y) = f(x, y)/(b - f(x, y))$ can be minimized at (x_0, y_0).

The first and second derivatives of $g(x, y)$ below.

$$\frac{\partial g(x, y)}{\partial x} = \frac{b\frac{\partial f(x,y)}{\partial x}}{(b - f(x, y))^2}, \tag{12.5.8}$$

$$\frac{\partial^2 g(x, y)}{\partial x^2} = \frac{b\frac{\partial^2 f(x,y)}{\partial x^2}(b - f(x, y)) + 2b(\frac{\partial f(x,y)}{\partial x})^2}{(b - f(x, y))^3}, \tag{12.5.9}$$

$$\frac{\partial^2 g(x, y)}{\partial x \partial y} = \frac{b\frac{\partial^2 f(x,y)}{\partial x \partial y}(b - f(x, y)) + 2b\frac{\partial f(x,y)}{\partial x}\frac{\partial f(x,y)}{\partial y}}{(b - f(x, y))^3}, \tag{12.5.10}$$

$$\frac{\partial g(x, y)}{\partial y} = \frac{b\frac{\partial f(x,y)}{\partial y}}{(b - f(x, y))^2}, \tag{12.5.11}$$

$$\frac{\partial^2 g(x, y)}{\partial y^2} = \frac{b\frac{\partial^2 f(x,y)}{\partial y^2}(b - f(x, y)) + 2b(\frac{\partial f(x,y)}{\partial y})^2}{(b - f(x, y))^3}. \tag{12.5.12}$$

Through Eqs. (12.5.9), (12.5.10) and (12.5.12), we obtain that

$$\begin{aligned}
&\left(\frac{\partial^2 g(x, y)}{\partial x \partial y}\right)^2 - \frac{\partial^2 g(x, y)}{\partial x^2}\frac{\partial^2 g(x, y)}{\partial y^2} \\
&= (b - f(x, y))^2[(\frac{\partial^2 f(x, y)}{\partial x \partial y})^2 - \frac{\partial^2 f(x, y)}{\partial x^2}\frac{\partial^2 f(x, y)}{\partial y^2}] - \\
&\quad 2(b - f(x, y))[(\frac{\partial f(x, y)}{\partial x})^2\frac{\partial^2 f(x, y)}{\partial y^2} + (\frac{\partial f(x, y)}{\partial y})^2\frac{\partial^2 f(x, y)}{\partial x^2}] \\
&\quad +4(b - f(x, y))\frac{\partial^2 f(x, y)}{\partial x \partial y}\frac{\partial f(x, y)}{\partial x}\frac{\partial f(x, y)}{\partial y}
\end{aligned} \tag{12.5.13}$$

If $f(x, y)$ is convex, it should satisfy

$$\left(\frac{\partial^2 f(x, y)}{\partial x \partial y}\right)^2 - \frac{\partial^2 f(x, y)}{\partial x^2}\frac{\partial^2 f(x, y)}{\partial y^2} < 0 \tag{12.5.14}$$

$$\frac{\partial^2 f(x, y)}{\partial x^2} > 0, \quad \frac{\partial^2 f(x, y)}{\partial y^2} > 0, \tag{12.5.15}$$

From (12.5.15), we can obtain that $\frac{\partial^2 g(x,y)}{\partial x^2} > 0$ and $\frac{\partial^2 g(x,y)}{\partial y^2} > 0$. Also, by using (12.5.14), we have

$$\begin{aligned}
&\left(\frac{\partial f(x,y)}{\partial x}\right)^2 \frac{\partial^2 f(x,y)}{\partial y^2} + (\frac{\partial f(x,y)}{\partial y})^2 \frac{\partial^2 f(x,y)}{\partial x^2} \\
\geq\ & 2\frac{\partial f(x,y)}{\partial x}\frac{\partial f(x,y)}{\partial y}\sqrt{\frac{\partial^2 f(x,y)}{\partial y^2}\frac{\partial^2 f(x,y)}{\partial x^2}} \\
\geq\ & 2\frac{\partial f(x,y)}{\partial x}\frac{\partial f(x,y)}{\partial y}\frac{\partial^2 f(x,y)}{\partial x \partial y}. \qquad (12.5.16)
\end{aligned}$$

Combining (12.5.16) and (12.5.13), it gives that $(\frac{\partial^2 g(x,y)}{\partial x \partial y})^2 - \frac{\partial^2 g(x,y)}{\partial x^2}\frac{\partial^2 g(x,y)}{\partial y^2} < 0$. Therefore, $g(x, y)$ is also convex in (x, y). □

Proof of Proposition 12.3.1 Since $f_1(\Delta_L) = q_M + q_L\alpha_L e^{-\beta_L \Delta_L}$, and $\frac{\partial^2 f_1(\Delta_L)}{\partial \Delta_L^2} = \beta_L^2 q_L \alpha_L e^{-\beta_L \Delta_L} > 0$, we can conclude that $f_1(\Delta_L)$ is convex in Δ_L, which implies that $g_1(\Delta_L)$ is convex in Δ_L from Lemma 12.3.1. Next we prove the convexity of $g_2(\Delta_L, \Delta_M)$. Notice that

$$\frac{\partial f_2(\Delta_L, \Delta_M)}{\partial \Delta_L} = \frac{m(\Delta_L, \Delta_M)}{(B - \Delta_L - \Delta_M)^2}, \qquad (12.5.17)$$

$$\frac{\partial^2 f_2(\Delta_L, \Delta_M)}{\partial \Delta_L^2} = \frac{q_L\alpha_L\beta_L^2 e^{-\beta_L\Delta_L}\alpha_M e^{-\beta_M\Delta_M}(B - \Delta_L - \Delta_M)^2 + 2m(\Delta_L, \Delta_M)}{(B - \Delta_L - \Delta_M)^3}, \qquad (12.5.18)$$

$$\frac{\partial^2 f_2(\Delta_L, \Delta_M)}{\partial \Delta_L \partial \Delta_M} = \frac{q_L\alpha_L\beta_L e^{-\beta_L\Delta_L}\alpha_M\beta_M e^{-\beta_M\Delta_M}(B - \Delta_L - \Delta_M)^2 + m(\Delta_L, \Delta_M) + n(\Delta_L, \Delta_M)}{(B - \Delta_L - \Delta_M)^3}, \qquad (12.5.19)$$

$$\frac{\partial f_2(\Delta_L, \Delta_M)}{\partial \Delta_M} = \frac{n(\Delta_L, \Delta_M)}{(B - \Delta_L - \Delta_M)^2}, \qquad (12.5.20)$$

$$\frac{\partial^2 f_2(\Delta_L, \Delta_M)}{\partial \Delta_M^2} = \frac{\alpha_M\beta_M^2 e^{-\beta_M\Delta_M}(q_M + q_L\alpha_L e^{-\beta_L\Delta_L})(B - \Delta_L - \Delta_M)^2 + 2n(\Delta_L, \Delta_M)}{(B - \Delta_L - \Delta_M)^3}, \qquad (12.5.21)$$

where $m(\Delta_L, \Delta_M) = \Lambda_s/\Lambda_m - q_L\alpha_L\beta_L e^{-\beta_L\Delta_L}\alpha_M e^{-\beta_M\Delta_M}(B - \Delta_L - \Delta_M) + q_H + (q_M + q_L\alpha_L e^{-\beta_L\Delta_L})\alpha_M e^{-\beta_M\Delta_M}$, $n(\Delta_L, \Delta_M) = \Lambda_s/\Lambda_m - \alpha_M\beta_M e^{-\beta_M\Delta_M}(q_M + q_L\alpha_L e^{-\beta_L\Delta_L})(B - \Delta_L - \Delta_M) + q_H + (q_M + q_L\alpha_L e^{-\beta_L\Delta_L})\alpha_M e^{-\beta_M\Delta_M}$.

Define $\phi(\Delta_L, \Delta_M) = (\frac{\partial^2 f_2(\Delta_L, \Delta_M)}{\partial \Delta_L \partial \Delta_M})^2 - \frac{\partial^2 f_2(\Delta_L, \Delta_M)}{\partial \Delta_L^2}\frac{\partial^2 f_2(\Delta_L, \Delta_M)}{\partial \Delta_M^2}$. From Eqs. (12.5.18), (12.5.19) and (12.5.21), the expression of $\phi(\Delta_L, \Delta_M)$ can be interpreted as

$$f(\Delta_L,\Delta_M) = \{\frac{2(q_H)}{\alpha_M e^{-\beta_M\Delta_M}}[-q_L\alpha_L e^{-\beta_L\Delta_L}(\beta_L+\beta_M)^2 - q_M\beta_M^2] - 2q_L\alpha_L\beta_L^2 q_M e^{-\beta_L\Delta_L} - q_L\alpha_L\beta_L^2 e^{-\beta_L\Delta_L}(B-\Delta_L-\Delta_M)[q_M\beta_M^2(B-\Delta_L-\Delta_M) - 2q_M\beta_M] - [q_L\alpha_L\beta_L e^{-\beta_L\Delta_L} - (q_M+q_L\alpha_L e^{-\beta_L\Delta_L})\beta_M]^2\}(\alpha_M e^{-\beta_M\Delta_M})^2(B-\Delta_L-\Delta_M)^2. \tag{12.5.22}$$

Notice that the sum of the second term and the third term of (12.5.22) is $-q_L\alpha_L\beta_L^2 e^{-\beta_L\Delta_L}(B-\Delta_L-\Delta_M)[q_M\beta_M^2(B-\Delta_L-\Delta_M) - 2q_M\beta_M + \frac{2q_M}{B-\Delta_L-\Delta_M}] \le -q_L\alpha_L\beta_L^2 e^{-\beta_L\Delta_L}(B-\Delta_L-\Delta_M)(2\sqrt{2}-2)q_M\beta_M \le 0$. The first equality and the second equality hold if and only if $q_M\beta_M^2(B-\Delta_L-\Delta_M) = \frac{2q_M}{B-\Delta_L-\Delta_M}$ and $\beta_L^2 e^{-\beta_L\Delta_L}(B-\Delta_L-\Delta_M)q_M\beta_M = 0$, respectively.

When $0 < q_L < 1, 0 < q_M < 1, \alpha_L \neq 0, \alpha_M \neq 0, \beta_L \neq 0, \beta_M \neq 0$, we always have that $\phi(\Delta_L,\Delta_M) < 0$. Therefore, $f_2(\Delta_L,\Delta_M)$ is convex in (Δ_L,Δ_M), which implies that $g_2(\Delta_L,\Delta_M)$ is convex in (Δ_L,Δ_M) from Lemma 12.3.1.

Recall that $S(\Delta_L,\Delta_M) = \lambda R_m h\left[g_1(\Delta_L) + g_2(\Delta_L,\Delta_M) + \frac{q_L\Lambda_m}{\mu_L - q_L\Lambda_m}\right]$, then $S(\Delta_L,\Delta_M)$ is concave in (Δ_L,Δ_M). Thus there exists unique (Δ_L^*,Δ_M^*) maximizes $S(\Delta_L,\Delta_M)$. Denote by $(\hat{\Delta}_L,\hat{\Delta}_M)$ the unique solution of $\frac{\partial S(\Delta_L,\Delta_M)}{\partial\Delta_L} = \frac{\partial S(\Delta_L,\Delta_M)}{\partial\Delta_M} = 0$. If (1) $0 \le \hat{\Delta}_L + \hat{\Delta}_M < B$, (2) $\frac{B-\hat{\Delta}_L-\hat{\Delta}_M}{c_h} > [q_H + (q_M + q_L\alpha_L e^{-\beta_L\hat{\Delta}_L})\alpha_M e^{-\beta_M\hat{\Delta}_M}]\Lambda_m + \Lambda_s$, and (3) $\mu_M > (q_M + q_L\alpha_L e^{-\beta_L\hat{\Delta}_L})\Lambda_m$, then we have $(\Delta_L^*,\Delta_M^*) = (\hat{\Delta}_L,\hat{\Delta}_M)$. Otherwise, the optimal (Δ_L^*,Δ_M^*) can be attained at the boundary of feasible region. □

Proof of Proposition 12.3.2 (1) If $q_L^e \cdot q_M^e \cdot (1-q_L^e-q_M^e) = 0$, it means that $q_L^e = 0$ or $q_M^e = 0$ or $q_H^e = 0$ $(1-q_L^e-q_M^e = 0)$. (1)-(i) If $U_L(0,0) < U_H(0,0)$ and $U_M(0,0) < U_H(0,0)$, then it is never wise to join L- or m- hospitals when all other customers join H-hospital. That is, $(q_L^e, q_M^e) = (0,0)$ is an equilibrium. The equilibrium conditions in (ii)-(Vii) of case (1) can be verified similarly, here we just prove the uniqueness of $q_L^{(1)}$, $q_M^{(1)}$ and $q_L^{(2)}$.

(a) Define $F_1(x) = U_L(x,0) - U_H(x,0) = P_H - P_L - q_{LM}q_{MH}P_H + h[W_h(x,0) - W_l(x,0)] = P_H - P_L - q_{LM}q_{MH}P_H + h(\frac{1-q_{LM}q_{MH}}{\mu_H - \Lambda_s + [x(1-q_{LM}q_{MH})-1]\lambda} - \frac{1}{\mu_L - \Lambda_m x} - \frac{q_{LM}}{\mu_m - q_{LM}\Lambda_m x})$, we have

$$\frac{dF_1(x)}{dx} = -\lambda h\left(\frac{(1-q_{LM}q_{MH})^2}{(\mu_H - \Lambda_s + [x(1-q_{LM}q_{MH})-1]\Lambda_m)^2} + \frac{1}{(\mu_L - \Lambda_m x)^2} + \frac{q_{LM}^2}{(\mu_m - q_{LM}\Lambda_m x)}^2\right) < 0,$$

which implies that $F_1(x)$ in decreasing in x, then there exists unique $q_L^{(1)}$ such that $F_1(q_L^{(1)}) = 0$.

(b) Define $F_2(x) = U_M(0,x) - U_H(0,x) = P_H - P_M - q_{MH}P_H + h[W_h(x,0) - W_m(x,0)] = P_H - P_M - q_{MH}P_H + h(\frac{1-q_{MH}}{\mu_H - \Lambda_s + [x(1-q_{MH})-1]\lambda} - \frac{1}{\mu_m - \Lambda_m x})$, which also decreases in x. Then there exists unique $q_M^{(1)}$ such that $F_2(q_M^{(1)}) = 0$.

(c) Define $F_3(x) = U_L(x, 1-x) - U_M(x, 1-x) = P_M - P_L - q_{LM}P_M - q_{LM}q_{MH}P_H + h[W_h(x, 1-x) - W_l(x, 1-x)]$, we have

$$\frac{dF_3(x)}{dx} = -\lambda h\left(\frac{(1-q_{LM})^2}{(\mu_M + [x(1-q_{LM})-1]\Lambda_m)^2} + \frac{[q_{MH}(1-q_{LM})]^2}{(\mu_H - \Lambda_s + [x(1-q_{LM})q_{MH} + (1-q_{MH})-1]\Lambda_m)^2} + \frac{1}{(\mu_L - \Lambda_m x)^2}\right),$$

which gives $\frac{dF_3(x)}{dx} < 0$. Then there exists unique $q_L^{(2)}$ such that $F_3(q_L^{(2)}) = 0$.

(2) If $q_L^e \cdot q_M^e \cdot (1 - q_L^e - q_M^e) > 0$, then we have $q_L^e, q_M^e, q_H^e \in (0, 1)$, where customers are indifferent in choosing hospitals upon their arrivals. Thus the equilibrium can be determined by solving $U_L(q_L^e, q_M^e) = U_M(q_L^e, q_M^e) = U_H(q_L^e, q_M^e)$. □

Proof of Proposition 12.3.3 It should be noted that $q_i^e \geq q_{i+1}^e$ if and only if $U_{L,i}(\mathbf{q}^e) \geq U_{L,i+1}(\mathbf{q}^e)$ in equilibrium. Notice that

$$U_{L,i}(\mathbf{q}) = R_m - P_L - q_{lh,i}P_H - hW_{l,i+1}(\mathbf{q}^e) \geq U_{L,i+1}(\mathbf{q}) = R_m - P_L - q_{lh,i+1}P_H - hW_{l,i+1}(\mathbf{q}^e)$$
$$\Leftrightarrow (q_{lh,i+1} - q_{lh,i})P_H + h[W_{l,i+1}(\mathbf{q}^e) - W_{l,i}(\mathbf{q}^e)] \geq 0.$$

If we have $q_{i+1}^e > q_i^e$, then it gives $W_{l,i+1}(\mathbf{q}^e) - W_{l,i}(\mathbf{q}^e) > 0$ since $q_{lh,i+1} - q_{lh,i} = \alpha(e^{-\beta\Delta_{i+1}} - e^{-\beta\Delta_i}) > 0$. Thus, we have $(q_{lh,i+1} - q_{lh,i})P_H + h[W_{l,i+1}(\mathbf{q}^e) - W_{l,i}(\mathbf{q}^e)] \geq 0$, i.e., $U_{L,i}(\mathbf{q}^e) \geq U_{L,i+1}(\mathbf{q}^e)$, which contradicts which the preassumption that $q_{i+1}^e > q_i^e$. Therefore, we have $q_i^e \geq q_{i+1}^e$. □

References

Anand, Krishnan S., M. Fazil Paç, and Senthil Veeraraghavan. 2011. Quality-speed conundrum: Trade-offs in customer-intensive services. *Management Science* 57 (1): 40–56.

Carlisle, A., and G. Dozier. 2001. An off-the-shelf PSO. In *Proceeding of the Workshop on Particle Swarm Optimization*. Purdue Indianapolis: School of Engineering and Technology.

Chen, Hong, and Murray Frank. 2004. Monopoly pricing when customers queue. *IIE Transactions* 36: 569–581.

Chen, Wuhua, Zhe George Zhang, and Zhongsheng Hua. 2015. Analysis of two-tier public service systems under a government subsidy policy. *Computers & Industrial Engineering* 90: 146–157.

Chen, Wuhua, Zhe George Zhang, and Zhongsheng Hua. 2016. Analysis of price competition in two-tier service systems. *Journal of the Operational Research Society* 67 (6): 897–910.

Chen, Wuhua, Zhe George Zhang, and X. Chen. 2020. On two-tier healthcare system under capacity constraint. *International Journal of Production Research* 58 (12): 3744–3764.

Gerber, H. U., and G. Pafum. 1998. Utility functions: from risk theory to finance. *North American Actuarial Journal* 2 (3): 74–91.

Gilbert, S. M., and Z. K. Weng. 1998. Incentive effects favor nonconsolidating queues in a service system: The principal—Agent perspective. *Management Science* 44 (12): 1662–1669.

Guo, Pengfei, Chris S. Tang, Yulan Wang, and M. Zhao. 2019. The impact of reimbursement policy on patient welfare, readmission rate and waiting time in a public healthcare system: Fee-for-service versus bundled payment. *Manufacturing & Service Operations Management* 21 (1): 154–170.

Hassin, Refael. 2007. Information and uncertainty in a queueing system. *Probability in the Engineering and Informational Sciences* 21 (3): 361–380.

Hua, Zhongsheng, Wuhua Chen, and Zhe George Zhang. 2016. Competition and coordination in two-tier public service systems under government fiscal policy. *Production and Operations Management* 25 (8): 1430–1448.

Kennedy, J. and R. Eberhart. 1995. Particle swarm optimization. In *Proceedings of ICNN'95-international conference on neural networks*, vol. 4, 1942–1948. IEEE.

Li, L., and Y. S. Lee. 1994. Pricing and delivery-time performance in a competitive environment. *Management Science* 40 (5): 633–646.

Milgrom, P., and J. Roberts. 1986. Price and advertising signals of product quality. *Journal of Political Economy* 94: 796–821.

Naor, Pinhas. 1969. The regulation of queue size by levying tolls. *Econometrica* 37: 15–24.

Qian, Q., and W. Zhuang. 2017. Tax/subsidy and capacity decisions in a two-tier health system with welfare redistributive objective. *European Journal of Operational Research* 260 (1): 140–151.

Sadat, S., H. Abouee-Mehrizi, and M.W. Carter. 2015. Can hospitals compete on quality? *Health Care Management Science* 1–13.

Sekhri, N., R. Feachem, and A. Ni. 2011. Public-private integrated partnerships demonstrate the potential to improve health care access, quality, and efficiency. *Health Affairs*. 30 (8): 1498–1507.

Wan, G., and Q. Wang. 2017. Two-tier healthcare service systems and cost of waiting for patients. *Applied Stochastic Models in Business and Industry* 33: 167–183.

Wang, J.J., Zhongping Li, and J. Shi. 2020. Hospital referral and capacity strategies in the two-tier healthcare systems. *Omega*, 102229.

Wang, Zhongbin, and Jinting Wang. 2021. Reimbursement policy in a healthcare system with priorities: Fee for priority versus bundled priority. *IMA Journal of Management Mathematics* 32 (3): 329–360.

Wang, Jinting, Zhongbin Wang, Zhe George Zhang, and Fang Wang. 2022. Efficiency-quality trade-off in allocating resource to public healthcare systems. *International Journal of Production Research* 60 (21): 6469–6490.

Zhang, Zhe George and Xiaoling Yin. 2020. Information and pricing effects in two-tier public service systems. *International Journal of Production Economics*, 107897.

MIX
Papier aus verantwortungsvollen Quellen
Paper from responsible sources
FSC® C105338

If you have any concerns about our products,
you can contact us on
ProductSafety@springernature.com

In case Publisher is established outside the EU,
the EU authorized representative is:
Springer Nature Customer Service Center GmbH
Europaplatz 3, 69115 Heidelberg, Germany

Printed by Libri Plureos GmbH
in Hamburg, Germany